Research for Development

The series Research for Development serves as a vehicle for the presentation and dissemination of complex research and multidisciplinary projects. The published work is dedicated to fostering a high degree of innovation and to the sophisticated demonstration of new techniques or methods.

The aim of the Research for Development series is to promote well-balanced sustainable growth. This might take the form of measurable social and economic outcomes, in addition to environmental benefits, or improved efficiency in the use of resources; it might also involve an original mix of intervention schemes.

Research for Development focuses on the following topics and disciplines:

Urban regeneration and infrastructure, Info-mobility, transport, and logistics, Environment and the land, Cultural heritage and landscape, Energy, Innovation in processes and technologies, Applications of chemistry, materials, and nanotechnologies, Material science and biotechnology solutions, Physics results and related applications and aerospace, Ongoing training and continuing education.

Fondazione Politecnico di Milano collaborates as a special co-partner in this series by suggesting themes and evaluating proposals for new volumes. Research for Development addresses researchers, advanced graduate students, and policy and decision-makers around the world in government, industry, and civil society.

THE SERIES IS INDEXED IN SCOPUS

More information about this series at https://link.springer.com/bookseries/13084

Maurizio Boriani · Mariacristina Giambruno
Editors

Architectural Heritage in the Western Azerbaijan Province of Iran

Evidence of an Intertwining of Cultures

Fondazione
Politecnico
di Milano

Springer

Editors
Maurizio Boriani
Retired, Former Member of Dipartimento
di Architettura e Studi Urbani
Politecnico di Milano
Milan, Italy

Mariacristina Giambruno
Dipartimento di Architettura e Studi Urbani
Politecnico di Milano
Milan, Italy

ISSN 2198-7300 ISSN 2198-7319 (electronic)
Research for Development
ISBN 978-3-030-83093-9 ISBN 978-3-030-83094-6 (eBook)
https://doi.org/10.1007/978-3-030-83094-6

This Springer imprint is published by the registered company Springer Nature Switzerland AG
The registered company address is: Gewerbestrasse 11, 6330 Cham, Switzerland

Contents

Foreword. Towards Protection, Conservation and Reuse of Architectural and Landscape Heritage

Maurizio Boriani and Mariacristina Giambruno

Abstract The book is about the history of Protection and Restoration in Iran, and particularly in Iranian Western Azerbaijan. A substantial part of the volume is dedicated to illustrating how the territory of the cities of Khoy and Salmas has been described in the reports of Western travelers. Finally, several essays are dedicated to the study of the Armenian and Assyrian churches still present in this area, evaluating the problems of their protection and conservation.

The current globalization processes also involve, indirectly yet strongly, the Cultural Heritage, in particular in the so-called emerging countries where the transformation of the territory is both faster and less controlled. This requires a reflection on what protection policies should be implemented in order to keep buildings, urban centres, and landscapes that have been preserved over the years but are now under danger by 'economy development' and 'modernization.' Iran cannot be regarded as a developing Country, experiencing a conflict between needs for modernization and economic difficulties due to the well-known events of the US sanctions. Even though the Country has a consolidated tradition in the field of architectural restoration, yet some criticisms remain, mostly about those episodes considered—right or wrong—of lesser importance. Criticisms deserve to be highlighted.

In fact, the conservation process of those sites has to confront with totally opposite forces: the need for modernization of a Country as well as of its people; the skills degree along with the upgrade to international standards for the technicians responsible for the safeguard of the Architectural Heritage; the disappearance, in many cases, of the communities that had realized it; the conflicts often arising from

Present Address:
M. Boriani (✉)
DAStU, Politecnico di Milano (retired 2016), Milan, Italy
e-mail: maurizio.boriani@polimi.it

M. Giambruno
DAStU, Politecnico di Milano, Milan, Italy
e-mail: mariacristina.giambruno@polimi.it

M. Boriani and M. Giambruno (eds.), *Architectural Heritage in the Western Azerbaijan Province of Iran*, Research for Development,
https://doi.org/10.1007/978-3-030-83094-6_1

the Cultural Heritage itself as representative of specific cultures. All factors often coming to evident conflict with the demand of safeguard and conservation.

We ought to realize, in fact, that the Built Cultural Heritage may be seen—for many reasons—in different ways according to different cultures, political interests, and to the knowledge shared on it. 'Heritage' is basically a relative concept even when it claims to be universal, like UNESCO does. The opinions about what should be preserved and how it should be done often look totally opposite to each other, both within a Country, and among operators from different cultural backgrounds, mainly when a specific social group assigns, to a particular group of buildings the task of establishing on a territory and its inhabitants, the ideological reasons of its own independence, if not even of its own dominance.

History is often written by winners, as recognized; hence it is obvious for them to try selecting, among their concerning, the history facts justifying their own behaviours in the past, in present time, and in the future. The physical inheritance, architecture in particular, being public object *pour excellence*, makes the material support for such a policy and, right in the Cultural Heritage policies, it is easy for intervention decisions to be made in line with that. If the recognition of an asset value implies a typically political decision, the safeguard and conservation practices of them, transformation interventions and even destruction actions are no else than consequences of those choices along with their supporting ideologies.

All of that implies decisions about what to preserve and hand over to future, and how to preserve it, particularly in case of limited resources to be devoted to Cultural Heritage. Preserving the Architectural Heritage means taking account of its authenticity and integrity, avoiding invasive and irreversible interventions, making rather use of non-destructive diagnosis techniques to assess the causes of degradation and instability, employing work methods and materials already tested: issues that often collide with the scarcity of available resources, as well as with the training of the restoration staff and the presumed need for the possibly fastest implementation of restoration works. Consequently, method carefulness tends to be regarded as delay and sophistry of technicians more interested to scientific publication of their performed studies, rather than to properly called restoration.

Moreover, the concept of authenticity in restoration theory and praxis, as recognized, had its start from a centuries-old debate developed until few dozens of years ago essentially in Europe, whereas those concepts not always have been shared in around the world. For different reasons, the main of which is that over nearly two centuries Industrial Revolution has made aware, in Western countries, of the irremediable gap between ancient and contemporary modes of production. Materials, working techniques, costs of labour and of building components have dramatically changed in comparison to the past, and any attempt to artificially revive ancient modes of production is seen by us like pale imitation, if not as bold falsification.

In such contexts, interventions in line with centuries-old traditions are still in common use. That is recovering the *quo ante* status, particularly where the buildings are undermined by partial though important collapses, but their remains are enough to let understand the original state. And today that happens mostly on buildings made with traditional materials and techniques. All of this, with no care of what authenticity will be left after such interventions.

Yet, technicians and cultural operators are not responsible for all wrong decisions! We ought to realize that in some cases the permanence of traditional construction techniques and of the related craftsmen skills makes it difficult correctly adopting new techniques, mostly when they are more expensive or difficult to learn.

It does not mean that modern materials and techniques are not used in those countries, mostly when they seem less expensive, more effective, or simply 'more modern'. Though it would be noticed that right in the Countries where traditional construction techniques are more in use than elsewhere, the negative effects of industrial materials and processing in restoration are not yet evident.

For Europeans that happen to work in these situations, adapting their own methods to the working context would be necessary. In fact, neither the needed material and human assets nor the long times often required for implementing the restoration works are sometimes available. Curiously, whenever works be carried out where the workmen cost, some actions that in highly developed countries would be expensive and might instead be easily performed.

As we know, archaeological research, restoration, and site enhancement strategies cannot be one-shot works but require continuous maintenance: weeds control as well as weather damage, surveillance, sites decorum and cleaning maintenance are all easy-to-do works where there is no need for particularly specialized workers. Yet, those works are seldom performed, and the responsibility for that has to be equally shared both on the side of the technicians who often do not forecast the sustainability—even economic—of their interventions in time, and on the side of the entities designated to manage the site after its rehabilitation, who should carefully plan the maintenance of buildings and work tools, which often happens rather for negligence than for lack of resources.

The above mentioned problems deal about safeguard and conservation of single monuments or of a clearly defined site of historical interest. Yet, the monumental heritage is rarely secluded from its establishing context.

The ongoing 'modernizations', typical of Countries in transition, often imply the loss of relationship between the historical building and the landscape or the surrounding urban context, and even more, change the historical landscape itself, which is seldom recognized as Cultural Heritage like the so-called monuments.

A new approach has for some time included the Heritage Conservation within the general issues of sustainable development, posing the problem of reconciling the current and necessary use of natural resources with the needs of the up-coming generations. By this point of view, mostly in the emerging Countries, keeping the traditional procedures in territorial government, and in particular of agriculture and environment, can make the difference. Traditional farmers are by definition 'maintainers of territory' and always played important activities to keep its natural balance.

Today, we are aware of the need for counteracting any short time exploitation policies of natural and cultural resources, paying great attention, instead, to their possibly in-coming long-term effects.

We are also aware that such purposes have to face the understandable aims to improve the economic conditions of peoples that have been poor for long time. Strong is the temptation for people to quickly improve their own economic conditions by intensely exploiting their available resources, which can be counteracted only with alternative policies, able to quickly improve situations, yet respecting the available resources.

Well-irrigated lands are very rare in Iran, since the territory is mostly a nearly desert plateau. An exception by this point is Western Azerbaijan itself, where the near important mountains that ranges in the West guarantees a great water supply, mainly in spring snowmelt. That caused since the most ancient times important canal works and storage of waters through dams and water basins. Thanks to that are due the flourishing orchards that dazzled the western visitors.

Aim of this volume is not dealing with the conservation of traditional agrarian landscapes; yet it is ought to remind that the historic watering system itself is one of the main monuments on earth, and that the permanence of people in the villages where it had been built and maintained is a crucial condition for its conservation. The recovery for new uses of the churches today abandoned, about which we are going to write in the final chapters of this publishing, could help rehabilitate the villages, also holding the youngsters there, particularly involved in activities that will attract visitors both from the Country and from abroad.

Another key issue in transition countries is that of tourism development as an economy source of great importance.

International tourism is often regarded, in such cases, like a *panacea*: preserving and enhancing the Cultural Heritage to attract visitors for economy increase seems to be the simplest solution to be adopted. A good solution because it preserves the heritage and strongly attracts well-off visitors from the world at large. And also it is good, at last, because there is no need for huge public resources, while the tourism supporting infrastructure (hotel, restaurants, shopping centres, etc.) can be implemented with private capital, even international.

Though it should be understood that good landscapes are the effect of good governance, each innovation in management and use of territories must be carefully assessed for its predictable impact.

Consequently, economic development shall be promoted together with socio-political development, complying with the principles of self-determination of peoples. The Western World can offer the history of its own mistakes and conquers, help the best energies of each Country through cooperation going along with the necessary reforms, yet being aware that it will be up to local people finding out a sustainable development way of their own.

The just over-mentioned issues—as in many other cases—are in background to the preservation of Western Azerbaijan Cultural Heritage. Azerbaijan: a centuries-old crossroads of cultures, where the great communication roads stretched through Caucasus, making it a kind of a bridge between Asia and Mediterranean Sea.

An impervious and mountainous land, rich in waters, among the most fertile in present-day Iran: 'In the large plains between the mountainous buttresses, there are delightful countryside cultivated with wheat, cotton, tobacco, rice, as well as castor, melons, pumpkins, and watermelons. Noteworthy in the field of the agrarian landscape architecture, as well as construction material, is the poplars growing, favoured by irrigated land that supplies building material in shape of long and thin elements jointed to form horizontal roofing. Agrarian economy has always played a decisive role on the territory: the market of Tabriz, for instance, can be reminded as one of the most important in West Asia (…). The rational land use has made the region surrounding the Urmia lake the best place where to live among all the plateau areas (in spite of its lack of fish and the marshy lands all around). Its population, thicker than in any other province, has been active and of some importance since old times' (Alpago Novello 1977).

This zone has changed—if not substantially—since the 1970s when Alpago Novello travelled through it; yet it still preserves a rich Architectural Heritage, produce of the peoples that in time inhabited it. Mosques, bazaars, caravanserais, but also rough earth traditional villages, and residual pieces of the agrarian landscape described by Alpago, intertwine with a still conspicuous heritage of Armenian religious buildings—even if often reduced to ruins—testifying of the dense presence of those people in the area.

Armenian presence in the area was already consolidated in XIII century when, thanks to more religious tolerance, the destruction of their towns in homeland, and the chances of commercial exchanges in Persian territory, Armenians established there in great numbers (Alpago Novello 1978).

In XIII century, Armenian communities existed in Tabriz, Marand, Khoy, Salmas, amd Maragheh (Abrahamian 1964).

From the Armenian presence itself, and from Architectural Heritage, its most tangible legacy, this volume takes its cue, and in particular from Alpago Novello's travels, along with his group of researchers in seek for the traces of Armenian architecture in Western Azerbaijan. Primary aim was checking the current consistency, in terms both of permanence and preservation state.

An idea of those 'nearly heroic' surveys can be drawn from the writing of Marco Brambilla, one of the members of those missions, reading the paper published here as *Adriano Alpago Novello and research on Armenian architecture in Iran.*

The results of the most recent research on that religious heritage are synthetized in the essay *Christian Churches in Salmas, Khoy, and Urmia Counties. Conservation problems and enhancement perspectives in the cultural and territorial context of Western Azerbaijan*, where the most relevant constructive techniques of the buildings are outlined, together with their current phenomena of decay and instability, and the safeguard and enhancement prospects to put on for their preservation.

The churches subject to our recent surveys are described in detail, with particular attention paid to their current preservation state, in *Directory of the Armenian churches of the Salmas and Khoy Counties*. Each of the 23 churches surveyed between 2016 and 2018 is identified as location, building period, and floor plan type. When available, descriptions from other scholars years ago are reported, in order to compare the current state with the previous one, and to assess whether decay and instability have gone further or to put in evidence recent restoration interventions.

A further picture of the characters of Armenian churches of the area is outlined in the essay *Studies in Armenian Architecture in Iran, Western Azerbaijan.*

As already said, Western Azerbaijan is the historical crossroads for connecting Asia to Mediterranean Sea and many travellers, for that reason, passed through it describing its features and the consistency of towns and monuments. The essay *The Regions of Khoy and Salmas in the Travelogues of Western Travellers between the 15th and 19th Centuries* reports about those journeys giving a detailed account of how Khoy and Salmas might have appeared between fifteenth and nineteenth centuries. The summary is supported by an *Anthology* of selected writings and reported in their original language, in appendix to this volume.

The roads network crossing the region was sustained by thick threads of infrastructural elements, many of which are still preserved today, enriching the group of the Architectural Heritage of these provinces. So, we chose to give account of them with the paper *Infrastructure for Trade Routes in Western Azerbaijan, Iran: Caravanserais, Bridges, and Fortresses* which documents and describes five caravanserais, urban and suburban, four ancient bridges, and three of the fortresses that presided the road tracks.

In order to point out the current situation of the architectural and landscape heritage of the area, its safeguard state and the historic-cultural context of the restoration interventions noted during the study campaigns recently carried on the religious Armenian Heritage, a history of restoration in Iran had to be retraced as well as the safeguard practices of Cultural Heritage in that Country.

The paper *Preservation of Historic Heritage Sites in Iran*, after a short analysis of the main material used in the construction of the Architectural Heritage, gives an account of Iranian legislation and safeguard management, then describing the origins of the restoration theory developed there, from the early 1900s to the present days, with particular attention paid to the contribution of European culture in the field.

The essay *The story of a National Heritage. The formation of the Iranian tools of preservation between Thoughts, Laws and Institutions* completes the picture of the institutions of protection of the Cultural Heritage in Iran. It describes in detail the history of theories and governance Institutions of Protection connecting them to the powers that ruled the Country, from Qajars (1797–1925) to Pahlavis, up to Islamic revolution and the current days. A focus completes the paper, on the effects in Western Azerbaijan of the organization of Heritage Protection, and of the consistency of the restoration implemented in the most celebrated monuments of the area.

The present book, therefore, is not an organic volume on the history of Protection or of Restoration in Western Azerbaijan, but a group of original studies contributing to put together the state and consistency of the articulated Historical Heritage; we still can find there within the protection of Cultural Heritage and the policies put in place in modern Iran. The aim of the book is to attract attention onto this important legacy of the past, testifying the fervent multiculturalism of the area, and to promote a more effective protection and a long lasting conservation.

Maurizio Boriani Architect, born in Milan, Italy, 1946.

Full professor in Architectural Conservation at the Politecnico di Milano (Department of Architectural Design—School of Civil Architecture)—retired in 2016.

Former Director of Department of Architectural Design at the Politecnico di Milano.

Member of the Teaching College of the Ph.D. in Architectural Conservation at the Politecnico di Milano.

Former President of the Scientific Board of Cultural Heritage Conservation Center of the Politecnico di Milano.

Research Co-ordinator on behalf of the UNESCO, World Bank and European Community, Italian Ministry of Foreign Affairs and other private and public agencies.

UNESCO Consultant for: Albania (2006–2009), Macedonia, Montenegro, Kossovo, Myanmar; expert in international cooperation on behalf of Italian Ministry of Foreign Affairs: China (2009–2012), Vietnam (2017–2019), Armenia (2019–2014), Pakistan (2011–2012), Myanmar (2012–2015).The opening parenthesis does not have a corresponding closing parenthesis in the sentence 'China (2009–2012)…'. Please insert the parenthesis in the appropriate position.done

Research Topics:

History of Architectural Conservation, Conservation and management of historic towns and archeological sites, Cultural landscape preservation and management, Gardens history and conservation, Cultural Heritage and International Cooperation. About those themes he is author or editor of about 240 papers and books.

Mariacristina Giambruno Architect, Ph.D. in Preservation of Architectural Heritage and graduated in Restoration of Monuments, she teaches at the School of Architecture, Urban planning, Construction engineering, and Graduate School in Architectural and Landscape Heritage (Politecnico di Milano).

Full professor in Architectural Conservation.

Member of Faculty Board of the Ph.D. program in Architecture, Urban Design, Conservation of Housing and Landscape until 2012, from 2019 Head of the Ph.D. program in Preservation of Architectural Heritage (Politecnico di Milano).

The teaching and research activity is oriented on the history and theories of Architectural Heritage, Conservation of Historical Heritage, Historical Centers Conservation and Management, and guidelines and management plans to preserve and promote Cultural Heritage.

In recent years, she has focused on emerging and transitioning countries, working and directing international projects in collaboration with organizations such as UNESCO, World Bank, AICS in Albania, Iran, Armenia, China, Pakistan and Vietnam.

In these countries, she investigates the role of Cultural Heritage as a "trigger" for sustainable development.

As a speaker at conferences in Italy and abroad, she held courses and workshops in Armenia, Vietnam, Portugal, where she is part of a research unit coordinated by the Lusìada University of Lisbon. She authored about 150 publications on the preservation of Historical Centers and Architectural Heritage.

Adriano Alpago Novello and Research on Armenian Architecture in Iran

Marco G. Brambilla

Abstract In studies of Armenian architecture, many extraordinary scholars contributed to the research, documentation, and exploration of Armenian architecture. Adriano Alpago Novello was undoubtedly one of the most accomplished experts of Armenian architecture. His contributions to this specialized field of studies will always remain milestones in the path of research on Armenian architecture. This essay will explore the academic and scientific aspects of the research programs implemented by Alpago Novello and the human element of this outstanding individual. The present article will further try to provide an insight into the challenging and exacting conditions of the field trips conducted by Alpago Novello, this author, and their team in the late 70s in rural North-Western Iran, a territory dominated mainly by the Kurdish population often in conflict with the central government of the time. In addition, due credit shall be given to other scholars who worked directly or indirectly with Alpago Novello and ultimately enriched the body of research on Armenian architecture.

Keywords Iran · Architecture · Armenians

There is a golden era in the history of research on Armenian architecture. Sometime in the early 60s through the mid-80s, there was a magical revival of international interest in the specifics of Armenian architecture by highly specialized and proficient scholars, architects, and art historians. This happened almost concurrently in Europe, more specifically in Italy, France, and Germany and to some degree in the USA. One of the pillars of this illustrious group was Professor Adriano Alpago Novello, a dear friend, a distinguished colleague, and an accomplished educator (Fig. 1).

This essay intends to illustrate not only the contributions of Alpago Novello to the field of studies of Armenian architecture in Iran but also the human aspect of this outstanding scholar, with whom I had the pleasure of working together on many occasions.

M. G. Brambilla (✉)
University of California, Los Angeles, USA

M. Boriani and M. Giambruno (eds.), *Architectural Heritage in the Western Azerbaijan Province of Iran*, Research for Development,
https://doi.org/10.1007/978-3-030-83094-6_2

Fig. 1 Adriano at the dome of the Shepherd's Chapel on the River Arax (photograph by the author 1976)

In Italy, two centers of research on Armenian architecture were established in Rome and Milan in the 60s. In 1967, with support from the Manoukian family, Alpago Novello established the Center for Armenian Studies in Milan (CSDCA). This organization would become an important center for studying and promoting the history of Armenian architecture. [*A/N: CSDCA, was also instrumental in publishing a series of monographs on Armenian architecture, Documents of Armenian Architecture, which became very popular and reached out to a vast audience*].

In Rome, Professor Geza de Francovich from the University of Rome encouraged and sponsored this scholarship, led by Professor Paulo Cuneo. These key figures promoted the study of Armenian architecture, generating interest and encouraging other scholars to join in the effort. It is thanks to their initiatives that a large group of architects, scholars, and students were introduced to this research program, ultimately contributing substantially to this field of studies. [*Prof. Aslan, Bolognesi, Maria Stella MarianiCalo, C. Fonseca, Manoukian, De Bernardi Ferrero, G. Boccali, T. Scalesse, and M. Boriani are just a few Italian scholars contributing to Armenian studies in Italy*].

At the same time, studies of Armenian architecture in Soviet Armenia, although well elaborated by various scholars, including Professor Murad Hasratian, Khalpakhchian, Aleksander Arami Sahinian, Ghafadarian, Armen Zarian, and others, were isolated from the rest of the world, with no dialog or interaction with the academic progress in the West.

A key individual was instrumental in facilitating a new environment of academic cooperation and ultimately a revival of interest in research on Armenian architecture. Armen Zarian (1914–1994), son of the well-known Armenian writer—Kostan Zarian, was a highly educated and successful architect in Italy, who decided to relocate from Rome to the Armenian SSR. His presence in Soviet Armenia, his ties to Italy, and the common cultural goals of the two countries were instrumental in facilitating a cultural and academic exchange between Armenia and Italy at a time when the Cold War was dividing everything Western from everything in the Soviet Union.

Many Italian scholars met with Zarian, became lifelong friends, and achieved something that no one could have imagined at the time: an intensive series of visits, interactions, joint ventures, and cooperation between architectural scholars in Armenia and abroad. [*A/N: The first visit to Armenia by Italian scholars occured in 1966 and then again 1967 led by P. Cuneo and T. Breccia Fratadocchi*].

Zarian facilitated several visits of Italian architects and academics to Armenia, which provided them with direct insight into the matters they were researching. They then extended their research into western Turkey, the historical Armenian highlands, and eventually to Iran.

This contribution focuses on the academic joint venture of the author with Adriano Alpago Novello in Iran.

In 1967, Adriano traveled to Iran on a short exploratory visit. Obviously, the most important part of the time was allocated to the major Armenian monuments of northern Iran: the monasteries of St. Thaddeus (Fig. 2) and St. Stephanos (Fig. 3).

Fig. 2 Monastery of St. Thaddeus (photograph by the author 1976)

Fig. 3 Monastery of St. Stepanos (photograph by the author)

This was most fortuitous, especially because at the time there was a general interest in expanding the Middle Eastern study programs in Italy. Istituto Italiano per il Medio ed Estremo Oriente (IsMEO), led by Giuseppe Tucci, was instrumental in bringing Italy and Iran together, and through the efforts of the Italian Ministry of Foreign Affairs, there was active support of the research programs of Italian scholars in Soviet Armenia and Iran [*A/N: While both groups of scholars from Milan and Rome visited Western Armenia, P. Cuneo concentrated his studies on Ani and the historic Armenian Highlands, while A. Alpago Novello expanded his research into Georgia, Iran, and Northern Syria*].

At the same time, Professor Geza de Francovich, director of the Institute of History of Arts at the University of Rome, had initiated a research program related to the history of Armenian architecture with the cooperation of Tomasso Breccia Fratadocchi, Enrico Costa, and Paulo Cuneo. These efforts were ultimately finalized and presented, not only in a series of publications and symposia but also in a major exhibition known as Armenian Medieval Architecture.

This exhibition, presented first in Rome in 1968, was subsequently exhibited internationally, thanks to the efforts of Adriano Alpago Novello,

Concurrently Adriano was preparing the organization of the first international symposium on Armenian architecture in Bergamo, Italy, in 1975 [*A/N: This symposium was organized jointly by the Politecnico of Milan, the University of Bergamo, and the Armenian Academy of Sciences*].

Until 1975, I was working with Paulo Cuneo on research material on Armenian architecture and its primary sources. Professor Giuseppe Zander, one of the most prominent scholars of architectural history and a key figure in the formation of the school of preservation of architectural heritage in Rome, Italy, introduced me to Adriano Alpago Novello in 1974. After this introduction, I became not only a contributor to the First International Symposium on Armenian Architecture (Bergamo 1975) but also one of its active coworkers and more importantly a close friend of Adriano.

In 1975, through joint efforts of Italian authorities and the local Armenian community, the Exhibition of Armenian Medieval Architecture was exhibited in Iran, home of one of the largest and most prosperous Armenian diasporas at the time.

Almost concurrently and in the same time period, I was asked to join the Department of History of Architecture at the National University of Iran (NUI) and to act as the chair of the department of architectural preservation at the School of Architecture at the NUI.

This event provided us with the opportunity to initiate a joint venture program between the National University of Iran and the Politecnico of Milan to study Armenian monuments of Iran.

Adriano lectured about the history of Armenian architecture to the students at both important schools of architecture in Iran: the National University and the Fine Arts Department of the University of Tehran. These were the first steps of Adriano's long-term involvement with Iran and served as the premise of our joint ventures for many years. Furthermore, the success of the first international symposium on Armenian architecture encouraged all those involved in this research program to further pursue these efforts. Ultimately, several other symposia were organized in Italy and Armenia. [*A/N: Subsequently five more symposia were organized in Italy and in Soviet Armenia*] The research on Armenian architecture in Iran was expanded and supported by Italian and Iranian academia.

In a magical way, during the mid-70s, everything seemed to be perfect for starting an extensive cooperation program between the Politecnico of Milan and the NUI. With the support of the Dean of the School of Architecture, NUI, Dr. Iraj Parvin, it was possible to formally initiate a cultural agreement between Italy and Iran, specifically with the universities of Milan, Turin, and Venice. [*A/N: Adriano gives a detailed summary of these efforts in his introductory notes in the volume published jointly by Politecnico of Milan and the National University of Iran "Ricerca sull" Architettura Armena # 17, Iran I. Milano 1977 and volume 18, Iran II, Milano, 1978*].

In this context, on behalf of the NUI, we were able to invite our colleagues from Italy on several occasions, and for the first time ever the lesser-known, secondary Armenian architecture of Iran became the subject of academic studies.

Our mission was actually vague. I had collected written documents regarding some 200 churches in Northern Iran, but in 1975, only a few had been known or visited, and none of them had been researched or published. The location and the conditions of these monuments were unknown, and no documentation was easily attainable. [*A/N: A major difficulty of identifying these churches was due to the changes of the toponymy. Names of villages were often changed during the previous centuries, and the most accessible source appeared to be oral traditions*].

Hence, our mission was to explore and to discover. This by itself was an adventure. In the 70s, it was not easy to travel through North-Western Iran, considered a "hostile" Kurdish-dominated territory, without the right connections on both sides. It seems unreal, and I am sure many of our colleagues did not realize that the Kurds, with active smuggling businesses and anti-government sentiments, had snipers and outlook posts all over the mountain passes, within which we were conducting our research program. If an unknown vehicle was seen and deemed "hostile", they were easy and likely targets. Thankfully our jeeps were well known by the local villagers, and thus, we had a safe passage throughout the Kurdish territories.

As such, we had to secure the official governmental permits and support from the local military, border patrol, etc., as well as the unofficial support of the Kurdish tribes, with whom we had established excellent relationships. In villages with no lodging and no food outlets, the only choice was to be hosted by a friendly Kurdish chieftain and to be known as a friend (Fig. 4).

Fig. 4 Typical tea break during the surveys inside Surp Sarkis Church in Fanai village (photograph by the author 1970s)

Fig. 5 Reaching our destinations on horseback (photograph by the author 1970s)

A significant part of our expeditions involved walking or, on occasion, traveling on horseback (Fig. 5). In a border town close to Turkey with literally no roads, we drove through the thorny hills on steep inclines to discover the Armenian cemetery of Charikh (Fig. 6).

When we learned about the churches of Darashamb (Fig. 7), we drove as close as possible, and then, we had to hike for a few miles through the hills in order to arrive at our destination. It is an important complex that is now easily accessible by road.

Fig. 6 Driving in ravines and hard to reach roads in Salmast (photograph by the author 1970s)

Our first mission was in 1976. Adriano was accompanied by Professor Gabriella Uluhogian, a linguistic expert from the University of Bologna, and Nice Vecchione (Fig. 8), a young architect, who did most of the photography and helped us with the architectural surveys. We traveled from Tehran to Tabriz on our way to Salmast. As fate would have it, we had to stay in Tabriz for a few days for necessary repairs to our cars, and thus, we visited the local churches, the community, and some of the unique monuments of the area, including the churches of Mujumbar (Fig. 9) and Sohrul (Fig. 10).

Salmast, at the time known as Shahpour, was the historical center of the local Armenian population since prehistoric times. However, for about 200 years, there was physically no city of Salmast. Sometime around 1722 the city disappeared or was abandoned. As the name was prominent and well known, *Salmas* became the name of the region, while a second city, Dilman, emerged as the leading town in the area. After the massive earthquake of 1930 and the total destruction of Dilman, the Iranian government, under Reza Shah, built a new city in the general area and called it Shahpour (the son of Shah). This city then was ultimately renamed Salmast after the Islamic Revolution of Iran.

Fig. 7 Church of Surp Asdvadzadzin in Darashamb (photograph by the author 1970s)

In Salmast during my previous visits, I had already been introduced to Ishkhan, the son of the local priest, and he was an expert beekeeper and extremely familiar with the region. He was friendly with Turkish, Kurdish, Persian, Armenian, and Assyrian communities and spoke all their languages.

Ishkhan became our principal guide and greatly contributed to the success of our missions by facilitating our travels to remote areas (Fig. 11).

Our typical day would start with a light breakfast of yogurt, flatbread, and tea (Fig. 12). A small, portable espresso machine was the only luxury item during our travels. Our work would start early at daybreak, and we would travel to the sites and the churches to be discovered, surveying, measuring, and photographing the monuments as much as possible.

Fig. 8 Gabriella and Nice at the St. Stepanos monastery with the caretakers (photograph by the author 1970s)

It was amazing how much our Italian colleagues could endure in an environment with minimal food supplies and water, basic human comforts, almost no access to showers or facilities, and being constantly harassed by mosquitoes, flies, and other critters while working under a scorching sun. The hardships of the time seem anecdotal today, but at the time, with no cell phones, no bottled water, or digital cameras and equipment, we were traveling on hard, wooden benches of our jeeps, driving through deserted landscapes for hours only to arrive in a housing with no hot water, rooms full of cockroaches, and no indoor bathrooms. Photographs were taken with 35 mm films, and surveys were done by hand and measuring tape. Food was scarce, mostly because of the weight that we had to carry. It was not unusual to have a lunch of an egg or two with some flatbread, as that was simply all the villagers had to sell or to give (Fig. 13).

When it was dark and work was impossible, we went to a prearranged Kurdish village, where the local chieftain generously hosted us in their "second" house, specifically built and reserved for visiting guests, apart from the main house of the landlord (Fig. 14). [*A/N: Typically the local Kurds have two houses within their property. The main house is always reserved for the family, while the second house is vacant and reserved for guests only*].

Fig. 9 Surp Hripsime Church in Mujumbar (photograph by the author 1970s)

These gatherings were certainly an extraordinary experience. We were a group of academics, tired and exhausted, sitting on the floor, and sipping tea while talking in several languages at the same time. We spoke Italian, Persian, Armenian, and communicated with some others in English, and our guide spoke with our hosts in Turkish, Kurdish, and, if necessary, Assyrian.

After hours of hard work in the sun, we were all tired, but we certainly enjoyed the generosity of our hosts, who prepared food and provided us with tea in exchange for vodka. At night, we all slept in the same room in beddings heavily infested with bed bugs, which were totally immune to all the bug-killers we had purchased (Fig. 15). It was an experience, by the way, that could not be repeated today, as now men and women are not permitted to sleep in the same room if they are not married.

Fig. 10 Unusual church of Surp Hovhannes in Sohrul (photograph by the author 1970s)

At the same time, we continued the study of Armenian churches in the Isfahan area. Although its major churches are now quite well known, the numerous churches in the outlands of Isfahan were barely explored. As part of the academic program at NUI, we had started a research initiative regarding the provincial churches of Charmahal, Peria, and other neighboring villages, around Isfahan. Adriano did not have the opportunity to visit them as much as we did in the Salmast region, but all the research material was shared and archived jointly between the universities in Iran and Italy.

A comprehensive report of our first expedition, including some of the initial work in the Isfahan region, was published in *Ricerca sull'Architettura Armenia* # 17, Iran I. Milano 1977 (Fig. 16).

In 1977, we conducted our second expedition to the region. Our research this time was more focused, as we had prepared a substantial amount of pre-expedition homework.

We had a larger team for this mission. We were joined again by Gabriella Uluhogian, her brother Haig, an architect, Nice Vecchione, as well as Professor Valeria Fiorani Piacentini (Fig. 17), who wrote an article about the fortress of Berdiuk (Fig. 18), which we discovered and documented during this expedition (Iran II 1978).

Fig. 11 Adriano, Nice, and Ishkhan at the fortress of Berdjuk (photograph by the author 1970s)

One of the first goals of this mission was to complete the work of the previous year and to look for unknown monuments. We had realized that our prior work did not properly cover all the monuments and their importance within the region.

It was fortunate that, by following whatever historical sources we were able to investigate and by carefully listening to the local oral history, we were able to re-discover the important church complexes of Darashamb and Sohrul as well as those in Fanai (Fig. 19) and Gris (Fig. 20) and Drishk (Fig. 21) and Khoy (Fig. 22) and Haftvan (Fig. 23), of course, the fortress of Berdiuk (Fig. 24).

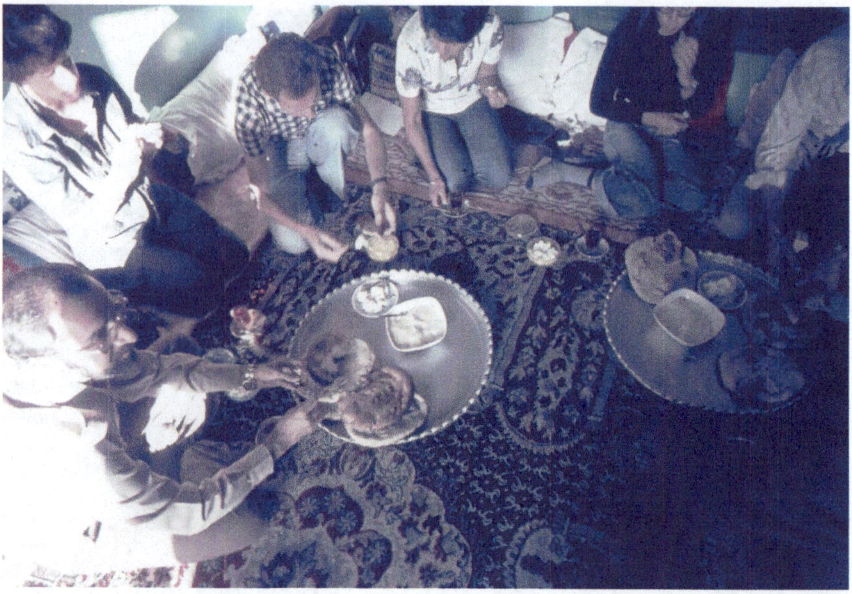

Fig. 12 Breakfast at our Kurdish hosts table (photograph by the author 1970s)

This was a particularly extensive expedition, with long stays in Kurdish mountains, always surrounded or accompanied by local Kurds, reaching into regions that were hardly ever explored.

The material collected during this survey was published in Iran II, Ricerca Sull'Architettura Armena, volume 18, Milan, July of 1978 (Fig. 25).

Besides the important discoveries of certain new structures, which were generally unknown until that time, this publication gave us an opportunity to initiate a typological study of the floorplans and structural techniques of all buildings, which remains a milestone in Armenian architectural studies.

While this book was being published in Italy, there was unrest in Iran. Ultimately, this would lead to the Islamic Revolution of 1979. The Kurdish-populated areas of North-Western Iran were in upheaval and in daily fights with the Persian military. Despite this situation, Adriano and I prepared a further program for our third expedition. However, at the last minute, the Italian team was advised that the situation in the region was not safe. Thus, this third mission (and then the fourth) was conducted without our Italian friends but with the help of our guide Ishkhan, his son, a few of the university students, and some local Kurdish and Armenian friends.

Fig. 13 Flat bread making in the fields in the mountains of Salmast (photograph by the author 1970s)

The material we collected was shared with Adriano, and an initial facsimile of Iran III was prepared. Alas, history changed our plans.

The Iran–Iraq war and the early formative years of the Islamic Republic of Iran were not conducive to the international research programs.

Research on Armenian architecture in Iran was on hold. Adriano, however, pursued his studies in Armenian architecture and published an important monograph on this subject.

Fig. 14 Typical Kurdish "great hall" with everyone sitting around the walls (photograph by the author 1970s)

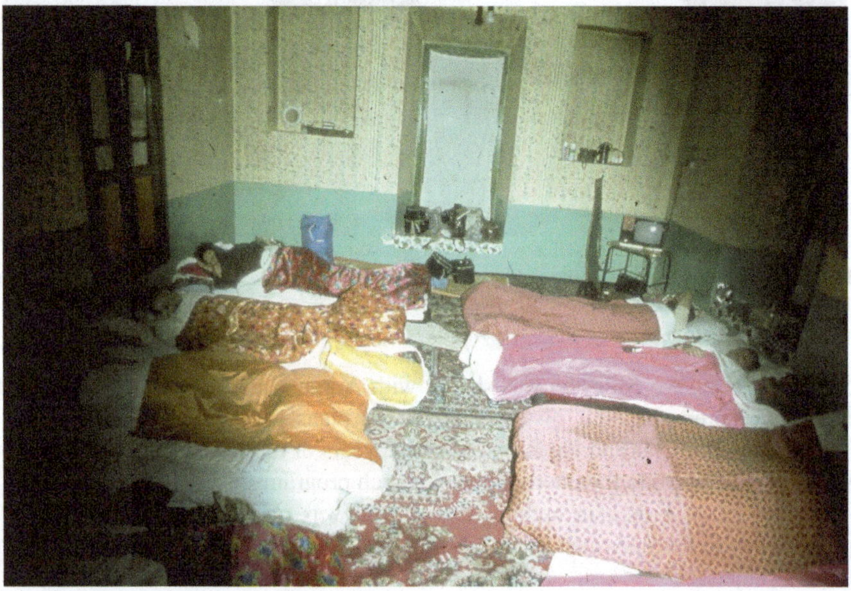

Fig. 15 Sleeping at night in the same "great hall" (photograph by the author 1970s)

Fig. 16 First volume of the research program

Fig. 17 Valeria Piacentini, the author, and our driver in the mountains at the Turkish border (photograph by the author 1977)

Fig. 18 Castle of Berdjuk (photograph by the author 1970s)

Fig. 19 Church of Surp Sarkis in Fanai village (photograph by the author 1977)

In the early 80s, I was living in Germany and often visited Adriano in Milan, working on diverse subjects of Armenian architecture. In the early 1980s, there was still considerable interest in matters related to Armenian cultural heritage in Italy, and Adriano was actively participating in many of these efforts.

In Milan, Herman Vahramian (1940–2009), who cooperated closely with CSDCA and thus de facto with Adriano and others, organized several conferences and programs in Milan with the support of Italian organizations as well as the local Armenian community. In Venice, Father Boghos Levon Zekyan founded a Chair in Armenian Language and Literature in 1976 at Ca' Foscari University's Asia and North African Studies Department. Further, he founded the Association of Padus Araxes and organized a series of summer courses on Armenian heritage in Venice, which has continued since 1986.

I saw Adriano for the last time in 1981 when he had already moved to Venice and Belluno. It was always a topic of our discussions to make sure that we would visit his paternal home in Belluno, and I would cook for the family. It never happened, but we met in Venice, and as always time was too short, with too many things to discuss, too many memories to cherish, and too many dreams to plan.

Fig. 20 Church of Surp Asdvadzadzin in Gris (photograph by the author 1977)

I must emphasize that I have known Adriano on two levels. The first is the scholar, the expert who had clear ideas of what needed to be researched, how, and why—and from whom I learned a lot. The second is Adriano the person, the father of a family that he cherished and loved beyond anything else.

As a scholar, he was one of the few who really understood the history of Armenian architecture and its intricacies and was willing to explore the secondary Armenian typologies of architecture. He was extremely open-minded and would listen to all of us, including myself, the younger, less experienced scholar, with my own ideas and suggestions. He would listen with particular interest to the villagers who shared with us mysteries, myths, and legends, and these would always attract his attention. He was methodical to the point that he could organize a clear vision within the confusion and multilayered information that we gathered during our missions. He brought a sense of order, discipline, and academic prestige to the entire undertaking. As such, it is not surprising that the volumes of Iran I and II are—after 40 years—still the points of reference for any research on the Armenian monuments of Northern Iran.

Fig. 21 Monastic complex of Surp Sarkis Zoravar in Drishk (photograph by the author 1976)

Fig. 22 Church of Surp Sarkis in Khoy (photograph by the author 1976)

Fig. 23 Surp Gevorg Church and the theater of Haftvan (photograph by the author 1976)

As an individual, he was the sweetest colleague I have worked with. I was always aggressive and impatient, and I always wanted more to be done in a shorter period of time, while Adriano was always composed and calm. I will always remember his quiet interventions: *Marco, ascoltami* (listen to me)…then he would point out what specifically I was doing wrong.

As part of our friendship and cooperative programs, Adriano (and sometimes the entire Italian team) often spent the night at our house in Tehran. But, we also spent many nights together in a third-rate motel room or in a simple Kurdish or Armenian house of a friend. I will never forget the tiny openable picture frame, with pictures of his family, that he carried with him at all times. Before going to sleep, he would put it next to his bed, say his prayers, and then say to me "*Buona Notte, Marco.*"

Adriano always appreciated life on many levels within the many cultures. He was a good listener, a mentor, and focused on the goals of the program he was working on. He was at home in Italy, of course, but he was also at home in Armenia, in Iran, and in Turkey. He would manage the daily difficulties without any problems and would always focus on the more important tasks at hand.

Fig. 24 Tower of the fortress of Berdjuk (photograph by the author 1970s)

Not surprisingly, both major Italian scholars of Armenian architecture, Adriano Alpago Novello, and Paulo Cuneo were also experts in Islamic architecture. This provided them with a broader perspective on Armenian architecture within the Islamic-dominated countries. In fact, both taught Islamic architecture in their later lives. More specifically, Adriano was appointed Professor of History of Muslim Art at the University of Ca' Foscari in Venice. He was modest, not controversial, and he was always happy to guide and to help. At the same time, he never compromised his stature as an expert on Armenian architecture, and he expressed his views in his books and his lectures as clearly as possible.

The specialized academic world as well as the Armenian community, both in Armenia as well as in the diaspora, respected and honored him with love and affection. He will always be a member of a pioneering group of Italian scholars who promoted a new vision of Armenian architecture in the world.

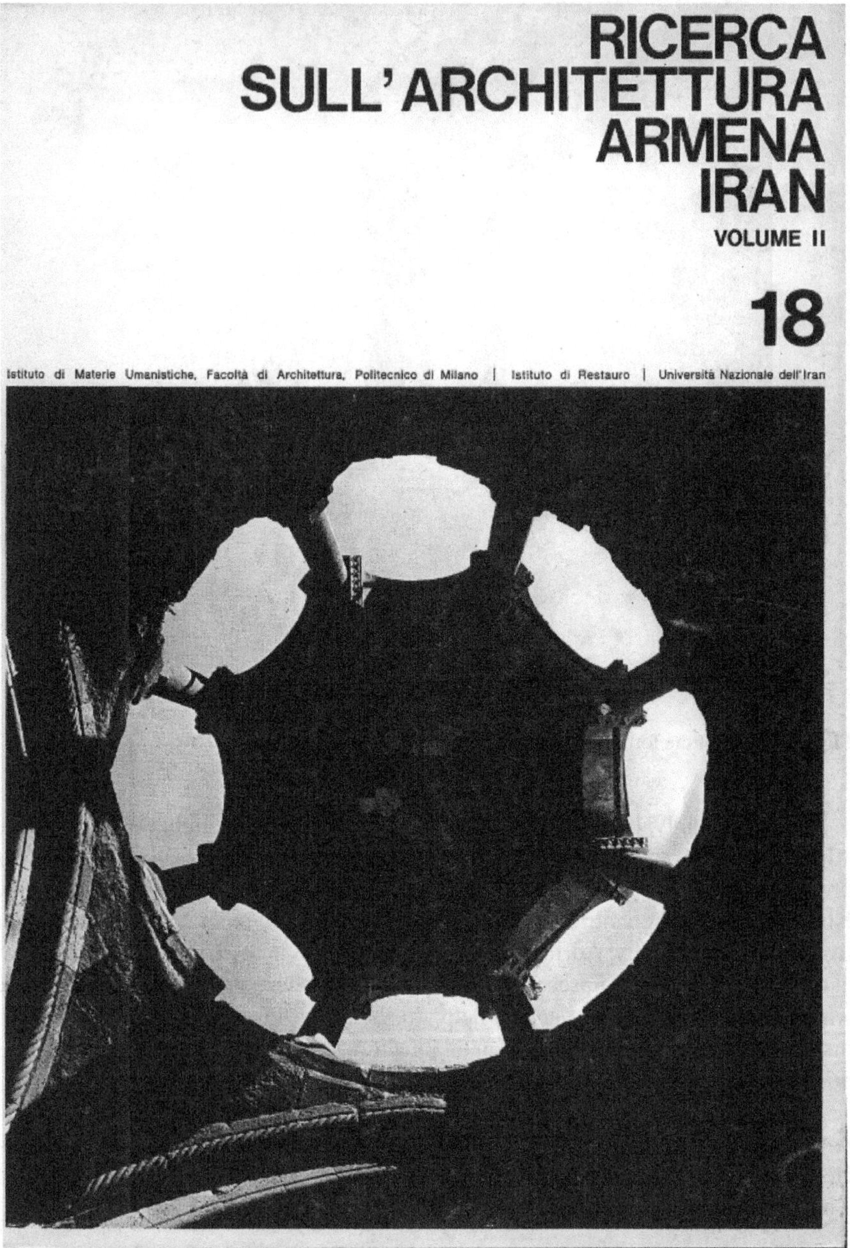

Fig. 25 Second volume of the research program

References

Architettura Medievale Armena (1968) In: De Luca (ed) Catalog of the exhibition. Palazzo Venezia, Rome, Italy
Cuneo P (1988) In: De Luca (ed) Architettura Armena. Rome
Documents of Armenian Architecture, (1988) Sorhul, vol 20. Oemme Edizioni, Milano
Ieni G, Zekyan L (eds) (1975) Atti del Primo Simposio Internazionale di Arte Armena. Bergamo 28–30 Giugno, San Lazzaro, Venezia
Kouymjian D (April 2018) Armenian studies past to present. Hrand Dink Foundation Publication, Istanbul
Novello A, Brambilla et al (1977, 1978) Iran I and Iran II, Milano
Novello AA (1987) The Armenians. Rizzoli, New York
Zekyan BL (ed) (1988) Quinto Simposio Internazionale di Arte Armena. San Lazzaro, Venezia

Marco G. Brambilla is a practicing architect and an architectural historian specializing in the history of Islamic and Armenian architecture. He studied art history and archaeology in Munich, architecture, in Paris and graduated from the School of Architecture in Rome in 1973 specializing in preservation of historic monuments.

He has taught architectural history and preservation of historic structures at the universities of Rome, Calabria and was the chair of the Department of Architectural Preservation, at the School of Architecture of the National University of Iran.

Since 1983 Marco Brambilla lives in Los Angeles, USA. He continues teaching history of Islamic and Armenian architecture in major Universities worldwide, Including Harvard, MIT, New York, Pennsylvania among others and is as an associate of Narekaci Program at the University of California in Los Angeles.

Preservation of Historic Heritage Sites in Iran

Marco Giovanni Brambilla and Bahram Ghadiri

Abstract The history of preservation of architectural heritage site in Iran is closely related to the development of archeological activities in the nineteenth century and the presence of a large number of European specialists in the country. In mid-nineteenth century, there was a revival of interest in Orientalism in general and the exploration of archeological sites in Iran in particular. Several French, German, British, and American experts moved to Iran, performed excavations on historic sites, advised the government about the notions of preservation and related laws, and contributed to the formation of a new generation of Iranian administrators and highly talented Iranian architects, who ultimately defined contemporary architecture in Iran in the twentieth century. The initial presence of European experts in Iran was heavily influenced by the French, who introduced the theories of Beaux Arts and Violet le Duc to Iran. In the late twentieth centuries, it was the Italian expertise that educated and formed a generation of qualified Iranians in the fields of conservation of all historical heritage values. This started with the direct involvement of Italians in restoring and preserving a large number of significant Iranian architectural sites as well as the educating process of the next Iranian experts. Furthermore, the evolution of preservation of heritage values in Iran was subject to massive political changes. The initial efforts started during the late nineteenth century during the Qajar period, upon the foundation of the Pahlavi dynasty by Reza Shah in 1925, the concept of preservation of historical sites and artifacts was restructured with a massive ideological program. They were then developed into a world class practice of conservation efforts on an international scale since 1972, only to be restructered again after the Islamic Revolution of 1978. This essay will explore the history, the parameters, the legislation, and the formation of the culture of preservation of historic heritage sites in Iran, from nineteenth century to present.

Keywords Iran · Restoration · Architecture

M. G. Brambilla (✉) · B. Ghadiri
Los Angeles, USA

M. Boriani and M. Giambruno (eds.), *Architectural Heritage in the Western Azerbaijan Province of Iran*, Research for Development,
https://doi.org/10.1007/978-3-030-83094-6_3

1 Introduction

With a history of construction that spans over millennia, it is not surprising that Iran also has a long-standing tradition of maintaining and restoring its architectural heritage.

However, the evolution of the culture of preservation of historical sites in Iran has been a complicated process. The country is vast; its climate and building materials are diverse, as is its ethnic mix and cultural and religious heritage.

Certain areas of Iran, especially in the north and west, are heavily prone to earthquakes. Other regions suffer from severe temperature changes: Temperatures may drop by 30 degrees or more during the night, thus creating material fatigue and difficulty controlling the environment. Environmental pollution, at least until recently, has not been a significant issue for many historical sites. The majority of Iranian historical monuments are either in smaller cities or in the countryside and are thus less affected by environmental pollution. This is not true for monuments in the larger city centers, like Tehran, Isfahan, and Tabriz.

Historical sites have also been subject to massive and rapid urban development programs. These programs, often promoted and initiated by the central government, were further aggravated by the private sector's speculative land-use projects. Reza Shah (1878–1944) implemented new urban development concepts similar to those in Europe to modernize Iran. These modernization concepts, including more extensive and broader streets, slash the city's historical fabric and change the traditional urban city centers forever.

This presentation will provide an overview of the history of preservation in Iran within the context of its gradual transformation from traditional techniques of restoring historic buildings to a contemporary scientific preservation approach of preserving historical heritage sites from the nineteenth century to date.

2 Building Materials

Stone is not a very common building material in Iran. However, it has been used both in ashlar and in rubble-type masonry construction. In the sixth century B.C., the Achaemenid kings' spectacular palaces in Persepolis were built with polished, cut stone (Fig. 1). There was already an expert technique for repairing cracks and failures in stones by specialized stonemasons (Tilia 1968, pp. 67–95). This expertise carried over centuries, as evidenced in stone repair works in the fourteenth-century Ilkhanid, Iran (Figs. 2 and 3).

Wood as a construction material was common in Gilan and Mazanderan at the southern shores of the Caspian Sea (Varjavi 2015, pp. 825–832.). It was typically used as part of the flat-roof support system and as a structural element to provide lateral support for buildings in brick or rubble stone construction (Fig. 4). In palaces or prestigious mosques, especially in the Safavid and Qajar periods, the supporting

Fig. 1 Persepolis, 515 BC, photograph by Andreh Marouti

Fig. 2 Walls of the citeadel of Sultaniyya, photograph by the author

Fig. 3 Restoration work on stones of Sultaniyya, fourteenth century, the largest brick dome in the Islamic world. Photograph by the author

Fig. 4 Isfahan, New Julfa, wood in roof framing and as a support element. Photograph by the author

Fig. 5 Isfahan, Ali Qapu (1597), decorated wood columns and ceiling. Photograph by Andre Marouti

columns would be elegantly finished with elaborately decorated capitals and artwork (Fig. 5). In simpler structures, the supporting posts were plain with simple capitals. Wood was also used as a ceiling cover, often painted and decorated (Figs. 6 and 7).

However, the Persian plateau's primary building materials were un-baked brick (*khesht*) (Fig. 8) or, alternatively, baked brick. A mixed construction technique took advantage of the best aspects of both materials (Fig. 9). Un-baked brick was used as the core of the construction, a cheaper, and more manageable task, and baked brick was used only as the exterior shell to provide better protection against natural elements. This technique was known in Iran as early as 1250 BC when the ziggurat of Tchogha Zanbil was being erected. By nature, un-baked brick requires more upkeep and maintenance. It is washed away by rain and water, erodes over time, and must be reconditioned regularly. Baked brick is more durable, but it subjects to the forces of nature: wind, rain, snow, and earthquakes; it also needs to be repaired, restored, and strengthened periodically. In traditional Iranian construction techniques, gypsum is often used as the main mortar element. It is quick-drying, easy to produce, and entails fast construction times. It is, however, subject to deterioration due to humidity and, as such, is not appropriate in humid climate zones. Gypsum mortar was also often used combined with lime or with clay.

All mud-brick buildings were planned to be maintained over time. They were thicker in volume, over and above the structure's need, to allow for erosion. After some mud-brick layers were washed away by rain or other natural occurrences,

Fig. 6 Isfahan, the House of David, decorated wood framed eves. Photograph by the author

Fig. 7 Isfahan, the House of Soukias, decorated wood ceiling. Photograph by the author

Fig. 8 Isfahan, New Julfa. Mud-brick as the main construction material. Photograph by the author

Fig. 9 Isfahan, New Julfa, mixed construction technique with mud and baked bricks. Photograph by the author

Fig. 10 Bam Citadel (fourth century BC through eleventh century AD). Massive mud-brick structure. Photograph credits, WikiCommons

another layer of mud would be added, thus recreating the original configuration and stability.

Due to their nature, mud bricks provide limited structural and architectural freedom. Typically, earthen architecture in mud bricks is limited to smaller structures of one or two stories. Still, there are examples of excellent uses of mud-brick in large-scale complexes, like that of the spectacular fortification complex of Bam, (Fig. 10) World Heritage Site (https://whc.unesco.org/en/list/1208). Baked bricks allowed master builders to plan and execute larger structures with impressive spans and high vertical volumes.

Nevertheless, throughout the ages, the builders of Iran were also conscious that these structures would require periodic repairs and maintenance. As such, preventive measures and pre-planning were an inherent part of any new construction. Preventing moisture accumulation and humidity in the structures and devices to drain all water away from the buildings was an integral part of the original projects. Furthermore, readily available building materials and traditional construction techniques contributed to the ease of building a facility and a longer lifespan for structures.

In the tradition of Iranian architecture, there is a unique phenomenon facilitating many brick buildings' future repairs. Being well-aware that buildings eventually would need to be maintained and repaired, and artisans and builders would incorporate various openings, *Moshtu*, essentially pre-arranged holes in the walls (Fig. 11). This would, in the future, allow the quick and straightforward installation of beams in these openings, thus creating a temporary scaffolding system to enable the craftsmen to perform their repair and restoration work. Moreover, in the building industry, masons, bricklayers, and artisans guaranteed that the maintenance work would be compatible with the original structure and warranted its survival in time.

Fig. 11 Moshtu in Iranian architecture. Opening left in the wall to allow the future scaffoldings for repair work. Photograph by Andre Marouti

3 Traditional Organization and Legislation of Preservation

While technically there was already a tradition of restoring and maintaining structures, since the ninth century, there was also a financial and legal support system to ensure the survival of most public buildings in Iran.

After introducing Islam in Iran, many public buildings, such as mosques, caravanserais, baths, bazaars, and others, became part of religious endowments. Based upon certain *hadiths* of the Koran, an institution, *Waqf*, is created to ensure the prosperity and the survival of many public buildings. This guaranteed an annual income dedicated to the upkeep of the structures. A trustee was appointed to spend the funds appropriately and to safeguard the facilities in perpetuity. The core element of these endowments was the continued financial support to enable the preservation of the bequeathed buildings. In this context, income from specific properties, businesses, and taxes were allocated as specific endowments; in theory, as long as the income was secure, the funds would be dedicated to the maintenance of the building (Nazari and Kakhaki 2018, pp. 239–257).

The tradition of caring for and preserving buildings was an inherently important factor in the history of preservation efforts in Iran. On the one hand, the *Waqf* system, the endowment system, provided funds for many structures to be maintained, and as many such monuments have survived over centuries. The ease of repairs, commonly available building materials, and artisans trained in a long tradition of construction techniques further facilitated the repairs and the conservation of the structures. On the other hand, the nomadic nature of the population often promoted the building of new structures instead of preserving existing ones. Furthermore, specific changes in religious inclination favored certain monuments while devaluing others. In general, though, whenever there was continuity in the political systems and the trade routes were safe, buildings were also well-maintained.

In the late nineteenth and early twentieth centuries, European scholars, archeologists, and art historians visited Iran frequently. Suddenly, there was a realization that there is immense architectural heritage in the country, dating from prehistoric times. This discovery pioneered a series of archeological excavations, research, and Iranian art and architecture studies. It was not surprising that at one point, it was noted that the preservation and structural restoration of many historical sites were necessary.

As is often the case, the first step was to create awareness in the private and public sectors. The presence of international specialists pointing out the importance of these monuments certainly contributed to the process. The entire development of the concept of preservation of historical sites was a complicated process. While critical figures in the administration supported these efforts, the population was more reluctant to join these initiatives.

The historical perception of what is valuable in a building, beyond its physical uses, is an educational and cultural process, constantly evolving and adapting to time's realities. What may appear as a fantastic monument to the visiting tourist or scholar is simply a building to the local inhabitants. The restoration and preservation of a historical site may or may not affect the quality of life of the local population, and hence, they were reluctant to support these initiatives. This was and is a common problem and one that has to be seriously considered. A consensus has to be reached between the local population, the administration, and the central government trying to preserve historical monuments and artifacts.

In any culture, the sentiment of pride of the ownership of historical heritage has to be perceived by both the central government and the population at large before groundbreaking practical efforts to save these monuments can be implemented. It is questionable if this occurred in Iran at any time in a meaningful, intensely perceived manner. Laws and regulations were slow to be implemented, and the only advantage was that a large body of monuments was already under the religious endowment care.

For the conservation efforts of these buildings, especially religious structures under *Waqf*—the Bequeaths Organization, a dialog had to be created between the secular administrators and the clergy. By nature, this is a tedious and laborious process that yields mixed results.

Ultimately, there were three essential milestones in creating awareness and preserving historic heritage values in Iran. The first was the formation of the National Monuments Council (NMC) in 1922. This was followed by the formation of the National Organization for the Preservation of Historical Monuments in

Iran (NOPHMI) in 1965. Finally, in 1985, the Iranian Cultural Heritage Organization (ICHO) developed from an organization into a joint ministry with tourism.

Furthermore, in the early twentieth century, the formation and the contribution of the "architect" were being defined.

In the modern sense, the profession of architect, as opposed to the traditional figure of *Mimar*, the master-builder, was developed in the 1940s. Arts colleges were formed where, next to architecture, painting and sculpture were also taught. This gradually developed into the school of architecture within the newly established University of Tehran. Its founders were almost exclusively from the School of Beaux Art in Paris. Three French nationals were its founding members (Godard, Siroux, Dubrulle), while the most influential Iranian architect, Mohsen Foroughi, was the first Iranian to graduate from Beaux Art. This French influence and its "modernism" would dominate professional education and architecture in Iran until 1960 when the School of Architecture at the National University of Iran was founded. This architecture school was heavily influenced by Italian architectural traditions, as most of its faculty had graduated from Italian universities.

4 Preservation of Historic Monuments in Iran in the Early Twentieth Century

The initial efforts of surveying and preserving Iranian historical heritage sites orig inate with several international archeologists, art historians, and practicing architects. The first wave of interest in Iran began in mid-nineteenth century, when European travelers, accompanied by graphic artists, visited Iran and described and illustrated many spectacular monuments. Upon their return to Europe, these sketches and rendering were widely published and contributed to the fashionable "Orientalism" in vogue at the time.

The second wave of European presence and activity in Iran is also attributable to various foreign archeologists. The first of many, between 1850 and 1852, William Kennett Loftus (1820–1858) initiated a series of excavations in Susa, the Achaemenid Empire's administrative capital (550 BCE–330 BCE), which was mentioned in the Bible and as such essential to the European elite of the time. Marcel Dieulafoy continued this work in 1884 with the generous support of Naser al-Din Shah (reign 1846–1896), who was personally interested in antiquities and was willing to sponsor excavations and archeological research. His collection eventually will become the first inventory of the first Iranian archeological museum, the Museum of Iran Bastan. The work of Dieulafoy was further advanced by Jacques de Morgan (1857–1924), who continued excavations in Susa.

Most importantly, he was also able to secure a monopoly for all archeological excavations in Iran for the French, which lasted over 30 years. In Iran, the activities of foreign archeologists would continue with Roman Girshman (1895–1979), who conducted excavations in Teppe Sialk, Tchoga Zanbil, and other sites in the

1930s. The prominent western archeological presence in Iran would continue into the twentieth century with David Stronach, Wolfram Kleiss, and many others.

The perception of the value of historical monuments and artifacts during these early periods, specifically during the Qajar period, is highlighted by archeological excavations and limited consideration for the actual preservation of either archeological sites or historical monuments. (https://iranicaonline.org/articles/conservation-and-restoration-of-persian-monuments).

With Reza Khan's rise to the throne of Iran in 1925, there was a new awareness of national identity and a policy to support archeological and conservation practices as further proof of the nation's glory. The political goal of Reza Shah was to promote a new Iranian identity and confirm a 2500-year continuity of the Iranian monarchy (Abdi 2001, N 105). In this context, Persepolis became a symbol of the new monarchy, and of course, its conservation was of the utmost importance. Furthermore, a unique architectural style was developed—de novo—that incorporated pseudo-Achaemenid architectural elements with modern European architectural trends. Many public buildings were built based upon an invented historical Iranian architectural style, including all significant public buildings, such as the National Bank (Fig. 12), the National Library, the Telegraph Office, the Military Academy, and many others.

In the early twentieth century, there was already an appreciation of Iranian monuments' importance and the necessity to safeguard and preserve them. In 1922, this general interest was formalized into ANJOMAN-E ĀṮĀR-E MELLĪ (AAM)—the National Monuments Council—the first Iranian organization with the specific goal to "promote interest in and to preserve Iran's cultural heritage" (Ṣadīq 2002, p. 83).

This effort was led by Moḥammad-'Alī Forūḡī, a friend of Reza Khan and a future prime minister, who continued his activities through 1934.

From its inception, the NMC invited a group of elite international experts to collaborate. These included the German Ernst Emil Herzfeld (1879–1948), the French André Godard (1881–1965) and Maxime Siroux (1907–1975), and the American Arthur Upham Pope (1881–1969).

Herzfeld moved to Iran in 1925 and stayed until 1934. He excavated the major Achaemenid sites of Persepolis and Pasargadae and prepared the first outline of the law of antiquities and the very first list of protected monuments in Iran. His archeological work in Iran was continued first by his assistant Friedrich Krefter and then by the American Erich Friedrich Schmidt (1897–1964). This list of 88 national monuments was published in 1934 by the NMC and was ultimately further expanded by Godard.

André Godard, an architect by profession, traveled to Iran to work for the government in 1929, soon after Reza Shah had established the Pahlavi monarchy. He stayed in Iran for 34 years, and his contributions to Iran's history and culture are extensive. He was one of the founders of the Fine Arts Department at the University of Tehran, of which he was the honorary dean. As a non-Iranian citizen, Godard could not be the "dean" of the newly founded architecture school. Instead, a prominent Iranian, Mohsen Foroughi, became the "administrative dean," while Godard was the "honorary president." In the early 1940s, the College of Architecture benefitted from

Fig. 12 The National Bank building in Tehran, representing a pseudo-Achamenid style. Photograph credits, WikiCommons

Fig. 13 The Iran Bastan Museum (1937). Designed by A. Godard, and it capitalizes on the images of Sassanian kings and their palace of Cteisiphone. Photograph credits, WikiCommons

the instructions of the French: Godard, Siroux, and Marcel Dubrulle. Gradually, a new generation of Iranian educators became the driving force of the school, among these, most significantly Mohsen Foroughi, but also Behzad Tagherzadeh, Khachik Yabloian, Mohammad Ali Heidarian, and Mohsen Moghadam. This created a new generation of prestigious Iranian architects, including Hushang Seihun, Ali Sardar Afkhami, Nader Ardalan, Kamran Diba, and Hossein Amanat greatly defined the architecture of modern Iran.

Godard published a *Journal of Archeology* and designed several modern landmarks of Iranian architecture, including the Iran Bastan Museum (Fig. 13) and the National Library in Tehran, and the Tomb Memorial of Hafez in Shiraz. But more specifically, he also advanced the efforts of architectural research and conservation in Iran. He studied the historical architecture of Iran and formed committees to preserve and restore Iranian monuments. These committees completed the first restoration efforts of the Shah Mosque in Shiraz, the Sheikh Lotfollah Mosque in Isfahan, and the Shah Mosque in Isfahan under the supervision of Godard. He also expanded the national monuments list, initially compiled by Herzfeld, by adding several more historical sites to this list, thus registering 247 historical monuments.

The work of Godard will remain a milestone in the history of architecture in Iran. Although he was not very open to dialog and exploring other possibilities in his restoration projects, he indeed spearheaded and caused many structures to be restored. Most probably, his most significant contribution to the preservation of historical sites in Iran is that he did pay attention to it rather than the actual quality of conservation projects he supervised.

Godard recruited another French national, Maxime Siroux, as a colleague. Siroux, also an architect, worked on several new architectural projects and worked extensively on numerous preservation projects with Godard, including the mosques of Qom and Yazd.

A unique phenomenon was the work of Sarkis Khatchadourian (1887–1947) in Iran (Katchadourian S.peoplepill.com). An Armenian born in Malatya, Turkey, from 1908 to 1911, he studied painting at the Accademia di Belle Arti di Roma and then from 1912 to 1914 at the National Higher School of Decorative Arts in Paris. At the recommendation of Arthur Pope, he came to Iran in 1929 and worked in Isfahan until 1934. He was the first to restore the mural paintings of Chehel Sotoun and Ali Qapu.

Next to these international experts, who contributed to and formed the early stages of the movement for the conservation of national monuments in Iran, others, although not directly involved in the preservation efforts, were instrumental in researching Iranian architecture and presenting it to the world.

Arthur Upham Pope and his wife Phillis Ackerman were invited to Iran in 1925 and resided there for many years. While he was not directly involved in preserving monuments, his enormous body of work regarding the history of architecture in Iran is presented to the world for the first time the impressive *A Survey of Persian Art from Prehistoric Times to the Present*.

Richard N. Frye (1920–2014), David Stronach (1931–2020), and Wolfram Kleiss (1930–2020) were among the luminaries of research on Iranian art and archeology and contributed enormously to emphasizing the importance of Iranian historical heritage both in Iran and abroad.

In the meantime, the NMC continued its work and, in cooperation with the Directorate General of Archeology, pursued issues related to national heritage sites. Oddly, it was concentrated more on creating new memorials and statues of Iranian cultural figures than actual preservation work. The first of many such memorials was the Ferdowsi Mausoleum in Tus, which was completed in 1934. Many others followed, including the mausoleum of Ebn Sīnā at Hamadān, Saʿdī mausoleum at Shiraz (Fig. 14), and the Nāder Shah mausoleum and museum at Maŝhad, as well as the mausoleum of ʿOmar Ḵayyām at Nishapur (Fig. 15). This was undoubtedly a political decision as it emphasized Iranian heritage by glorifying certain key figures in Iranian history.

The NMC was ultimately restructured in 1965 when, by royal decree, the Sāzmān-e mellī ḥefāẓat-e āṯār-e bāstānī-e Īrān (the National Organization for Preservation of Historical Monuments of Iran, or NOPHMI) was established within the Ministry of Culture and Arts. This marked the beginning of a new era of historic conservation in Iran. In prior decades, preservation of historic structures was accomplished to the best of the ability of those involved. Local architects, community leaders, or supervisors would interpret a given structure's historical element and determine the means and techniques for its conservation. This often resulted in significant changes to the original design, demolishing parts of buildings, and generally performing work without scientific guidelines. Traditional masons and artisans carried out all

Fig. 14 Sa'di Mausoleum in Shiraz (1950–52). Photograph credits, WikiCommons

Fig. 15 Tomb of Omar Khayyam in Nishapur. Built in 1963 by Houshang Seihun. Photograph credits, WikiCommons

preservation work, and their intervention was based upon their own experience, the ease of execution, and the budget to complete the project. There were no specific or clear laws regulating historic buildings' preservation in the private sector or religious endowment charters.

With the formation of the NOPHMI, there was a systematic approach to preserving historic buildings. These efforts were multifaceted and comprehensive. There were clear by-laws, an organizational structure with regional offices in each significant province, a precise mechanism, and a system of prioritizing what to preserve with considerable financial resources. There was a new system and guidelines for all preservation work. This included the necessary studies and research, documentation, evaluation, and preparation of a restoration project package. At the core of the NOPHMI, there was the Technical Committee. This is the body that approved every conservation project in the country. Clear guidelines were given to the regional offices and any other organization intended to perform restoration/conservation work on any structure. A format was created with which all applicants must comply. This included a historical evaluation of the building and essential documentation, including surveys, as-built plans, photographic data, a description of damages or perils to the building, a program of preservation complying with international guidelines, a time frame, a budget, and a completion date. This general concept and methodology were later adopted by the newly configured, newly structured Iranian Cultural Heritage Organization in 1985.

The committee reviewed all the projects proposed for a fiscal year and prioritized the budget with the proposed scope of work. If a building needed urgent repairs or temporary measures to prevent its collapse, that was a priority. If a structure required long-term planning, in-depth surveys, studies, or photographic documentation, this was considered a priority and approved. The entire system was based upon a long-term plan (typically five years). A system of reporting to the central body was established. All projects had to provide monthly reports about their progress, and a system of checks and balances ensured that funds disbursed were allocated to the appropriate progress of the work.

At the same time, Iran entered the international community of preservation of historical heritage. The Charta of Athens and Venice (www.icomos.org/en/charters-and-texts/179) was adhered to by all Iranian authorities to implement conservation work. The government's commitment to the conservation of national heritage contributed significantly to adapting many new laws promoting historical heritage sites, similar to those adopted in other progressive countries.

It cannot be ignored that the NOPHMI, notwithstanding its mistakes, provided outstanding service to the field of preservation of historical sites in Iran. The entire effort of this organization was also self-serving. The Iranian government cared for its historical heritage sites, but the NOPHMI was also used as a public relations program to project the Pahlavi era's image. There has been an ongoing critique of the preservation efforts of the Pahlavi period, claiming that the main emphasis of these programs was pre-Islamic monuments, Ilkhanid and Safavid palaces, and public buildings, like caravanserais, baths, and bazaars, but not many Islamic structures. This would become a point of contention in the post-revolution years, as many suggested that the NOPHMI was less concerned with religious establishments and more so with secular buildings.

5 Preservation of Cultural Heritage in Iran and the Italian Contribution

With a rising economy and a powerful organization (NOPHMI) in place, it was only natural that the next step would be to improve its quality of restoration and conservation efforts. Not surprisingly, this brought the governments of Iran and Italy together with mutual long-term benefits. In the early 1960s, Italy was embarking on exploring and defining the essence of historical conservation. Piero Gazzola (1908–1979) was one of the first Italian pioneers in the field. Nevertheless, at the same time, the theory of preservation of monuments was widely advanced by R. Bonelli, C. Brandi, G. Carbonara, G. De Angelis D'Ossat, and R. Pane, and later by G. Zander, E. Galdieri, and others. It is not surprising that after the Charta of Athens of 1931, the subsequent groundbreaking guidelines for preserving historical monuments were staged in Venice in 1964. By the 1970s, Italy had already started, organized, and implemented what were to become the general guidelines and the norms for preserving historic buildings for decades to come.

At the same time, in Italy, there was a sentiment of outreach to the Middle East. With perfect timing, Giuseppe Tucci became the president of IsMEO (Istituto Italiano per il Medio ed Estremo Oriente) and initiated an active dialog with Iranian authorities (Puoti 2009/2010). This Italian/Iranian interaction began first with archeological work and then extended into a large-scale involvement of Italian experts in Iran in other disciplines. Many scholars describing Italians' work on the restoration of monuments describe preservation projects in detail but ignore a critical factor. The Italian experts' involvement in Iran was not limited to the individual rehabilitation of certain structures. Instead, it was a turning point in the entire approach to preserving heritage sites in Iran, with long-lasting effects and traditions.

The total effort of the Italian presence in Iran was not to complete a proper restoration project. It was a learning and educating process. The most advanced preservation theories and technologies available in the west were integrated with traditional Iranian techniques and experiences to formulate a unique strategy and a methodology to safeguard historical monuments. The most advanced restoration theories were applied to create a particular, custom-made approach to safeguard Iranian monuments beyond simple preservation. This was research, documentation, study, exploration, and education.

What the Italian team brought to the country were not just specialists. They brought Iran the culture of preservation. Students and potential architects were provided with the opportunity to visit Italy and learn the required skills and expertise from the most qualified experts of the time. They promoted architectural restoration courses in the primary schools of architecture in Iran and established a series of desirable student exchange programs in Italy. Over the years, this would become a key element of the culture of architectural preservation in Iran. In fact, most individuals teaching or working in Iran today are graduates of schools of architecture in Italy.

The most critical Italians contributing to these efforts are well-known. Still, they deserve to be mentioned again as they created a great tradition of appreciating, studying, and preserving architectural monuments not only in the Middle East but worldwide. They set the example of working with prehistoric mega-monuments, like Persepolis and delicate and intricate Islamic structures, like Isfahan's Safavid palaces. As the visionary and overall leader of these efforts, Tucci was supported on the technical end by De Angelis D'Ossat, a considerable authority on architectural preservation matters at the time. Giuseppe Zander acted as the director of the restoration projects in Iran and published an essential volume regarding the contribution of the Italian experts to the preservation of historical sites (Zander 1968). Eugenio Galdieri, the project architect of many of the restoration projects in Iran, committed his heart and soul to understand the complexities of Islamic architecture in Iran and identifying a methodology to safeguard them. Starting with groundbreaking archeological excavations led by Umberto Scerrato, the Italian group assumed responsibility for the preservation of the Persepolis complex in 1965. The meticulous studies and detailed work of Giuseppe Tilia and his scholar wife Ann Britt created an example of how to work, study, and preserve a pre-historical monument in the world.

Soon after that, the preservation of Safavid structures in Isfahan was initiated with the direct involvement of Eugenio Galdieri. As the project leader, the in-depth studies of past Iranian building technologies allowed Galdieri to devise modern preservation methodologies of historical monuments within the Iranian building tradition. His restoration projects at Ali Qapu (Fig. 16), Chehel Sotun, Hasht Behesht (Fig. 17), and the Congregational Mosque of Isfahan (Figs. 18 and 19) are remarkable achievements, which were honored internationally and are to date the benchmark of quality for any restoration project in Iran. The list of Italians contributing to these efforts is not limited to a few individuals. Soon after Galdieri started working in Isfahan, Rosario Paone joined the Italian team and led an impressive restoration and study project in Pir e Bakran. When the Iranian government had given up the most spectacular brick dome structure of the Islamic world —the Mausoleum of Uljaitu in Sultaniyya—(Fig. 20) as doomed to be lost, Tucci introduced another Italian expert restoration work, Piero SanPaolesi, to Iranian authorities. Together, they were able to convince the Iranian authorities that the dome of Sultaniyya could be saved and was worth the effort. Thus, one of the most important monuments of the Islamic world was saved from destruction (http://www.Sultaniyya.org).

The quality of these restoration projects was beyond anything that had occurred in Iran and set an international benchmark of quality for all similar restoration/preservation efforts, not only in Iran but worldwide. The actions of the Italian group were not limited to architectural experts in the preservation of buildings. Italians brought Iran the culture of properly surveying monuments, researching and studying them, and proper documentation. They also insisted that all projects should be evaluated for short-term dangers and have long-term projects prepared for their

Fig. 16 Isfahan, Ali Qapu (1597), one of the most complicated restoration project led by the Italian team and especially E. Galdieri. Photograph credits, Andre Marouti

preservation. They brought in experts to repair and restore mural paintings, wood structures, plasterwork, and other elements. They provided scholarships for Iranian students to continue their education in Italy and then return to Iran to preserve an immense historical patrimony. They gave the Iranians the prestige of being at the forefront of architectural preservation worldwide.

This chapter would not be complete without mentioning key Iranian individuals who assisted the Italian group and were instrumental in carrying out Iran's preservation efforts independently. Probably, the most critical individual to be mentioned is Bagher Shirazi, a graduate of the School of Architecture at Tehran University, and he was one of the first Iranians to attend the School of Restoration in Rome. Upon returning to Iran, he became the director of the regional office of the NOPHMI in Isfahan. He was in direct contact and collaboration with the entire Italian team of IsMeo. He was the primary and the most critical counterpart to the Italian group and greatly facilitated the enterprise, sharing the Aga Khan prize with Galdieri. (https://www.akdn.org/architecture/project/ali-qapu-chehl-sotoun-hasht-behesht).

After the Islamic Revolution, Shirazi became the director of the NOPHMI for a short period. When the NOPHMI was restructured in 1985 and the new ICHO was formed, he was its executive vice president and a leader of all efforts to preserve Iranian cultural heritage sites. For many years, he prevented the destruction of historical sites, provided guidelines to new, emerging specialists, and advocated the need

Fig. 17 Isfahan, Hasht Behesht. Another major project by the Italian preservation experts in Iran. Photograph credits, Andre Marouti

Fig. 18 Isfahan, the courtyard of the Jami/Congregational Mosque. One of the most spectacular mosques in Iran spanning centuries of construction and additions. Photograph by the author

Fig. 19 Isfahan, the Jami Mosque. Interior. Photograph by the author

Fig. 20 Sultaniyya. The mausoleum of Uljaiytu (1314). Photograph by the author

for preservation, funding, and research to the government. Shirazi's impact on the conservation of historical sites in Iran cannot be ignored. He was instrumental in the creation of the School of Restoration at Farabi University in Isfahan. He founded the prestigious ASSAR publication, which to date is the main venue for architectural studies and essays in Iran. Upon his retirement, he founded the Iranian chapter of Icomos and continued teaching and lecturing in all significant Iranian universities.

Mohsen Foroughi, a prominent, French-schooled Iranian architect, was a knowledgeable force in the community and a dominating member of the Technical Committee of the NOPHMI. Firouz Bagherzadeh and Ezzat Negahban, both from an archeological background, were extremely helpful in monitoring and preserving archeological sites. Fat'hollah Mojtabai and Karim Pirnia, icons of Iranian architectural studies, always provided help, directives, and suggestions to implement preservation projects throughout Iran better.

6 Preservation of Historical Heritage Sites After the Islamic Revolution

In 1979, Iran witnessed one of the most dramatic changes in the entire spectrum of government, administration, and legislation of the twentieth century. The monarchy was abolished, and Islamic law became the prevailing legislative instrument. A new religious, governmental structure was established, and the new Islamic Republic of

Iran redefined every aspect of daily life. It was natural that this would affect Iran's cultural life and reformulate the values of its historical past.

Post-revolutionary societies always struggle to re-evaluate their perspectives on values based on their history. In Iran, the process was further aggravated by the Iran and Iraq war of 1980–1988. The urgency of the war efforts and myriad internal religious and social issues were of urgent importance, and thus, appreciating the nation's cultural heritage became a secondary issue. To further complicate the situation, there were other disorienting elements in the population's minds and perception. Westerners carried out the entire effort of discovering the treasures of Iran. These efforts, while advocating the values of Iranian arts worldwide, ultimately served primarily to enrich European and American museums.

However, at this stage, the people of Iran were free from all western influences. In this context, the power and the impact of the west were generally considered injurious; as such, it is evident that the values of preservation of historical heritage promoted by the west were by default also degrading and needing to be denounced the fact that the international community contributed to Iranian arts' glory, and preserving many Islamic and non-Islamic monuments in Iran was irrelevant. The second point of ambiguity in people's minds was that the Pahlavi era sponsored the preservation of historical monuments considerably. At this point in time, though, the Pahlavi regime was considered evil; thus, it is logical to assume that anything they sponsored or initiated is also evil and must be repudiated.

This created a period of uncertainty related to identifying Iranian society's historical heritage values under the Islamic Republic of Iran. People perceived that the historical heritage sites were tied to the previous regime and, as such, could not be good. In the heat of the early days of the revolution, the preservation of historical sites was not the primary concern of the government or the population. In this challenging time, many individuals, either as private citizens or as part of the administration, took it upon themselves to save cultural heritage sites and increase awareness of their importance both in the minds of the people and within the administration. Shirazi, Hodjat, Dr. F. Tehrani, Taghizadeh, and F. Bagherzadeh are just a few of this elite group.

Furthermore, several other factors contributed to an already complicated and delicate situation—the change of regime and administration promoted by necessity the appointment of new individuals to sensitive government positions. New governors, mayors, and political appointees gained certain jurisdictions with no prior knowledge about historic heritage values. Thus, it is not surprising that the same responsible parties would approach preserving historical heritage with a liberal, cavalier attitude. The mayor of Hamadan at the time proposed to use the city center, the historic site of the eighth century B.C. ancient city of Ekbatana, the cradle of Iranian civilization, as a new bus stop terminal. Others tried to destroy the structures of Persepolis as it was "connected" to the Pahlavi era.

As Iran became the free Islamic Republic, all previous laws and regulations became questionable and had to be verified by the new government. With the establishment of the new Republic, new social requirements surfaced. Among these, the practice of large-scale Friday prayers was and is of utmost importance.

These, however, required large spaces to accommodate the attendees' ever-increasing numbers, including prayer halls, spaces for ablution, bathrooms, offices, and storage spaces. Thus, in a revolutionary fever, local authorities changed or added to historical monuments, like the work done in the Vakil Mosque of Shiraz to provide for these new requirements, without much concern for the historical integrity of the heritage sites.

Urban development, an ever-present peril to all historical sites worldwide, assumed an unprecedented influence in Islamic Iran. After the war with Iraq, and encouraged by the clergy, the country's population exploded, and thus, new housing requirements in urban settings became urgent, necessary, and lucrative. City centers with historic fabrics and structures inevitably paid the price.

The transition from the turbulent war with Iraq to a relative stabilization of Iran's situation spans ten years. During the Iran–Iraq war, many historical sites in Iran were bombarded by the Iraqi military. The extreme problem of preservation of historical sites during the Iran/Iraq war has not been addressed sufficiently. During this war, the historical world heritage site of Tchogha Zanbil, a three thousand-year-old structure, the historic cities of Isfahan, Kashan, and Susa were bombarded by Iraqi forces. UNESCO, in its minimal efforts to address the issue, initiated a campaign to protect historical structures during wars. As always, their intervention and input were more theoretical than adequate. However, Iranian scholars and specialists on a local level took advantage of the international attention to forge ahead of the mechanisms that would eventually safeguard these monuments.

In 1985, three years before the Iran–Iraq war ended formally, NOPHMI was abolished and replaced with the Iranian Cultural Heritage Organization ICHO. Upon finishing the war with Iraq, a period of relative prosperity dominated Iranian society. The government's efforts were able to focus on matters not related to survival, martyrs, and other urgent issues resulting from the war.

Gradually, Iran gained a new appreciation for its past historical values, and the efforts of preservation multiplied. Many new universities with schools of architecture were established, and most of them provide courses on architectural conservation. The structure of reaching out from the capital into the regions via regional offices expanded and offered valuable, on-site working capability. Many preservation experts emerged, most of them from the Italian school, and provided high-quality preservation work at an international level. Among these Abdollah, Jabal Ameli, who, as the head of the Isfahan regional office of the preservation efforts, continued the work started by Galdieri, Akbar Taghizadeh Assl, was the director of the office of ICHO in Azerbaijan, and was an essential contributor to the efforts of architectural conservation. Mohammad Mehriar did outstanding work in the preservation of the Temple of Anahita in Hamadan and the conservation of the Bam complex. Dr. Farhad Tehrani and his colleagues contributed to the restoration of the bazaars of Kashan and Tabriz (Figs.21 and 22), and their group of architects on the project won the Agha Khan Award.

The expanded efforts of architectural preservation created its problems. There were simply not enough specialized human resources in the public sector to complete

Fig. 21 Bazar of Tabriz. One of the best preservation projects of historic sites in Iran and winner of the Agha Khan Award. Photograph by Amir Hossein Younessi

Fig. 22 Bazaar of Tabriz. Interior dome structure. Photograph by Amir Hossein Younessi

all the preservation projects. With that in mind, a structure was created, and independent architects and contractors were encouraged to bid and work on historical monuments. In theory, this would be a workable scenario, but many contractors who were not qualified were awarded specific projects with specific budgets. The results were almost always less than perfect.

The integration of the ICHO within a new ministry, Tourism and Preservation of Heritage Sites, gave more importance to the ICHO but created specific local and political problems. With leaders like Mehdi Hodjat and Saiyed Mohammad Beheshti, the scope of the ICHO expanded, and its impact was significant.

Today, Iran actively participates in the international forum of architectural preservation and is an active member of Icomos. Shirazi, who formally founded the Iranian chapter of Icomos, spearheaded Iran's efforts into the international community. Iran was supposed to host the Icomos General Assembly of 2011. Sadly, it was cancelled at the last minute due to restrictions imposed upon Iran by the west.

Internally, Iran continues its preservation efforts that are more driven by individuals than by overall structure. In the recent years, Iran and its cultural enterprises have suffered dramatically from the international sanctions imposed upon the country. Tourism has been reduced to an insignificant level, and difficulties with trade, currency exchange, and travel bans have significantly isolated the country. The exchange of ideas, scholarly interaction, and intellectual interchange have been reduced to the bare minimum. Iranian scholars have limited access or the possibility to interact with their global counterparts. Students have difficulty in obtaining travel visas or permission to enroll in European or American universities. In a way, Iran has been forced to define its existence within its confines.

7 Restoration in Azerbaijan in Northwestern Iran and the Historical Heritage of Armenian Sites in Iran

Soon after the NOPHMI was established in 1965, it created two special consultancy groups to help the organization provide quality preservation projects. Qualified Italian experts of IsMEO already led the prestigious projects of Persepolis and Isfahan. This covered the central and southern regions of Iran. Thus, the NOPHMI assigned six projects in the northern provinces of Iran to the two most influential architectural schools in Iran, both led by Italian-educated younger academics and preservation experts. This resulted in the universities being the consultants and contractors to the NOPHMI and providing professional preservation services.

In this context, the National University of Iran's School of Architecture was awarded the contract for the conservation work on the congregational mosques of Saveh (Fig. 23) and Qazvin (Fig. 24) and the Mausoleum of Uljaitu in Sultaniyya (Fig. 25). These projects have initially been led by Prof. Reza Kassai and later continued by the author from 1975 to 1979. The University of Tehran's Department of Fine Arts was awarded the restoration and preservation projects of the three

Fig. 23 Saveh, The Jami Mosque (twelfth–sixteenth century). Photograph credit, WikiCommons

Fig. 24 Qazvin. The Jami Mosque (ninth–sixteenth century). Photograph credits, WikiCommons

major Armenian monuments on Iranian soil: the monasteries of St. Taddeus (Fig. 26) and St. Stepanos (Fig. 27) and S. Sarkis's church in Khoy (Fig. 28). These projects were led by M. Falamaki, another specialist educated in Italy, whose work was continued mainly by Mohammad Hassan Mohab'Ali. All work proposed and directed by the above consultants was reviewed by the Technical Committee of the NOPHMI and often included international experts and advisors. The Armenian monuments' restoration efforts in Azerbaijan included input from the Armenian community and other specialists, among them Dr. Armen Haghnazarian, who completed his thesis on St. Taddeus and founded the prestigious organization of Research on Armenian Architecture.

It must be acknowledged with gratitude that while other neighboring countries with an abundance of Armenian heritage sites were more inclined to destroy them, the Iranian authorities then and now invested in and facilitated the preservation of Armenian monuments on Iranian soil. After creating the Islamic Republic of Iran and its turbulent initial years, the central government of the Islamic Republic renewed its commitment to warrant the preservation of Armenian architectural heritage sites. Thanks to Iranian authorities' efforts, the monasteries of St. Taddeus and St. Stepanos became UNESCO World Heritage Sites (Unesco.org. *Armenian Monastic Ensembles of Iran*), and the government still assists financially and with expertise, the preservation efforts of Armenian monuments in Iran. This reflects the policies of the Islamic Republic of Iran to preserve the heritage of minority cultures, which inherently are part of Iranian culture. Many Armenian monuments were included in the roster of

Fig. 25 Sultaniyya, interior decorations, rendering by the author

Fig. 26 St. Taddeus Monastery (68–1811). Photograph credit Andre Marouti

Fig. 27 St. Stepanos Monastery (62 AD–sixteenth century). Photograph credit Andre Marouti

Fig. 28 St. Sarkis Church in Khoy (twelfth century), photograph by Maurizio Boriani

protected and historic monuments after the Islamic Revolution. The government has continuously funded and helped their preservation efforts at all times.

Two singular cases in this respect must be mentioned. First of all, due to the efforts and financial support of local Armenian communities, the Chapel of the Shepherd (Fig. 29) was restored and saved from destruction (Rohani 2015). Second, the small but significant church of DzorDzor (ninth–fourteenth century) (Fig. 30), which was doomed to be inundated as a result of the construction of a nearby dam, was dismantled and reassembled at a higher elevation due to a joint effort between the Iranian authorities and the Armenian community in a perfect example of goodwill and joint venture.

In more recent years, Armenian churches' minor architectural structures have also been included in Iran's protection programs for historic sites. A significant discovery in recent years is the frescos of the church of Haftvan (Figs. 31, 32 and 33), which were restored by the efforts of ICHO. Many of these secondary structures are now included in the list of Iran's historical monuments, and the local government sponsors their preservation efforts. Notwithstanding political and social restrictions and sanctions imposed upon Iran by the west, there is significant synergy at the local level and a series of efforts to pursue conservation and research.

The Salmast Heritage Association has promoted and funded various expeditions to these sites by local architects and Italian experts. Scientific research material has thus been collected, evaluated, and used in part in this publication. Intergovernmental

Fig. 29 The Chapel of the Shepard, (sixteenth–nineteenth century). Photograph credit, WikiCommons

Fig. 30 The Chapel of DzorDzor (ninth–fourteenth century)... photograph credit, WikiCommons

Fig. 31 St. George Church in Haftvan (thirteenth–twentieth century)... photograph by Maurizo Boriani

accords have been created and renewed between Iranian and Italian universities, and it is more than likely that this cooperation will become more fruitful in the years to come. It is imagined that soon, with a joint effort by private and public sectors, local academia, the ICHO, and the Armenian Diaspora, the research and preservation of Armenian monuments in the region will be further preserved and integrated within the cultural horizon of Iranian/Armenian vision.

8 Conclusion

A short review of the background of historical preservation in Iran emphasizes a few critical points.

The effort of exploring and later preserving historical heritage sites in Iran was initiated primarily by Europeans, and a specific self-serving element cannot be dismissed. It was then part of the early Pahlavi era policy to validate the monarchy and legitimize its continuity. Both were ideologically tainted and would be controversial during later years. The establishment of the Islamic Republic of Iran was initially concerned with the Iran–Iraq war and many social internal issues. Nevertheless,

Fig. 32 St. George Church in Haftvan (thirteenth–twentieth century)… photograph by Maurizo Boriani

Fig. 33 St. George Church in Haftvan (thirteenth–twentieth century)… photograph by Maurizo Boriani

even before the end of the war, it embarked upon a renewed revival of architectural preservation.

There is no doubt that political aspirations and public relations efforts aside, the heritage preservation work of Iran during the 1960s and 1970s, with the assistance of Italian experts and specialized Iranians, catapulted Iran to the forefront of international preservation practices. After the revolution of 1979 and after a period of turbulent post-revolutionary confusion, the Iranian government fully adopted the international preservation community's rules. It made a considerable effort to safeguard its historical inheritance.

Once again, it seems that the cultural ties between Iran and Italy are being renewed, and more cooperation programs are being implemented (https://en.irna.ir/news/83558993/Iran-Italy). The Iranian government's support in preserving the heritage of minority cultures in Iran is yet another significant proof of the government's commitment toward a progressive view of the preservation of cultural sites worldwide.

The economic restrictions imposed upon Iran have caused further isolation of the country. Iranian professionals have difficulty in interacting with the international community and participating in the more extensive conservation policies worldwide. A network of international experts is not able to contribute to the preservation and research efforts in Iran. New technologies, construction techniques, and materials used in architectural preservation are not accessible to Iran. The cultural and intellectual isolation of any country prevents the opportunity to provide the best possible historical preservation endeavors for what currently is considered as the historic heritage of humanity.

Iran is committed to safeguarding its multifaceted historical heritage. It is important to note that the very first list of historical monuments compiled by Godard contained 88 monuments; Herzfeld updated this list to 247 sites in 1932. By the end of the Pahlavi era, this list had grown to 1633, while today, there are 32,708 historical sites registered. Significantly, many of these are related to minority cultures, modern architecture, and intangible heritage manifestations.

References

Abdi K (2001) Nationalism, politics and the development of archaeology in Iran. Am J Archaeol 105

Grigor T (2013) The king's white walls. Modernism and bourgeois architecture in culture and cultural politics under Reza Shah, 2013, Routledge

http://www.icomos.org/en/charters-and-texts/179-articles-en-francais/ressources/charters-and-standards/167-the-athens-charter-for-the-restoration-of-historic-monuments, http://www.international.icomos.org/charters/venice_e.pdf

https://www.akdn.org/architecture/project/ali-qapu-chehl-sotoun-hasht-behesht, http://www.Sultaniyya.org

Nazari F, Kakhaki AS (2018) Saz-va kare Tarikhiye Hefze Moghufat dar Iran (The working mechanism of historical preservation of bequeathed structures in Iran). Iranian Archeol Res, pp 239–257. NBSH.2018.12450.1541/10.22084. https://iranicaonline.org/articles/conservation-and-restoration-of-persian-monuments

peoplepill.com › people › sarkis-katchadourian

Puoti F (2009/2010) Il Centro Restauri dell'Is. M.E.O. Il ruolo di Eugenio Galdieri nel Restauro Italiano dei Monumenti in Oriente, Tesi di Laurea, Anno Accademico 2009/2010, Facoltà di Lettere e Filosofia, Corso di Laurea in Storia e Conservazione del Patrimonio Artistico.

Rohani SD (2015) Ph.D, student of Urban Planning and Design of Tehran Azad University. Restoration and rehabilitation of the world heritage site of Chupan Church. Int Arch Photogram Rem Sens Spat Inf Sci XL-5/W7. In 25th international CIPA symposium 2015, 31 Aug–4 Sept 2015, Taipei, Taiwan

Ṣadīq Ī (2012) ANJOMAN-E ĀṮĀR-E MELLĪ. Encyclopædia Iranica II/1:83. http://www.iranicaonline.org/articles/anjoman-e-aar-e-melli-aam-the-national-monuments-council-of-iran-established-in-1301-s. Accessed on 30 Dec 2012. https://whc.unesco.org/en/list/1208

Tilia AB (1968) A study on the methods of working and restoring stone and on the parts left unfinished in Achaemenid architecture and sculpture, East and West, vol 18, no 1/2 (March–June 1968):67–95. Rome IsMeo

Varjavi MS (2015) Identification of the Mazandaran Indigenous architecture concepts by studying the style of the wooden structures in the Northern Parts of the Country: with sustainable architecture approach. J Appl Environ Biol Sci 5(10S)

Zander G (ed) (1968) Travaux de Restauration de Monuments Historiques en Iran. IsMeo, Rome

Marco Giovanni Brambilla is a practicing architect and an architectural historian specializing in the history of Islamic and Armenian architecture. He studied art history and archaeology in

Munich, architecture, in Paris and graduated from the School of Architecture in Rome in 1973 specializing in preservation of historic monuments.

He has taught architectural history and preservation of historic structures at the universities of Rome, Calabria and was the chair of the Department of Architectural Preservation, at the School of Architecture of the National University of Iran.

Since 1983 Marco Brambilla lives in Los Angeles, USA. He continues teaching history of Islamic and Armenian architecture in major Universities worldwide, Including Harvard, MIT, New York, Pennsylvania among others and is as an associate of Narekaci Program at the University of California in Los Angeles.

The Story of a National Heritage: The Formation of the Iranian Preservation Tools Between Thoughts, Laws, and Institutions

Panteha Karimi

Abstract Many significant factors characterize a historical era and its thought about the inheritance of its predecessors. The interpretation of this heritage and its attributed values can reflect the sociopolitical circumstances of the society and the stance of its intellectuals. The relationship of the Iranians with the millennial history of their country and their inherited heritage has been and continues to be a complex historical subject that has undergone crucial alterations since the latter half of the nineteenth century. The present study aims to review important and decisive moments as well as the role of influential figures in forming the modern criteria and institutions for the preservation of Iranian heritage. The essay is divided into two main sections: the first section focuses on Iranian cultural policies in Qajar, Pahlavi, and Islamic Republic eras. The second section addresses the West Azarbaijan Christian heritage by analysing the restoration projects of St. Thaddeus church, St. Sarkis church, and Dzordzor chapel.

Keywords Iran · Preservation · Restoration · Qajar · Pahlavi

1 Introduction

The Persia described by the travelogues of the nineteenth century was a country in decline, which was just the shadow of its glorious past during ancient empires and Safavid Kingship. Iranian cities and villages covered with ruins of past epochs became a symbol of a decadent civilization known through classical sources as the oriental counterpart of Greece and Rome.

The French Eugène Flandin wrote in the Orientalist fashion:

> Throughout Persia, from north to south, from east to west, it is the same devastation; the ruins fall on top of each other ... to see the villages in ruins, the houses in dust, it seems that people no longer need shelter. Civilization is extinguished, the social state disappears, the nation is

P. Karimi (✉)
PH.D program in Preservation of Architetctural Heritage, DAStU, Politecnico di Milano, Milan, Italy

M. Boriani and M. Giambruno (eds.), *Architectural Heritage in the Western Azerbaijan Province of Iran*, Research for Development,
https://doi.org/10.1007/978-3-030-83094-6_4

dying; such is reality, such is Persia. (Flandin 1851, p. 353) [A/N: All the translations in the present study from French and Persian texts to English are done by the author.]

Encountering with the imperialist West, the first systematic archaeological excavations, and the formation of a nationalist and reformist political opposition to the Qajar monarchy are considered as some of the decisive events of the nineteenth century, which affected the notion of historical Heritage in Iran. Consequently, Iranian heritage was studied, discovered, exported, and preserved by new laws and institutions since the nineteenth century, and this is where our story of national heritage begins.

2 Qajar Dynasty (1797–1925)

Aqa Mohammad Khan founded Qajar Dynasty in 1979, after years of turmoil following the fall of Isfahan and the end of the Safavid era in 1722. Qajar monarchs reigned for more than a century under the constant pressure of the Russian Empire and Britain in searching for more political influence, economic encroachments, territory, and market for their nascent industry. However, Iran was never officially colonized and could preserve its national sovereignty. Qajar kings had to seek alliances and support between the Russian Empire, Britain, and other Western countries such as France in a complicated game of power to balance the foreign presence. In this tumultuous time in the history of Iran, European diplomats, ambassadors, and army officers arrived at the Iranian court with different aims such as "hunt[ing] for knowledge and antiquities" (Mousavi 2012, p. 148).

Based on biblical and classical texts, Persians were "considered part of the Western Foundation myth" (Ansari 2012, p. 26). Thus, the interest in Iranian antiquities was also motivated by research on European narratives of descent. In addition, in the time of high colonialism and imperialism, the study of antiquities was an excellent opportunity for foreign presence in a country with immense geopolitical importance due to its vicinity to the precious colony of Britain, i.e. India, as well as due to the interest of the Russian Empire in Iran.

Therefore, while the chess pieces of the "Great Game" were played, and "The rivalry between France and England in the Middle East was a constant, archaeology was one of the fields in which it was exercised" (Gran-Aymerich 1999, p. 358).

Henry Creswick Rawlinson was one of the first key figures of this era. Lieutenant Rawlinson was an army officer, an orientalist, and scholar who landed on the eastern coast of Iran in 1833 from India as a British committee member. He was sent to Tehran to train Iranian military troops defending the country against Russian threats. In 1835, he was chosen as a military advisor and assistant to the governor of Kurdistan by Mohammad Shah Qajar (Shah means King in Persian) and later came into contact with the Sassanian and Achaemenid monuments of Iran such as Bisotun and Taq-e Bostan (Rawlinson 1898). Rawlinson, with much effort, succeeded in transcribing the inscriptions of Bisotun by visiting the site several times. He made significant progress in the decipherment of the Persian cuneiform script in 1857, which revolutionized "the conceptual framework of archaeology" and revived interest in Iranian antiquities (Mousavi 1996, p. 3). He left Iran in 1839. In 1851, he participated in the British archaeological activities in Nineveh while keeping his archaeological interest in Persia and its "abundance of antiquities" (Rawlinson 1989, p. 64). When the British returned to Iran as the Turkish–Persian frontier commission and began the first systematic archaeological excavations, Rawlinson offered assistance to the committee and obtained £500 from the parliament to fund the Susa excavations (Loftus 1898, p. 356).

William Kenneth Loftus, Henry Churchill, and Colonel William were responsible for these excavations at Susa, which lasted two campaigns between 1850 and 1852.

At Susa, Loftus and Williams found the Apadana Palace (or rather the biblical Shushan palace), along with many interesting artefacts such as inscriptions, terracotta figurines, bricks, and ancient coins (Curtis 1993). The results of Susa excavations, which are astonishing for today's canons, were disappointing based on nineteenth century standards. Rawlinson affirmed in this regard that Loftus "had turned the mound of Susa topsy-turvey without finding much" (as cited in Curtis 1993, p. 15).

Therefore, the British preferred to concentrate on Mesopotamian sites by sending Loftus to Nineveh and Warka and letting the French discover and export the antiquities of Susa with the arrival of the Jane and Marcel Deiulafoy in 1884.

The archaic monuments belonging to the Achaemenid and Sassanid eras always attracted the attention of Westerners who visited Iran. James Morier (1809–11), William Ouseley (1810–12), and Robert Ker Porter (1817–18) provided valuable observations and details related to Iranian monuments, especially the Achaemenid archaeological sites of Persepolis and Pasargadae, in the first half of the eighteenth century (Mousavi 2013, pp. 4–5). These studies were completed by French scholars such as Charles Texier, Eugène Flandin, and Pascal Coste a few years later.

Charles Texier, an architect and art historian, arrived in Iran in 1839 to complete his studies on Asian monuments. Texier wrote in *Description de l'Arménie, la Perse et la Mésopotamie*:

> As early as October 1836, after traveling through different parts of Lycia, I proposed to the Minister of Public Education to continue my excursions to the east of Asia and to go and identify the ancient and modern monuments of Persia to create the most complete set of Asiatic monuments we had ever gathered. (Texier 1842, p. 3)

The French architect visited the most important cities of Iran, such as Tabriz and Isfahan, as well as the Achaemenid and Sassanid archaeological sites and created 108 plans, sections, details, and views of Iranian monuments. A substantial part of his work concentrated on pre-Islamic sites and, more importantly, Persepolis, to which he dedicated 53 of his drawings (*see* Fig. 1). Texier aimed to add more "technical elements" to what was known about Persepolis for determining the main form of the Achaemenid ruins in a more precise way and restoring them "to their old state" based on "the elements that still existed" (see Fig. 2) (Texier 1842, p. 61). Texier also paid considerable attention to the Achaemenid and Sassanid rock reliefs such

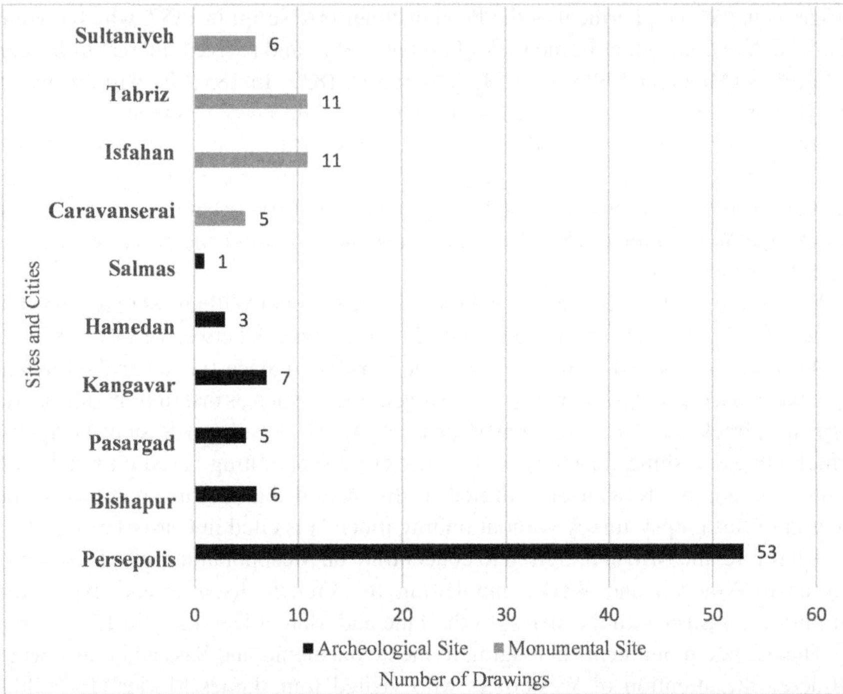

Fig. 1 Quantitative analyses of Texier Drawings published in "*Description de l'Arménie, la Perse et la Mésopotamie*"

as Naqsh-e Rostam and Naqsh-e Rajab and reported the cuneiform writing of these monuments (Fig. 2).

In the same years, the Academy of Beaux-Arts ran a competition and chose the painter Eugene Flandin and architect Pascal Coste to accompany Comte de Sercey's French diplomatic mission to Iran. Flandin wrote about this in *Voyage en Perse*:

> We were given the mission to explore Persia in all directions. We had to collect detailed information which would present an interest related to philology and history. Above all, we were assigned to study the monuments as well as the ancient and modern civilization of these people who, under the name of Mede, Parthian, or Persia, had successively gone through phases with such diverse influences on their arts, customs, traditions. (Flandin and Coste 1851, p. 8)

From 1839 until 1841, Flandin and Coste visited most of the Iranian territory and studied its historical heritage, executing "468 architectural drawings, 86 drawings of sculptures" and had transcribed "35 rock inscriptions" according to the account of *l'Académie des Beaux-Arts.* (Flandin and Coste 1851, p. 492) Three hundred seventy of these drawings are now available online through the digital collection of the New York Public Library. A quantitative analysis of these drawings confirms the importance of archaeological sites belonging to the Persian pre-Islamic empires (*see* Fig. 3). In the work of Flandin and Coste, similar to Texier's works, Persepolis has

Fig. 2 Harem Palace of Persepolis and its Ideal Reconstruction: Texier, Charles. "*Description de l'Arménie, la Perse et la Mésopotamie*", 1842, p. 596

undeniable importance, and this special attention in the first half of the nineteenth century is an early premonition of the future role of Persepolis in the maturation of the culture of preservation in Iran.

The travel account of Flandin and Coste, titled *Voyage en Perse*, was published in 1851. It is a magnificent work of systematic documentation of Iranian monuments and archaeological sites.

In 1867, another volume dedicated to the drawings and studies of the Pascal Coste was published, entitled *Monuments Modernes de la Perse*, which aimed to publish the unprinted drawings of the French architect dedicated to Iranian architectural heritage. The book focuses mainly on the Safavid monuments of Isfahan, to which 54 drawings, plans, sections, and details were dedicated out of the 71 drawings of the book (Coste 1867).

Pascal Coste, in 1879, donated 6100 of his drawings to the *bibliothèque de la Ville de Marseille*, where they are still preserved.

The study undertaken by the French can be considered the premise for the scientific study of the Iranian historical heritage, which was performed by other Western scholars such as Herzfeld, Pope, and Godard in the first half of the twentieth century.

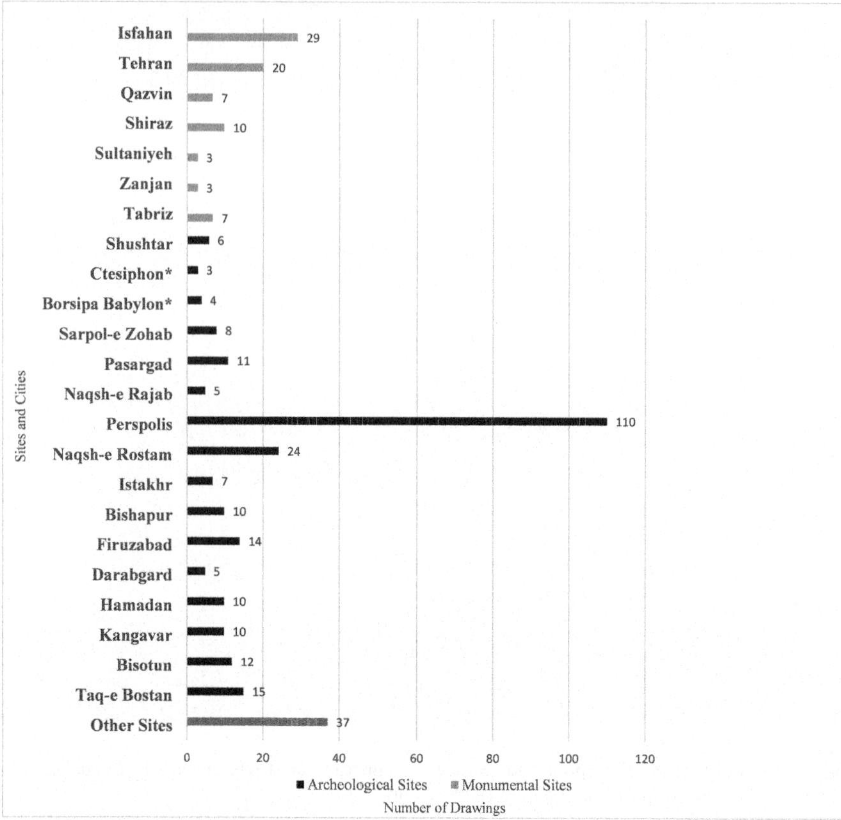

Fig. 3 Quantitative analysis of Flandin and Coste Drawings published in "*Voyage en Perse*"

The interest of the French in Iranian monuments and archaeological sites took a significant turn with the excavations initiated by Marcel and Jane Dieulafoy at Susa in 1884.

During two excavation campaigns between 1884 and 1846, the Deiulafoy couple found relics of great interest, which were mainly from the Achaemenid era of the site. "The presentation of the finds including the frieze of lions, the archers, and the monumental bull capital, opened on 6 June (1888) by President Carnot at the Louvre, was the striking proof" of the archaeological interest of Susa (Chevalier 2018, p. 43). These remarkable discoveries motivated the French government to obtain the exclusivity of the excavations in Persia, considering "in a country where the political weight of Great Britain is indisputable. [archaeological activities could] represents France and ensures her cultural and political influence." (Gran-Aymerich 1999, p. 362).

The negotiations of the French with the Qajar court resulted in stipulating an agreement between the two nations in 1895 and forming the *Délégation Scientifique Française en Perse* in 1897. According to this agreement, Persia granted the exclusive right to conduct excavations throughout the national territory, except in places of religious interest, to the French, who had to share half of the discoveries with the Iranian government. France also committed that "Precious artifacts such as gold and silver-ware, if found, shall be the strict property of the Sublime Government of Iran" (Hojjat 1995, p. 150).

The direction of the French delegation "was not left to chance: the Shah of Persia having very clearly expressed his rejection of Marcel Dieulafoy [because of the exportation of all antiquities despite the agreement], the director of the Missions Office, Xavier Charmes, addressed J. de Morgan, who since 1892 has been doing wonders at the head of the Antiquities Service in Egypt." (Gran-Aymerich 1999, p. 362).

De Morgan wrote about Dieulafoy's campaigns:

> The results of this expedition were very remarkable from an artistic perspective. Very beautiful pieces brought to light [that] enrich our Louvre Museum today. However, history had gained nothing from new explorations. Elamite Susa was still buried in complete oblivion. (de Morgan 1902)

Therefore, De Morgan's excavation campaigns at Susa aimed to shed light on The Elamite era of Susa, while the Monopoly granted the committee significant financial resources and more freedom of action. The change in circumstances in favour of the French is evident from the comparison of Jane Dieulafoy's memoirs with De Morgan's writings. For example, it is sufficient to compare the humble house built on the site by Dieulafoy with the Islamic era bricks found in the excavations and the imposing fortress built by De Morgan in just five months which still today looks over the plains of Susa (*see* Fig. 4) (De Morgan 1902; Dieulafoy 1997, p. 155).

Susa remained the centre of the French excavations, leaving the rest of the territory quasi unexplored, which aroused criticism against the French archaeological

Fig. 4 Shush Castle, Built by De Morgan in the late 1890s, photo by: Ninaras, https://commons.wikimedia.org/wiki/File:Shush_Castle_and_the_ruins_of_the_ancient_city_of_Susa_04.jpg, accessed on: 1 June 2021

committee. On the other hand, the methods of Dieulafoy and De Morgan, even for the canons of the time, were questionable and irreparably damaged the stratigraphy of the site. Especially, De Morgan's special technique to dig as much as possible to find valuable objects resulted in destructing earthen architectural remains (Abdi 1994).

Neither Dieulafoy nor De Morgan respected the contract stipulated with Qajar monarchs and unscrupulously exported everything which came to light. There is no sign of wondering about the rights of Iranians on their patrimony in the memoirs of Jane Dieulafoy, who also takes the merit of the difficult enterprise of "taking out of the hands of the Iranians half of the treasure, which was considered as the share of Iran by the order [of the Shah], the value of which is inestimable" (Dieulafoy 1997, p. 294).

This attitude was no longer tolerated when the first Iranian parliament was created in 1906 after years of opposition to the absolute monarchy. The abolition of the French Monopoly became a national matter of great interest after 1920 and boosted the formation of Iranian legislation for the preservation of the national heritage.

The second half of the nineteenth century was a crucial period for Iranian archaeology since, apart from the western interest, the Iranian aristocracy became interested in collecting antiquities that were found in small-scale and unscientific excavations (Shahmirzadi 1988, p. 135). Naser-al-Din Shah, who reigned from 1848 to 1896, was so fond of antiquities that he established the first Iranian Museum in Golestan Palace in Tehran. De Morgan writes in this regard:

> [The Persian King] was desirous of enriching the Royal Cabinet, … [so he] had reserved half of the discoveries. He loved the memories of the old days of Persia, collected them in his palace, and enjoyed visiting them (Morgan 1902).

Additionally, Houtum Schindler, who was an engineer and lived in Iran for 30 years, confirms the existence of this royal collection for which the term Museum was used for the first time in Iran:

> The *Shahanshah* [king of kings] has permitted some foreigners to dig at [some] mounds. It is a pity that these ancient artifacts are being taken away from this land. It would be a good idea to put everything like bricks, seals, etc., in the *Shahanshahi* museum. (as cited in Abdi 2001, p. 54)

The initiative to establish a museum in the Naseri era is usually attributed to the effect of the Shah's visits to European museums during his royal travels. Considering his initiatives during almost five decades of his reign, the aforementioned statement seems too simplistic. It should be noted that when the ruler was only 19 years old and "had just ascended to the throne two years earlier … ordered the photography of Persepolis motifs to the Frenchman Jules Richard in the first half of the 1850" (Adle 2000). Naser-al-Din Shah also attempted to excavate in Khorhe in 1859 and 1892. Therefore, the King was not ignorant of the value of antiquities but certainly lacked the sensitivity for interpreting such patrimony as the national heritage.

After the assassination of Naser-al-Din Shah by the revolutionary forces in 1896, Mozaffar al-Din Shah, the new king, signed a new agreement granting exclusive excavation rights to the French during a trip to France in 1900. He "granted France the possession of all … discoveries in Susiana [although] the clauses remained the same for the northern provinces" (De Morgan 1902). The article that obliged the French to deliver precious metal objects to the Iranian government was also modified, and the French established that "the discovered gold and silver items must be weighed by a French official, with the knowledge of the local governor, and once their weight is determined, their price must be remitted by the French embassy to the Iranian government" (Hojjat 1995, p. 154).

French scientific invasion of Persia with such questionable terms resulted in the Iranian intelligentsia's discontent. Zein-al Abedin Maraghei, the Iranian writer and reformist, writes in his diary:

> Through the intermediary of the French plenipotentiary ambassador, the Monopoly of the discovery of antiquities in Shushtar and Hamadan was given to a French company. Although the Iranian people were unaware of such concerns, those who knew the evil outcome of such actions were adrift in a sea of sorrow and pain [to see] that all those boundless treasures of our ancestors, which our motherland had long preserved in her bosom for the Iranians and were worth millions, had been given away in exchange for a European benediction. (as cited in Hojjat 1995, p. 151)

In fact, the local Iranian population saw the foreign presence and the exportation of Iranian patrimony as a serious concern in a veil of mystery. For instance, in the case of Susa excavation which was close to the holy shrine of Daniel, an official letter from the religious authorities of Khuzestan was sent to the ruler of the city to object to the activities of the French:

> They [the French] are digging in the graves of people who have been buried thousands of years ago, whose religion was believed by believers during their lives. They take out the spells that our prophets used to bury for the health of Khuzestan from the depths of the earth. Which diseases will destroy our land? It has been proven that whenever the *farangi* [foreigners in Persian] set foot in Khuzestan, the first signs of divine wrath are revealed, followed by descending terrible heavenly calamities. (Dieulafoy 1997, p. 172)

The values attributed to the archaeological findings in this letter are far from modern and scientific fundamentals of archaeology, but still, there is the notion of "inherited heritage" which is expressed to legitimize the protestation.

Another example of the same interest was reported in the Loftus travel account. The guardian of the Daniel shrine informed Loftus about a "black stone" which was considered "as a powerful talisman against the plague, hostile invasion, and other evils." Although the local population prevented its exportation by the westerners in 1812, the sculpture was exploded by gun powder during a treasure hunting few years later. However, "In order to preserve the fragments intact, they were collected together and secretly built into a pillar in the veranda of the tomb" (Loftus 1857, pp. 415–417). This episode is an example of how antiquities of thousands of years ago were perceived by the local population and, above all, it is noteworthy that special values were attributed to such heritage although they were very different from scientific canons.

These were few examples of the reactions of the local population to exporting Iranian archaeological patrimony. However, it is interesting to ponder what was Iranian archaeological heritage meant for Iranians over the centuries.

Ali Mousavi is one of the few scholars who considered this issue regarding the Persepolis through studying the incised sketches, graffiti, and inscriptions carved in stones on the Achaemenid ruins of Persepolis from the third century to the fifteenth century by Sassanid rulers as well as the kings of the Islamic era. He concluded:

> These are testimonies to the inspirational effect of the ruins on new generations of Iranian rulers associating themselves with a long line of predecessors. It is clear that notwithstanding any loss of specific historical knowledge about the precise identity of the Achaemenid builders of the monuments at Persepolis, a crucial essence of its symbolic value was handed down over the centuries. This essence promoted Persepolis as a place of spiritual resonance in Iranian traditions of noble greatness, which transcended several social-political changes. (Mousavi 2012, p. 93)

With the affirmation of archaeology as a scientific discipline, Persepolis was no longer considered as the spiritual place with suggestive notions like fugacity of the life or as a palace of the mythical protagonist of *Shahname* named Jamshid. Instead, it became a monument built by the great Darius and his successors and, above all, an important source of scientific documentation for historical studies. [*A/N: Persepolis still today is known as Takht-e Jamshid (Throne of Jamshid) among Iranians since, before the XIX century, it was believed to be the palace of Jamshid, who is a mythical king of Shahname with clear similarities with the historical personage of the great Darius.*]

The development of archaeology led to rediscovering the Iranian past based on the facts. Iranian history began to be rewritten according to scientific canons, which resulted in the wrong conclusion that the Iranians were unaware of their pre-Islamic history before archaeological activities and European scientific research on merit. However, the reality is more complex and articulated in order to be exhausted with such simplicity, which was explained irrespective of the historical memory of a nation before the arrival of the Europeans. The fact is that "the Iranians had not forgotten their past, [but] they just remembered it differently" via myth and poetry, as in the case of Persepolis (Ansari 2012, p. 27).

3 The Fall of the Qajar Dynasty

The French archaeological Monopoly was not a unique case of Iranian concessions to foreign governments in the second half of the nineteenth century. Several monopolies, contracts, and licenses were signed by Qajar kings with Russian, British, and other European governments since 1872.

Ervand Abrahamian summarizes the plight of the Qajar era very clearly:

> Military defeats led to diplomatic concessions; diplomatic concessions produced commercial capitulations; commercial capitulations paved the way for economic penetration; and

economic penetration, by undermining traditional handicrafts, was to cause drastic social dislocations. (Abrahamian 1982, p. 52)

The general discontent with the economic status, invasive foreign presence, and formation of a new class of educated Iranians, as well as the rise of Iranian Nationalism, led to the Constitutional Revolution of 1905–1911, which "signified the rise of a new national community which was conscious of itself and its deprivations" (Amanat 2017, p. 421). With the formation of the Iranian constitution in 1906, the first parliament in Iranian history was established, which endowed the country with a constitutional law signed a year later by Muzaffar al-Din Shah before his death.

Between 1905 and 1921, the constitutional movement and the constitutional revolution were a period of great importance in Iranian history. The occupation of the country in 1907 by the Anglo-Russian Convention was a sign of the dissent of the two countries, which felt threatened by the liberal ideals of the constitution.

The parliament was forced to close several times due to immense difficulties and threats. It is not surprising that no crucial measure was taken during this period to preserve the historical and artistic heritage of the country. The main three measures taken during this era were as follows:

- One of the first laws to preserve the historical heritage of the country was drafted by parliament as a sub-chapter of the Municipal law in 1907. In this law, establishing "libraries, museums and preserving and restoring mosques, schools, and antique buildings" was entrusted to the municipalities. (Islamic council Parliament Research Center, https://rc.majlis.ir/fa/law/show/90099, accessed on: 10 April 2021)
- Based on the law regarding the Ministry of Education, Endowment, and Antique Beaux Industry, in Sep. 1910, the aforementioned tasks were entrusted to the Ministry of Education by adding the responsibility of "acquiring antiquities and supervising excavations," "preserving ancient monuments," and "preserving precious objects and others properties of sacred buildings and other places containing national entrusted patrimony" (Islamic council Parliament Research Center, https://rc.majlis.ir/fa/law/print_version/90121, accessed on: 10 April 2021).

  In the same years, the Minister of Education founded the Office of Antiquities. However, there is no information about the exact date of its foundation as well as its responsibilities.
- In 1917, the first Iranian national museum, known as *Maaref* Museum, was founded to "collect Iranian historical heritage and artifacts" so that Iran "as well as other civilized countries could have a complex of historical works and artifacts of their progeny" (Mostafavi 1955, p. 26). This museum with its 270 objects was later transferred to the collection exhibited at the *Iran-e Bastan* Museum inaugurated in 1927.

4 Reza Shah, the First Pahlavi

Colonel Reza Khan entered Tehran with his troops in February 1921 and proclaimed himself the commander in chief of the armed forces following a coup. During the next four years, *Sardar Sepah* was able to affirm his position as the only person capable of handling the complicated political situation of the country while foreign pressure and internal revolts threatened the nation. In 1925, Reza Khan proclaimed himself as the Shah and chose the surname Pahlavi for his dynasty with an explicit annotation to the pre-Islamic history of the country. The coronation of Reza Shah Pahlavi was the beginning of a new and different era in Iran.

Abbas Amanat writes about the 15-year reign of Reza Shah:

> Reza Shah's authoritarian rule, boosted by oil revenue and a consolidated military, helped centralize the country, create modern administrative and educational institutions, co-opt the old elite, nurture a nationalist ideology, and conduct a relatively independent course of foreign policy. These were achieved at the expense of democratic aspirations as well as individual and political freedoms which were at the core of the constitutional experience. (Amanat 2017, p. 502)

Reza Shah's secular nationalist ideology had been inherited from the nineteenth century Iranian intellectuals (i.e. enlightened thinkers) who played a vital role in the constitutional movement. The encounter with westerners in the nineteenth century influenced the political and economic situation of the nation and created perceptions of backwardness and inferiority among Iranian intelligentsia in front of modern western nations. "Torn between a desire to resist and the impulse to appease foreigners in their midst", the Iranian sense of self changed and their relation with foreigners was shaped (Matthee 2004, p. 337).

Nationalism was one of the first answers to this dilemma, which embraced the "European racialism" towards Arab Semites and Islam, blaming the decline of the country in the nineteenth century on the advent of Islam in Iran in 761 AD (Zia-Ebrahimi 2014, p. 1052). The hostility towards Islam and the glorification of ancient Persian empires as an ideal of purity of the Iranian race and culture spiked during the reign of the Pahlavi and shaped the Pahlavi's cultural policy. On the other hand, Iranians' propensity towards Europeans' racial theories such as the Aryan race created the possibility of claiming the equality of Iran with the west, which became an essential part of the Iranian national identity.

The Aryan myth aroused interest in the field of art history and architecture with Josef Strzygowski's famous book *Orient oder Rome?* published in 1901. The Austrian scholar, who had overcome Eurocentrism of his time, suggested that the source of the Western architecture be sought in Iran as the cradle of the Aryan culture. Despite the questionability of this claim, the Iranian intelligentsia adopted this hypothesis "to cross temporal boundaries in search of their authentic national origins in pre-Islamic, specifically Achaemenid (ruled 559–331 BC), antiquities of Iran" (Grigor 2005, p. 26). Achaemenid heritage was also the font of inspiration for the neo-classic architecture of the Pahlavi era, and buildings such as the National Bank of Tehran were inspired by the archaeological site of Persepolis.

The secular nationalism adopted by Pahlavi had two pillars for its legitimation: "the narrative of historical descent" from pre-Islamic *Shahanshahs* and the "myth of progress and emancipation" from foreign intrusion and backwardness (Ansari 2012, p. 17). Reza Shah was pictured as the great saviour of Iran from all maladies and the indisputable leader of the country towards progress and modernity. However, he was surrounded and supported by a group of elite politicians who played a crucial role in the transformation of the country towards a modern nation-state (Devos and Werner 2017, p. 4).

Some of these politicians who were interested in preserving and revitalizing historical monuments established the Society of National Heritage in 1923 (Afsar and Mousavi 1977, p. 60). Although this Society was a private institution, its founders were among the most important and influential men of the time: Abdolhossein Teymurtash (the first court minister of Pahlavi), Hassan Pirnia (historian and four-time prime minister), Mohammad Ali Foroughi (scholar, writer, and the first and the last prime minister of Reza Shah), and Hassan Mostowfi (five-time prime minister of Qajars and one-time prime minister of Reza Shah) were between the founders of the institute (Grigor 2004).

The main objectives of the Society were listed as follows:

1. Cultivating the public interest in Iranian scientific and historical-artistic heritage, preserving the fine arts and handicraft as well as their old style and techniques
2. Establishing a museum and library in Tehran
3. Registering and cataloguing the monuments and artworks considered as important national heritage
4. Registering the precious collections in the property of museums or libraries owned by the government or national institutes. The society will seek to take advantage of these collections to benefit the population through teaching in schools, public exhibitions, and speeches.
5. The Society will ask the government to set up a suitable place for the museum and library and hire specialized personnel. In addition, the government will be required to ratify a law to preserve the national heritage.

Anjoman-e Athar-e Melli (Society of National Heritage) committed that once the objectives were achieved, it would hand them over to the Antiquities Office in the Ministry of Education and contribute to its maintenance (Afsar and Mousavi 1977, pp. 60–62). With the foundation of this Society and the ascend of Reza Shah in 1925, a new chapter began for the preservation of Iranian heritage, and all the aforementioned objectives were achieved in the next few years.

One of the first steps of the Society was to invite Arthur Upham Pope and Ernest Herzfeld in 1925 to deliver speeches for influential politicians, deputies, and diplomatic guests. Pope's speech was on the importance of Iranian artistic and historical heritage, which "deeply influenced the listeners", especially Reza Khan, then prime minister, who later assumed the role of the honorary president of the Society, considering the importance of preserving ancient monuments "to revive national consolidation and pride" (p. 64).

Pope affirmed that "there was no civilized country without collections of Iranian works of art," underlined the Iranian situation where, despite the fact that the country was the "mother and inspiring force in creating Islamic art, was still without an important collection of its most significant artworks" (ibid). Pope judged the art of the Qajar era as having no artistic value and praised the era of the Safavid Heritage as the last era of splendour of Iranian art. Arthur Upham Pope and Phyllis Ackerman had an important role in promoting Iranian art and architecture, and the publication of "*A survey of Persian Art*" was their greatest contribution. Furthermore, Pope founded the American Institute for Persian Art and Archaeology in 1928 and was its director for 24 years.

Ernest Herzfeld was another key figure of this era. The eminent German archaeologist arrived in Iran for the first time in 1905. Two years later (1907), he presented his Ph.D. dissertation on Pasargadae. Herzfeld was a man with many outstanding qualities and extensive knowledge. He studied architecture and had considerable experience in archaeological activities. He was an "excellent draftsperson and surveyor," "an expert in Islamic art", and not surprisingly the "first full professor of Near Eastern Archaeology in the world" (Hauser 2012, pp. 290–293).

Herzfeld had visited Iran several times before being invited by the Society of National Heritage to hold a speech in August 1925. In his speech, he highlighted the need "to establish appropriate regulations" for preserving monuments and "taking into account the excavation and discovery of antiquities" (Herzfeld 1925a, b, pp. 43–44). Herzfeld's proposals were in line with the programs of the Society, which have asked Herzfeld to prepare an inventory of the Iranian "Ancient Monuments." Eighty-nine monuments and archaeological sites were chosen by Herzfeld, and the results of his research were published in the first publication of the Society as "*A Brief Inventory of the Historical Heritage and Edifices of Iran.*"

Nearly half of the monuments chosen by Herzfeld were of archaeological interest and belonged to the pre-Islamic era, while the other half were religious and civil buildings built after the advent of Islam (*see* Fig. 5). Considering the selected monuments, it is possible to deduce greater attention towards the Achaemenid, Sassanid, Seljuk, and Safavid monuments since Herzfeld considered these four eras as the moments of maximum splendour of the Iranian art and architecture (*see* Fig. 6) (Herzfeld 1925a, b, p. 42).

The Iranian government ratified the decree for Herzfeld's employment as a professor of "ancient languages, ancient Iranian history, historical geography, and expert of antiquities" on May 24 in 1928 (Session: 247, Detailed accounts of the parliament on Thursday, May 4, 1928).

Firouz Mirza, Herzfeld's friend and Finance Minister, explained to parliament that Iran needs Herzfeld to educate students and teachers and write "the history of Iran to be read by our children." The minister explained that the Iranian government should hire Herzfeld since Iran "should not always resort to a foreign country [to know] about the great honours of [its] ancestors." On the other hand, the opponents of this decree saw emergencies in other sectors, which could lead to the progress of the country, and asked for "saving this country and this nation from poverty and misery, and then going back to our ancient works." It is interesting to note how progress

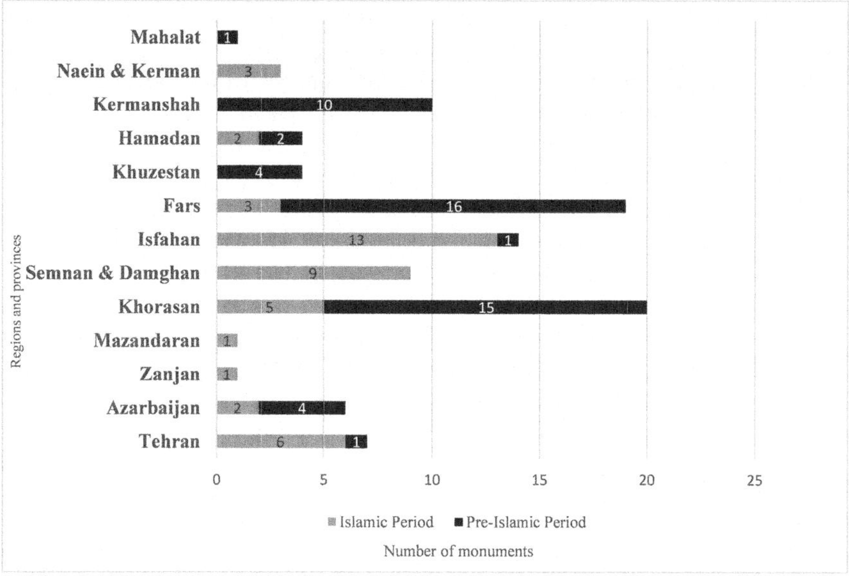

Fig. 5 Quantitative analysis of The Archaeological sites and Architectural Monuments proposed by Herzfeld to be registered in the Inventory of Iranian National Heritage in 1925, categorized by the region and by the era of belonging of the monuments; pre-Islamic and Islamic period. (Herzfeld indicated the Safavid royal complex of Isfahan as one item in his inventory, but for better analysing this inventory, Isfahan's Safavid royal complex is considered as ten monuments)

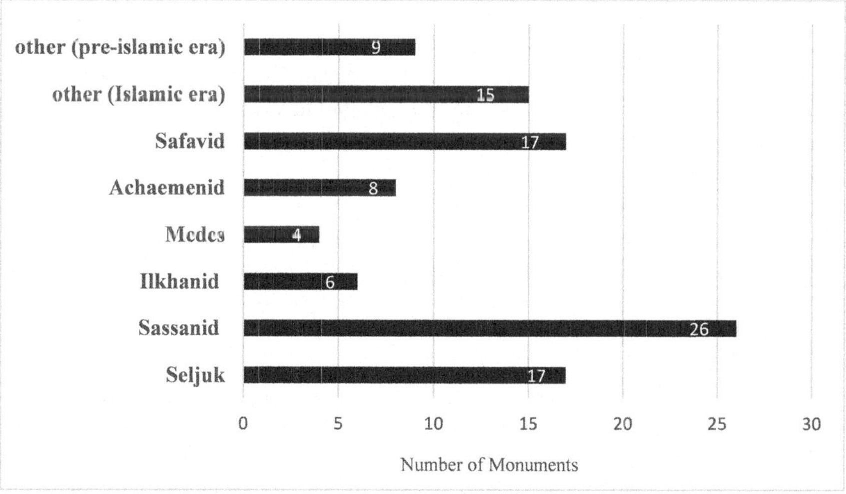

Fig. 6 Quantitative analysis of the Archeological sites and Architectural Monuments proposed by Herzfeld to be registered in the Inventory of Iranian National Heritage in 1925, categorized by the era of belonging

and modernization were considered urgent matters contrary to the preservation of Iranian heritage (ibid).

[*A/N: The digitalization of the Iranian parliament archive offers a unique opportunity to see the detailed accounts of each session of the Iranian parliament since its foundation.* https://www.ical.ir/ical/fa/fa/MashroohMozakerat].

Despite the teacher's role Herzfeld's main task was to be the advisor to the government in archaeological activities. Additionally, he had an important role in proposing the law of antiquities (the ratification of this law would also permit Herzfeld to excavate at Persepolis) and he served as a "strong weapon" to the exclusive presence of the French in the field of archaeology (Mousavi 2005, p. 456).

In the same years, the abolition of the French Monopoly in archaeology was finally achieved. Mohammad Taqi Mostafavi, the author of one of the first books on the history of preservation in Iran, reports two decisive episodes for the annulment of the French Monopoly. The first episode is related to Reza Khan's visit in 1924 to the fortress built by De Morgan and his negative impression of the massive French fortress in the Iranian national territory. The second episode included the telegraph of Farajollah Agh Avali (one of the army commanders and a co-founder of the Society of National Heritage) employed in state affairs near Susa, who wrote to the Reza Shah in 1927: "I pray his excellency to do not allow our identity documents to be exported out of the country" (Mostafavi 1955, p. 37).

Indeed, overemphasizing Reza Shah's role is not exhaustive to explain the abolition of Monopoly, which lasted for 27 years. With the Constitutional Revolution and the formation of the Iranian parliament, various concessions of the Qajar era were examined by the deputies. It should be noted that a bill was proposed to parliament in 1911, which criticized the French Monopoly and demanded the prohibition of excavations without the permission of the government throughout the national territory (Yazdani 2001, p. 26). The French ambassador in Iran intervened to prevent the ratification of this law in 1913, and the Iranians had to wait another 14 years for the Monopoly to be cancelled on 18 October 1927, after lengthy negotiations with the French government.

Ève Gran-Aymerich writes in this regard:

> In the political situation of Iran between the two world wars, the results of the 1927 negotiations were a form of victory for France, which, against the growing influence of Germany, managed to retain an important archaeological mission and prompted the creation of an Antiquities Service, the management of which, as in Egypt, would be retained for a long time. (Gran-Aymerich 1999, p. 368)

According to the new contract, the Iranian government committed to hire a Frenchman as the head of the antiquities office for at least 20 years and the French government committed to abolish the archaeological Monopoly and limit the French excavations in Susa and Susiana. Thus, André Godard arrived in Iran at the invitation of the Iranian government in 1929.

The abolition of the aforementioned Monopoly was the cause of interest for the British, the Germans, and the Americans. In Germany, even a campaign was advertised in specialized magazines that supported the Iranians' right to cancel the French Monopoly (as cited in Gran-Aymerich 1999).

It should be noted that one year after the inauguration of the University of Tehran, the first modern university of Iran in 1934, a three-year course called "*athar-e Bastan*" (ancient monuments) was taught to young Iranians who wanted to become archaeologists. Nevertheless, archaeology remained an almost exclusive domain for western presence for many years.

5 The Legislative Measures

In addition to the French Monopoly, the clandestine excavations and trade of antiquities that had been in vogue for many years threatened Iranian historical heritage in the first half of the twentieth century. Therefore, the first code concerning the preservation of national heritage, entitled the "New code of Antiquities approved by cabinet ministers" ratified in 1924, was mainly a solution and a controlling tool of illegal excavations (Gholamdoost 2013, pp. 12–13).

There are many interesting points in the 29 articles of this regulation that are reviewed in the following:

1. Two macro-categories called "underground relics" and "on earth relics" are created to describe better the objects included in the code.
2. The permission of the Ministry of Education is required for exporting any manuscript books, paintings, vases, and pottery (even if in pieces), elaborate metals, carpets, embroidered cloth, elaborate or painted wood, pieces of plaster or elaborate stone, statues, and pieces of bones or pre-Qajar coins. If such permission is not requested, the objects in question will be confiscated and sent to the *Maaref* Museum. If the objects belonging to the aforementioned categories do not belong to the government and are not considered essential for the country's history and are over 50 years old, the owners must pay 20% of taxes to the government in the case of exportation.
3. If ancient objects from endowed religious mausoleums and ancient madrasas such as the *Kashi* (Iranian tiles) are found, they will be confiscated and returned to their place of belonging. The objects will be brought back to the *Maaref* museum only if the latter is unknown.
4. Objects belonging to kings or anything bearing a King's stamp or signature cannot be exported and must be bought and delivered to the *Maaref* Museum.
5. It is forbidden to perform excavations in Iranian territory without the official permission of the Ministry of Education. In addition, if any objects are found without the excavation permit, they will be confiscated and handed over to the *Maaref* Museum.
6. Twenty-five percent of the objects found in excavations belong to the government, but this share rises to 75% in the case of metal objects such as gold and silver. The objects must be catalogued with the presence of representatives of the Ministry of Finance and Education. Then, everything will be packed up

and sent to the Ministry of Education, where the share of the government is subtracted.

7. Owners of statues (complete or damaged), stones carved with Iranian inscriptions, and vases with "strange and Surprising" shapes must make a cast of these objects and deliver them to the *Maaref* museum before exportation.
8. When important historical antiquities from a national and scientific point of view are found, the government can cancel the excavation permit and independently continue excavations to compensate for the caused harm to the excavator.
9. If the excavations are not executed according to scientific principles, and without necessary care, the excavation permit will be cancelled.
10. The relics of antiquities owned by the government cannot be sold unless the proposal is ratified in the parliament.
11. If this regulation is not respected, there will be one-year imprisonment as a penalty.

The 1924 law was the first legislative attempt to organize clandestine excavations and the exportation of the founded objects. However, it is evident that the thought and criteria regarding the archaeological findings were not fully matured. This claim is supported by Article 16, which states the "strangeness" of the shape of an object is a criterion that obliges the owner to create a cast before its exportation.

It is interesting to note that architectural fragments such as *kashi* (Iranian tiles) decorations were cited as works of art while the preservation of architectural monuments was only mentioned to prohibit excavations on the foundations, and there was no measure to preserve the historic buildings. The importance attributed to the architectural fragments confirms that the detachment and exportation of these elements was a concrete problem and the few known cases of the exportation of *mihrabs* or *kashi* decorations are not sufficient to exhaust the threats imposed by the antiquity market to monumental heritage in this period. On the other hand, the extensive modernization and urbanization of Iranian cities had not begun yet. Therefore, there were still no significant concerns about preserving the historical buildings.

Six years after the aforementioned code, the first law of antiquities was ratified on 4 November 1930. Mohamad Ali Foroughi, the minister of foreign affairs, presented the law to the parliament and asked for its ratification as soon as possible since "The heritage of our [Iranian] ancestors," which is the "source of dignity and pride of this nation" should be "collected and preserved." Furthermore, Foroughi highlighted the importance of the "scientific and historical" benefits which could be derived from this initiative as well as the financial benefits for the government, which aimed at the "material and spiritual" use of the historical heritage (Afsar and Mousavi 1977, pp. 63–64). Foroughi was a crucial figure in the maturation of Pahlavi's cultural policy and had an essential political role.

The law of antiquities had 20 articles. A few months later, an executive regulation of another 20 articles was released. The ratification of this law is certainly an important moment for the preservation of historical heritage. This law aimed to create and define a bureaucratic process within which a "national heritage" was identified, registered "in the inventory of national heritage" and "preserved and supervised." The two main responsible authorities were the Ministry of Education and the Office of Antiquities. The law defined the responsibilities of each authority and the owner of the national heritage as well as the duties of the excavators and the government for preserving the national heritage. The different matters and notions of this law are summarized by the author from its original draft for a better understanding:

1. **The definition of national heritage***:* "All works of art, buildings, and places which were built in Iran until the end of the Zand Dynasty (1779), whether movable or immovable, are considered as national heritage and placed under the supervision of the government." These antiquities should have specific "historical, scientific, and artistic" values and be regarded as highly important for "the Iranian national history."
2. **The preservation procedure**: An "inventory of known and identified, national heritage" should be "created, published, and always updated by the government." The Ministry of Education is in charge of identifying the national heritage. Each item of the inventory should have two descriptive dossiers with photographs. One dossier should be kept in the National Heritage Office, and the other should be given to the owner. The head of the Antiquities Office should identify the National Heritage and propose it to the minister of education. To create a dossier for an immovable property, the following items must be compiled: name, geographical position, name of the builder, construction date, general plan, and details of decorations, as well as other valuable details for future studies. It is also suggested to create casts of inscriptions if there are any. It is even suggested to record the local legend about the building. Regarding the restorations, the restoration work, date of execution, and the restorers' names should be described. The restorations must be recorded in the dossier of the monument even when the Antiquities Office performs them. The Antiquities Office is responsible for preserving the monuments. Preserving monuments could be entrusted to municipalities in the case of city monuments.
3. **The ownership of the national heritage**: If the national property belongs to a private owner, the owner can keep his/her rights as owner. Before registering the property, he/she will be informed of the will to include this property in the list of national heritage, and he/she can appeal to this decision, which will be taken over. When the property is registered in this list, the owner has no right to oppose the measures taken for preserving the Heritage, and no financial contribution will be requested from the owner. If the owner wants to "restore and renovate", the project related to the planned works must be delivered to the Ministry of Education. The execution of this project must be performed by experts chosen by the Ministry of Education. In addition, the owner must provide a guarantee until the end of the works. If the owner wants to sell such a property, he/she must notify the government so that the government, if it deems it necessary, can buy it before other purchasers. If the owner has lands registered in the list, he/she cannot hinder the excavations even if he/she is given compensation for

damages. The owner of relics cannot export that object if the government prefers to purchase it.

4. **The prohibited acts**: Destroying and damaging a national heritage is prohibited. Any act in the vicinity of the monuments which may "damage the foundations or modify the appearance" is also prohibited. Additionally, it is forbidden to appropriate or sell materials or elements of a registered building without the permission of the Ministry of Education. Furthermore, it is forbidden to dig for antiquities without acquiring permission from the government, even when digging on one's own land. Illegally excavated items will be confiscated, and the excavators will have to pay a fine. For some monuments and sites of national interest, "special areas" could be identified by the government, in which it is forbidden to build or plant trees, dig wells and Qanat, and build cemeteries. To create these "forbidden zones," the Ministry of Education will clearly communicate the prohibited operations by paying compensation for damages to eligible people.
5. **Excavations and digging**: "Excavation and exploration for extracting antiquities is the right of the government, and the government is free to exercise this right exclusively or transfer it to scientific institutions, individuals, or companies." The excavations are divided into two categories. If the excavation aims to find a national heritage and conduct scientific research, it is considered as scientific excavation and only allowed for scientific institutes (Two years later, scientific excavation was better defined in a separate article entitled "Excavation is scientific when it aims to find documents to investigate the Iranian civilizations and their relations with other civilizations"). On the other hand, if the excavation aims to trade in antiquities, it is considered as a commercial excavation. Commercial excavations are prohibited in registered places and buildings. The committee of ministers will grant permission for excavations to scientific institutes in registered sites. For unregistered sites, a permit will be given by the Minister of Education. The permit for excavations is granted upon presenting an excavation program that identifies the area and the schedule for performing the work. The head of the Antiquities Office or the representatives have the right to visit and supervise the site, as well as the found objects.

   At the end of each excavation season, the excavation report, including the plan and the map of the excavations with descriptions of the findings, should be delivered to the head of the Antiquities Office. Further, this report should include the drawings of the buildings and found objects and describe the excavation campaign results in detail.

   The share of the government in scientific excavations should be preserved in museums and collections of the government (It became more specific two years late: antiquities discovered in scientific excavations are pure historical documents and must be preserved in national collections and could not be sold. Exports in scientific excavations are exempt from any levies and taxes.) The share of the government in commercial excavations should be evaluated and, if adequate, should be preserved in museums; otherwise, it could be sold by auction (The auction will be eliminated in 1932 code).

6. **The share of the government in excavations**: The government has the right to select ten objects found in the excavation and divide the rest of them into equal parts with the excavator. These regulations are not valid in the case the buildings or parts of buildings that belong to the government.

The Iranian law of 1930 is probably inspired by the French law related to preserving historical monuments that was drafted in 1913; the classification of national heritage, the government supervision procedure, the rights of the owners of the registered monuments, the two defined macro-categories, movable objects, and non-movable architectural monuments are some of the similarities with the French legislation.

This law was an important legal measure seeking to legitimize and organize the preservation of architectural monuments, archaeological sites, and historical artworks. Through this law, historical and economic values were attributed to the national heritage. Unfortunately, it is more concerned with a single monument rather than on the historical fabric of the cities (Moradi 2018, p. 27). The most serious shortcoming of this law was that it excluded the monuments of the Qajar era, thus neglecting 130 years of Iranian history. Reza Shah demolished an important number of palaces and buildings built by the Qajar kings for political rivalry with the dynasty, which preceded him, with the claim of urbanization and modernization. Fourteen years later, the Iranian parliament approved the inclusion of Qajar buildings in the list of national monuments in case of their public utility, which helped preserve some of the Qajar monuments (Mehdizade and Hanachi 2016, pp. 8–9).

The law of antiquities was an effective legislative tool for restoring and preserving Iranian monuments, which were formerly entrusted to kings, regional leaders, and endowment authorities. However, the funding of the necessary project for preserving the national monuments was not considered in the law of 1930. To remedy this shortcoming, the "*Oqawf*" (Islamic endowment) law was drafted four years later, upon which a part of the earnings of the endowments must be dedicated to the restoration of the monuments. The *Oqawf* funds were not sufficient for all the projects across the country. In fact, little had been achieved until 1940, apart from the urgent repairs of a few historic buildings and projects for safeguarding the Safavid monuments in Isfahan (Mostafavi 1955, pp. 47–50). The financial problems of the national monuments were finally addressed in 1940 when part of the budget of the Ministry of Culture was dedicated to the restoration of historical monuments, excavations of archaeological sites, and other necessary measures. The funds provided by the aforementioned two laws, as well as peoples' contributions and financial aids by municipalities and the King, in almost 20 years since 1930, financed numerous restoration projects throughout the country (ibid).

6 Archaeology Office

During the Qajar era, Iranian monuments were in a severe state of decay caused by years of lack of maintenance, earthquakes, and wars. There are several travelogues of the nineteenth century which describe Iranian cities covered with the ruins of the majestic monuments of the past eras. Here are some pages of these descriptions which depict Iran in the nineteenth century:

> Such was the Chehar Baqh (a street in Isfahan) in the plenitude of its fame. But now, what a tragical contrast! The channels are empty, their stone borders are crumbled and shattered, the terraces are broken down, and the parterres are unsightly bare patches. The trees, all lopped and pollarded, have been chipped and hollowed out or cut down for fuel by the soldiery of the Zil, the side pavilions are abandoned and tumbling to pieces, and the gardens are wildernesses. Two centuries of decay could never make the Champs Elysees in Paris, the Unter der Linden in Berlin, or Rotten Row in London look one half as miserable as does the ruined avenue of Shah Abbas. It is in itself an epitome of modern Iran. (Curzon 1892, p. 39)

> We went to Soltaniyeh from Zanjan in two days, the monuments of which ... are extremely interesting ... I do not know why this city, which used to be so prosperous and prestigious, should be so ruined and devastated now ... The reason according to what I've been told by inhabitants of these ruins is the negligence of the government and the internal turmoil of the country. (Jaubert 1943, p. 107)

> Out of 250 mosques in Tabriz, mentioned by Chardin, the ruins of only three are visible. (Porter 1821, p. 222)

> No man can enter Persia without remembering he is about to tread on a land with a long line of native princes, cities, and towns, and fertility. … But of all these towns, villages, and structures, the erections of so many different ages and generations of men, few remain of any kind that are not sunk in ruin or furrowed with decay. Where there were once cities, hamlets, and cultivated fields, there is vast solitude now without any houses, huts, trees, blades of grass, or many miles. (Porter 1821, as cited in Mathee 2016, p. 5)

> No scholar knows that the rule slipped out of the hands of the Parsi people by the oppression of time, and Muslims, Arabs, Afghans, Mongols, Tartars, and different groups have constantly been encouraging plunder and tumult by ruling with greed and covetousness now after 1200 years… Thereby, what at first was "Iran the mark of Paradise" (irān mino-niśān) should now be called "Iran the ruined" (irān verāṇ). (Manekji 1865, as cited in Sheffield 2013, p. 34)

The abovementioned testimonies can epitomize the critical state of the preservation of Iranian monuments at the dawn of the twentieth century. With the Constitutional Revolution in 1906 and the rise of Iranian Nationalism, a new age began for the cultural heritage of the past epochs. The foundation of the Office of Antiquities to preserve Iranian national monuments dates back to this period. However, there is no exact information about its responsibilities or the exact date of its foundation. In the following, the opinions of three scholars in this regard are reported (Figs. 7 and 8).

Mostafavi, the head of the Archaeology Office, in his account of the history of the preservation of Iran published in 1955, refers to the opinions of two employees of the Office of Antiquities and says that the organization was founded in 1297 SH (1918–1919) even if the desire to found this office dated back to the time of the

Fig. 7 Flandin and Coste, Mosquée Chân-Khodâh-Bendèh, à Sultânyèh [Sultânièh]. The New York Public Library Digital Collections. 1851–854. https://digitalcollections.nypl.org/items/510d47e2-9059-a3d9-e040-e00a18064a99, accessed on: 5 may 2021

constitution and the then Minister of Education Sani-ol Douleh who was minister from 1915 to 1916 (Mostafavi 1956, p. 23).

Keramatollah Afsar, in his book "*Preservation of Antiquities in the Pahlavi Era*", reports a piece of an Iranian magazine from 1926 which affirms the formation of the Office of Antiquities by the Minister Sani-ol Douleh. Additionally, he stated that the responsibility of this office was limited to supervising the excavations and acquiring taxes from objects found at the historical site of Rey before they were exported to Europe (Afsar and Mousavi 1977, p. 40).

Mehdi Hojjat, in his doctoral thesis titled as *Cultural Heritage in Iran: Policies for an Islamic country*, collected unpublished archival material and concluded that the Office of Antiquities had been already active since 1910, four years after the Constitutional Revolution (Hojjat 1995, pp. 164–170).

Therefore, the Office of Antiquities existed before the Pahlavi era and the foundation of the Society of National Heritage in 1923. With the ratification of the law of antiquities in 1930, the responsibilities of this Office were defined, and the preservation and restoration of architectural and archaeological heritage was assigned to this organization.

With the foundation of *Farhangestan-e Iran* (Academy of Iran) in 1935 to purify and protect the Persian language from foreign languages such as Arabic, the Office

Fig. 8 Flandin and Coste, Tabriz. Mosquée ruinée. The New York Public Library Digital Collections. 1851–1854. https://digitalcollections.nypl.org/items/510d47e2-9052-a3d9-e040-e00a18064a99, accessed on: 5 may 2021

of Antiquities became the Archaeology Office in order to use Persian words in the name of the organization.

André Godard was the head of the Office of Antiquities and then the Archaeology Office from 1928 to 1953. Godard is one of the most influential figures of this era. He was the designer and director of the first Iranian archaeological museum called *Iran-e Bastan* Museum. In addition, he played a crucial role in founding the first Iranian architecture faculty as part of the University of Tehran in 1940. During his long career, Godard examined and studied Iranian monuments and archaeological sites in collaboration with Yedda Godard (Godard's wife) and published the results of these studies in prestigious journals such as *Athar-e Iran: Annales du Service Archeologique de l'Iran.*

During the directorship of Godard, important restoration projects were executed, and local traditional architects played an important role in preparing and implementing these projects. Traditional architects, or rather the *Memars* (architect in Persian), were the only experts who were acquainted with Iranian architectural methods of construction. Their knowledge was inherited from former generations of *Memars* through long-term practice and apprenticeship.

Before the foundation of universities in Iran, architecture was a profession linked to spirituality and faith. In other words, the architects' attempts to achieve expertise were a kind of spiritual austerity, and all stages of their work were followed with spiritual and religious accuracy (Mohammadi 1992, pp. 10–12).

In Iran, "architecture for centuries had been a craftsman's profession which was taught from father to son or from teacher to apprentice through teaching on the construction site" (Ahmadian and Sharifian 2014, p. 1161). The master was the spiritual father of the disciple and "the apprentice had to pass through different levels of skill before becoming the master of the masonry" (ibid).

André Godard wrote about *Memar*s of Isfahan:

> Now, there are skilled masters in Isfahan who can well renovate the tiling decorations of old buildings, even in cases where only a tiny trace is left, they can reconstruct them again without any mistakes, as if they were contemporary to the period of construction of the building. (Godard 1998, p. 465)

Despite this affirmation, Godard shared the reductive opinion of his time regarding the traditional architects:

> His only scientific document was a paper tube which he held with pride, and in his opinion, all the sciences of the world were gathered in it. In this old paper tube, some old designs of arches and vaults as well as some formulas were gathered … But, that time has passed. Now, the mason is just a mason. Nowadays, there are talented Iranian architects, painters, sculptors, and artists who have completed their higher education in these fields, in both Iran, European and American countries. (ibid.)

Notwithstanding Godard's negative opinion regarding the knowledge of traditional architects, neither he nor any other Western architect who had studied the history of Iranian architecture could intervene to preserve Iranian monuments. The restoration projects were all executed by Iranian *Memars*. Mostafavi, in his book titled as *Attempts to serve National Heritage* (1955), lists the personnel of the regional offices of the Archaeology Office and notes that each branch had an "architect for historical buildings" (Mostafavi 1955, pp. 88–90). Even today, restoration of Iranian monuments is strongly dependent on the knowledge of traditional architects who dedicate their lives to understanding the Iranian art of construction. Some of these architects such as Hossein Maarefi had a significant contribution to preserving the Iranian architectural heritage. The book "Architects and Traditional Restorers of Isfahan", published in 2008, is a precious source for a better understanding of the *Memar*s important role in the field of conservation and restoration in Iran. (Sajadi Naini 2008).

However, Godard was right to point out the beginning of a new era in which traditional *Memars* were replaced by new architecture graduates. The ancient knowledge of Iranian architecture was utterly unprepared by encountering with modern and Western architecture. With the establishment of the faculty of architecture of the University of Tehran in 1940 based on Western models, new architects were educated to contribute to the modernization of Iran, while passing on the Iranian architectural tradition was not among the priorities. On the other hand, the lack of Iranian architectural written sources such as treatises or books did not facilitate university teaching. Therefore, the syllabuses were mostly based on the French model (Bidhendi 2009, pp. 5–20). Thus, the new architecture graduates could not compete with the local *Memars* in terms of knowledge about traditional architecture. Hence, restoration projects were entrusted to the *Memars,* while the construction of new buildings was given to the university graduate architects (Rahiminia et al. 2016, p. 7).

When the Office of Antiquities began to intervene to preserve the architectural monuments, the Iranian built heritage was in a severe state of decay. Therefore, the interventions were mainly concerned with structural consolidation and reconstruction of the collapsed elements. Based on Mostafavi's account of some of these interventions, modern materials such as iron beams and reinforced concrete were widely used. For instance, Mostafavi writes about the restoration of Safavid buildings in Isfahan "with structural metal elements and buttresses" or the restoration of the dome of the Khaje Rabi tomb in Mashhad with "reinforced concrete" to maintain its "original composition and shape" (Mostafavi 1955, pp. 54–60). Restoration projects to bring the building to the original state were not limited to religious or civil monuments still in use by the population. For example, Niasar Fire Temple was "systematically repaired" by reconstructing the dome and renovating the bricks (Mostafavi 1955, pp. 100–101).

Reconstructions, as part of the restoration projects, were not limited to less known monuments. In the same years, Herzfeld and Krefter, during their excavations in Persepolis in 1930, reconstructed a large part of The Harem of Xerxes for "quarters of the expedition and a future museum" (Schmidt 1939, p. 7). Mostafavi, the Iranian supervisor of the archaeological committee directed by Herzfeld in Persepolis in 1932, was apparently cautious in criticizing the reconstructions by Herzfeld so as not to contradict the consent given by Reza Shah in his visit to the site. However, he reports the opinion of some visitors who judged the reconstruction as "an improper act" which damages the "original and ancient form" of Persepolis. Additionally, he writes about his perplexities regarding the reconstruction of the Small Tachara Palace (which had to be rebuilt before the Harem Palace was chosen) since it would have irrevocably damaged the landscape of Takht-e Jamshid (Mostafavi 1975, pp. 9–10).

Unfortunately, no reports were prepared for the restoration projects before 1965. Therefore, it is not easy to evaluate the conducted restorations. However, based on the extant results of a few case studies, it could be deduced that the stylistic restoration for revitalizing and bringing the monument to the original state was the horizon of expectations. Even the use of terminology in Mostafavi's book in 1955 confirms this desire. Terms such as "repair and renewal", "renewal and completion", or "revitalization and renewal" are used to describe the interventions performed on the historical monuments.

7 Mohammad Reza Shah, the Second Pahlavi

The allied forces of Britain and Russia invaded Iran in 1941, and Reza Shah agreed to leave the country to his son, Mohammad Reza Shah, as the condition of the withdrawal of foreign occupation. The post-war period beginning with the occupation of Iran led to severe social and economic problems as well as more political freedom.

The unfinished Constitutional Revolution was reassembled through the movement of the Iranian Oil Nationalization with the leadership of Prime Minister Mosaddeq. The Shah was forced to flee to Italy and return to his throne just after the coup of 1953, which was supported by American and British agents and resulted in the overthrow of Mossadeq's government. Abrahamian divides Mohammad Reza Pahlavi's reign after the 1953 coup into two periods: the period of consolidation of power from 1953 to 1963 and the period of socio-economic development from 1963 to 1977.

Not surprisingly, during the second period, in 1965, the National Organization for the Conservation of Historical Monuments of Iran (NOCHMI) was founded. The foundation of NOCHMI was the beginning of modern and scientific restoration and conservation practices in Iran. NOCHMI educated and trained experts and enhanced the institutional fundaments of preservation in 1965–1979 period, known as the golden era in the field of preservation of historical heritage. In collaboration with international institutions and organizations such as UNESCO, IsMEO (Istituto Italiano per il Medio ed Estremo Oriente), and Iranian universities, NOCHMI introduced international methodologies and conservation criteria to Iranian experts, and Iran adhered to international charters during this period (Galdieri and Afsar 1992).

In particular, the collaborations between NOCHMI and IsMEO have been fruitful and had an essential role in introducing modern principles and methods of intervention in Iran. IsMEO was founded in 1933 by the philosopher and politician Giovanni Gentile and Indologist and Tibetologist Giuseppe Tucci with the aim of "promoting and developing cultural relations between Italy and the countries of central, southern and eastern Asia." (Gnoli 1983, p.4) As a country of history and culture and above all almost free from colonialist entanglements, Italy was one of the main political allies of the Shah. Accordingly, the cultural collaborations and the arrival of the IsMEO committee were supported by the political relationship between the two countries.

IsMEO's presence in Iran began with archaeological activities in Sistan in 1959 but soon have extended to include significant restoration projects in Achaemenid sites of Fars with the direction of Giuseppe Tilia and Safavid monuments of Isfahan with Eugenio Galdieri as the director. The archaeological site of Achaemenid Persepolis and Safavid palaces, mosques, and urban monuments of Isfahan constituted the most important and symbolic historical heritage for Iranians as the emblem of Iranian pre-Islamic and Islamic culture. Thus, the intervention of IsMEO in two of the most significant Iranian Heritage sites attests to the importance of these interventions and caused their vast resonance through the years as one of the first's restorations projects to be executed according to scientific and international criteria of the Venice Charter.

8 National Organization for the Conservation of Historical Monuments of Iran

The first meeting of the Superior Council of the National Organization for the Conservation of Historical Monuments of Iran (NOCHMI) was held in October of 1966.

According to the statute of the organization, the 70 members of this council were ministers, heads of universities, professors of archaeology, architecture, and history, members of National Heritage Society, and other influential politicians. The policies of this organization, presided over by Prime Minister Hoveyda, were drafted in 11 articles as follows (Mehran 1977, pp. 10–21):

1. The organization should publish a comprehensive list of historical monuments and archaeological mounds and organize special programs to raise the awareness of the Iranian people about the importance of the Iranian heritage.
2. Modifications in historical buildings should be strictly prohibited, and restorations should be performed only by experts. The restorations should be based on the original style and historical methods of building construction so that each building could represent the architectural style, art, and craft of its contemporary artists.
3. The organization should collaborate with universities and scientific institutes to train technical experts to assume in future projects.
4. For each of the historical monuments, two types of buffer zones must be specified. In the special buffer zone of historical buildings, any operation which damages the foundations or causes alterations in the appearance and landscape of the building should be prohibited. In the general buffer zone, which limits will be determined by the needs of the building, a special criterion should be defined for constructing and developing operations in order to preserve the historical landscape and harmony of the surroundings. Special regulations must be defined for determining the buffer zones for some buildings which form a unit together such as Naqsh-e Jahan Square and the buildings inside and around it. New construction criteria in cities such as Isfahan, Shiraz, and Qazvin should be established in such a way that the historical character of the city and the architectural style of its buildings would be preserved at least in the historic and ancient core of the city.
5. Each monument must have an identity dossier containing its architectural plans in order to represent the era of the construction, decorations, and architectural characteristics. These dossiers must be published as books and essays to be available to a vast audience.
6. The organization should collaborate with tourist organizations to train guides for historical sites.
7. Ancient and historical inscriptions must be photographed, and their casts should be made to be kept in a special museum. Collecting and publishing the inscriptions as well as translating the books and essays related to the subject should be encouraged.
8. The archaeological mounds of the country should be preserved, and any destructive operations should be prevented. A buffer zone must be set for each of them to prohibit any operations within its limit.
9. The organization should collaborate with *Sepahian-e Danesh* (troops of knowledge) to identify historic sites and villages.

10. The organization should cooperate with the *Oqawf* organization for the restoration of the monuments which are financed by *Oqawf* funds.
11. The organization should raise funds from private and government institutions and organizations for its activities.

A few months later, NOCHMI drafted a technical training program for four different categories of experts as a part of the future restoration/conservation committees (Mehran 1977, pp. 25–29). These categories were explained as follows.

1. *Experts of Archaeology*: Excellent students of archaeology must be selected and attracted to join the organization. NOCHMI should provide scholarships for these outstanding students to pursue their studies abroad. They should be introduced to archaeological committees in Iran. The organization should provide favourable conditions for hiring archaeologists.
2. *Engineers*: NOCHMI should contact engineering and architecture students in their last year of university to employ them in the field of restoration and conservation of monuments. There should be a possibility for interested students to gain scholarships for continuing their studies abroad in the field of restoration of historical monuments. These students should be hired in the organization.
3. *Traditional architects*: A traditional architect should be hired in each region to work under the engineers' supervision. Traditional architects must be instructed on restoration methods and criteria in construction sites by the responsible engineers. Then, the architects will be responsible for guiding the workers and artisans on the site.
4. *Chief artisans and workers*: Workshops and laboratories must be created in different regions depending on the historical importance of the area to train the artisans specialized in working with wood, ceramics, plaster, painting, etc. The artisans must be instructed under the chief artisan's supervision and should perform a period of practical internship. The organization must hire workers and artisans who properly perform their duties.

The division of responsibilities between different experts, the definition of the hierarchy on the site, and the will to involve and hire new graduates were important factors which could greatly influence the process of preservation of the Iranian built heritage. NOCHMI founded the Center for Education, Research, Restoration, and rehabilitation of Historical Monuments to educate expert personnel in 1974. This centre aimed to educate archaeologists, architects, restoration experts, and artisans. To this end, the students of archaeology and architecture were selected and provided with different courses and internships in 12 months to acquire specialization in the restoration of monuments. Additionally, technical courses were offered to artisans and technicians to gain necessary expertise.

A program for the restoration and conservation of historic buildings was drafted. Accordingly, the first steps for the proper preservation of the monuments were defined: (Mehran 1977, pp. 30–32)

- Groups including an engineer, a traditional architect, and a photographer should be formed to visit the monuments, realizing surveys and photographs to identify the necessary interventions. A work program should be designed according to the information gathered for differentiating urgent interventions from long-term interventions.

- In each region, a traditional architect and an engineer should be hired. In the case of a lack of staff in the region, two experts must be hired elsewhere and sent on missions.
- A central council must be formed to supervise the work in progress in the national territory. Then, a central technical office and a technical council will be created. The technical council will consist of the directors of the faculties of the architecture from the University of Tehran and *Melli* University, the head of the archaeology institute, a building engineer, two restoration experts, and a representative of the Ministry of Culture.
- A descriptive dossier must be prepared for each monument. Restoration projects and works performed to preserve the monument must be added and written in these dossiers. The files must be collected in the central archive in Tehran as well as the local archives.
- The responsibilities of the regional councils for the restoration of monuments and their relationship with the main office must be defined.
- Other organizations such as municipalities, local police, and the Waqf Organization must be informed about historical monuments and the potential risks which threaten them. NOCHMI should collaborate with these and other organizations to preserve the Iranian heritage and the respective buffer zones of the monuments.
- A sign must be installed to attract the attention of visitors to sites registered in the list of national monuments. Trained guards should be employed where the site requires.

NOCHMI will publish the monthly *Agahiname* magazine on the activities performed by the organization, found a specialist library in Tehran, establish 17 provincial offices in the first ten years of its activity, and organize two national seminars in 1969 and 1975 in Tehran and Isfahan. These events led to the formation of specialized discourse in the field of restoration and conservation of Iranian cultural heritage.

The 1970s were the critical years for the maturation of restoration methods and criteria in Iran, which could mainly be due to the significant rise of oil revenue resulting in the augmentation of the budget of the Ministry of Art and Culture. The scale of intervention on the national territory after the foundation of NOCHMI was impressive in comparison with the previous period. In the first five years, 370 monuments were subject of restoration, and ten provincial technical offices were created with the necessary technical staff such as engineers, cartographers, traditional architects, and artisans (Masoumi 1977, p. 191). After ten years of activity, 470 restoration projects were executed on monuments, and 17 provincial offices were created (Mehran 1977, p. 125).

The important role of the NOCHMI is also evident from the number of monuments registered in the Inventory of National Heritage. From 1941 to 1964, only 101 architectural and archaeological sites had been registered, while from 1964 until

Regitsration's period	Total	Tappe	Monument
First Pahlavi (1931-1941)	366	13	353
Second Pahlavi (1941-1965)	101	14	87
Second Pahlavi (1965-1979)	1175	464	711

Fig. 9 Number of Monuments (archaeological and architectural sites) and Tappe (mounds) registered in the Iranian National Heritage Inventory during different periods

1979, almost 1200 monuments were chosen as national heritage, which clearly show the radical change in the cultural policy of the country (*see* Fig. 9).

Despite the development of national preservation organizations, the 60s and 70s were also years of the country's unbridled development, modernization, and globalization (Rouhani 2011, p. 1022). Iranian cities were subject to development plans, and the old historic fabrics were dismantled to build modern multi-store buildings or create boulevards and streets. De Angelis d'Ossat, in his speech, *Città Crocifisse*, in Tehran's seminar on the preservation of historic buildings and fabrics in 1971, explains how modernization had damaged cities such as Mosul, Cairo, or Italian cities such as Palermo and was then the major threat for Iranian cities:

> Urban problems and issues must be answered by finding quick but decisive and studied solutions, and it is sure that … the crucifixion of ancient centers [by constructing streets] not only does not eliminate the problem but also annihilates the cities. (de Angles d'Ossat 1971, p.6)

Hamdan, Semnan, Kerman, and Kashan are cities that de Angelis mentioned as victims of the modern vertical streets, by which the old fabric of the cities had been mutilated. The Italian historian of architecture affirms about Kashan:

> In this arid city at the edge of the desert, which is made of clay and mud, with its meandrous alleys and narrow passages, has created a valuable and reasoned texture against the climate, … [Notwithstanding, recently] north-south and east-west streets which are wide sixty meters have been constructed. this practice, like in other cities, has been done without the slightest study and without much result … [this kind of intervention] goes against architectural values. (de Angles d'Ossat 1971, pp. 10–11)

These urban interventions carried out in a distorted perspective of modernization symbolize a lack of cultural maturation and sensibility towards historic urban fabric. Moreover, if compared with European Theories of the 60s and 70s, concerning the measures necessary to safeguard the historical cities (Carta di Gubbio 1960; Charter 1964; Amsterdam Declaration 1975), these interventions could epitomize the interpretation of the concept of the monument, as the single building of significant historical-artistic interest while neglecting the environmental context. It also demonstrates the negligence towards the documentary and social value of historical urban fabrics, apparently "minor," while they constitute precious testimonies of the civilizations.

The preservation of Iranian historic cities in their entirety and not just their major monuments remained one of the main challenges in Iran also after the Islamic Revolution of 1979.

The Islamic Revolution of Iran and the consequent eight-year war with Iraq slowed down the activities of the NOCHMI. After the 1979 revolution, the Ministry of Art and Culture was abolished, but NOCHMI remained active. Based on the archival documents, it could be concluded that most of the provincial experts continued to operate until 1985 and also after the abolition of the organization in the same year. In 1985, a new organization called the Iran Cultural Heritage Organization (ICHO) was founded and inherited what was created by NOCHMI over the years.

From 1965 to 1979, the National Organization for the Conservation of Historical Monuments of Iran succeeded in creating the first modern organization for the preservation of national heritage through an efficient bureaucratic system that was well related to different components of the national organization in different stages. NOCHMI created the link between the academic institutions and the restoration and conservation sites which did not exist before.

9 Iran Cultural Heritage Organization

The Islamic Revolution of Iran in 1979 brought a radical change in the political and cultural status of the country. Political repression, human rights violation, the discontent of westoxification, and unsuccessful socio-economic and cultural policies are just some of the causes for the end of the Shah's dictatorship in Iran. Abrahamian describes the 1979 revolution as follows:

> The revolution came because the Shah modernized on the socioeconomic level and thus expanded the ranks of the modern middle class and the industrial working class, but failed to modernize on another level, the political level; and that this failure inevitably strained the links between the government and the social structure, blocked the channels of communication between the political system and the general population, widened the gap between the ruling circles and the new social forces, and, most serious of all, cut down the few bridges that had in the past connected the political establishment with the traditional social forces, especially with the bazaars and the religious authorities. (Abrahamian 1983, p. 427)

Based on the glorification of the ancient Persian empires and the secularization of the Iranian culture, the cultural policy of Pahlavi had reached its peak in the years before the revolution. The 2500-year celebration of the Persian Empire in Persepolis in 1971 and the Changing of the Solar Hijri Calendar to the Imperial Calendar in 1975 were the culmination of such policies.

The cultural policy of the Islamic Republic after 1979 was a response to Pahlavi's denial of the Islamic identity of Iranians and sought to revive Iranian Islamic traditions by refusing Western values.

The Islamic Revolution caused the immediate cessation of most international cooperation in the field of cultural heritage and the departure of Western committees from Iran. However, due to the Iran–Iraq war and the complexities of the post-revolution era, it took several years to define a new governmental organization in the new regime. Meanwhile, NOCHMI was not dissolved and remained active until 1985, when it was replaced by the Iran Cultural Heritage Organization (ICHO). Most

Iranian experts, such as directors of provincial offices, maintained their positions after the revolution and continued to operate on the field to preserve the Iranian heritage, especially during the war when many monuments and cities suffered from war damages. The directorship of NOCHMI after 1979 was entrusted to Dr. Shirazi, who was the head of Isfahan's technical office before the revolution. Shirazi, known as the father of modern conservation in Iran, was one of the founding members of the ICHO in 1985. In the following, some of the articles of the statute of ICHO are reported to shed light on the objectives of the new organization (Islamic Council Parliament Research Center, https://rc.majlis.ir/fa/law/show/91528, accessed on one march 2020):

- ICHO is responsible for preparing, organizing, and implementing research programs on the testimonies of the past eras. Surveying and identifying sites, mounds, monuments, and historical complexes as well as registering them in the inventory of National Heritage are ICHO responsibilities. In addition, it is responsible for preparing a comprehensive list of monuments and the archaeological map of the country as well as conducting archaeological research and scientific excavations.
- Conducting research on traditional arts and providing the necessary facilities for their preservation, revival, and growth should be sponsored by ICHO.
- The ICHO should be involved and consulted in the urban and provincial development plans concerning the cultural and historical sites, as well as the historical fabric of the cities and villages, to prevent any damage to historical monuments.
- ICHO should determine a buffer zone for the registered monuments and set particular criteria for any intervention inside these buffer zones.
- The establishment, development, and management of museums in the country are entrusted to ICHO.
- ICHO should establish a centre for collecting cultural heritage documents (e.g. scientific reports, maps, and photographs of buildings) to provide scientific and research services in the field of preservation of monuments.
- ICHO should introduce cultural heritage values at national and international levels by publishing a collection of studies and research via using audio-visual and mass communication means.
- ICHO is also responsible for organizing and implementing educational programs to prepare trained and expert human resources related to cultural heritage in coordination with other relevant organizations.

The formation of ICHO was an important milestone for preserving architectural heritage, which did not have an immediate impact. The major boost arrived almost ten years after establishing ICHO and the election of Mohammad Khatami as the president of Iran in 1997 for two consecutive periods. Khatami started a period of reforms. According to Wastdnige (2014):

> Khatami's concept of Dialogue among Civilizations ... came from a perception of Iranian civilizational weight and importance in the world, much in the same way that the Shah also sought to use similar cultural standing, albeit with different reference points which emphasized the Islamic and pre-Islamic identity of Iran. (pp. 5–6)

During the Khatami government, the architect Mohammad Beheshti was chosen as the fourth director of ICHO in 1997 and held the position for the next seven years, which is the most prolonged period of presidency of ICHO until now. In this period, a growing number of monuments were registered in National Heritage Inventory due to the stability of the ICHO and the importance attributed to cultural patrimony in the Khatami's government. This trend continued during Ahmadinejad's government from 2005 onwards but with much more complications caused by the directorship of Cultural Organization by political figures instead of experts.

The years after the Islamic Revolution are almost unexplored by specialized literature dedicated to historical and artistic heritage. The Inventory of National Heritage offers the possibility to better understand the circumstances of cultural heritage over the years. To study and analyse the Inventory of National Heritage according to the political situation of the country and the year of foundation of important organizations such as NOCHMI in 1965 and ICHO in 1984, eight different periods have been created:

1. 1932–1941: First Pahlavi reign
2. 1941–1965: Second Pahlavi, before the foundation of NOCHMI
3. 1965–1979: Second Pahlavi, from the foundation of NOCHMI until the Islamic Republic Revolution
4. 1979–1985: Islamic Republic, NOCHMI years of activity after the revolution
5. 1985–1989: Islamic Republic, the foundation of ICHO in 1985
6. 1989–1997, Islamic Republic, Rafsanjani as the president
7. 1997–2005, Islamic Republic, Khatami as the president
8. 2005–2009, Islamic Republic, Ahmadinejad as the president.

Given the propaganda of the Pahlavi and the radical ideological changes after the revolution, the advent of Islam was chosen as the decisive moment to characterize and classify the monuments registered in the National Heritage inventory:

a. Before 550 BC: before the foundation of Achaemenid Empire
b. 550 BC to 651 AD: the era of the Persian empires such as Achaemenid and Sassanid dynasty
c. 651–1797 AD: from the advent of Islam until the foundation of Qajar Dynasty
d. 1797–1925: Qajar
e. 1925–1979: Pahlavi
f. Coeval: monuments which were contemporary to the period of their registration
g. Unknown.

In the following, the results of analysing the national heritage inventory from 1935 to 2009 are reassumed in a table based on the corresponding period and the number of monuments registered in each era (*see* Fig. 10).

It is interesting to note from the data reassumed in Fig. 10:

- The percentage of monuments which belonged to the Persian empires between the 550 BC and the advent of Islam in 651 AC (including the Achaemenid and Sassanid monuments) decreases after the 1979 due to the cultural propaganda of

Period of Resgistration	Total	before 550 BC	500 BC - 651 DC	651 DC - 1778 DC	Qajar	Pahlavi	coeval	unknown
1932-1941	366 (4% mound)	23 (6%)	83 (23%)	236 (64%)	12 (3%)	1	0	11 (3%)
1941-1965	101 (14% mound)	13 (13%)	15 (15%)	58 (57%)	8 (8%)	0	0	7 (7%)
1965-1979	1175 (39% mound)	367 (31%)	140 (12%)	467 (40%)	119 (10%)	36 (3%)	4	42 (4%)
1979-1985	80 (20% mound)	17 (21%)	4 (6%)	42 (53%)	14 (16%)	4 (5%)	0	1 (1%)
1985-1989	28 (20% mound)	6 (21%)	1 (4%)	16 (57%)	5 (18%)	0	0	1 (5%)
1989-1997	182 (2% mound)	3 (2%)	3 (2%)	68 (37%)	94 (52%)	0	9 (5%)	5 (3%)
1997-2005	1148 (35% mound)	2091 (18%)	719 (6%)	3836 (33%)	2770 (24%)	689 (6%)	107 (1%)	1273 (11%)
2005-2009	13198 (40%	3730 (28%)	2205 (17%)	3645 (28%)	2236 (17%)	652 (5%)	20 (0.1%)	704 (5%)

Fig. 10 The registered Monuments and Archaeological sites in the Inventory of National Heritage, classified by the era of foundation and the period of registration

the Pahlavi's era. Not surprisingly, during the reign of Reza Shah, the percentage of pre-Islamic monuments in the National list was at its highest point and constituted 23% of all the registered monuments.

- The number of Qajar monuments registered in the national inventory significantly increased after the revolution and the overcoming of Pahlavi's anti-Qajar policy. It should be considered that in the meantime, the temporal distance with the Qajar period had also increased.
- Despite the cultural propaganda of Pahlavi and Islamic republic governments, the monuments of the Islamic period from the advent of Islam in 651 until the beginning of Qajar period have always constituted an important part of the National Heritage inventory.
- The pre-historic archaeological sites acquainted importance from the foundation of NOCHMI, which could be related to the presence of foreign archaeological committees and also because the prehistory was out of the domain of cultural propaganda of Pahlavi era.
- The remarkable increase in the number of the sites registered between 1996 and 2004 is a significant sign of the formation of a cultural policy after almost two decades since the Islamic revolution in 1979. It should be noted that no single monument was registered between 1990 and 1996.

It is noteworthy that the abovementioned analyses are not comprehensive to explain the situation of Iranian cultural heritage after the Islamic Revolution. In fact, many other factors should be considered and studied by a detailed account of ICHHTO activities during almost 42 years. The archive of ICHHTO (ICHO became ICHHTO during the years) is definitely a precious source of information in this regard. A general glimpse through the list of documents indicates that although more and more monuments were registered after Khatami's government, the number of restoration and conservation projects diminished over the years; from 1979 until 1985, while NOCHMI was still operative, there are almost 2000 reports of restoration campaigns and studies catalogued in the list of ICHHTO archive. However, there are fewer than 1000 documentations in the eight years of Khatami's government.

ICHO, as a Research and Executive Organization, was part of the Ministry of Culture and Higher Education until 1993, when it became part of the Ministry of Culture and Islamic Guidance. In 2003, ICHO was merged with the Iranian Tourism Organization and became the "Cultural Heritage and Tourism Organization" operating under the president's direct supervision. Three years later, in 2006, other major changes took place in the structure of the organization. The Handicrafts Organization was merged with ICHTO, and the "Cultural Heritage, Handicrafts, and Tourism Organization" (ICHHTO) was established.

The performance of the organization has provoked considerable criticism, especially after merging with the Tourism Organization in 2003 and becoming a state-owned company headed by political figures instead of experts. ICHHTO became an independent ministry in 2019, which could be a solution for many problems of an organization that had lost its executive power by merging with other organizations and suffered from severe financial and management problems (Beheshti 2019).

10 Conclusion

Since the first law of antiquities in 1930 until the foundation of ICHO in 1985, the Iranian historical heritage underwent significant difficulties such as uses and abuses of political propaganda and the constant threads of development and modernization. The notions attributed to the inheritance of the past had been radically changed through the years to adopt national and political dimensions and scientific values. This long journey is well-presented if we confront the first law of antiquities with the statue of NOCHMI first and ICHO after. The monuments and archaeological findings from treasures to be exported became important scientific materials to be studied, surveyed, and preserved. While preserving few major monumental sites such as the Safavid royal complex of Isfahan or the Achaemenid Persepolis raised some concerns among the elites at the beginning of the twentieth century, historic private houses or entire villages were considered as a national heritage for preservation after almost four decades. Thus, while the circle expanded to include every testimony of different dimensions of the Iranian civilization, the thought towards the inherited patrimony of past ages matured, and their conservation was entrusted to trained experts and modern institutions.

Several Iranian architectural and archaeological heritage have been nominated as UNESCO World Heritage since 1979: Persepolis (1979), Meidan Emam of Esfahan (1979), Tchogha Zanbil (1979), Takht-e Soleyman (2003), Pasargadae (2004), Bam and its Cultural Landscape (2004), Soltaniyeh (2005) and Bisotun (2006), The monastic ensembles of St. Thaddeus and St. Stepanos, and the Chapel of Dzordzor (2008), Shushtar Historical Hydraulic System (2009), Tabriz Historic Bazaar Complex (2010), Sheikh Safi al-din Khanghah and Shrine Ensemble in Ardabil (2010), The Persian Garden (2011), Gonbad-e Qabus (2012), Masjed-e Jame of Isfahan (2012), Golestan Palace (2013), Shahr-e Sokhte (2014), Cultural Landscape of Maymand (2015), Susa (2015), The Persian Qanat (2016), Historic City of Yazd (2017), and Sassanid Archaeological Landscape of Fars Region (2018).

Selecting monuments such as nine gardens in different regions as the Persian Garden or the eleven traditional irrigation systems as the Persian Qanat as well as nominating the entire city of Yazd and Bazar of Tabriz, in recent years, can epitomize the liberation of the Iranian Heritage from the grandeur of Achaemenid and Sassanid Empire and the splendour of Safavid era to finally valorize the Iranian traditions and culture.

Scientific literature is rarely dedicated to restoration campaigns and projects of Iranian architectural heritage. In the following to provide some examples of executed restoration projects in Iran, St. Thaddeus and Dzordzor chapel restorations are studied in detail along with the restorations of St. Sarkis church in Khoy.

11 Restoration and Conservation of the Armenian Churches in West Azarbaijan

Armenians and Christian minorities lived in the West and East Azarbaijan provinces of Iran for centuries. The existence of nearly 100 churches in West Azarbaijan confirms the high density of a Christian population in past eras (Mohammadi 2005). Unfortunately, most Christian monuments in Azarbaijan are currently in ruins as a result of the immigration of the local Christian population. However, important Armenian churches such as St. Thaddeus and St. Stepanos still stand and attest to the historical presence of the Armenians in the region.

In 2006, St. Thaddeus Church, St. Stepanos Church, and the Chapel of Dzordzor were proposed to be registered as UNESCO World Heritage Site. UNESCO accepted the nomination of the three Christian ensembles in 2008 and motivated this decision by the following explanation:

> The monastic ensembles of St. Thaddeus and St. Stepanos as well as the Chapel of Dzordzor are the main heritage of the Armenian Christian culture in Iran. They were active over a long historical period, perhaps since the beginning of Christianity and certainly since the 7th century. They have been rebuilt several times, either due to regional socio-political events or natural disasters (e.g., earthquakes). Until today, they have remained in a semi-desertic environment in keeping with the original landscape. The nominated property consists of three monastic ensembles of the Armenian Christian faith. … They were created by the Armenian religious culture over the course of its history and its establishment in the north-east of present-day Iran (provinces of East and West Azarbaijan). Each zone centers on a major architectural property which is the main reason for its nomination and includes several

> complementary elements, either built or in ruins: annex chapels, cemeteries, and villages, which were associated with the monastic way of life in the past. (ICOMOS 2008, p. 52)

Regarding the Christian heritage of the West Azarbaijan, it should be noted that 30 churches were registered in the Inventory of Iranian National Heritage until 2006 (*see* Fig. 11).

The ICHHTO archive and library, located in Tehran, preserves a precious collection of reports and preliminary studies of the restoration projects conducted in Iran since 1965 when the NOCHMI was founded. The following table in Fig. 12 summarizes and lists the documentation on Armenian churches of West Azarbaijan, preserved in ICHHTO archive. As can be noted, some churches such as St. Sarkis and St. Thaddeus have been subject to multi-year campaigns of restoration and conservation (*see* Fig. 12).

As mentioned before, Iranian restoration projects have rarely been the subject of scientific publications. Therefore, to examine the methods and criteria of restorations conducted at Armenian heritage in West Azarbaijan, St. Thaddeus, St. Sarkis, and Dzordzor churches and chapel have been selected as case studies. The selection was motivated by the historical and architectural importance of the sites as well as the importance of their multi-year restoration campaigns. In order to summarize the documentation, collected at the ICHHTO archive, a summary sheet is created for each monument. These case studies are not exshaustive to draw conclusions about the Iranian restoration practices. Still, they are reported here to inspire future reflections on the reality of restoration projects in Iran.

11.1 Summary Sheet of the Restoration and Conservation Campaigns of St. Thaddeus Church

Location: Maku, Qare Kilisa Village, West Azarbaijan.

Experts and collaborators: Mansour Falamaki, Hedaiati, Etehadi, Marten, Dr Sarafraz, Latif Abolqasemi.

Responsible Institution: *Sazman Melli-e Hefazat az Asar-e Bastani* (NOCHMI), University of Tehran (Faculty of Fine Arts), UNESCO Consultancy.

Duration of campaigns: 1972–1986, 1990–1995, 2001.

The registration number of the consulted documents at the ICHHTO archive: 501 (1973), 502 (1975), 504 (1975), 505 (1971), 507 (1973), 508 (1974), 510 (1974), 566 (1976), 5667 (1976), 519 (1360).

Documentation attached to the reports: Photos, reports of the restorations, contracts, and stipulated agreements.

	Church	Location	Date of Registration	Construction period	Number of Registration
1	Monastery of Saint Thaddeus (Armenian)	Chaldoran	1956	Seventh century	405
2	St Serkis church (Armenian)	Khoy	1968	Sixteenth century	822
3	Marsergiz Church (Assyrian)	Urmia	1975	Sassanid	1122
4	Mariokhne Church of Jamalabad (Assyrian)	Urmia	1983	Sassanid	1648
5	Church of Our Lady (Assyrian)	Urmia	1983	First century BC	1649
6	Church of Karabakh (Assyrian)	Urmia	1983	Ilkhanid-Safavid	1650
7	AftarKhane Church (Assyrian)	Salmas	1998	Safavid	2073
8	Margivargiz church Hazrat-e Maryam church	Urmia	1998	Qajar	2209
9	Mahlazan Church (Armenian)	Khoy	1999	Safavid	2389
10	Fanai church (Armenian)	Khoy	2000	XV-XVI century	3380
11	Church of Qeris (Armenian)	Khoy	2000	Qajar	3381
12	Petros Polis Church (Assyrian)	Urmia	2001	VIII century	4790
13	Qadimi Church	Khoy	2001	Ilkhanid	4847
14	Old Vank Catheral	Salmas	2001	Qajar	4839
15	Old Armenian Church (Armenian)	Salmas	2001	Qajar	4840
16	Old Church of Ghezeljeh (Armenian)	Salmas	2001	Safavid-Qajar	4841
17	Haftvan Church (Assyrian)	Salmas	2001	Safavid	4842
18	Mar Yaghub Church (Assyrian)	Salmas	2001	1952	4843
19	Dzordzro chapel (Armenian)	Maku	2002	XIV century	6157
20	Mahlam Church	Salmas	2002	1880	6158
21	Aftarkhaneh Church (Armenian)	Salmas	2002	1890	6159
22	Sapourqan church (Assyrian)	Urmia	2002	Qajar	7387
23	Adeh Church (Assyrian)	Urmia	2002	Qajar	7388
24	Mavana Church	Urmia	2002	Qajar	7389
25	The old church of Marserkis (Assyrian)	Salmas	2002	Qajar	7532
26	Margivargis Church (Assyrian)	Salmas	2002	XIII century	7859
27	Tappeh church Tappeh Si	Maku	2003	First century BC	10039
28	St George Church	Urmia	2005	Qajar-Safavid	14295
29	Martouma Church (Assyrian)	Urmia	2005	VII century	14287
30	Margabriel church	Urmia	2005	XVIII	14977
31	Tappeh church	Urmia	2005	First century BC	15939
32	Mar Quriaqus (Assyrian)	Salmas	2006	Ilkhanid	16392

Fig. 11 Churches of West Azarbaijan registered in the Inventory of National Heritage until 2006

<table>
<tr><th>Churches in
West Azarbaijan</th><th>Responsabile</th><th>Reports of Restoration
project of ICHHTO archive</th></tr>
<tr><td rowspan="2">St Stepanos</td><td>Faculty of fine arts</td><td rowspan="2">1977, 1978, 1982, 1983,
1984(5), 1985, 1986(2)</td></tr>
<tr><td>NOCHMI</td></tr>
<tr><td rowspan="4">St Thaddeus Monastry</td><td>Faculty of fine arts</td><td rowspan="4">1973(2), 1974(4), 1975, 1976(2),
1977, 1978, 1979, 1980(2),
1981(2), 1982(3), 1983(2), 1984(5),
1985 (2), 1986(2), 1990, 1991,
1993(3), 1994, 1995, 2001</td></tr>
<tr><td>NOCHMI</td></tr>
<tr><td>Unesco Consultant</td></tr>
<tr><td>ICHO</td></tr>
<tr><td rowspan="3">Surp Serkis Church</td><td>Faculty of fine arts</td><td rowspan="3">1974, 1975, 1976-78, 1979,
1981(2), 1982, 1983, 1984,
1985, 1987, 1989, 1990,
1993, 1994, 1995</td></tr>
<tr><td>NOCHMI</td></tr>
<tr><td>ICHO (ICHHTO)</td></tr>
<tr><td>Dzordzor Chapel</td><td>ICHO (ICHHTO)</td><td>1984(2), 1988, 1989, 1990,
1991, 1993(2)</td></tr>
<tr><td>Markhne Church</td><td>NOCHMI</td><td>1980, 1983, 1985</td></tr>
<tr><td rowspan="2">Maar-Sergiz Chucrh</td><td>NOCHMI</td><td rowspan="2">1978, 1981, 1986, 1987-1991</td></tr>
<tr><td>ICHO (ICHHTO)</td></tr>
<tr><td>S. Maryam Church</td><td>NOCHMI</td><td>1983</td></tr>
<tr><td>Nane Maryam Church</td><td>NOCHMI</td><td>1984(2)</td></tr>
<tr><td>Qare Bagh Church</td><td>ICHO (ICHHTO)</td><td>1990</td></tr>
<tr><td>Chapel of Sandokht</td><td>ICHO (ICHHTO)</td><td>1990, 1991, 1992</td></tr>
<tr><td>Dare Sham Church</td><td>NOCHMI</td><td>1983(3)</td></tr>
<tr><td>Mahlazan Church</td><td>NOCHMI</td><td>1981(2), 1982(2)</td></tr>
<tr><td>Petros Poliz Church</td><td>NOCHMI</td><td>1984</td></tr>
<tr><td>Fanai, Darshamb, Bardivak Complex</td><td>National University of Tehran</td><td>1978</td></tr>
<tr><td>Saharqand Church of Marand</td><td></td><td></td></tr>
<tr><td>Reports of restoration projects in the churches of the province</td><td>ICHO (ICHHTO)</td><td>2002(2)</td></tr>
</table>

Fig. 12 Documentation of ICHHTO archive regarding the studies and interventions executed on West Azarbaijan churches

11.2 Origins of the Monument

St. Thaddeus church, founded in the seventh century, underwent significant transformations in the thirteenth and nineteenth centuries. This church is an important religious centre for Armenians since it is considered the burial place of St. Thaddeus, the Thaumaturgic apostle. In the 1970s, nearly 6000 pilgrims visited the church, and the site is considered as the spiritual centre of the Armenians in Iran. The church has two main buildings, one in black stone, which is the oldest construction, and one in white stone, which was built in the Qajar era in the first half of the nineteenth century (*see* Fig. 13).

Fig. 13 St. Thaddeus, Edouard Utudjian, report for UNESCO, 1970, ICHHTO Archive, report number: 505

11.3 The Restoration Project

- In 1970, an agreement was signed between National Organization for the Conservation of Historical Monuments of Iran (NOCHMI) and the faculty of Fine Arts of Tehran University to perform "the study and restoration project of the St. Thaddeus church in Qare Kilisa". The project was stopped for two years due to lack of funding, and the restorations resumed in 1972. Article 9 of this agreement mentions the obligation to collaborate with Edouard Utudjian, the UNESCO adviser who was a French Armenian architect, according to the agreement signed with UNESCO in 1966. In the contract, the NOCHMI was stipulated as the "employer" and the University of Tehran mentioned as the "consultant engineer" and "project manager".
- The objective of this project is described as follows.

"Preparing a restoration project for the reinforcement and structural consolidation and revitalization of the building … (where) at least five architecture students will perform their internship. This project helps revitalize the Iranian cultural heritage and the Armenian architectural tradition. More importantly, it will help develop tourism" (report number 501, Qare Kilisa 1972).

- In 1966, an agreement was signed between Iran and UNESCO. Accordingly, Iran had to consider several monuments as its priority for restoration: the Sassanid monuments in Fars, the Jame Mosque and Naqsh-e Jahan Square in Isfahan, and the Tomb of Oljeitu in Soltaniyeh and Qare Kilisa in Maku. (Mehran 1976, p. 245). With the help of UNESCO's councilors, preservation of the selected historical heritage aimed to help the development of tourism in the country.
- According to this agreement, Eduard Utujdian came to Iran in 1969 to work as a consultant in the restoration project of Qare Kilisa or better St. Thaddeus church. Utudjian had studied Armenian architecture before his mission to Iran and published *Les monuments arméniens du IVe siècle au XVIIe siècle* in 1967.

12 Research Campaign

- In the preliminary project, extensive research was conducted on historical maps and drawings, manuscripts, and ancient essays about the church, in order to define the evolution and modification of the building in different eras. Additionally, a typological study of the church in confront of other churches was performed.
- Different surveys were performed in the interior space, the external environment in the south and southwest, the central courtyard, and the western courtyard, where there were ruins of other buildings.
- The chapels in the east and southeast of the church were studied.
- Petrographic analysis of the black church and the white church was performed. The stones were analysed in the laboratories of Tehran and Cologne to identify their characteristics and find a remedy for the degradation of the white stone.
- The ground on which the monument was founded was analysed to identify the critical points of the foundations and better understand if structural needs motivated the construction of a subsequent portion of the building.
- The foundations and possible damages caused by humidity were studied.
- Analysis of the roofs and bricks as well as the structural analysis was performed to identify the deterioration source. Then, the damages and their causes were divided into different categories.
- Several tests were performed on the walls to recognize the construction system.
- The project aimed to organize the surrounding environment. Therefore, in the preliminary project, the relationship of the church with its surrounding environment over the years was studied along with the historical irrigation system of the place.
- Three architects and civil engineers as well as a petrography expert and an archaeologist were hired for the project.

12.1 *Structural and Material Consolidation*

Before starting the work, metal and wood scaffoldings were installed in interior and exterior spaces (*see* Figs. 14 and 15). The restoration project included:

- Containing and arresting the progress of damages with temporary solutions such as wood shoring and some permanent solutions
- Removing the roof of the black church, numbering the roof stones, and performing general consolidation of the building and the underlying walls with reinforced concrete beams (in some portions)
- Reinforceinment and dehumidification of the foundations of the church and the underlying walls (see Fig. 16)
- Finding suitable measures to restore stones in a severe state of decay
- Removing the roof of the white church, cleaning the roofing stones, and replacing broken stones with new ones if necessary
- Reinforcing the foundations, walls and vaults of the building (see Figs. 17, 18)
- Removing a part of the flooring to rebuild suitable flooring
- Performing emergency restorations with recognizable materials in three chapels around the church, one of which is the Sandokht chapel. Based on the photos, it seems that they were in an intense state of decay and that they were reconstructed mainly with original materials (*see* Fig. 19).

12.2 *Future Measures for Using the Monument*

Considering the mass pilgrimages to the church and the importance of the site, three different types of buffer zones were proposed:

1. The large protected area: A large surrounding area was defined in the mountainous landscape around the church where any intervention could alter the relationship of the church with the landscape. In this area, necessary services planned to be provided for the visitors in the Qare Kilisa village.
2. Archaeological buffer zone: In this area, no intervention was allowed, and archaeological excavations had to be performed to better define the underground relics. Any intervention in this area was defined dangerous for integrity and harmony of this historical complex.
3. Architectural/Ancient buffer zone: This area defined and characterized traditional and local architecture, where any intervention had to be supervised by authorities.

Unfortunately, no plans and maps were attached to the documents to show the buffer zones. Providing more details about its impossible, although, from the reports of the last few years of restoration campaigns, it seems that the desire to establish buffer zones was not immediately actualized since the experts continued to emphasize the importance of having a buffer zone around the monument.

Fig. 14 St. Thaddeus, Wooden Props, faculty of Fine Arts of Tehran University, 1974, ICCHTO Archive, report number: 502

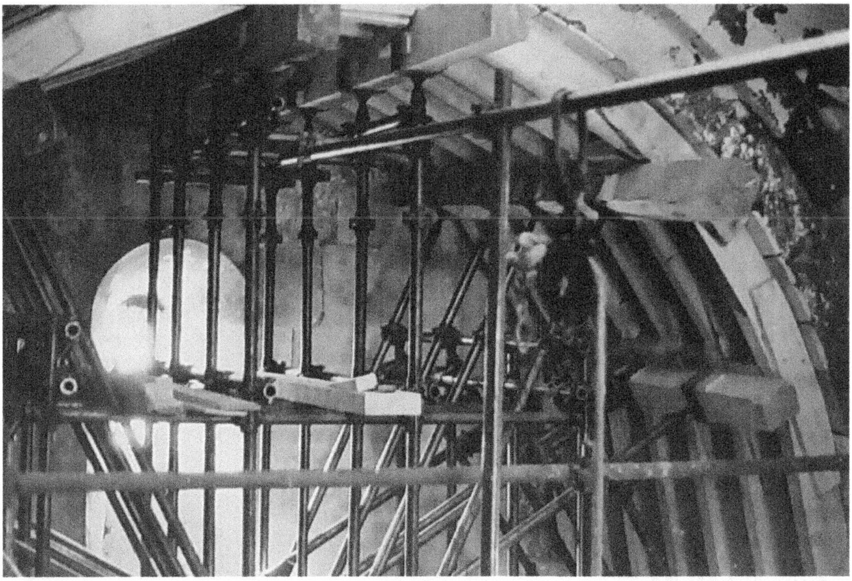

Fig. 15 St. Thaddeus, Scaffolding before the beginning of the Restoration project, faculty of Fine Arts of Tehran University, 1974, ICCHTO Archive, report number: 502

In the document prepared for the Nomination Qare Kelisa as UNESCO Heritage in 2007, four different protected boundaries are defined: (*see* Fig. 20) (UNESCO 2008, pp. 96–97).

1. The core zone: "No allocation is permitted. Construction, digging wells or subterranean canals, tree plantations, excavation, earth filling or civil engineering operations are prohibited unless they are undertaken by the ICHHTO."
2. The buffer zone: "It is permitted to carry out traditional farming in existing farms and on agricultural lands within their boundaries at surface level" and "all the installation should be taken under permission and monitoring of ICHHTO."
3. The landscape buffer zone: "It is permitted to continue to carry out agricultural or farming activities" but "industrial installations causing environmental pollution, mining, or any activities altering the structure of the mountains or damaging the general landscape are forbidden."
4. The expansion zone of the villages: "Enclaves are allocated to future expansion of villages" and "are highly recommended that all constructions in the expansion areas be carried out in respect to local traditional architecture."

In the UNESCO document, it is written: "The first nominated zone is the monastic ensemble of St. Thaddeus and the Chapel of Sandokht. This zone corresponds to a total area of 40.16 ha. It is situated in a buffer zone of 310.98 ha and a larger landscape zone of 1438 ha."

Fig. 16 St. Thaddeus, Reinforcement and dehumidification of the foundation from the interior space, faculty of Fine Arts of Tehran University, 1976, ICCHTO Archive, report number: 5667

Fig. 17 St. Thaddeus, Original stones of the central vault, faculty of Fine Arts of Tehran University, 1976, ICCHTO Archive, report number: 5667

Fig. 18 St. Thaddeus, Top layer of concrete on the vaults, faculty of Fine Arts of Tehran University, 1976, ICCHTO Archive, report number: 5666

12.3 Characteristics and Criteria of the Restoration Project

The following points could be deduced from the photos and explanations in the reports:

Fig. 19 St. Thaddeus, Dismantling, numbering and reconstructing north chapels, faculty of Fine Arts of Tehran University, 1976, ICCHTO Archive, report number: 5667

- Using original materials available on the site as much as possible in order not to alter the authenticity of the church
- Preserving decorations in their original state
- The reconstructed parts were limited to the minimum possible to maintain authenticity of the monument. If it was necessary to dismantle the stones or repair

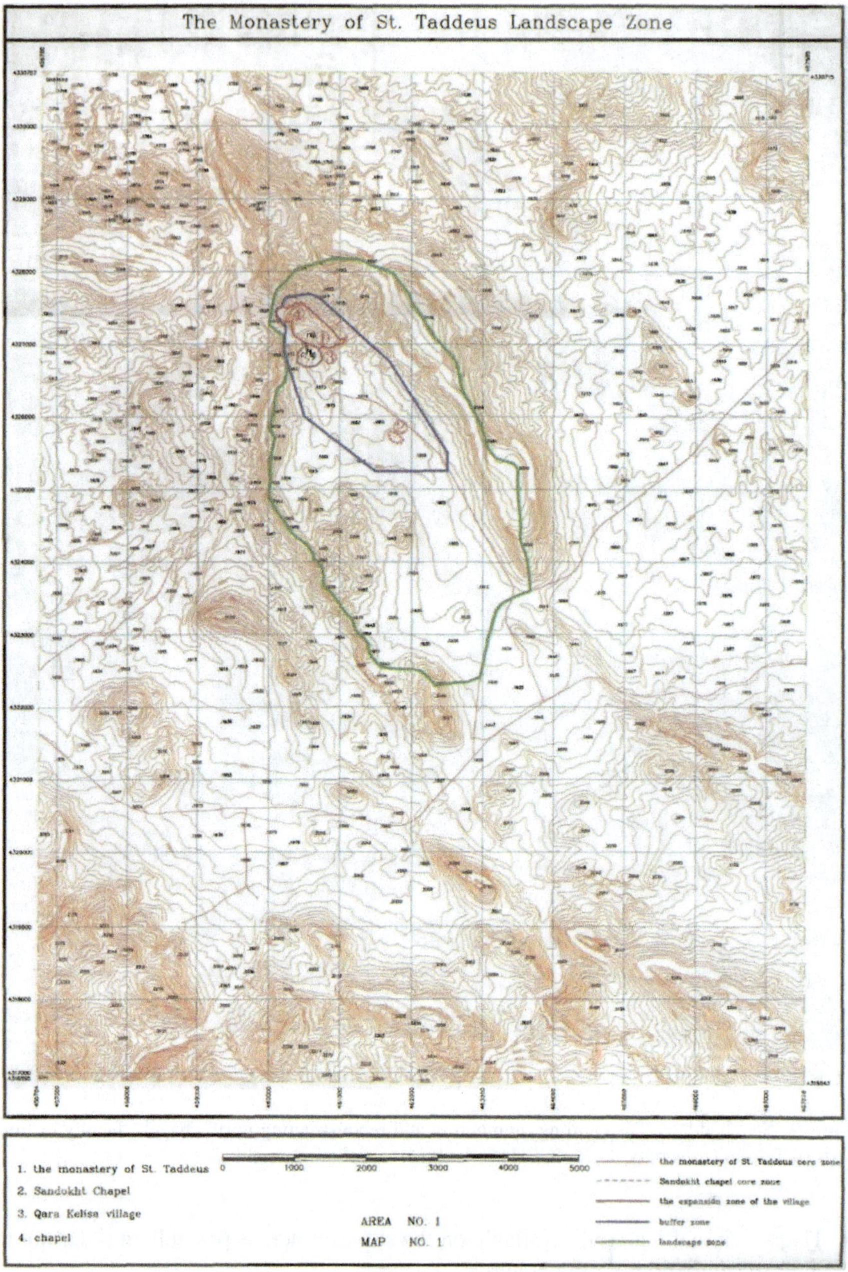

Fig. 20 The monastery of St. Thaddeus Landscape zone, UNESCO documentation, 2008, https://whc.unesco.org/en/list/1262/documents/ accessed on: 13 may 2021

structural damages, the stones were numbered and returned to their original position.

- Using modern materials such as reinforced concrete beams was considered as an obvious solution to structural problems.
- The 1976 report is very interesting, which describes a meeting between site managers and important figures of Iranian preservation such as Foroughi and Mohammad Karim Pirnia to decide how to strap beam the black church. Episodes like this explain how carefully such projects were performed.

12.4 Advisory Body Evaluation (ICOMOS)

> "The major restoration works undertaken since the 1970s thoroughly respected ground plans, available data, and materials. The stone came from the same quarries. Interior restorations and stabilization works on the edifice were performed in a completely satisfactory spirit of authenticity, which did not rule out some isolated problems such as the pointing of joints with cement, which is promoting the appearance of salt deposits. The large number of low relief and painted mural decorations were authentic. The buildings of the inner courtyard of the monastery were restored in a spirit in keeping with the original by reference to early documents." (ICOMOS 2008, p. 56)

12.5 Summary Sheet of the Restoration and Conservation Campaigns of St. Sarkis of Khoy

Location: Khoy, West Azarbaijan.

Experts and collaborators: Ebrahim Heidari, Latif Abolqasemi.

Responsible institution: Sazman Melli-e Hefazat az Asar-e Bastani (NOCHMI), Faculty of Fine Arts (University of Tehran).

Duration of campaigns: 1975–1995.

The registration number of the consulted documents at the ICHHTO archive: 554 (1975), 555 (1976–78), 5666 (1976), 5667 (1976), 556 (1979), 557 (1981), 558 (1981), 559 (1982), 560 (1983), 561 (1984), 562 (1983), 563 (1987–88), 564 (1989–91), 565 (1991–93), 4424 (1995).

Documents attached to the reports: Photos, reports of the restorations, official letters, and architectural drawings.

12.6 Origins of the Monument

St. Sarkis Church is an ancient monument located in the northwest of Khoy in Qamsal Square. The church was registered as a National Heritage with the registration number of 822 in 1980. The exact date of its construction is not known, but historical books estimated it dates back to IV and IX century. The present building seems to belong to Ilkhanid era based on its architectural characteristics and a stone inscription dated to 1120 AD. However, significant repairs were made in the Safavid period (Sistani 2008). The name of the church could be because of Sarkis Gabadokian, who was one of the missionaries of Christianity in the third century AD and was killed by Shapur, a Sassanid king (ibid).

12.7 The Restoration Project

- In a visit by Latif Abolqasemi, representative of the University of Tehran in 1975, the general state of the church was described as unsatisfactory due to severe deterioration caused by humidity, degradation of the materials, and inclination of the walls.
- The construction of a new urban street was mentioned as a disturbing element for the surrounding context of the church.
- The severe state of structural and material degradation is visible in the collapsed vaults and the high humidity of the external walls. Latif Abolqasemi wrote in his report that the state of the building was quite critical and needed significant restoration and consolidation works (*see* Figs. 21 and 22).
- The project lasted for almost 12 years due to financial problems and lack of necessary funds. The Armenian Diocese had contributed some funds to the restoration campaigns.

12.8 Research Campaign

- A program of historical studies and research including archaeological surveys was drawn up by the Tehran University of Fine Arts. However, there are no references to archaeological research in the executive reports.

12.9 Structural Consolidation

- Wooden props were installed as the first measure and remained attached to the building for almost ten years before being removed (*see* Fig. 23).

Fig. 21 St. Sarkis Church before the restorations, faculty of Fine Arts of Tehran University, 1975, ICCHTO Archive, report number: 554

Fig. 22 St. Sarkis Church, cracks caused by settlement and thrust, NOCHMI, 1979, ICCHTO Archive, report number: 556

Fig. 23 St. Sarkis Church, wooden props, faculty of Fine Arts of Tehran University, 1975–77, ICCHTO Archive, report number: 555

- A canal in bricks and concrete was built around the eastern side and the apse's side to eliminate humidity of the building (*see* Fig. 24).
- The foundations on the south side were reinforced and interconnected to reinforce the stability of the building.
- New and wider foundations were built for the pillars (*see* Fig. 25).
- The decayed stones of the pillars were replaced to support the roofing weight.
- A new pavement was made with a layer of concrete and a layer of brick on top of the concrete. (It is interesting to note that in the 1362 SH report (1983–84), the person in charge suggested stone as a flooring material since it was more in tune with the stone walls. However, he stated that the NOCHMI board had to make the decision. Based on the photos, it could be deduced that NOCHMI has opted for brick pavement.)
- The arches were scaffolded to prevent further collapse and remained attached untill the termination of the work.
- In the early 1970s, a second roof was constructed over the original one, and the load of this new structure damaged the original roof and caused infiltration from the top. In 1983, the new structure was removed, the original brick vaults were restored, and the roof was repristinated with some new elements (*see* Fig. 26).

12.10 Restorations and Reconstructions

- The collapsed parapets of the building were rebuilt with "the original shape and size" (*see* Fig. 27).
- Internal parts of the church such as the niches were restored by reconstructing the parts in ruins. The reconstruction of the arches of the apse was the most significant element reconstructed.

Some elements of the apse's arches were in place. More details were revealed about the original form and shape of the apse based on the "collapsed stones found on the floor and the remaining elements on the opposite side." However, these testimonies were not sufficient to define the original form in detail and with certainty. Therefore, it was decided to resort to a "typological comparison with other churches" and reconstruct the apse arches in a more probable form and shape. The original stones on the site were used in the reconstruction process, and the missing pieces were created by considering the original pieces. (report n. 564) (*see* Fig. 28).

12.11 Summary Sheet of the Restoration and Conservation Campaigns of Dzordzor Chapel

Location: Maku, West Azarbaijan.

Fig. 24 St. Sarkis Church, reinforcement and dehumidification of the foundation, faculty of Fine Arts of Tehran University, 1975–77, ICCHTO Archive, report number: 555

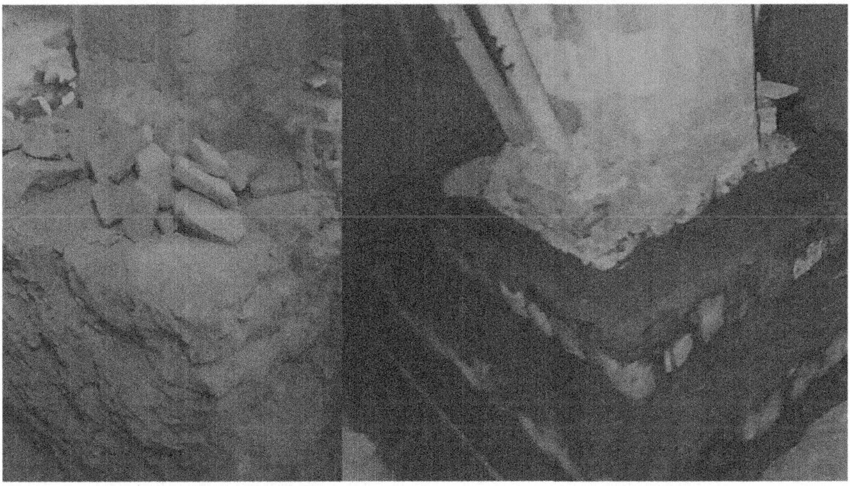

Fig. 25 St. Sarkis Church, the foundation of pillars before and after the restoration, NOCHMI, 1981, ICCHTO Archive, report number: 557

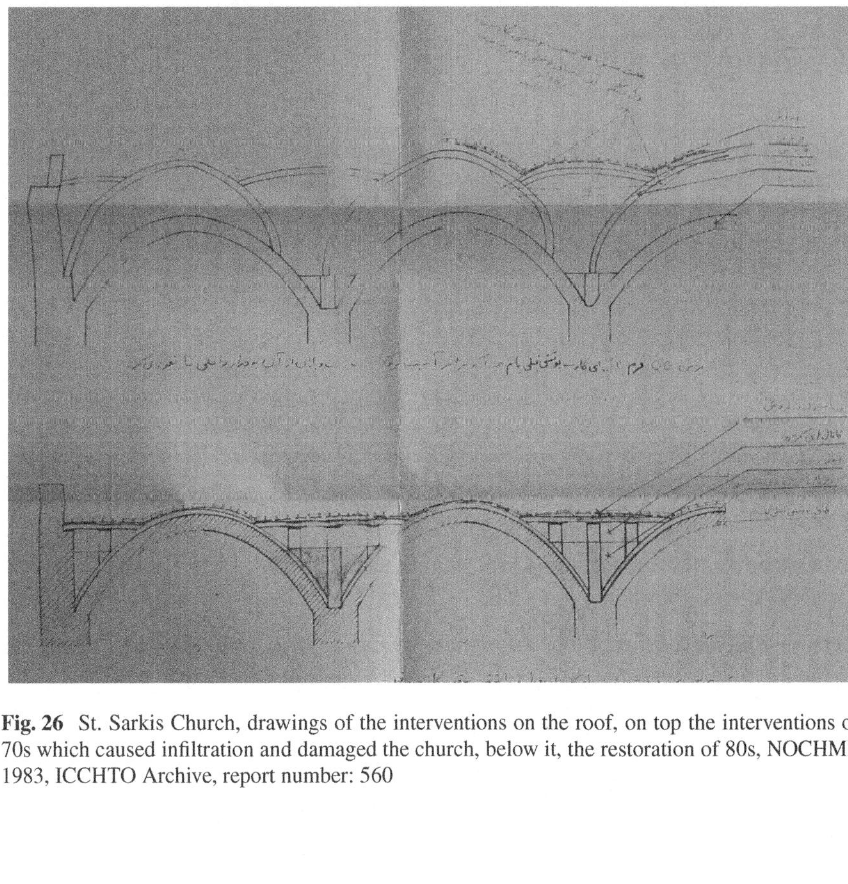

Fig. 26 St. Sarkis Church, drawings of the interventions on the roof, on top the interventions of 70s which caused infiltration and damaged the church, below it, the restoration of 80s, NOCHMI, 1983, ICCHTO Archive, report number: 560

Fig. 27 St. Sarkis Church, reconstruction of collapsed parapets, NOCHMI, 1987–88, ICCHTO Archive, report number: 563

Fig. 28 St. Sarkis Church, reconstruction of the arches of the apse, ICHO, 1989–92, ICCHTO Archive, report number: 565

Experts and collaborators: Ebrahim Heidari, Bahman Kargar, Moslem Mohammad Zade, Sadr-aldin Shirdel.

Responsible institution: The National Organization of Cultural Heritage *(Sazman-e Miras-e Farhangi-e Iran).*

Duration of campaigns: 1984–91, 1993.

The registration number of the consulted documents of the ICHHTO archive: 501 (1973), 502 (1975), 504 (1975), 505 (1971), 507 (1973), 508 (1974), 510 (1974), 566 (1976), 5667 (1976), 519 (1360).

Documentation attached to the reports: Photos, reports of the restorations, and architectural drawings such as plans and sections.

12.12 Origins of the Monument

The Dzordzor Chapel was built in the nineteenth century by the Archbishop of St. Thaddeus Church named Zakaria, who descended from one of the great family and owners of the area. "The Chapel of Dzordzor is the remains of a much larger monastic ensemble, which disappeared at the beginning of the seventeenth century" (ICOMOS 2008, p. 55).

12.13 The Restoration Project

- The construction of the *Baran* dam in 1987 exposed the chapel to the risk of being waterlogged. After several reports from the regional technical office regarding the risks of such operations, the aforementioned office was given 25 days to take the necessary measures and safeguard the monument. Accordingly, the chapel was dismantled stone by stone and relocated to a new emplacement. The reconstruction of the chapel lasted for three months (*see* Figs. 29 and 30).
- The chapel was relocated 600 m to a location 114 m higher.
- Experts of stone restoration were invited from Shiraz to evaluate the project. The administrative work was performed by the local staff, who performed the work with a 10th of the proposed budget while living on-site with immense difficulty.
- The stones were dismantled and numbered. Additionally, the thickness of the mortar joints was measured and written (*see* Figs. 31 and 32).
- An extensive photographic campaign was conducted with video footage before and during the dismantling operations. Videos and photos were considered as primary materials which could be consulted in the reconstruction process.
- Before starting the project, the bulldozers from the dam project dug almost four metres around the chapel and entirely ruined the archaeological stratigraphy. The reports showed the impossibility of collecting the required data from the archaeological research due to this inconvenience.
- Patterns were made from the arches to be able to reconstruct them accordingly to the original form (*see* Fig. 33).
- Research campaigns were performed around the monument to find the collapsed stones. Several stones were found, and after dividing them into different categories, 52 stones were recognized as the original stones of the chapel.

Fig. 29 Chapel of Dzordzor, before the relocation and restoration, ICHO, 1988, ICCHTO Archive, report number: 116

- Some of the stones found on the site belonged to the collapsed dome of the building, which helped define some characteristics of the original dome such as shape. At the end, the church dome was reconstructed. The original stones were considered as templates for the new white limestone to be used for reconstructing the dome (*see* Fig. 34).

12.14 Research Campaigns

- The report about the history of the monument along with the historical documents concerning the building was collected with the help of the Armenian diocese council.
- Typological analysis of the small chapel was performed together with the identification of the type of stones and type of mortars used in the construction. The stones were basalt, travertine, and lime.
- A geologist was employed in the project.

Fig. 30 Chapel of Dzordzor, before the relocation and restoration, ICHO, 1988, ICCHTO Archive, report number: 116

12.15 Structural Consolidation

- Reinforced concrete beams were inserted into the building, and a concrete floor was created in the new construction site with the same slope as the original site.

12.16 Characteristics and Criteria of the Restoration Project

- The new limestone used for reconstructing the dome was bright white, which is in stark contrast with the original material and can provide a good recognition criterion. However, the recognition criteria do not seem to worry the experts. Instead, the report implies that it was crucial to use materials similar to the original ones. Thus, limestone was chosen, and it was specified that it would have the same colour as the rest of the church over time (*see* Fig. 35).

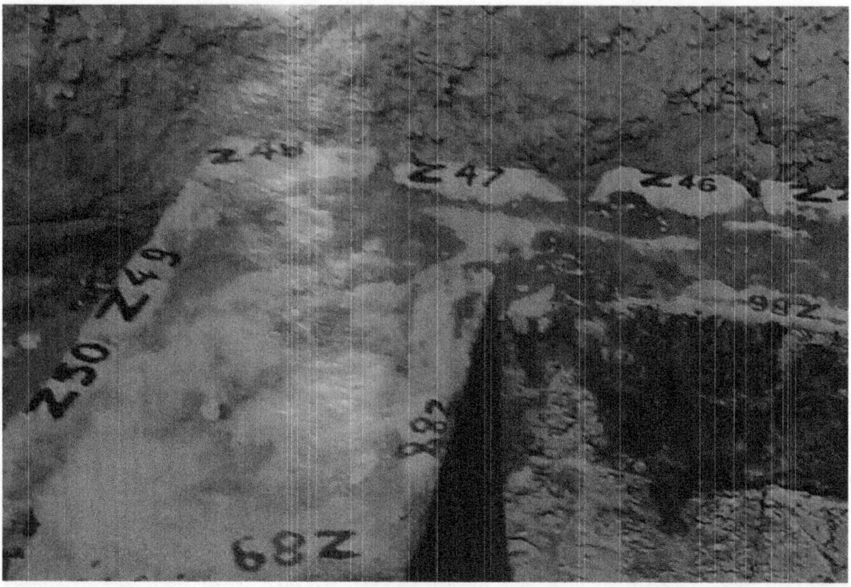

Fig. 31 Chapel of Dzordzor, coding and detachment of the stones, ICHO, 1988, ICCHTO Archive, report number: 247

12.17 Advisory Body Evaluation (ICOMOS)

"The authenticity of a building which was initially in ruins, before being dismantled, moved, and rebuilt, must be carefully verified, as this type of action is a priori and not acceptable in terms of outstanding universal values. ICOMOS asked the State Party for details related to this operation, which was necessary for constructing a dam. A specific document including a detailed plan and elevation views of various operations were provided by the State Party. It appears that moving and reconstructing the Chapel of Dzordzor were performed with great care at each stage. Each stone in place was numbered level by level before removal and subsequently put back into its initial position. The upper part of the edifice, including the dome, was in ruins at the time of the operation. The stones scattered on the ground were collected, and the original assembly was reconstituted wherever it was possible, as in the case of the cornice at the dome base. The reconstruction of the dome required the most plausible interpretation possible on the basis of the remaining stones and their shape, the architectonic study of the whole structure, comparative studies based on other edifices in the Armenian religious heritage, and the opinion expressed by the Armenian prelacy. Of the 1548 cut stones required for reconstructing the chapel, only 250 were not from the original building. The colour of the stones was slightly lighter in tone, which indicated that to the visitor the stones which absolutely conform to the state of the original when it was moved, and the stones which have been added. The base for the foundations should be adapted as the slopes ofvf the original location and

Fig. 32 Chapel of Dzordzor, two of the drawings included in the report, ICHO, 1988, ICCHTO Archive, report number: 247

Fig. 33 Chapel of Dzordzor, assembling the arches with use of wooden patterns prepared before the detachment, ICHO, 1988, ICCHTO Archive, report number: 247

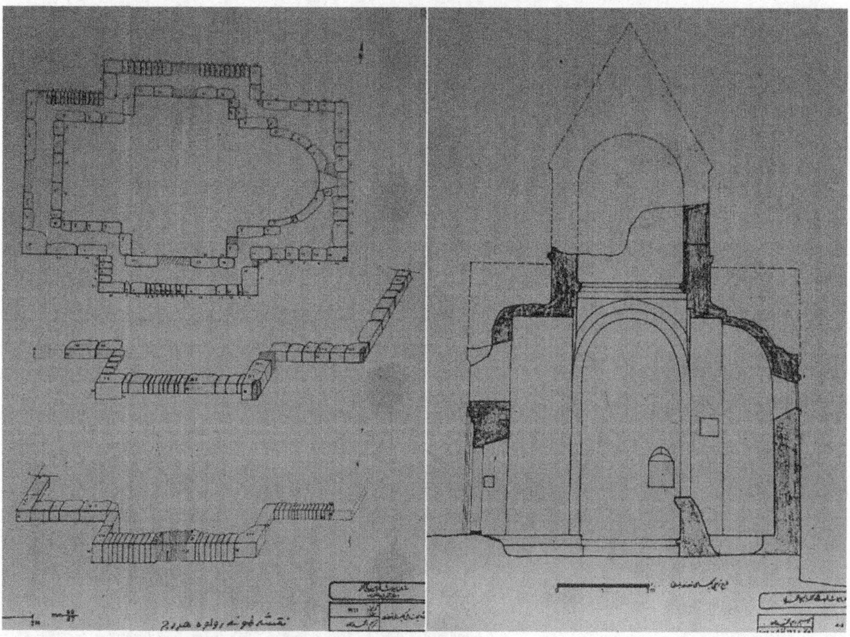

Fig. 34 Chapel of Dzordzor, final part of the project with the reconstruction of the dome with new stones, ICHO, 1991, ICCHTO Archive, report number: 576

Fig. 35 Chapel of Dzordzor, assembling the stones with use of new stones in case of damaged or lost pieces, ICHO, 1988, ICCHTO Archive, report number: 247

the location of the rebuilt chapel were not the same. ICOMOS, considers that in the difficult conditions arising from the fact that Chapel of Dzordzor was initially in ruins, and the need to move it to prevent it from disappearing beneath the dam reservoir waters, reconstruction of the building led to an acceptable technical and architectural reconstitution. This criterion alone would not be satisfactory. However, there were other factors which argue in favour of acceptance. The aim was to evaluate the authenticity of a set of three monasteries, which were the vestiges of a tradition and the last conserved testimony of the Armenian culture in the region. The history of the region was marked by dramatic events, which resulted in repeated destructions of both human and natural origins as well as a cultural tradition of reconstruction under the control of the episcopate of the Armenian church. Moving the Chapel of Dzordzor respects this tradition" (ICOMOS 2008, p. 56).

References

Abdi K (2001) Nationalism, politics, and the development of archaeology in Iran. Am J Archaeol 105(1):51–76

Abdi K (1994) One hundred years of excavations in Susa. J *Miras-e Farhangi* (ICHO), pp 87–105. [in Persian]

Abdul Rasool Ahmadian, Mehrdad Agha Sharifian Isfahani (2014) Fetvat of Guilds, Iranian and Islamic business model. In Proceedings of the third conference of the Iranian Islamic model of progress, Iran, pp 1153–1173. [in Persian]

Abrahamian E, Revolutions IBT (1982) Princeton studies on the near East. Princeton University Press, Princeton

Adle S (2000) Khorheh, the dawn of Iranian scientific archeological excavation. Tavoos Q J (3–4):226–239. [in Persian]

Afsar K, Mousavi A (1977) Preservation of antiquities in the Pahlavi era. Ministry of Culture and Arts, General Directorate of Preservation of Antiquities and Historic Monuments, Tehran [in Persian]

Amanat A (2017) Iran: a modern history. Yale University Press, New Haven & London

Ansari AM (2012) The politics of nationalism in modern Iran, vol 12, n 4. Cambridge University Press, New York

Beheshti (2019) It was a mistake to merge the three areas of cultural heritage, handicrafts and tourism, an interview with Mohammad Beheshti by Samane Zeinolabedin, Newspaper Irna article, 2019, www.irna.ir/news/83533469/. Accessed on 12 May 2021. [in Persian]

Bidhendi MQ (2009) An overview of the written sources of the history of Iranian architecture and the method of research. Golestan Honar Bahar J 15:5–20. [in Persian]

Carta di Gubbio (1960) Convegno Nazionale per la Salvaguardia e il Risanamento dei Centri Storici, Gubbio 17–18–19 settembre, 1960, https://www.ancsa.org/la-storia-e-larchivio/la-prima-carta-di-gubbio-1960/. Accessed on 10 May 2021

Chevalier N (2018) France and Elam. In Álvarez-Mon J, Basello GP, Wicks Y (eds) The Elamite World. Routledge. https://www.routledgehandbooks.com/. https://doi.org/10.4324/9781315658032-4. Accessed on 16 Feb 2021

Coste PX (1867) Monuments Modernes de la Perse. A. Morel, Paris

Curtis J (1993) William Kennett Loftus and his excavations at Susa. J Iran Antiq 28:1–55

Curzon GN (1892) Persia and the Persian question. London, Longmans, Green & Company

De Angelis d'Ossat G (1971) Le Citta' Crocifisse. Asar J 22–23:6–12. [in Persian]

de Morgan J (1902) La Délégation en Perse du Ministère de l'Instruction Publique. E. Leroux, Paris

Devos B, Werner C (2017) Culture and cultural politics under Reza Shah, The Pahlavi State, New Bourgeoisie and the creation of a modern society in Iran. In Devos B, Werner C (eds). Routledge
Dieulafoy J (1997) In Susa: diary of excavations 1884-1886, In Farahvashi I (trans) (from Á Suse, Journal des fouilles 1884–1886). Tehran University [in Persian]
Flandin E, Coste P (1851) Voyage en Perse de MM. Eugène Flandin, Peintre, et Pascal Coste, Architecte, attachés à l'Ambassade de France pendant les années 1840 et 1841. Paris, Gide et J. Baudry
Galdieri E, Afsar K-A (1992) Conservation and restoration of Persian monuments. Encyclopaedia Iranica. https://iranicaonline.org/articles/conservation-and-restoration-of-persian-monuments. Accessed on 12 May 2021
Gholamdoost M (2013) The role of the law of antiquities and commercial excavations. In: Hassanzadeh Y, Miri S (eds) Proceedings of 80 years of Iranian archaeology. Pazineh Pubblishers, pp 11–43
Gnoli G (1983) Nel Cinquantenario dell'IsMEO, discorso tenuto a palazzo Brancaccio il 16 Febbraio 1983, Istituto Italiano per il Medio ed Estremo Oriente, Roma
Godard A (1998) The art of Iran. In: Habibi B (trans). University of Shahid Beheshti publishers [in Persian]
Gran-Aymerich È (1999) Archéologie et politique française en Iran: convergences et contradictions (1881–1947). J Asiat 287(1):357–374
Grigor T (2004) Recultivating "good taste": the early Pahlavi modernists and their society for national heritage. Iran Stud 37(1):17–45
Grigor T (2005) Of Aryan Origin(s), Western Canon(s), and Iranian Modernity, in Repenser les limites: l'architecture à travers l'espace, le temps et les disciplines. In: Thomine-Berrada A, Bergdol B (eds). Publications de l'Institut national d'histoire de l'art, Paris
Hauser SR (2012) Herzfeld, Ernst i. Life and work. Encyclopædia Iranica XII/3:290–293. Available online at http://www.iranicaonline.org/articles/herzfeld-ernst-i-1. Accessed on 12 Mar 2012
Herzfeld E (1973a) A brief list of the Monuments and Historical buildings of Iran. Collection of old publications of the Society, series of publications of the National Heritage Society, (published originally in 1925), pp 13–28. [in Persian]
Herzfeld E (1973b) National Monuments of Iran. Collection of old publications of the Society, series of publications of the National Heritage Society, (published originally in 1925), pp 30–44. [in Persian]
Hojjat M (1995) Cultural heritage in Iran: policies for an Islamic country. Ph.D. thesis, University of York
ICOMOS (2008) Advisory body evaluation (ICOMOS), Armenian monastic ensembles (Iran) No 1262, file:///C:/Users/pante/Downloads/1262-ICOMOS-1542-en%20(8).pdf. Accessed on 5 June 2021
Jaubert PA (1943) Voyage en Arménie et en Perse, 1805–1806. In Hedayat M (ed). Taban Publishers [in Persian]
Loftus WK (1857) Travels and researches in Chaldæa and Susiana; with an account of excavations at Warka, the "Erech" of Nimrod and Shush, "Shushan the palace" of Esther, in 1849–52. Robert Carter & Brothers, New York
Mahkouri J (1953) The history of Urartu and Urartian inscriptions in Azerbaijan. Institute of Studies, Studies and Scientific, Artistic and Historical Exploration of Ancient Iran [in Persian]
Masoumi G (1977) Archaeology of Iran, from 1848 until 1977. Ministry of Art and Culture [in Persian]
Mathee R (2004) Sympathy and enmity: nineteenth-century Iranian views of the British and the Russians. In: Harder H (ed) Looking at the coloniser: cross-cultural perceptions in Central Asia and the Caucasus, Bengal, and Related Areas. Ergon Verlag, Beate Eschment, pp 311–338
Mathee R (2016) From splendour and admiration to ruin and condescension: Western travellers from the Safavids to the Qajars. Br J Persian Stud 54(1):3–22

Mehdizadeh S, Hanachi P (2016) The role of Western architects and archaeologists in the restoration of Historical monuments of Iran in the Pahlavi period 1925–1979. J Fine Arts Archit Urban Plan 21(3, Fall):5–14 [in Persian]
Mehran M (1977) Ten years of service to the national organization for the preservation of antiquities of Iran. Art and Culture ministry [in Persian]
Mohammadi AAK (1992) Fetvatname of masons. J *Saffeh*. Publication of the Faculty of Architecture and Urban Planning of Shahid Beheshti University, Second Year, no 5, pp 11–15. [in Persian]
Mohammadi BK (2005) Allocation of the Armenian historical churches in Western Azerbaijan. Peiman J 31 [in Persian]
Moradi AM (2003) The process of preservation in Iran since the establishment of the Antiquities Office until the Islamic Revolution. Haft Shahr J 1(11, Spring):14–28 [in Persian]
Moradi AM (2018) The evolution of conservation and restoration interventions in Iran until the Islamic Revolution of Iran. Publisher the author [in Persian]
Mostafavi M-T (1975) Guardianship of soil. J Hist Anal XXI year, special number, Tehran [in Persian]
Mostafavi M-T (1955) Serving the national heritage and hoping for the future. Published by The Archaeology Office, Tehran [in Persian]
Mousavi A (1996) Early archaeological adventures and methodological problems in Iranian archaeology: the evidence from Susa. Iranica Antiqua J 31:1–17
Mousavi A (2012) Persepolis, discovery and afterlife of a world wonder. De Gruyter
Mousavi A (2013) The history of archaeological research in Iran: a brief survey. In: Potts DT (ed) The Oxford handbook of ancient Iran. Oxford University Press
Mousavi A (2005) Ernst Herzfeld, politics, and antiquities legislation in Iran. Gunter and Hauser, pp 445–75
Municipality Law of 1286 SH, Islamic council Parliament Research Center, https://rc.majlis.ir/fa/law/show/90099. Accessed on 15 Apr 2021
Naini MS (2008) Architects and traditional restorers of Isfahan, Center of Isfahanology and the home of nations and Isfahan Municipality Cultural and Recreational Organization [in Persian]
Porter RK (1821) Travels in Georgia, Persia, Armenia, Ancient Babylonia, during the years 1817, 1818, 1819, and 1820. Longman, Hurst, Rees, Orme, and Brown
Rahimnia R, Gharaati M, Zamanifard A (2016) A narration of Iranian architect: a survey on the historical evolution from mason to architect. In: Conference: Terra 2016 (XIIth world congress on earthen architecture), Lyon, France
Rawlinson G (1898) A memoir of Major-General Sir Henry Creswicke Rawlinson. Green and Co., London, New York Longmans
Rouhani B (2011) Development and cultural heritage in Iran: Policies for an ancient country. In: ICOMOS 17th general assembly, 2011–11–27/2011–12–02, Paris, France
Schmidt EF (1939) The treasury of Persepolis and other discoveries in the homeland of the Achaemenians. University of Chicago Press, Chicago
Session: 247, Detailed accounts of the parliament on Thursday, June 13, 1928, Islamic council Parliament Research Center. https://www.ical.ir/ical/fa/Content/4_artmajles. Accessed on 5 Mar 2021
Shahmirzadi SM (1988) A brief overview of the evolution of archaeology in Iran. Asar J 12, 13, 14:133–160
Sheffield D (2013) Iran, the mark of paradise or the land of ruin? Approaches to reading Two Parsi Zoroastrian Travelogues. In: Sharma S, Micallef R (eds) On wonders of land and sea. Harvard University Press, Cambridge, MA, pp 14–43
Sistani IA (2008) St. Serkis church. Peiman J 44 [in Persian]
Statute of the Cultural Heritage Organization 1367 SH, Islamic council Parliament Research Center, https://rc.majlis.ir/fa/law/show/91528. Accessed on 1 Mar 2020
Texier C (1842) Description de l'Arménie, la Perse et la Mésopotamie, Typ. de F. Didot frères

The Declaration of Amsterdam 1975, Congress on the European Architectural Heritage 21–25 Oct 1975. https://www.icomos.org/en/and/169-the-declaration-of-amsterdam. Accessed on 10 May 2021
UNESCO (2008) Armenian monastic ensembles of Iran nomination documentation. UNESCO. https://whc.unesco.org/uploads/nominations/1262.pdf. Accessed on 14 May 2021
Venice Charter 1964, International charter for the conservation and restoration of monuments and sites, https://www.icomos.org/en/179-articles-en-francais/ressources/charters-and-standards/157-the-venice-charter. Accessed on 10 May 2021.
Wastnidge E (2014) The modalities of iranian soft power: from cultural diplomacy to soft war. Polit J 35(3–4):364–377
Yazdani M (2001) Documents of archeological delegations in Iran 1875–1966. Publications of the National Organization for historical documents of Iran, Research Institute
Zia-Ebrahimi R (2014) "Arab invasion" and decline, or the import of European racial thought by Iranian nationalists. Ethnic Racial Stud 37(6):1043–1061

Panteha Karimi Architect, Iranian, Ph.D. student in Preservation of Architectural Heritage, Department of Architecture and Urban Studies, at Politecnico di Milano. Her research focuses on the protection of Architectural Heritage in Iran, particularly on the Italian contribution in this field.

The Regions of Khoy and Salmas in the Travelogues of Western Travellers Between the Fifteenth and Nineteenth Centuries

Maurizio Boriani

Abstract In the article, a series of excerpts from the travsselogues dealing with the territory north-west of Lake Urmia in Iran, particularly the cities of Khoy and Salmas and the Armenian people who lived there, are considered. The purpose of this research is to document the territorial, economic and social events which characterized this region over time through the textual sources of Western travellers; therefore, unless strictly necessary, the excerpts documenting the ideological, religious, and historical events only (political struggles, wars, invasions, etc.) will not being considered. The most significant texts in English, French, and Italian languages, available on the Internet, were selected starting from the main bibliographic sources: 115 texts (some of them in several volumes) were studied, and 64 authors (diplomats, missionaries, militaries, orientalists, artists and architects) were identified who provided information on the Khoy and Salmas region, ranging from the fifteenth to the early twentieth century. Section 5.1 accounts for the descriptions left to us by Western travellers who visited the city of Khoy. The city was the seat of the Governor of the County of the same name, was equipped with an important bazaar and fortified by clay walls, with an outermost "European" circle. Khoy was therefore the object of particular attention from diplomats and military, especially French and English, with the aim of evaluating the economic potential and the strength of the defences. During a short period of occupation by Russia, there was the opportunity to make a survey of the city and its surroundings that documents the layout of the city and its territory in an exceptional way in the early decades of the nineteenth century. The Governor's palace and its suburban villa, a prosperous Armenian community and an important Islamic sanctuary were also points of interest for orientalists, archaeologists and architects, who reported interesting descriptions, images and reliefs. Section 5.2 regarding the first Western travellers through Salmas area explains that their main interest was a bas-relief from the Sassanid period, carved on a rock near a village a few kilometres south of Salmas. The different descriptions of this bas-relief and the accompanying drawings and photographs allow you to have interesting information on the emergence of Western interest in Persian archaeology. Salmas and the villages of the countryside were largely inhabited by Christians, Armenians and Assyrians,

M. Boriani (✉)
DAStU, Politecnico di Milano (retired 2016), Piazza Leonardo da Vinci 26, 20133 Milano, Italy

M. Boriani and M. Giambruno (eds.), *Architectural Heritage in the Western Azerbaijan Province of Iran*, Research for Development,
https://doi.org/10.1007/978-3-030-83094-6_5

starting from the first decades of the nineteenth century: for this reason, it was the object of numerous visits by religious, Catholics and Protestants, who founded some missions there. The travel reports describe in great detail these communities and the relationships they had with the local Islamic population.

Keywords Persia · Iran (West Azerbaijan) · Travellers · Khoy · Salmas · Armenians

In this article, a series of excerpts from the travelogues dealing with the territory north-west of Lake Urmia in Iran, particularly the cities of Khoy and Salmas, are considered. The purpose of this research is to document the territorial, economic and social events which characterized this region over time through the textual sources of Western travellers; therefore, unless strictly necessary, the excerpts documenting the ideological, religious and historical events only (political struggles, wars, invasions, etc.) will not being considered. Also, regarding the ethnographic aspects, only the information related to the specific area of this study was selected neglecting the more general texts dedicated to the entire Persia. At the same time obviously, the geographical and historical context cannot be overlooked and it is necessary to study and understand the events and locations described by travellers.

The most significant texts in English, French, and Italian languages, available on the Internet, were selected starting from the main bibliographic sources, first of all those published in Encyclopedia Iranica (ENCYCLOPAEDIA IRANICA 2012: Andreeva 2016; Bast 2012; Bedrosian 2020; Hodjat 1995; Nasiri-Moghaddam 2016; Richard 1990; Tissot 1994; Touzard 2016), and also other general inventories regarding Middle East, Armenia and The Silk Road (Vatoogian 1974; Bedrosian-Ashtiani 2004; Salzani and Tötösy de Zepetnek 2016; Badalyan 2019; Waug and Lee 2020; Wilmshurst 2000). Among the chosen texts, there are also works by authors who originally wrote in other languages (Dutch, German, etc.), but it was preferred to use the versions in English, French, and Italian, in order to make it easier to understand the relationships to a wider audience. If translations into English of texts originally written in other languages were available online, the English language was chosen, because a large number of people understand it. However, it should be noted that the original text does not always correspond literally to the translated one.

A single exception was made to include the fundamental text of Evliya Celebi in the anthology, originally written in Turkish, but also available in English: even though he is not a strictly Western author, the information that can be obtained from him, with respect to the geographical area of our interest, significantly enreach the situation relating to the seventeenth century for which the travel stories that have affected the area of Khoy and Salmas are, as we shall see, infrequent.

While for the reports prior to the eighteenth century all the news (generally brief) from the various travellers were referred, for the subsequent period and, in particular for the nineteenth century, only the most significant and less repetitive texts were chosen, leaving out those which reported only brief notes on the specific themes studied in this paper.

As a conclusion to the research, 115 texts (some of them in several volumes) were studied and 64 authors were identified who provided information on the Khoy and Salmas region, ranging from the fifteenth to the early twentieth century. As Internet continues to be enriched day by day, we assume that texts by other authors or other texts by already selected authors may gradually be found in future. However, it is believed that the documentation collected and analysed up to this moment is amply sufficient to provide the information necessary for the research. The proof lies in the fact that these texts are present in almost all the bibliographic inventories used. For reasons essentially dependent on the linguistic knowledge of the author of this article, the texts of Armenian, Arab, Turkish and Persian authors have not been systematically studied, for which the commitment of specialists is required. A recently published essay (Moradi 2020) partially covers this gap, especially with regard to Arab and Persian travel stories prior to the nineteenrh century.

Finally, an attempt was made to keep the texts in their original language or in the most complete version translated into English and available online: those texts, especially the oldest ones, sometimes sound archaic to us and not free from grammatical errors or linguistic expressions which were later dropped/disused. The geographical names have been kept as they were written by the various travellers, while minor grammatical corrections have been made when dealing with obvious writing or printing errors. For the English language, the original version of the author (British or American) has been kept. Among the selected authors, 38 reported news about the town and the Khoy countryside, 21 wrote about Salmas and 15, while not specifically describing the two cities, gave general and noteworthy information on housing, on the travel support infrastructure, on the agricultural landscape and on other ethnographic themes. In the attached anthology, we also wished to publish the most significant information available on the passage of the Araxe river and on the remains of the town of Julfa (entry point into Persia for many travellers), as well as on Lake Urmia, around which gravitate our two cities.

The most ancient texts written by Europeans, if we exclude Xenophon, Strabo and other Greco-Roman geographers, date back to the thirteenth/first half of the fourteenth century and are produced by religious (Andrew of Longjumeau, Dominican; Ascelinus and Simon of San Quentin, Dominicans; Giovanni di Pian del Carpine, Franciscan; William of Rubruck, Franciscan; John of Monte Corvino, Franciscan; Odorico di Pordenone, Franciscan; Giovanni di Marignolli, Franciscan): all were sent by the Pope or by their respective religious orders to make contact with the new Mongol rulers of central Asia and attempting a first missionary action (Vanzan and Bernardini 2012). However, the courageous attempt was not very successful, since in 1304 Islam was proclaimed the state religion of the empire; however, it helped keeping alive the many Christian communities that, from the very beginning, had settled in the area. The information found in these writings is very general and specifically referred to the missionary purposes of travellers: lacking direct references to the area of our interest, these texts have not been taken into consideration.

The only layman who in the thirteenth century has left sufficiently ample news of his travels is famous Marco Polo, a Venetian merchant, regarding the second journey made with his father and uncle between 1271 and 1295 (Marsden and Wright 1948;

Ramusio 2015). In his report, the information that can be found in relation to the area of our interest is very poor, and this cannot always be considered first-hand. In this specific case, Marco Polo does not explicitly mention Khoy and Salmas and only dedicates a few lines to the town of Tabriz itself. However, the mere mention of a geographical area, a town or an economic activity that took place in a specific territory, is already a testimony of their importance, at international scale.

More significant for these times is instead the rich cartographic apparatus left to us by Arab and Persian authors, especially for the fact that in their maps the cities of Khoy and Salmas are often specifically indicated, a sign of their political-military, economic or religious importance. (Galichian 2007, pp. 88, 94, 96, 98, 103, 106). We will briefly deal with these news in the chapters specifically dedicated to the two cities (Fig. 1).

A resumption of contacts is instead documented by the travels made in the fifteenth–seventeenth century above all by various Venetian merchants (who often had also received diplomatic posts, more or less official); to these were also added some French, Dutch and English travellers. Starting from the eighteenth century, writings by specialists (orientalists, archaeologists, geologists, botanists, architects and artists) began to appear, who had observed and described the territory visited above all through their specific disciplines, without neglecting references to more general issues of interest.

With the first decades of the nineteenth century, thanks to the conquest by Russia of the sub-Caucasian region, the travel reports of missionaries, orientalists and above all diplomats and soldiers became much more frequent, the latter involved in activities of political mediation and intelligence, but also consultancy to support the modernization of the Persian economic and military apparatus.

From the middle of the nineteenth century, studies on the complex religious issues of the area finally appeared (and became increasingly frequent), in particular by Protestant and Catholic missionaries, obviously interested in proselytism and in the living conditions of Christian populations of the region.

For the entire time period examined, Tabriz was the most important point of attraction for the trafficking and diplomats who reached Persia by land (Fig. 2). However, we can say that the town had alternating fortunes, since it was a disputed land for several centuries: conquered by the Arabs when it was nothing more than a village, it was rebuilt by the Abassids in 791, and after a devastating earthquake, it was the capital of Ilkhanid Empire from 1270 to 1305, of the *Ak Koyunlu* dynasty from 1469 to 1502 and of the Safavids from 1502 to 1548. Conquered and devastated by the Georgians, the Mongols and the Ottomans over the centuries, it suffered another tragic earthquake in 1721. For these reasons, despite being the natural transit point for those coming from the north and headed to the centre of Persia, could not always be a point of political or commercial attraction, especially since the Portuguese discovered the sea route to the East through the Cape of Good Hope. It came back into vogue in the nineteenth century, with the conquest by Russia of the Khanates of Yerevan and Nakhichevan in 1828, once again becoming a strategic, productive, commercial and military point, within the great Anglo-Russian competition for the control of Persia.

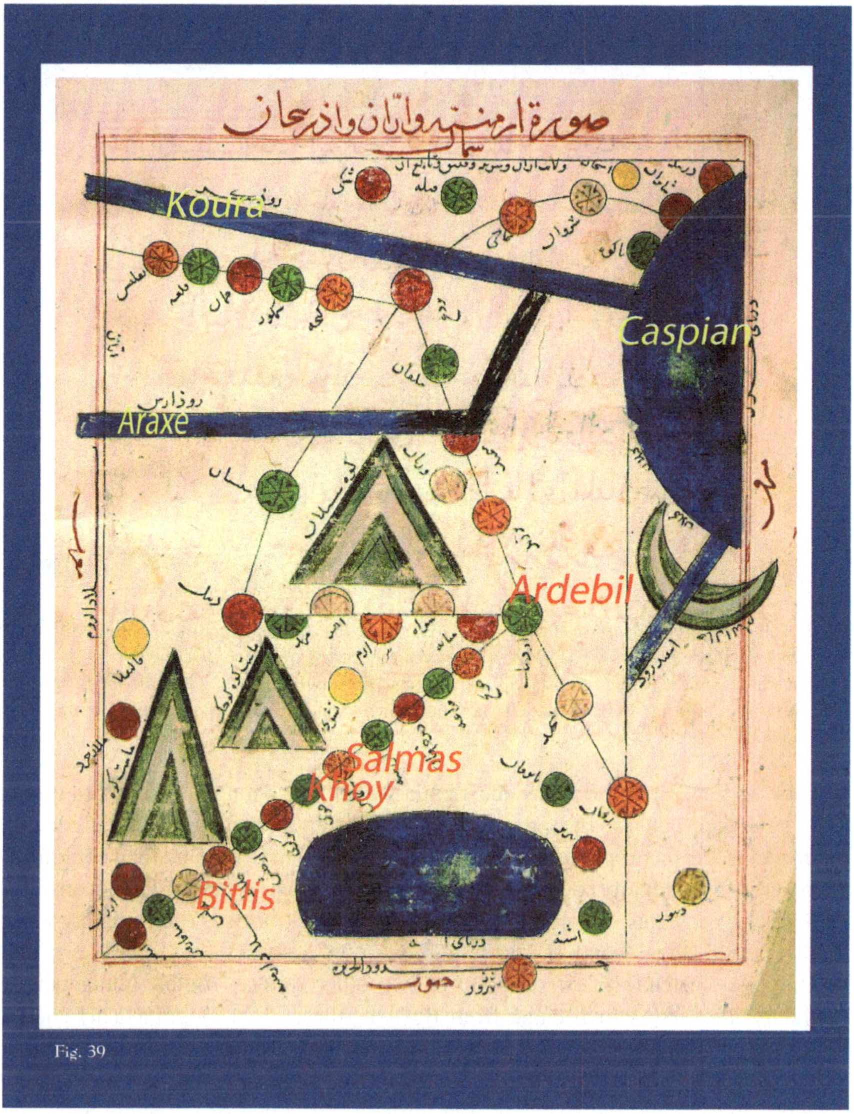

Fig. 1 Al-Istakhri's Map *Surat Arminya va Arran va Adharbeijan*, 1325 copy of a 950 CE original Map. Top, Araxe river flowing into Koura, that flows into Caspian Sea. Centre, the two peaks of Ararat. Bottom Lake Urmia. The diagonal marked by small circles represents the road from Ardebil to Bitlis. The towns of Khoy and Salmas are marked (Galichian 2007)

Fig. 2 John Pinkerton. *Map of Persia*, 1818—Detail of the Azerbaijan Region (Pinkerton, J., *A Modern Atlas, from the Lates and Best Authorities, Exhibiting the Various Divisions of the World with its chief Empires, Kingdoms, and States; in Sixty Maps, carefully reduced from the Larges and Most Authentic Sources*. 1818, Philadelphia, Thomas Dobson Edition.—Wikimedia Commons)

Khoy and even more Salmas and Urmia were close enough to Tabriz (a couple of days' journey) but far from the main line of communication, which included the entrance into Persia proper to Julfa, by ferry on the Araxe River, and then continue to Marand, Sofian and Tabriz. This was the preferred transit route for Europeans, also made easy by a series of caravanserais and some brick bridges, built mainly in the Safavid period (Figs. 3 and 4).

Khoy was instead the arrival point of two other caravan routes, minor but not secondary: the first, which came from Bayazit (the current Dogubayazit) entered Persia at Maku (capital of a semi-independent khanate), was not practised by Europeans; the second, which came from Bitlis and Van in Turkey, with entry into Persia at *Qotur*, went down the river that ran through the valley of the same name and reached Khoy, home to an important bazaar. These itineraries were chosen mainly because they gave the opportunity to visit Van, an important Armenian town which also preserved some monuments with cuneiform inscriptions from the Urartu period.

All these routes could start in Constantinople/Istanbul and reach Trabzon (ancient Trebizond), or Batumi, Georgia, sailing along the southern shore of the Black Sea. Khoy could be reached crossing mountain ranges with passes of 2000–2300 m of altitude through the first itinerary via Erzurum and then Erevan. Or, through the second way, via Tbilisi, Dilijan, Erevan (Kleiss 2016). Erevan was a fundamental point of attraction for Europeans, since a fair road connected it to Erzurum and in its

Fig. 3 The Ferry on the Araxe River near Joulfa (Binder 1887)

Fig. 4 A.F. Stahl, Routenkarte von der Russishen Grenze *nach Tabriz und Kaswin, 1903* (West part), Original scale 1: 840,000. The roads connecting Julfa, Marand, Sofian and Tabriz are represented, as well as the detour from Sofian to Khoy and Salmas (ANVARI 2015)

vicinity there were two important stops: the great monastery of Echdmiadzin, seat of the Armenian Catholikos and Mount Ararat, of biblical memory. With the Russian conquest of the town, the interest in these streets increased significantly, also for the technical improvement of the roads and logistical support for travel.

From Erevan, one could travel along the Araxe river valley (another mythical border river between the Roman Empire and Sassanid Persia), stop in Nakitchevan and enter Persia by crossing the Araxe by ferry near ancient Julfa, the Armenian town destroyed by Abbas the Great to make scorched earth between the Ottomans and Safavids and to deport to Isfahan the Armenian merchant community that populated it. From there, the caravan led to Marand and Tabriz. The one just described, for example, was the path followed by Chardin in 1673 (Chardin 1686, Chap. 1, pp. 247/288) and by Dubois de Montperreux (up to Julfa) in 1834 (Dubois de Montpéreux 1839–43, Chap. 3, pp. 275/331 and Chap. 4, pp. 5/20).

Those who were not interested in this route, from Erevan could, alternatively, reach Van and enter Persia, as mentioned, via Bayazit or *Qotur*. This meant, however, crossing the Kurdish territory, considered dangerous due to the scarce control that the two neighbouring empires exercised over this people and, in particular, due to the frequent assaults by marauders.

Khoy was, however, considered an important town for its Bazaar, and Salmas attracted the interest of those who wanted to know the Christian, Syriac and Armenian communities, who lived in the surrounding villages (*Khosrova* and *Haftvan* in particular). In general, the two locations could be reached, with a detour of 80 km about halfway between Marand and Tabriz, near Sofian. In other cases, it was preferred to stop in Tabriz for some time and organize the entire circumnavigation of Lake Urmia. In this case, you could take the opportunity to visit also the town of Urmia, the mythical homeland of Zarathustra, its thriving Christian communities and the rich surrounding countryside.

The journey by caravan from Trabzon to Tabriz was about 900 km long and involved 35–40 days in winter and up to 4–5 months in summer (in the summer, the possibility of grazing animals in the prairies along the path: time was obviously a secondary variable). Those who could afford to travel on their own could be even faster, although it should be noted that the official, diplomatic or scientific missions that they were, could not exempt themselves from visiting the Governors or local leaders or make detours and prolonged stops to study the geographical aspects, geological, naturalistic, economic, archaeological and architectural of the places crossed. As an example, Texier, architect and archaeologist who runs between the end of August and the end of October 1839, the itinerary from Trebizond to Tabriz without passing through Erevan, 9 days stop in Erzurum, 5 days in *Hassan Kalé*, one in *Ani*, 2 in Kars, 8 in Van, one in Khoy and one in Salmas. Altogether it takes 57 days, dedicating the stops to studies and surveys of the main monuments encountered (Texier 1842, pp. 229 and following). Shiel, an English military and diplomat, completed the only Tabriz-Van section of a longer route through Kurdistan in 8 days in 1836, producing a beautiful and precise map (Shiel 1838, pp. 54/55) (Fig. 5).

Muller-Simonis, travelling in 1888, takes 14 days from Tbilisi to Khoy, stopping for a couple of days in Erevan and passing the Araxe at Julfa (Muller-Simonis 1892, p. 602).

Finally, another route to Persia was possible: through Syria, it was possible to reach, Mosul or Kirkuk by land, in present-day Iraqi and from there reach Tabriz by entering Persia from the South West at *Kermanshah*. This route, apparently easier because it did not involve crossing high mountains, was originally the one preferred

Fig. 5 Detail of the itinerary followed by J. Shiel in the section between Khoy and Salmas (Shiel 1838)

by European merchants and diplomats; however, it was gradually less frequented with the decline of Baghdad following its conquest and sack by the Mongols in 1258, repeated in 1401 with Tamerlane and the Ottomans in 1534. This was however the preferable overland route to point directly to Isfahan, bypassing Tabriz. The itinerary made a comeback in the nineteenth century, especially due to the interest of the missionaries for the strong presence of Christian communities in the territories of present-day Iraq, as well as for the nascent demand for independence from the Kurdistan region, which attracted European military and diplomats. A good description of the different itineraries that can be followed in Persia, in particular of the two routes from Trabzon and from Tbilisi to Tabriz, is found in the text of Curzon (Curzon 1892, pp. 38/41).

The turning point in the relations between Europeans and Persia was certainly that of the conquest of Constantinople by the Ottomans in 1453. The way to the east was already known at least from the travels of Marco Polo and other missionaries who in the thirteenth century intended to get in touch with the Mongols and reach China along the Silk Road or by sea starting from the Persian Gulf (in 1219 the conquests of Genghis Khan begin). The Hundred Years War in Europe (1337–1453) brought about a setback, with the exception perhaps of the Spaniard Ruy González de Clavajo who, in the embassy in Samarkand on behalf of the king of Spain, passed through Tabriz in 1404 and, above all, for what concerns us, prepared a precise written account of his journey, also briefly describing the town of Khoy (Clavijo 1928, p. 148).

On the contrary, it was after the fall of Constantinople that European (and especially Venetian) interest in Persia became evident. It was a matter of evaluating how much the enmity between Persians and Ottomans could have benefited Europeans and which commercial prospects could be opened through agreements with the new oriental lords. It was the turn of the travels of the Venetian ambassadors, with official credentials: Caterino Zeno (1471), Ambrogio Contarini (1474–77), Giosafat Barbaro (1474–78) and Giovanni Maria Angiolello (1482). He was an adventurer who even assumed important administrative positions, first in Istanbul and then in Tabriz itself.

In fact, these missions did not lead to great results, both due to the logistical difficulties of a political-military collaboration at such a great distance and due to the fact that the Genoese and Venetian rulers and merchants deemed more convenient and easier to find an agreement with the Ottomans, if only for the fact that to reach Persia it was essential to pass through the territories controlled by them.

The discovery of America (1492) and the sea route to the East via the Cape of Good Hope (1487), the unrest following the fall of the Timurid dynasty (1506) and the Protestant Reformation (1517) once again diverted European interest for Persia, which instead resumed with the rise of the Safavids and, in particular, of Shah Abbas the Great (1587–1629). Then, it was the turn of the English merchant Newberie (1581), and the English preacher John Cartwright (1593), followed in 1620 by an official Vatican embassy who brought printing presses with Arabic characters to Isfahan. The European interest was paid by the Persian shahs, who had realized, after the terrible defeat in the battle of *Chaldiran* by the Ottomans in 1514, of their military inferiority and, above all, of the need to update the modest and backward artillery they had with the contribution of the Europeans.

With the seventeenth century commercial relations consolidated: Adam Olearius (1635–1639), Jean Chardin (1664–66 and 1671–74), John Baptista Tavernier (1670), Paul Lucas (1699–1703), and Cornelis de Bruyn (1703) were merchants who not only visited Persia for business, attracted by the splendour of the new capital, Isfahan, but reported at the same time on the political, economic and social conditions of the empire, also publishing for the first time views of the main cities and monuments visited during their travels. Moreover, the Shah Abbas had promoted, as already said, the construction of a dense network of caravanserais and bridges all over the country, and in particular to support the road that from Julfa reached Tabriz and, from there, Isfahan (Kleiss 2016, p.96).

A new slowdown in relations occurred during the first half of the eighteenth century, with the devastating earthquake that razed Tabriz to the ground (1721), the Afghan invasion of Persia (1722) and, above all, the competition from sea travel organized by British and Dutch commercial companies, much safer and cheaper than those by land. This also led to the decline of the road system and caravanserais, promptly noted by travellers of the eighteenth and nineteenth centuries (Steensgaard 1974; Drouville 1825, p. 101), who began to reach Persia by sea and to focus their attention on Isfahan and Shiraz (Fig. 6).

At the end of the century, another protagonist appeared in Persia: the Russia of Catherine the Great obtained control of Georgia (1783), which voluntarily granted itself, considering that it was unable to resist Persia alone. Russia would soon become

Fig. 6 The "road" to Marand (Jackson 1906)

the main player in the area: on October 2, 1827, General Ivan *Paskevich* entered Erevan, and the following year, with the peace treaty of *Turkmenchay*, Persia would be obliged to cede the khanates of Erevan and Nakhichevan, to the southern border with the Araxe. Among the clauses of the treaty, Russia was authorized to encourage the migration, to the newly conquered territories, of Armenians residing in the Persian and Ottoman empires. Many important cities (Erzurum, Van, Kars, Khoy in particular) and large part of the surrounding countryside were abandoned by what was the most active part of the population of those territories, especially dealing with trade and handicraft production. Of this great migration, there are, as we will see later, various testimonies from European travellers.

With the Russian conquest of the sub-Caucasian region, the interest of Europeans in Persia increased significantly. First of all, geopolitical issues came into play: Russia and England intended to defend their Asian conquests and Persia was at their borders (the Russian Empire to the north, the British-held India to the east); therefore, it appeared important to be able to control and influence Persia, a state too weak at that time to be a danger, but an important possible ally and base for operations.

Secondly, the Armenian question took on new vigour: the conquest of the khanates of Erevan and Nakhichevan by Russia and the consequent policy of encouraging immigration in this territory of the Armenian communities dispersed in the empires of the Ottomans and the Persians, provoked two consequences that are only apparently contradictory. On one hand, Russia set itself up as a defender of the Armenian people, exploiting the common Christian religion, and on the other hand the Armenians, concentrated in an increasingly homogeneous territory from the ethnic point of view, began to claim their autonomy, if not a true Armenian independent state. The Catholic Church and the Protestant Churches also began to compete with each other to assist the Christian communities that remained in the two great empires, sending missionaries, founding missions and carrying out diplomatic activities, especially in Persia, traditionally more tolerant towards Christians and in particular Armenians than the Ottomans.

The role of France in this situation was particularly significant: the Napoleonic attempt to stipulate a Franco-Persian alliance in an anti-English function practically failed (General Gardane's Mission—1807/9). In fact, the primary purpose was to gather information on Persia and its military strength; see: "Vous ne manquerez point de lui [General Gardane] faire connaître que le principal but de son voyage est de s'instruire de la situation de la Perse et de la marche des gouverneurs qui y existent, province par province, et de reconnaître à combien se montent ses forces—Napoleon's letter to Talleyrand (Vinson 2009—2/3). Instead, following the mission, an intense policy of scientific research developed in the archaeological, linguistic and ethnographic fields that made the French language the main vehicle of nineteenth century Orientalism: the missions of Eugène Flandin and Pascal Coste would be one of the most famous and celebrated results (Flandin and Coste 1854).

Finally, the opening of political and economic relations between Western Europe and Persia, contemporary to the development of historical and archaeological research that characterized the nineteenth century, favoured the sending of scientific missions and the carrying out of studies, the first excavations and the first restorations, especially in the sites of greatest archaeological interest in present-day Iraq and Persia mentioned in biblical and Greco-Roman sources. These missions, protected by their respective governments, would have determined a real competition in acquiring favours and permits from the local rulers, which ended in 1895 with the recognition by the Persian State of the French monopoly on all the archaeological activities to be carried out in its territory. The agreement was broken by Persians in the 1930s (Tissot 1994, p. 238; Mousavi 2005, p. 447).

How did they travel? All authors are extremely critical about the quality of Persian roads: according to Maclean and Browne "The roads are seldom made, and are merely the result of being constantly trodden down by passing caravans" (Maclean-Browne 1892, p. 65). Since they are often at a lower level than the surrounding fields, when these are irrigated they turn into real muddy waterways. In theory, it would be up to the individual villages to take care of maintenance but no one controls and pressures for this to happen. "The only possible vehicle in Persia is the horse" writes Filippo De Filippi, physician and zoologist, one of the few Italians who in the nineteenth century left us travel memories in Persia (de Filippi 1865, p. 123).

Apparently, Persia is the only country in the world where, after paying customs taxes, a traveller does not have to spend anything else, since he is considered a guest of the country and therefore can sleep and eat at the expense of the richest man of the villages and of the cities he crosses. This obviously applies to groups travelling with letters of introduction from sovereigns and governors who are presented with official credentials: a little for hospitality and most likely, because the necessary food supplies are paid for by the population of the countryside. However, this hospitality is not free of charge: the local elders expect to be reciprocated by gifts, which are rarely less valuable than the received services (de Filippi 1865, pp. 129/131).

Fig. 7 The bridge over the Qotur River, south of Koy, in a recent photograph. The abundant parts reconstructed in recent times are evident. (*Photo* Boriani 2018)

In fact, a large number of caravanserais are available to travellers which, at the distances of one day's journey from each other, allow to rest and cool down during the day or spend the night. Those who do not bring with them the necessary can buy it at reasonable prices in attached shops. I write in theory, since, as travel reports often note, caravanserai are generally poorly maintained or even in ruins, especially those in isolated places outside major cities. The reason lies in the fact that outside the town, in general, this service is provided at the expense of sovereigns, local governors or wealthy landowners, in order to acquire religious merit and that, upon their death, no one, not even the heirs, feels in duty to maintain these buildings properly. The same is true for brick bridges too: it is not uncommon for travellers to be forced to ford a river as the originally built bridge falls into disrepair (e.g., the *Kathun Bridge* south of Khoy—Muller-Simonis 1892, pp. 117/118) (Fig. 7). We have many descriptions of the caravanserais, together with those of the caravan journey; both constitute one of the most frequent *topoi* of the journey to the East. One of the first is that of Tavernier (Tavernier 1678, pp. 105/107), which clearly describes its most picturesque aspects. A more recent description is that of Drouville, which illustrates well the reasons for the construction of these buildings and their subsequent decay:

> Les caravanserais forains sont très différent de ceux des villes, tant par leur construction que par leur usage ... Ces édifices sont pour la plupart très-vieux, délabrés, sans gardiens, et servent assez souvent de retraite aux brigands. Schah-Abas-le-Grand en avait fait construire

Fig. 8 External facade and internal courtyard of a *tschaparkhané* (Persian post-stage) (Binder 1887)

> de fort beaux sur tous les points de la Perse. Ils furent pendant son règne, et même encore au commencement des guerres civiles, assez bien entretenus, mais depuis ce temps ils sont presque tous tombes en ruine, et comme on néglige beaucoup les réparations en Perse, il est à présumer que dans quelques années il n'en restera plus de vestiges. (Drouville 1825, pp. 100/101)

Diplomats and travellers on official missions on behalf of their governments (military, scientists, archaeologists, etc.) by far preferred to be hosted by local lords, also because this guaranteed a privacy that was impossible to have if you resided in a caravanserai, subjected to the curiosity of the local population and the insistence of street vendors. As far as we are concerned, it allows us to have detailed descriptions of the visiting ceremonies, food and drink, local customs and, in particular, the houses in which one was hosted. Some of these descriptions, which concern the Khoy area, are mentioned further on and are published more extensively in the appendix to this volume. [A/N: *If the reader is interested, he can refer to the texts cited in the bibliography regarding the main towns that are outside the field of research of this work: in particular Tabriz, Qasvin and, above all, Isfahan].*

Since the journey on horseback was the norm (the use of slower camels mainly concerned goods and therefore caravans), it was essential to have frequent changes of horses: this was possible in the villages, buying them from the farmers; however, there was also a service of real post offices,which in Persian were called *tschaparkhané [Chapar Khaneh].*

De Filippi gave us a description:

> un piccolo cortiletto chiuso fra mura di fango, nelle quali mura sono scavate piccole nicchie che servono da mangiatoia pei cavalli, una stanza terrena, ed un'altra superiore alla porta d'entrata costituiscono quasi dovunque i soli quartieri per gli alloggiamenti.

> [A/T: a very small courtyard enclosed by muddy walls, where small niches are carved out to serve as horse feeder; a ground floor room and another one, over the entrance door, almost everywhere constitute the only quarters for accommodation.]

Binder left us two photographs showing one that perfectly matches De Filippi's description (Binder 1887, pp. 75, 79) (Fig. 8).

In rural villages, partly out of necessity, partly out of choice, it was preferable to be accommodated in the normal homes of the locals, in particular those of the

Fig. 9 Nasi village, bed on the terrace of a traditional house (Binder 1887)

village chief or a mayor (generally there was a coincidence between the two figures). This allows us to have also descriptions of the common houses of the villages. All travellers observe that the building material par excellence in Persia is raw earth mixed with straw, which is used not only for the houses of rural and urban villages, but also for the palaces of magnates, mosques and walls, both border and fortification ones (Figs. 9 and 10).

As for the houses in the villages, these are common for all the ethnic groups found in the north-west of Persia: they are one-storey high houses, partly underground, with a vast central room, in the middle of which an oven is dug (the *tandour*), to which corresponds a vacuum in the ceiling, to let the smoke out (Fig. 11). This ground area is covered with carpets, and a few small niches in the walls contain the few objects of the family that cannot be hung. One access door and no windows. Next to it, there is the stable, often communicating with the inhabited rooms. The oven is used to bake bread and to heat during the long and harsh winters. The only fuel is the excrement of cattle and camels, which are dried in the sun in tall piles during summer. The roof is flat and can be used during summer to work, eat and sleep.

Seen from the outside villages made up by these buildings are almost unrecognizable, as they are not only partially embedded in the ground but also of the same colour as it. The only visible elements are the gardens (in the fertile plains) or a few isolated groups of trees. De Filippi describes these villages very well, adding a notation of particular interest:

Notre palais à Khatibâba.

Fig. 10 Traditional house in Khatibaba village (Mullers-Simonis 1892)

> while in Europe the most prosperous villages are along the main roads, in Persia it is perfectly the opposite. Along those lines that are called main roads, because they are more travelled by travellers, the villages generally have the appearance of the most squalid poverty, while in the most segregated corners ... there are the most populous and flourishing villages. (de Filippi 1865, p. 140)

The reason is that the streets of caravans are also beaten by armies and processions of kings and nobles, who impose serious corvée and "donations" in kind on the villages they cross.

Isabella Bird also pauses to describe in a similar way the houses of the Persian villages; however, she adds a precise description of the construction systems which deserves to be reported in the appendix (Bird 1891, p. 148).

A precise description of the *tandour*, the bread oven, comes to us from Flandin (Flandin and Coste 1854, pp. 128/129) and above all from Muller-Simonis, who also gives us two sketches in which he represents a vertical section of the oven and an image of a grinder for wheat (Muller-Simonis 1892, pp. 123, 126).

On the other hand, a curious fact concerns the scarce attention that travellers paid to the architecture of the Armenian and Assyrian churches in the region, contrary to what happened to churches and convents in proper Armenia. Of course, the quality of the architecture of the latter is incomparable in respect to the general modesty of the religious buildings in the Khoy and Salmas area; however, the authors who mention them are rare (Smith and Dwight 1834, p. 357, Anaclerio 1868, p. 59, Wilson SG 1895, pp. 90/91, Aubin 1908, pp. 57/58), in comparison with the attention raised by the modest rural houses, described by many also in detail. Texier and Ker Porter, who dedicated so many studies and surveys to the churches of Greater Armenia and Flandin/Coste who gave us a rich documentation on the archaeological areas,

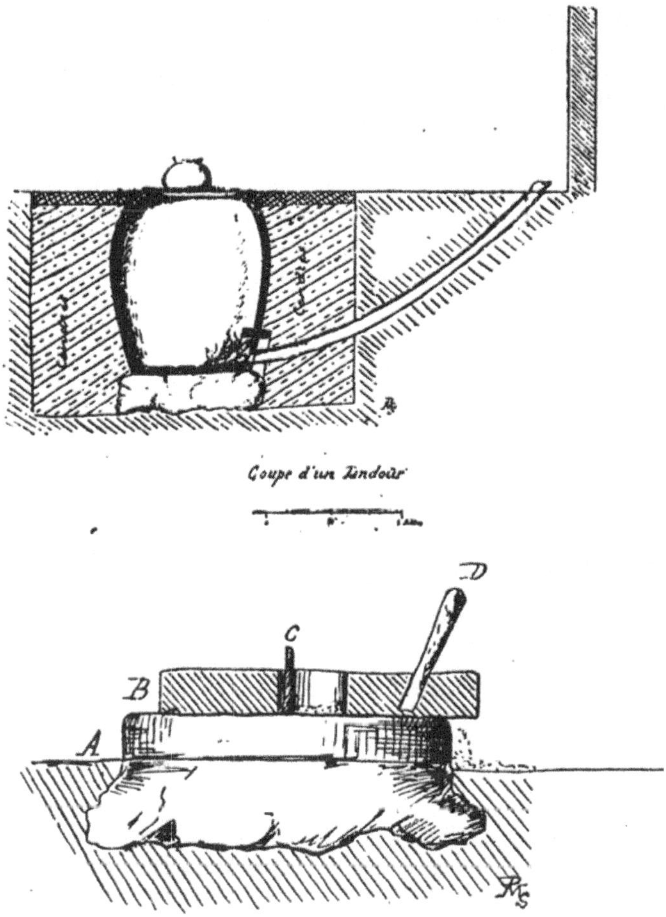

Fig. 11 Vertical section of a *tandour* oven and sketch of a hand grinder (Muller Simonis)

mosques and palaces of Persia, are completely silent on the churches of Khoy and Salmas, ccrtainly of a lower architectural level, but in any case not negligible, at least in the case of the S. Sarkis of Khoy, the Mar Givargis of Khosrova, the S. Khatch of Malhazam or the S. Gevorg of Haftvan.

Muller-Simonis is the only author who gives us a general description of minor churches:

> Les églises de ces pays sont toutes très pauvres, et on doit louer le zèle des habitants quand elles sont tenues à peu près proprement. Toutes ont une porte ridiculement basse où ne peut passer qu'une personne à la fois, en se courbant en deux. L'on donne cet usage comme un signe de l'humilité qui doit caractériser le chrétien ... mais j'avoue que je crois beaucoup plus à une préoccupation d'un tout autre genre. Dans les villages, l'église est la seule construction un peu solide. Sa terrasse peut servir de camp retranché; les femmes et les enfants peuvent se réfugier dans l'intérieur, et si la porte est bien basse et bien étroite, elle peut être barricadée

solidement et à peu de frais. Dans des pays sans cesse exposés aux incursions des brigands, cette opération doit, et surtout a dû se pratiquer assez souvent; c'est à la préoccupation de la faciliter le plus possible que je suis tenté de rapporter l'usage dont je parie. (Muller-Simonis 1892, pp. 162/163)

While for Western travellers to Persia the appearance of the villages looks modest, if not miserable, the enthusiasm with which one looks at the agrarian landscape of this part of Iranian Azerbaijan is shared. The appreciation for the verdant valleys of Khoy and Salams is also justified by the arid mountainous territory that has to be crossed to reach the two cities from the north. One of the *topoi* of travel reports is that of the appearance of the fertile plains once travellers reach the last mountain pass that hides the view. [Reference to the anthology of texts in this regard is published in the appendix]. Here, it is enough to cite the impressions of an Englishman, a Frenchman and an Italian:

We crossed over a ridge of trachytes and descended by *Khâni Berin*, re-ascending amid hills of conglomerate and igneous rocks, from whence we obtained our first view of the fertile plain of *Salamast* … with the lake of *Urumiyah* beyond. There are moments which never slip from a traveller's memory, when, after a long journey on a heated or monotonous plain, a range of mountains with their anticipations of cool waters and refreshing breezes come into sight, or when, fatigued with mountain-toil and travel, a plain, smiling with gardens and villages, and full of promises of delicious repose, presents itself to his delighted vision. (Ainsworth 1841, p. 59) (Figs. 12 and 13)

Après avoir franchi [le] passage, qui forme le point réel de séparation entre la Turquie et la Perse proprement dite, nous entrons dans la plaine de Khoie. L'aspect du pays change totalement. Nous nous trouvons comme par enchantement transportés au milieu des jardins tracés avec art, des champs bien arrosés, d'une population active, industrieuse et qui parait au-dessus du besoin. (Texier 1842, p. 27)

Quasi ogni villaggio è ombreggiato da alberi fronzuti, addensati, in filari, in boschetti, ora all'aperto ora nei chiusi, è circondato da frutteti arruffati ma rigogliosi, da aiuole verdeggianti, da campi ubertosi, bruscamente contornati nel deserto. (de Filippi 1865, p. 138)

[A/T: *Nearly each village is shaded with leafy trees, gathered in lines, in groves, both in the open and in fenced areas. It is surrounded with ruffled but thriving orchards, and verdant beds, fertile fields, abruptly surrounded with the desert*] (Fig. 14).

1 Khoy in the Descriptions of Western Travellers

European authors' detailed descriptions about the town of Khoy begin to appear in the fifteenth century. However, the town dates back to ancient times, and its importance is evidenced by its inclusion in many Iranian Azerbaijan maps created by Arab and Persian cartographers between the tenth and fourteenth centuries. In a map that represents Iranian Azerbaijan, with Lake Urmia in its centre, *Ibn Hawqal*, a famous Arab tenth-century geographer, identifies Khoy and Salmas as prominent cities (Anvari 2015) (Fig. 15). Several maps place the town along the important road that connected Ardebil and Bitlis (Galichian 2007). Khoy, indicated as *Khawi*, is mentioned by several Arab and Persian geographers: *Yâkût* (*Yaqut Al-Hamawi*, 1179–1229), *Kazwini* (*Zakarya Al-Qazwini*, 1203–1283), *Mustawfi* (*Hamdallah Mustawfi*, 1281–post-1339–40).

> To the north-east of Salmas lies *Khawi*, pronounced Khoi, on a stream that flows north to the river *Aras* (Araxes). *Khawi* was a strongly fortified and flourishing town according to *Yâkût* and *Kazwini*, surrounded by fertile lands and famous for its excellent brocades. There was also a spring here which had the reputation of being hot in winter and cold in summer. Mustawfi says that the enceinte of its town walls measured 6500 paces, and that its people were a white skinned race like the *Khatai* (Chinese); eighty villages were of its dependencies. (Le Strange 1905, p. 166)

This was the great British orientalist Guy Le Strange's summary in *The Lands of the Eastern Caliphate: Mesopotamia, Persia, and Central Asia, from the Moslem Conquest to the Time of Timur* regarding how the ancient geographers recorded the town of Khoy.

In 2000, Gabriella Uluhogian published a 1691 map that was commissioned by a Bolognese scientist and created by an Armenian from Constantinople. It documents the territorial distribution of the Armenian Church throughout the territory of "Greater Armenia". Khoy's significance is indicated by its being highlighted, along with *Khram*, as an Azerbaijan town with the caption "Monastery of the apostle *T'umay* (Thomas), bishopric". This is a clear sign of the town's importance to the

Fig. 12 Orchard gardens on the edge of the village of Akhtekhané (From drone, Sina Yavarinia 2018)

Fig. 13 Agricultural landscape near the village of Mahlam (Salmas County). (*Photo* Boriani 2017)

Armenian church and its superiority to Salmas, which is also represented on the map, but considered less relevant (Uluhogian 2000).

The first written information of European origin dates back to the early fifteenth century: Khoy "lies in a plain encircled by many orchards and corn lands, these again being surrounded by extensive grazing pastures irrigated by many streams and conduits of water". This is Ruy Gonzalez de Clavijo's description of the town at the beginning of the fifteenth century (Clavijo 1928).

This image will be the leitmotif of the descriptions that follow. The most enthusiastic, as we can see, will be that of James Morier: "The first view of the plain of *Kiwi*, from the summit of the pass in the mountains, is sublime" (Morier 1812, p. 298).

It is no coincidence that the Khoy plain still has very rich vegetation. Indeed, this seems to be even more noticeable to the ancient travellers after they crossed the arid mountains that block access to the town. What De Clavijo notes above all is the abundant water (another rarity in Persia), which is a function of the streams and artificial canals (conduits). There are not the famous Persian qanats. They are open channels fed by the melting snows from the Kurdish mountains to the west of the plain (Fig. 16). At the end of the nineteenth century, Muller-Simonis provides a detailed description of the water conduit system in the villages on the Urmia plain, a geographically similar area south of Khoy:

> La distribution des eaux est réglée par un fonctionnaire spécial appelé *Myr-Ab* (en persan, grand maitre des eaux). Ce fonctionnaire est nommé par le seigneur de l'endroit; il a

Fig. 14 Landscapes of the plateau on the road to Gris. (*Photo* Boriani 2018)

> sous sa juridiction, soit un seul village, si celui-ci est suffisamment important, soit les différents hameaux situés sur un même cours d'eau. Dans le pays d'Ourmiah l'usage des eaux d'irrigation est libre jusqu'au commencement de Mai; à partir de ce moment, chacun irrigue à tour de rôle. Un particulier veut-il arroser hors tour, le *Myr-Ab* lui vend ce droit d'eau supplémentaire. Cette vente ne doit avoir lieu que sur présentation d'un bon signé par le chef du village; mais, sans aucun doute, il est des accommodements. Parfois le village comme corps achète à un autre village situé au-dessus de lui le droit d'irrigation pour une période donnée. Dans les villages à population mixte, le vendredi, jour de repos des Musulmans, est réservé aux Chrétiens et le dimanche aux Musulmans. (Muller-Simonis 1892, p. 134)

The second detail that De Clavijo provides is that the town's population is largely Armenian. This is confirmed by subsequent visitors up to the 1820s. As we know, after the Russian conquest of the Yerevan khanate, a large portion of the Armenian population was encouraged (and sometimes forced) to migrate to Russian territory, leaving historical Armenia (Erzurum, Kars, and Van in Ottoman territory, as well as Khoy and Salmas, which were then part of Persia). The Europeans visiting Khoy after the 1828 Treaty of *Turkmenchay* regularly document the new demographics. The large majority of the town's population was Muslims, who lived primarily within the walled area. The remaining Armenian population lived in three outer streets west of the fortified centre, which is recognizable in the Russian map drawn by *Kolokov* and *Yai Shenikov* in 1827 (Mehryar et al. 2000). It is the first scientific survey of the town.

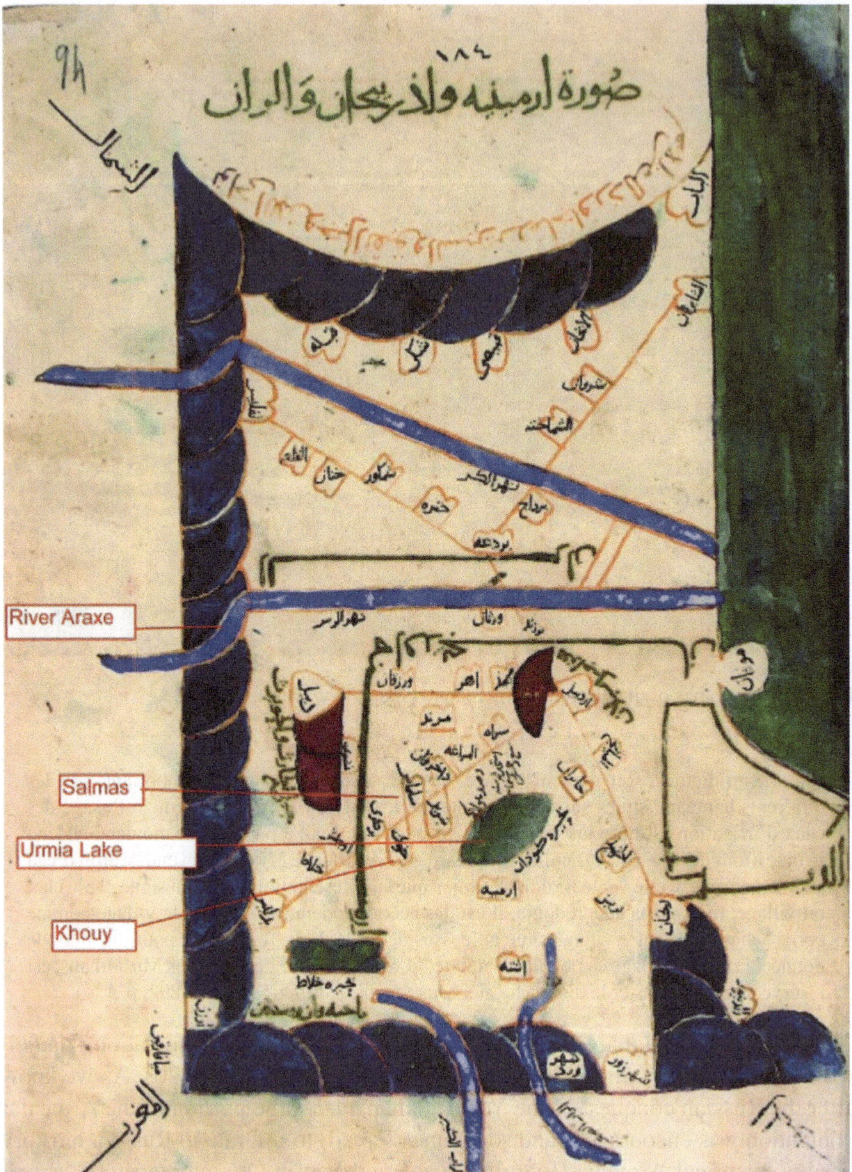

Fig. 15 Ibn Hawqual, *Map of Armenia, Azerbaijan and Arran.* Copy dated 1086 CE (the original around 973 CE) (Galichian 2007)

Fig. 16 Firuraq, a large orchard garden (from drone, Sina Yavarinia 2019)

The second traveller, Caterino Zeno, has left us significant information from his visit to Khoy between 1471 and 1472 during *Ak Khoyunlu* rule (1468–1514). According to Zeno, Khoy was a big town in ancient times, as evidenced by the likely visible large ruins. Zeno hypothesizes, without basis, that it was an ancient Armenian town, which he calls *Artasata*, the site of *Artaxata* (*Artašat*). This ancient capital of Armenia is identified in the 1920s through archaeological excavations on a hill near the present *Khor Virap* monastery.

Much more information, however, is provided by another Venetian traveller, Giovanni Maria Angiolello (1451–1525), who appears to have been in Persia between 1499 and 1525. His story, published by Giovan Battista Ramusio, includes the following:

> Questa città prima era rovinata, ma poi che *sciech Ismael* ... ha cominciato a rifabricarla e n'ha rifatta una gran parte: e fra l'altre cose è stato fatto un gran palagio, il quale con vocabolo persiano è detto *Doulet chana*, che vuol significare la casa graziosa. Questo palagio è tutto murato di mattoni, grandissimo, con un arin *[Harem, A/N]* tutt'insieme; dentro vi sono molte sale e camere, ed è fatto in un volto, come sarebbe dire in un solaro, e ha un bellissimo e gran giardino. Ha poi due porte con due magnifiche corti degnamente fabricate, e quest'entrate sono simili a due chiostri di convento di frati; avanti la porta che sta verso ponente vi sono tre torrioni fabricati in tondo, e ciascuno d'essi volge otto passa, e d'altezza sono da 15 o 16 passa. Questi torrioni sono fatti di corna di nanfroni cervi, e si giudica che nel mondo non ne siano altretanti; e appresso i Persiani queste cose sono riputate molto magnifiche, onde per magnificenza hanno delle corna di quelli animali murato tutti questi tre torrioni, però

che tutte quelle montagne sono alpestre e piene di salvaticine; e sultan sciech Ismael porta il vanto co' suoi baroni d'aver ammazzati tutti li detti animali. (Ramusio 1978–88, p. 1141)

[A/T.:

This town, short time ago, was in ruins; but when Sciech Ismael succeeded to the throne he began to rebuild it, and has now restored the greater part. And, among other things, a large palace has been built, which in the Persian language is called Douler Chana, signifying "pleasant abode." This palace is all walled round with bricks, and is of great extent, with an Arim [Harem] all together; inside there are many halls and chambers, and it is built in one vault – that is to say, with one flooring; and it has a large and magnificent garden. It has two gates, with two fine courts, beautifully decorated, and these entrances are like two cloisters of a convent of friars. Before the gate which looks west are three round turrets, each of them eight steps in circumference, and about fifteen or sixteen steps high. These turrets are built of the horns of Namphroni stags, and it is considered that there are none like them in the world. The Persians also consider these things very magnificent. Therefore for magnificence they have built these three turrets of the horns of these animals, as the mountains are rocky and full of game].

This is the only document stating that there were originally two additional towers similar to the famous *Shams Tabrizi* tomb tower. He also indicates that the town had commercial links with India to which it exported a substance, obtained from a local root, that was used to make the colour crimson (Figs. 17, 18 and 19).

A probably dated 1536 Matrakçi miniature gives a rough idea of what Khoy—the only Persian town, as well Tabriz—which is illustrated in *Matrakçi's* substantial treatise (which will be discussed later), might have looked like (Nasuhu's Silahi called *Matrakci* 1976, and Balilan and Jafari 2013). Angiolello and *Matrakçi* suggest that

Fig. 17 The two Shams Tabrizi towers in Khoy that survived at the time of James Morier's voyage (Morier 1818)

Fig. 18 The only tower of Sham-e-Tabrizi today surviving in an old image (JayHani and Zali, 2015). (*Photo* Boriani 2018)

Khoy played a very important role in Persia between the late fifteenth and early sixteenth centuries. This role appears to have faded in the first half of the eighteenth century, however. There are almost no travel accounts about Khoy until the early nineteenth century. The exception is that of *Evliya Celebi*, the famous Ottoman traveller who, of course, cannot be counted among European voyagers.

Fig. 19 A detail of the tower of Sham-e-Tabrizi with the deer antlers still in place. (*Photo* Boriani 2018)

The lack of information about Khoy might be the result of the continual wars between the Ottomans and Safavids that characterized the sixteenth century, precisely to contend with the current Iranian Azerbaijan. Moreover, the transfer of the Persian capital to Qazvin and then to Isfahan, Shah Abbas's scorched earth policy regarding the Ottomans, the deportation of wealthy Armenian merchants to Isfahan, and the major earthquake that destroyed Tabriz in 1721 strongly influenced the shrink of news. Khoy, a frontier town, suffers a great deal of physical, social, and economic harm. Tabriz maintains its role as a magnet for the traffic from Trabzon and Tiflis but only as a stop on the route to Qasvin and, above all, Isfahan. Thus, travellers can bypass Lake Urmia.

A decisive factor in the area's commercial decline, however, is the Portuguese discovery of the possibility of reaching Persia and, in general, the Far East by rounding the Cape of Good Hope (1497). For about a century, the Portuguese maintain their dominance over the naval routes to Persia and India and the land bases along the African coast and Arabian Peninsula in order to facilitate navigation and trade. The attempts of the British, with the help of the Russians, to open an alternative route to the north, across the Caspian Sea, are in vain (Curzon 1892, Vol. II, Chap. 29, pp. 533–545).

Don Juan of Persia (a Persian whose given name is *Uruch Beg* [about 1560–1604]), who converted to Christianity, interrupts the silence from European sources.

He writes about his travels in the Khoy area in no very good Spanish. According to Don Juan, Azerbaijan Province, with Khoy and Salmas as its main cities, has a population of 135,000. He states that the people are "not very civilized". He also describes them as warlike. He spends half of the year encamped in the countryside.

Evlyia Celebi, who passes through the town in 1655, provides a detailed account of the region. He describes the fortress in *Qotur*, which controls the mountain pass on the border between Turkey and Persia on the road connecting Bayazit and Khoy. He describes the castle as one of the strongest in Persia. It has withstood the fierce attacks of the Ottoman sultan, who climbed the Qotur valley after devastating Western Azerbaijan (including Khoy and Marand). Celebi describes Khoy under Ottoman control. He states that the Sultan temporarily resides there, and all the state civil and religious authorities have a presence. Khoy Castle has a rectangular plan, and the perimeter of the walls is seven hundred steps with two gates. The moat is not very deep. Within the walls are approximately one hundred houses and a mosque. The town, which has been sacked several times, has seven thousand houses and seventy mosques, including a congregational mosque, as well as two public baths, seven caravanserais, and a thousand shops with gardens. *Celebi* makes no mention of city walls but calculates the perimeter of the built-up area as ten thousand steps. The climate is favourable, and the fruits grown there are famous for their sweet flavour. In the Khoy district, there are one hundred eighty villages whose main inhabitants are Salafist Sunnis. Curiously, he does not mention the Armenian population (Celebi 1834, pp. 148–149).

It was not until the early nineteenth century that 1813new travellers' accounts of Khoy became available (Pierre Amedé Jaubert 1805–1806, James Morier 1808, Gaspard Drouville 1812–1813, Macdonald Kinneir , and Joseph Michel Tancoigne 1813). Not surprisingly, almost all are diplomats during the Napoleonic wars.

The Europeans' renewed attention to northern Persia stems essentially from Napoleon's involvement of the Shah in an anti-Russian and, above all, anti-English alliance. This is of great interest to the Persians, who have recently lost control of Georgia. Moreover, Napoleon aimed gaining Persian support for the possible invasion of British India. Of course, the issue is the profitability of this alliance. The secret instructions that Napoleon sends to Tayllerand, the French foreign minister, to be passed on to Jaubert are clear:

> Vous ne manquerez point de lui faire connaître que le principal but de son voyage est de s'instruire de la situation de la Perse et de la marche des gouverneurs qui y existent, province par province, et de reconnaître à combien se montent ses forces. Il doit, s'il lui est possible, parcourir les bords de l'Araxe et pousser jusqu'aux frontières russes. Il prendra des renseignements sur les évènements passés, et enfin sur tout ce qui peut me mettre à même de connaître bien le pays. Il sera très réservé, cependant, en causant avec les ministre et l'Empereur (le Shah), il dira que je veux entrer en alliance avec lui et lui offrir des secours. (Vinson 2009, p. 3)

In 1807, after a long preparation phase (*Treaty of Finckenstein on April 27, 1807*), Napoleon sends an important seventy-people mission headed by his aide-de-camp, Claude Mathieu de Gardane. In addition to the civilians and diplomats are thirteen military people (including cartographers and artillery experts), a doctor, an architect, and three missionaries. Tancoigne and Dupré were also a part to the mission. Drouville would undertake there another mission in 1812. The mission was not successful, especially because of Napoleon's indecision and imprecision. To Gardane's embarrassment, France signed a treaty with Russia during this same period. However, that introduced Persia to French society and lead to the development of an important phase in the French orientalists' study of Persia a few decades later (Fig. 20).

A trade route, the Silk Road is now essentially a sea route, especially after the opening of the Suez Canal in 1869. Then, it was essential to understand the disputed territory between the Russian Ottomans and Persians, i.e., Georgia, Armenia, and Azerbaijan of Tabriz. Those territories were no longer of commercial significance; however, they were of military and diplomatic great importance.

Clearly, the British have not watched the French initiatives passively. Diplomatic missions are sent to oppose Napoleonic politics. James Morier, John Macdonald Kinneir, and William Montheith, British diplomats in Persia, write about their travels and experiences. Thus, Khoy is often mentioned in French and English relations. Despite having lost its commercial importance, the town is still on the front lines. On the border between Turkey and Persia, it has been a disputed area between the two powers several times.

Fig. 20 *Persian envoy Mirza Mohammed Reza-Qazvini meets Napoleon at Palazzo Finckenstein to sign the Treaty of Finckenstein on April 27, 1807* (François Mulard—https: // creative commons.org/publicdomain/mark/1.0/deed.en accessed February, 25, 2021)

The first French visitor of the town is Pierre-Amédée Jaubert, who enters Khoy in 1806 on his preparatory journey for General Gardanne's mission:

> La ville de Khoi, qui n'est pas fort ancienne, a des fortifications régulières. On n'y voit pas un grand nombre de mosquées ni de maisons considérables; mais les rues sont ombragées d'arbres, et l'on y trouve un assez beau caravanserai spécialement réservé pour les marchands. (Jaubert 1821, p. 148)

Jaubert estimates the town population at approximately twenty-five thousand. According to him, they considered themselves to be "of Tatar origin". This is quite strange because, according to many other contemporary travellers (Drouville in 1812), Khoy still seems predominantly Armenian. It was possible for Jaubert to be referring to the ruling class, specifically the provincial governor, whose residence he visited. He describes it as being located in a pavilion surrounded by a garden with trees: "symétriquement alignés, des bassins de marbre et des jets d'eau" (Jaubert 1821, p. 149). Stained glass windows illuminate the pavilion, which is slightly raised. The front is open. This is certainly the *Dagh Baghi*, a Persian garden that is indicated on the 1827 Russian map of Khoy. The map is now very degraded but still readable. It is the subject of a recent study that focused on its restoration (Jayhani and Zali 2015).

J. M. Tancoigne, attaché to the Gardanne Mission, also dedicates a few lines to Khoy. He describes the city as having high brick walls, tree-lined streets flanked by canals, and mostly one-storey clay houses. It suffered an earthquake (which Tancoigne writes about in 1807 just after it happened) and has a population of twenty-five thousand. They are attracted by the fertile countryside, which is among the most productive in Persia (Tancoigne 1820, Vol. I, p. 67; the English translation, published in 1820).

James Morier, a British diplomat, publishes a memoir (two volumes in 1812 and 1818) of his visits to Persia (in years 1808, 1810, and 1814 to 1816). Morier's accounts include the usual enthusiasm about the Khoy plain: "The first view of the plain of *Kiwi* is sublime ... The plain ... was the richest tract that we had seen", but his main attention is on the city fortifications:

> Khoy is surrounded with a wall, and with towers of a different construction to any which we had remarked in the fortified towns in Persia ... There are four gates which are of stone, and very superior to most of those that I had noticed elsewhere. (Morier 1812, p. 299)

Morier provides additional interesting information. He states that the town has twenty mosques and six public baths; however, he makes no mention of the bazaar, which previous travellers have found interesting. He estimates that there are 10,000 houses with 15,000 inhabitants, mostly Armenians. Muslims instead live in a secluded neighbourhood. The strange relationship between the number of houses and inhabitants could be explained by the devastating 1870 earthquake in Tabriz and the escape from the town (as documented by Jaubert) to avoid the fallout from the succession wars during Shah Agha Muhammad Khan's reign (1794–1797) (Morier 1812, p. 298).

Morier provides a detailed description of the governor's great garden, which was built by Hossein Khan at the end of the seventeenth century:

> [A] fine alley of chenar [Platanus orientalis] trees, which leads up to a pleasure-house, now falling into decay, built on the elevation of six terraces, from each of which falls a beautiful cascade, conducted by qanats from the neighboring mountains. On the right and left is a wood of fruit trees of every sort and description, with a fine crop of grass at their roots. (Morier 1812, p. 298)

During his second trip, Morier evaluates the quality of the new European fortifications that are under construction. He considers them imperfect:

> Attempts have been made to fortify Khoi according to rule, upon a plan drawn out by one of General Gardanne's officers. It has cost the Prince large sums of money … He now has determined to complete the fortifications, and we found some hundreds of workmen employed under the direction of a young Persian, by name *Mirza Baiker*, who acquired a superficial knowledge of engineering, from both the French and English officers. (Morier 1818, p. 305)

Gaspard Drouville (Fig. 21) has an even more negative opinion of Khoy's new European fortifications. A cavalry officer in the service of the Tsar, Drouville can be considered an expert. He writes:

> Il n'existe dans toute la Perse que deux villes qui soient fortifiées dans un système européen, Khoï et *Abas-Abad*; il est impossible de trouver rien de plus mauvais. Tout ce qu'on y a

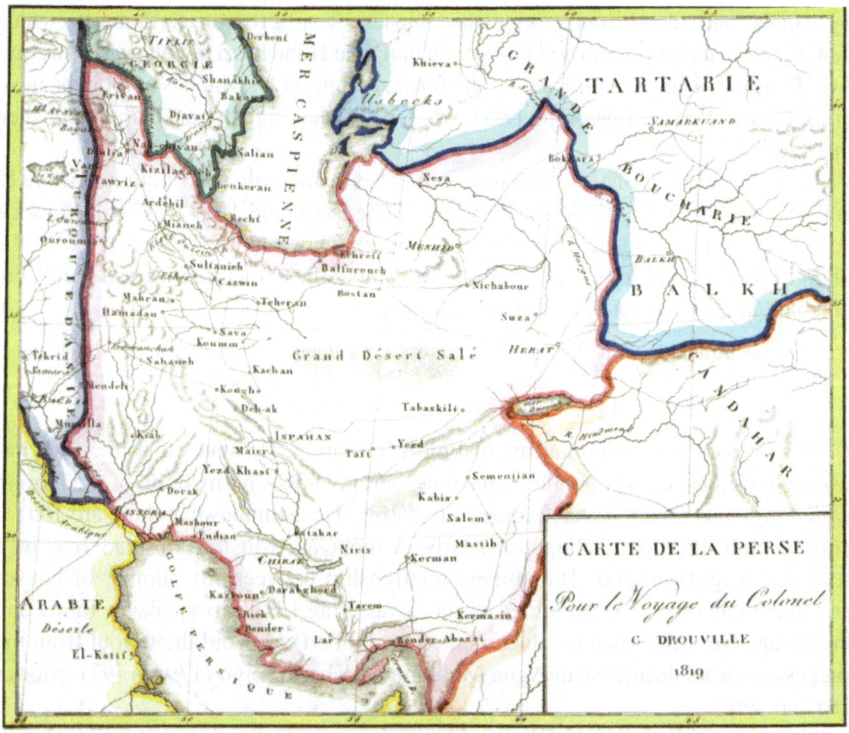

Fig. 21 The map of Persia prepared in 1819 for the mission of Colonel Drouville (Drouville 1825)

> fait, consiste en de mauvaises courtines, très-imparfaitement flanquées par de petits bastions si étranglés, et d'un saillant si aigu, qu'on aurait de la peine à y manouvrer deux pièces de canons. Ajoutez à cela que le tout est sans revêtemens, à une toise près, sans chemins couverts, ni glacis, ni palissades. Ce que l'on a bien voulu décorer du nom de contrescarpe, n'est d'aucune défense, et l'on n'aurait pas beaucoup plus de peine à la descendre que les degrés d'un escalier un peu élevé. Ces mauvais ouvrages furent construits par un officier anglais, soit-disant ingénieur et qui était au service de la Compagnie des Indes. Peut-étre fut-il, faute d'autres, considéré comme le Vauban de l'Asie. (Drouville 1825, p. 54)

It is worth noting that Drouville, as a good Frenchman, attributes the blame for the poor execution of the project to an English *soit-disant* expert in fortifications.

The Khoy region becomes interesting again after Russia's conquest of the Khanate of Yerevan (1828 Treaty of *Turkmenchai*). Iranian Azerbaijan and the eastern territories of the Ottoman Empire (Kars, Erzurum, Van) enter the Russian political-military orbit. The start of the modernization of the road infrastructure in Russian Armenia facilitates trade with Tabriz and, consequently, Khoy. The new political situation attracts not only diplomats but also a new group of travellers who are interested mainly in the geographical, naturalistic, historical-artistic, and archaeological aspects of the region. Among the visitors stand the French Charles Texier, who visits Persia in 1839, and Eugène Flandin, who travels there with Pascal Coste between 1839 and 1841.

Among the first travellers to enter Khoy shortly after the end of the Russo–Persian war are several Englishmen: Macdonald-Kinneir (in 1813), Fowler (1829), Alcock (1829), and Ballie-Fraser (1836). Except for Macdonald-Kinneir, one of the Shah's soldiers since 1803, the men declare themselves independent visitors. We might well guess they were intelligence officers in disguise.

Fowler focuses on the fortifications. He declares the city gates "respectable". He states that the moat, which is dry, is crossed by crumbling bridges supported by slender poles that can be dismantled in five minutes in case of danger (Fowler 1841, p. 242). Fowler also testifies to the depopulation of the area surrounding Khoy because of the Armenian migration to Russia. This has saddened the town's governor because the Armenians are well off and could pay high taxes.

Alcock too reports the ongoing migration. He states there are reports that the Armenians in the Russian territory are not treated much better than they were by the Persians. He adds that Khoy's houses have been damaged by war and the Russian occupation of the town (Algock 1831, p. 103). Colonel Shiel, an English soldier, does not describe Khoy; however, he provides some interesting information about the *Qotur* valley. The village of the same name, on the border with Turkey, depends on Khoy. It has a fort and a raw earth fort that Shiel considers a "tolerable mud-fort" that the Kurds of the *Hakkari* tribe had sacked a year earlier. They are formally dependent on Van's *Pashalik*. The village is made up of about 60 houses that are inhabited mainly by Armenians, who are dirty and in rags. Although they appear to be poor, they own a great deal of cattle. The descent to Khoy takes place through a wide valley that further widens into an immense prairie. This last section of the route, which is controlled by a Kurdish tribe, is the most dangerous stretch between Van and Khoi (Shiel 1838, p. 57) (Fig. 22).

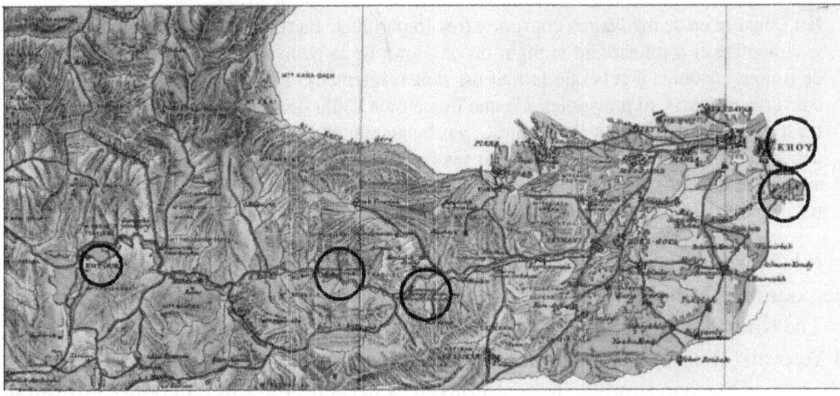

Fig. 22 *Map of the Turco-persian frontier made by Russian and English officiers in 1849–1855.* Detail of the valley of the Qotur river (Library of congress—Wikipedia commons)

Curiously, another English soldier, Robert Macdonald, does not write about Khoy's fortifications. Instead, he describes the governor's palace and a garden outside the city (the *Bagh Daghi* mentioned by Joubert). Like the other houses in the city, the palace is built of raw earth; however, it is elegantly painted, sumptuous, and classy. The garden, with beautiful fountains, is in a state of disrepair; however, there are still beautiful trees and rose hedges. It is surrounded by a high earthen wall. The *Bagh Daghi* also appears poorly maintained, and the palace is in ruins. In Macdonald's precise, description is a new detail. The rose gardens, which were once considered the most beautiful in Persia, are populated by nightingales (Macdonald 1859, p. 126) (Figs. 23 and 24).

Ballie-Fraser does not provide new information other than the fact that the city gates are closed at sunset. Texier, the French architect who is also an archaeologist, unearths *Hattusa*, the ancient Hittite capital, near Ankara. His descriptions of Armenia, Persia, and Mesopotamia (1842–1845) inaugurate a new series of publications in which the iconographic apparatus is not only abundant but also of an excellent technical level. The plans, sections, elevations, decorations, and architectural details are documented in beautiful plates, many in colour. In Persia, Texier is clearly focused on Persepolis and the great monuments of Tabriz and Isfahan. However, a plate is dedicated to an idealized representation of Salmas' famous Sassanid bas-relief. It is accompanied by a fairly detailed description of the state of conservation:

> Ces rochers sont formés d'un calcaire bleu, un peu schisteux, et par conséquent difficile à travailler, s'exfoliant à l'air, en un mot, très-peu propre à recevoir une sculpture. Aussi le bas-relief de Selmas a-t-il sensiblement souffert. On peut néanmoins reconnaitre que ce monument date de l'époque des princes sassanides. Son exécution est assez grossière: les figures manquent principalement de modelé, elles sont seulement découpées sur la pierre en silhouette. (Texier 1842, pp. 33–34)

To visit the bas-relief, Texier stops at Khoy and Salmas, which he describes in detail over several pages. This is the first such description.

Fig. 23 A Qajar period image of the Dagh Baghi palace and garden in Khoy (Jayhani and Zali 2015)

Fig. 24 Another Qajar period image of the Dagh Baghi palace and garden in Khoy (Jayhani and Zali 2015)

Texier enters Persia from Van via the caravan route that connects Bayazid and Khoy. His main objective is Salmas' Sassanid bas-relief. Khoy is the first Persian town that he visits. This might explain why he dedicates more pages than previous travellers to it. After crossing an inaccessible, almost uninhabited territory, Texier reaches Khoy on 14 October, 1839 and stays between Khoy and Salmas until 19 October. The Khoy plain appears after Texier crosses a difficult hill and a narrow valley:

> Nous nous trouvons comme par enchantement transportés au milieu des jardins tracés avec art, des champs bien arrosés, d'une population active, industrieuse, et qui parait au-dessus du besoin. Ce premier sentiment fut tout à fait en faveur des Persans. (Texier 1842, p. 27)

Even the town seems pleasant at a first glance:

> L'aspect de la ville de Khoie ne démentit pas l'idée que nous nous étions faite à la vue de ses environs. Elle est entourée d'un retranchement en terre, construit d'après les lois de la fortification moderne ... Après avoir franchi le pont-levis, nous entrâmes dans une rue bien alignée, et plantée des deux cōtés d'une allée de platanes qui sont constamment rafraichis par des ruisseaux d'eau vive. (Texier 1842, p. 28)

However, the urban landscape is less attractive: "c'est que nulle part les maisons n'ont de fenêtres sur la rue, et qu'on ne voit de part et d'autre que de longs murs de briques dans lesquels sont percées de petites portes". What is striking is the commercial activity in the bazaar (Figs. 25, 26, 27 and 28):

> Je fus surpris de l'activité qui régnait dans le bazar; et ce qui me frappa surtout, c'est que pas une des cinq cents boutiques qui le composent n'était à louer, tandis qu'en Turquie, dans les bazars des villes les plus commerçantes, il y a toujours six boutiques sur dix, et quelquefois plus. (Texier 1842, p. 28)

In addition, caravanserais occupy a large area of the city and many Armenians who are wealthy traders still live in the city. The *Shia Muslims* are particularly involved in public functions. As expected, the mission visits the provincial governor, and this provides an opportunity to visit the palace:

Fig. 25 The south gate of the city and Khoy Bazaar before the restorations. (*Photo* Moradi 2020)

Fig. 26 The south gate of the city and Khoy Bazaar after the restorations. (*Photo* Boriani 2019)

> Nous entrâmes par un kiosque orné de peintures dans une grande cour oblongue, plantée de rosiers et d'arbustes, et arrosée par des canaux d'eau courante. A l'autre extrémité de la cour s'élevait un kiosque dont la façade, supportée par des colonnes, était ouverte sur le jardin. (Texier 1842, p. 29)

After leaving Khoy, Texier visits the two famous *Shams-e Tabriz* tomb towers (the third has apparently already collapsed), which are covered with animal horns. He mistakenly believes them to be the minarets of a vanished mosque. On the road to Salmas, he describes a wide tree-lined three kilometre-long avenue. Along it are palaces and gardens (one was certainly the *Dagh Baghi*) that, for the most part, appeared to be in ruins. Not much farther is a six-arch bridge across a river that Texier refers to as *Atou*r. This is the *Khatun* Bridge across what is now known as the *Qotur* River.

Texier leaves Dilman (see the chapter dedicated to the Salmas region) on 19 October, 1839, and Eugène Flandin enters Khoy on 15 January of the following year:

> Devant nous s'étendait une longue muraille crénelée dont le ton brun tranchait fortement sur la blancheur de la neige; beaucoup d'Arbres placés derrière surmontaient ses créneaux, mais, du reste, aucune maison, ni même aucun minaret n'accusaient là l'existence d'une ville. (Flandin and Coste 1854, p. 135)

Fig. 27 An image of the Khoy Bazaarrecently restored. Note the old shop not yet modernized. (*Photo* Boriani 2018)

Flandin's description of the city walls adds a few more details than Texier's. He mentions wide moats and bridges that are martial and defensive structures of European origin. Following the Gardane mission, some officers have offered suggestions on the topic; however, they were only partially accepted after several years. The Azerbaijan governor, Prince *Abbas-Mirza*, has made changes despite his lack of knowledge of military science. Flandin renders a harsh judgement: "ces travaux, tout en valant mieux que ceux qui sont dus aux simples maçons du pays, étaient insuffisants pour protéger la ville contre un siège régulier fait par une armée européenne" (Flandin and Coste 1854, p. 136). He shares a similar opinion about the city inside the walls. He describes the usual single-storey houses without windows overlooking the public streets. He states that the mosques have no minarets; they have domes that rise just above the regular houses. With the exception of some central tree-lined

Fig. 28 Another image of the Khoy Bazaar with an old shop not yet modernized. (*Photo* Boriani 2018)

and canal-flanked streets, the city streets are narrow and winding. There are an estimated twenty thousand inhabitants, which include many Armenians. Flandin does not mention the city's bazaar, caravanserai, or economic activity, which is considered to be in decline because of the 1827–1828 wars.

A theme in the descriptions of the city of Khoy is the fortifications. Juste Pichon, another soldier on the Flandin–Coste mission, offers a harsh criticism, which can be considered definitive:

> Dans l'après-midi nous fûmes visiter les remparts, qui ne sont autre chose que deux murs de circonvallation, flanqués de tours de distance en distance. Ces murs sont très élevés sans avoir de bonnes bases; aussi n'est-il pas rare à Khoi, comme dans les autres villes de la Perse, de voir tomber lambeaux par lambeaux les choses dans lesquelles surtout ils devraient mettre tout leur salut. Les fortifications de Khoi, qui sont bien renommées en Perse, me firent la

plus grande pitié. Nous fûmes visiter ensuite la caserne, vieux amas de ruines, où se sont retirées quelques femmes depuis que les soldats logent chez eux. (Pichon 1900. p. 46)

Pichon's negative opinion cannot be attributed to his character because, a few lines later, he provides an enthusiastic description of Khoy's bazaar: "Je fus réellement étonné des bazars que nous visitâmes. Il sont grands et sont sous de vastes voûtes assez régulièrement construites. La hardiesse de ces voûtes m'étonna beaucoup et était de beaucoup supérieure à celles d'Erzeroum" (Pichon 1900, p. 47) (Fig. 29).

An engineer and geographer, Xavier Hommaire de Hell, another French technician, is less well known than Texier and Flandin. However, he provides a great deal of information. Some of it is new and very interesting. De Hell enters Persia from the Qotur valley and gives a precise description. Upon approaching Khoy, he is impressed, as has been everyone else, by the beauty of the landscape. He adds a small technical note. The water from the *Qotur* is collected through "nombreaux canaux étagés les uns au dessus des autres". Towards evening, the peasants come home from the city, and the atmosphere is pleasantly animated. The houses are clean. The façades include glazed verandas that overlook gardens located both in front of and behind the buildings (Hommaire de Hell 1855, Vol. 2, p. 524) (Figs. 30 and 31).

Upon entering Khoy, de Hell notices that the European fortifications are in poor condition and that the bazaar and many houses have been damaged by an earthquake

Fig. 29 Khoy's Caravanserai in an aerial photo. The main arm and the three transverse arms are recognizable, as well as the three courts of the caravanserai (from drone, Sina Yavarinia 2019)

Fig. 30 The inner courtyard of Raffi's birthplace in Payadjuk (Alice Navasargian, *The Immortals*, 2012)

Fig. 31 The inner courtyard of a Persian princely house (Binder 1887)

that occurred five years earlier. Reconstruction is slow. The exception is a well-designed baked brick mosque, whose construction has already reached the base of the dome. Within the walls are 2800 houses, and beyond, in the suburbs, there are 3700. Only about fifty Armenian families remain. Many soldiers (four thousand infantry and cavalry and four hundred gunners) would also reside in the town. However, at the time of his visit, they are employed outside the city so that the city gates are not even guarded by sentries. De Hell also observes that the bazaar and city gates close at sunset. Two hours later, a triple drum roll announces the curfew (Hommaire de Hell 1855, pp. 526–528).

Khoy receives a brief mention by Lady Mary Sheil, the first woman to write about her journey to Persia. She travels between 1849 and 1852. Although she stays only a few hours, she has a negative opinion of the city's fortifications: "We only stayed to rest and breakfast at *Khoee*, that important and defenseless city, open and ready to be seized upon by any invader" (Shiel ML 1856, p. 288).

The opinion of the French aristocrat Arthur de Gobineau, who is appointed secretary of the French Legation in Tehran in 1855, is more extensive. Despite his "*Essai sur inégalité des races humaines*" (1853), which makes him one of the precursors of white supremacism, he is a great admirer of ancient Persia (as Aryan). He gives a brief but interesting account. According to de Gobineau, Khoy is a rich, charming trading town that is famous in Persia for slipper manufacture. The bazaar has a beautiful, distinctive black and white marble door. It is worth noting that this is the only opinion about the architectural quality of the Khoy bazaar, which has been generally underestimated by previous travellers except Jules Pichon. The town surroundings are cultivated with beautiful tree plantations, a rare feature that de Gobineau ascribes to the governor, who is a royal prince. De Gobineau describes him as a highly skilled soldier and intelligent administrator (de Gobineau 1859, p. 510).

In a later account, Colonel Stuart, an English soldier, notes that Khoy is surrounded by a double raw earth wall, and the city and the wall are in disrepair. The outer wall has curtains, salient, and bastions, with battlements, moats, and covered roads. The inner wall is high and flanked by towers. Between the two curtains are gardens and shacks. This is evidence of the survival of the old Persian walls into the mid-nineteenth century. It is also a nonjudgemental description of the quality of the work. The reason could be the fact that Stuart attributed the project to another Englishman, Colonel Montheith (Stuart 1854, pp. 128–129).

An important piece of information confirms de Hell's claim about the change in the population characteristics of the town. According to Stuart, Khoy has four to seven thousand families (i.e., houses). This is compatible with De Hell's data and other travellers' estimations of twenty to twenty-five thousand inhabitants. However, only one hundred families are Armenians, who were removed to a suburb just west of the walls. This suburb is clearly recognizable on the 1827 Russian map. An irregular layout of alleys converges on a central wide stretch of roads surrounded by a wide border of orchards. Stuart regards the bazaar as one of the best in Persia, second only to that in Shiraz. Built of baked brick, it is tall and spacious, and the internal road is covered by arched domes. The shops are generally small. The exception is an octagonal shop with a fountain in the centre (Stuart 1854, p. 130).

Moritz Wagner, a German geographer and naturalist who visited Khoy in 1843, also described the fertile plain surrounding the town. He ascribes the fertility to the intelligence and industriousness of the rural population because the area would be a desert without artificial irrigation. He is critical about the city walls:

> The town of *Choi* is almost buried under the growth of the surrounding foliage, and you are scarcely aware of its existence, till you are within the place, which is encompassed by an

> indifferent mud wall, an effectual protection against Koordish forays, but absolutely useless against European field-pieces [field artillery]. (Wagner 1856, pp. 77–78)

Wagner then adds that two buildings emerge: the bazaar, which appears larger than would be expected in a small town, and the caravanserai, that are located in two large courtyards. One courtyard, which is surrounded by trees, has a fountain. In his opinion, the town proximity to the Russian and Turkish borders is an advantage for trade. He states that the bazaar is not as splendid as that in Tabriz. However, a variety of valuable goods, such as silks, carpets, pewter, leather works, and articles of European origin, are on display (Wagner 1856, pp. 78–79). A dozen years later, Wagner's evaluation will be confirmed by Ussher, who will write about a small town with many traces of bygone greatness and a busy, crowded bazaar noted for sock and slipper manufacture.

Diplomats, the military, and orientalists frequent the Khoy region in the period following the Treaty of *Turkmenchay*. There are also accounts from Christian missionaries in Urmia and Salmas, which have a large, albeit minority, concentration of Nestorian, Assyrian Catholic, and Armenian communities.

The first Christian missionary to describe the town of Khoy in modern times seems to be Horatio Southgate, a US Episcopalian bishop who worked mainly in the Ottoman Empire. He makes several trips between the Nestorians and the other Christian communities in Turkey, Mesopotamia, Armenia, Kurdistan and Persia between 1836 and 1838. The observations about Khoy do not include religion. The first impression is related to the situation of the Persian roads. According to Southgate:

> The reader may generally understand, when I speak of roads, simple footpaths, such as may be seen across pasture grounds in New-England. The country over which they pass is, for the most part, in the state in which nature left it. (Southgate 1840, pp. 325–326)

He states that there are excellent masonry bridges that are often in poor condition, thus causing the traveller to wade through the waterways. During some seasons, this is not easy or safe. The access road to Khoy, already described by other visitors, is no exception. It is lined with willows that "once must have been beautiful" but are in poor shape. The road is crossed by canals, via small passages three or four feet wide. The countryside is beautiful. However, the Urmia countryside is more attractive and has a wider variety of products.

Southgate rates Khoy's caravanserai as excellent. It has an upper floor with rooms reserved for prayer. According to Southgate, curiosity of the population about foreigners is not as annoying as that in Salmas and Urmia because Khoy is a border town. Thus, it is not uncommon to meet Turks with fezzes or turbans. Another bit of information from Southgate concerns the bazaar. The brick structure is covered with vaults and domes. It develops on a single line that extends from the city gate to the city centre. It is intersected at short distances by short roads that lead to the caravanserai. In the bazaar, Southgate meets a very kind, hospitable Persian. He is well dressed and therefore of a relatively higher class. However, he cannot understand that the reason for the trip of the religious man could be the desire to know the area (Southgate 1840, pp. 328–329).

The spate of visits following the Russian conquest of Yerevan and Naktchevan ends with these authors. A new phase will open between the end of the nineteenth century and the beginning of the twentieth. It is the period in which Khoy and the Tabriz region enter modernity. A new group of travellers, Christian missionaries (Catholics and Protestants), will engage in fierce competition. Thus, it seems appropriate to summarize the state of the town prior to this last phase. The two beautifully drawn 1827 Russian maps accurately describe many of the sights reported by the visitors.

There are several maps of Khoy. The most detailed are the two drawn by two Russian army officers, *Kolokov* and *Yai Shenikov*, in 1827 (Mehryar and Alii 2000, pp. 73–80) along the military occupation of the town during the Russo-Persian war. There is also *Matrakçi's* famous 1536 map (Matrakci 1976, p. 26b]). There is another map, "Iran and Ottoman Borders: Khoy Fortress," which *Amir Houshang Anvari* published during the Qajar period (Anvari 2015, p. 263).

Balilan and Jafari (2013) studied the Matrakçi map (Fig. 32). It has a south-up orientation because the *Qotur* River, represented with a bridge (which can be only the *Khatun* Bridge) over it, flows south of the town of Khoy. The Shams-e Tabrizi mausoleum (represented on the right side) is west of the current historic centre. The sixteenth-century city, which is rectangular, is surrounded by crenellated walls with some towers. The walls appear to be interrupted in several places. In two locations where there is no wall, there are streams. However, in other cases, the interruptions are inexplicable. They might be sections that were destroyed during the Turkish–Persian war. Along the interior of the long sides of the walls are rows of trees, which, by their shape, appear to be poplars because cypresses do not grow in this part of Persia. The map also shows a minaret and the *Shams-e Tabrizi* mausoleum with two minarets flanked by an isolated tower (the *Shams-e Tabrizi* tomb tower). Attached to the mausoleum is a large garden area (perhaps also a cemetery). In the centre of the city is a walled palace.

The map published by Anvari has a north-up orientation (Fig. 33). It schematically represents the walled city, which has a rounded shape. It includes a detailed description of the surrounding area. A dozen villages are noted. Several of them bear a name: *Rabāt* (the Armenian quarter), *Zoraq, Qashqablaq, Shabani, Daqbagi, Ravian*, and *Pajukan*. The water system (*Qotur* River and *Qale Spring*), roads, the bridge over the *Qotur*, and the entrance to the *Qotur* valley, which has a caravanserai, as well as the mountains that surround the plain, are indicated. Some villages appear to be surrounded by gardens. Other crops are grown just outside the walls to the west and southwest. Near a grove, an *Emamzade* (a saint's mausoleum—it is likely to be *Shams.e Tabrizi's* tomb) is indicated. Another mausoleum, *Emamzade Sheikh*, which is also near a grove, is in the lower part of the map. The representation of the crops suggests that they could be groves and rice fields, as mentioned by some travellers. Away from Khoy, at the southwestern edge of the map are two other important landmarks. Along the river near *Qotur* village is a caravanserai. Beyond a hill further down is an indication of a large complex labelled "barracks". Curiously, those buildings are never cited by travellers, although some of them have seen other barracks, which, instead, are in the immediate vicinity of the city.

Fig. 32 Khoy city in an image from *Chronicle of Stages of Campaign of Iraq and Persia* (NASUHU'S SILAHI (MATRAKCI) 1976—Wikipedia Commons). The map is oriented with the south at the top and the Qotour river with the bridge and the mountain pass that separates the Khoy plain with that of Salmas are recognizable

Fig. 33 *Map of the city and castle of Khoy* in an anonymous map from the Qajar era (Anvari 2015—Archive of the documentation and diplomacy history center of the ministry of foreign affairs of the Islamic Republic of Iran)

More reliable and even richer sources of information are the two Russian maps, which constitute an accurate topographical survey. The first represents the town and approximately twelve kilometres east–west and nine kilometres north–south of the surrounding area. The original scale is approximately 1: 27,000. The second map, which represents the walled city and its immediate surroundings, is at a scale of approximately 1: 6000 (Figs. 34 and 35).

The map of the territory shows that gardens surround the double circle of city walls and a large village to the west. There is also a cemetery. This is certainly what would remain of the area inhabited by the Armenian population that did not migrate to the Russian territories. A fair number of other external settlements are reached by rural routes. The map indicates the names of almost all the settlements. There are often contiguous small green areas. It also shows the connecting routes that go up the *Qotur* valley towards Bitlis, the northern road to Maku, the northwestern road to Marand, and the western road to *Thessuj*. However, these are simple trails. A perfectly straight tree-lined road starts from the south gate of the city. It runs just more than three kilometres to the *Qotur* River, interrupting on its banks. A small graphic sign suggests the bridge mentioned by many travellers, even though they said it was not in

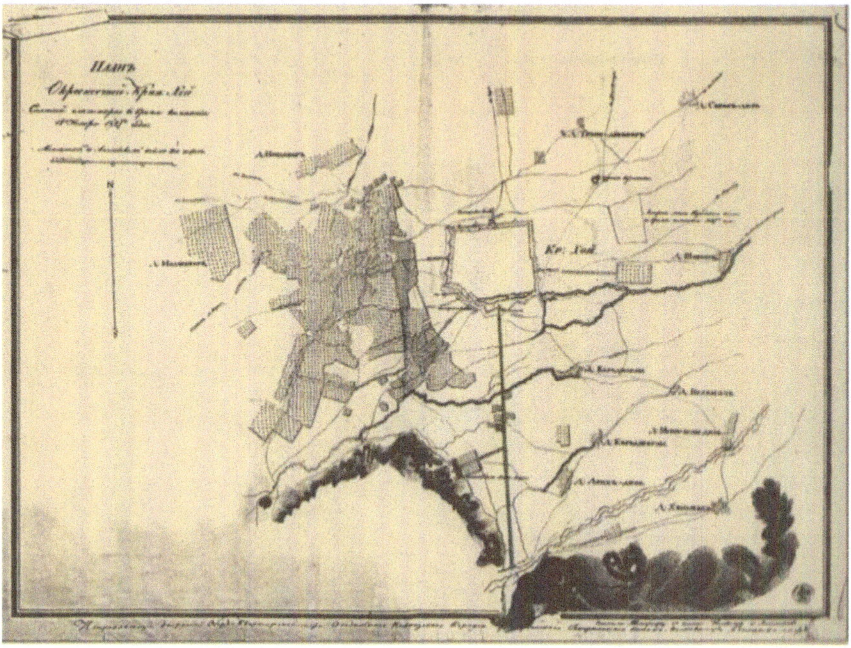

Fig. 34 Kolokov and Yai Shenikov, *Suburb of the Khoy City*, Fort and Gardens, 1827 (Archive of the Documentation and Diplomacy History Center of the Ministry of Foreign Affairs of the Islamic Republic of Iran)

use as confirmed in Dupré's 1819 account (p. 245), which indicates that the river was forded. Beyond the river, the route loses its regularity and continues in the direction of Salmas. Halfway through the avenue, several small gardens, which are mentioned by Texier, overlook the street. A little further on is a tree-lined four-hundred-metre deviation to the west that leads to a garden. This is the *Bagh Daghi*, which has been mentioned by many sources.

The second Russian map illustrates Khoy and provides some interesting details. The two city fortifications are represented. The internal fortification seems to be completely preserved despite reports of its deterioration from lack of maintenance or destruction during the frequent wars in which the town was involved. It is an eleven-foot-thick earthen wall (Wilson SG 1895, p.83) reinforced by crenellated towers with a triangular or circular plan, approximately 30 m tall as almost all travellers describe. Outside of those the "European" ramparts can be seen, which almost all regard as inadequate military architecture. Between the two walls to the north, east and west sides, there is a twenty-metre-wide path in relief. To the south, the area widens. You can find large buildings that should be the military barracks and accessory buildings for armour and services. Along the exterior of the rampart there is a moat about one hundred metre wide, lacking water (evidently too scarce to be used for defense) (Fig. 36).

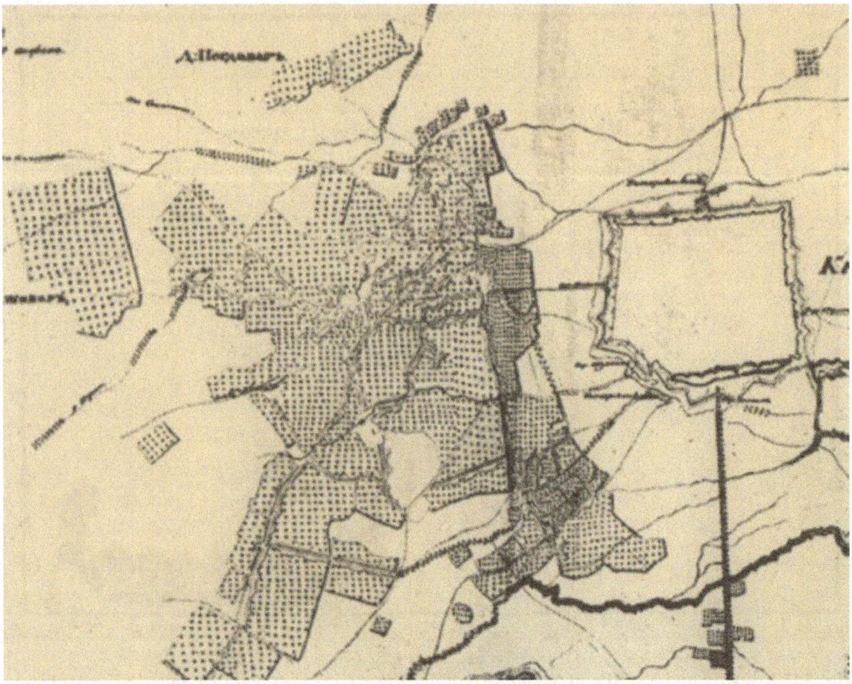

Fig. 35 Kolokov and Yai Shenikov, *Suburb of the Khoy City, Fort and Gardens*, 1827. Detail of the Armenian quarter to the west of the walls. (Archive of the Documentation and Diplomacy History Center of the Ministry of Foreign Affairs of the Islamic Republic of Iran)

The inner city is described in two detailed levels. The streets are clearly indicated: several tree-lined streets intersect perpendicularly, as reported by many visitors. The city blocks are recognizable; however, the distribution of the regular houses is not indicated. The water of Qotur flowing from southwest through the city is also evident. Yet, where it exits is less clear. The Matrakçi map shows it at the extreme north of the west flank. However, it does not appear on the Russian map, thereby suggesting the possibility of an exit to the north at one of the city gates (Fig. 37).

Two buildings are represented in their plan, albeit in brief. The bazaar is shown along with the two annexed caravanserais and the governor's palace. The bazaar is accessible directly from the older walls through the monumental gate that survives. It develops in a long vaulted corridor and ends in a large open space where the courtyards of the caravanserais are recognizable. A short annex extends westwards. It is connected to the crossroads that run east to west through the city.

The governor's palace is enclosed by a wall with turrets. The south entrance is accessed through a garden area. Two large courtyards articulate the buildings. To the north, there is another more private garden. The palace also has a north door, which is manned by a guardhouse and leads directly to the ancient walls. The map corresponds to Texier's description of the building. There is a relief and a view of

Fig. 36 Kolokov and Yai Shenikov, *Fort of Khoy*, 1827 (Archive of the Documentation and Diplomacy History Center of the Ministry of Foreign Affairs of the Islamic Republic of Iran)

an internal courtyard created by Pascal Coste in 1840 during Eugène Flandin's trip (see: "Eastern elevation of *Saray i sardi* by Pascal Coste", Marseille, Bibliothéque Municipale, ms 1132 fol. 33). There is no full correspondence between the two reliefs. The level of detail is not the same. The Russian map is at urban scale, while the French map is at architectural scale. However, it must be said that the French map was created thirteen years later, after the wars and a major earthquake.

The second Russian map has one last detail. It is a schematic of a stretch of the ramparts and moat. It includes a detail of one of the internal towers. Curiously, it does not highlight mosques, churches or other public buildings. This indicates the possible military focus of the map. With very few exceptions, the aforementioned visitors also do not report of churches or mosques. This suggests that these buildings, which must have been there, held no social, religious, or architectural interest for the travellers (Figs. 38, 39, and 40).

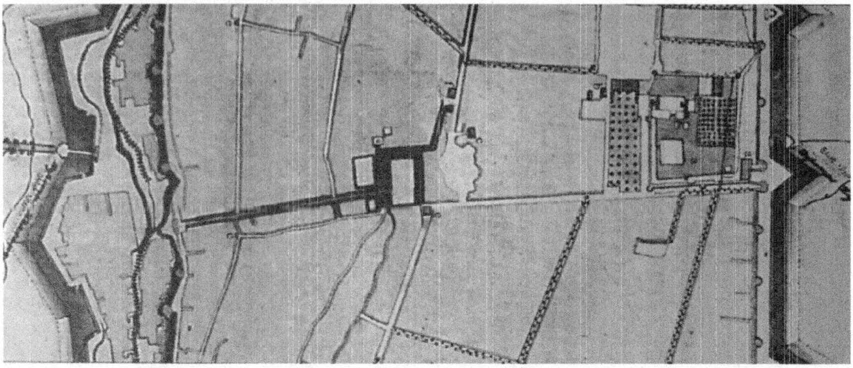

Fig. 37 Kolokov and Yai Shenikov, *Fort of Khoy, Castle and Bazar*, 1827—Detail (Archive of the Documentation and Diplomacy History Center of the Ministry of Foreign Affairs of the Islamic Republic of Iran). The map is oriented to the west: observe the historical arm of the Bazaar and the open door in the old walls. Beyond these are the military barracks that some visitors talk about

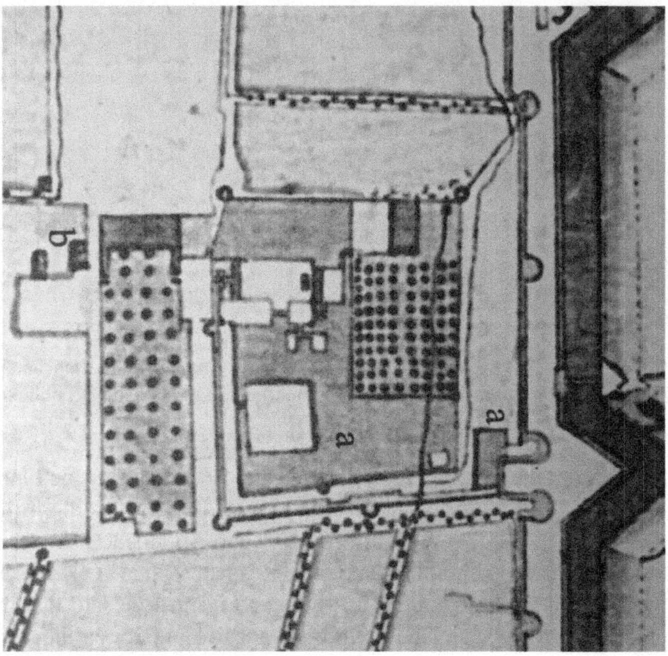

Fig. 38 Kolokov and Yai Shenikov, *Khoy Castle*, detail of 1827 Map with the Castle and its gardens—Detail (Archive of the Documentation and Diplomacy History Center of the Ministry of Foreign Affairs of the Islamic Republic of Iran)

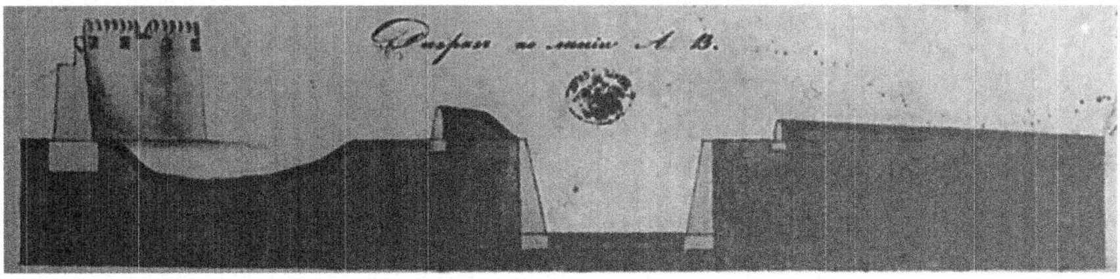

Fig. 39 Kolokov and Yai Shenikov, Cross section of the Khoy wall and moat, detail of the 1827 map (Archive of the Documentation and Diplomacy History Center of the Ministry of Foreign Affairs of the Islamic Republic of Iran). Note the detail of the old earth walls and their poor state of conservation

The years following the mid-nineteenth century represent a period of decadence and internal struggle for Persia. It ends, at least in part, with the ascent of the Pahlavi dynasty (1925) following a coup of Reza Khan, who overthrew the last Qajar Shah in 1921. During 1870 and 1871, a famine resulted in two million deaths from starvation. The ensuing rampant corruption and mismanagement of state resources led to the 1906 short-lived Constitutional Revolution, whose epicentre was Tabriz, by far the most modern town in Persia at that time. The 1907 Anglo–Russian agreement, which stipulates the two powers' respective zones of influence over Persia, likely represents the lowest point in Persian state power.

It is clear that in a period of such upheaval, a renewed interest in the situation in Iranian Azerbaijan helps attract new visitors, mainly diplomats, soldiers, and especially missionaries, towards the end of the century. Even though the travel reports of this period do not provide new information on the layout of Khoy town and his territory, yet they attest to the social and economic climate.

Weeks is an anomaly. An American painter and orientalist, he has left many genre scenes of Morocco, Turkey and Persia. He provides little information about his independent travels through Persia in 1870. However, two pieces of information should be highlighted. During his visit to the town, there is a procession of Armenians, a sign of a certain degree of freedom and tolerance that they enjoy at this time. The sanitary conditions in the town, however, seem rather precarious (Weeks 1896, p. 37).

A few years later, military, geographer and archaeologist Sir Charles Wilson writes a travel guide. Perhaps the first of its kind about Persia, it provides essential information, such as the main roads, travel times, and main cities. It is interesting to note the presence of many rice fields (which are documented in the Russian map) and indications of the main road connecting Khoy to Julfa (what remains of it, on the Araxe) Van, and Urmia. There are two roads to Tabriz via Marand and *Tasuj (Thessuj)*. The bazaar is regarded as one of the best in Persia (Wilson CW 1895, p. 226).

About ten years later, Edward Browne, an English Iranologist, left a small report of his three days in Khoy. He mentions the "beautiful little city of *Khúy*" with "pleasant gardens of poplars, willows, and fruit trees, and fields planted with cotton". He talks

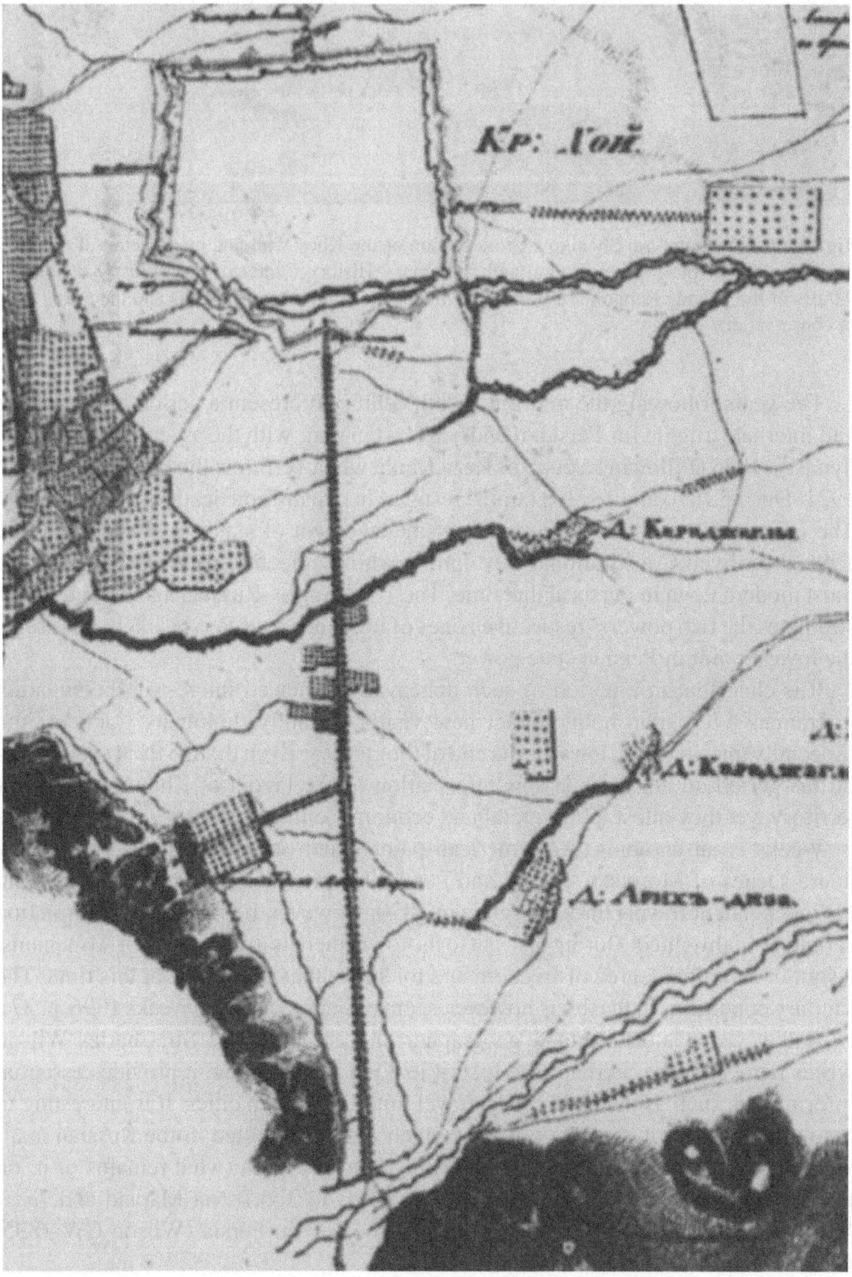

Fig. 40 Kolokov and Yai Shenikov, The road with the row of trees leading to the Dagh Baghi garden and the Qotur river—Detail (Archive of the Documentation and Diplomacy History Center of the Ministry of Foreign Affairs of the Islamic Republic of Iran). The pass that connects the Khoy plain with that of Salmas is clearly recognizable

with local personalities, who are interesting guides to the town but whom he considers very superstitious (Browne 1893, p. 50).

During the same years, a few missionaries visit the town. The first is Paul Muller-Simonis, who travels through Western Azerbaijan between 1888 and 1890. An Alsatian Catholic priest who studied medicine and engineering, he is very involved in social work at home. The first distinguishing feature of Khoy is that, unlike Turkish cities, the view from afar does not include minarets or buildings of a certain height. There is a row of treetops behind a long wall. From a distance, the raw earth fortifications seem impressive; however, when approaching, they appear insignificant. They are falling into disrepair because of a lack of maintenance. The caravanserai has a "primitive" appearance. The windows are missing; only the frames remain. Nonetheless, the town is considered one of the most beautiful in Persia because of its wide and regular laid-out tree-lined streets that are flanked by canals.

When Persia and Turkey were powerful states, the town was one of the most important commercial hubs, with roads connecting it to Erzurum (via Bayazid), Van (through the *Qotur* pass), and Erevan (via Nakhitchévan). According to Muller-Simonis, Khoy is clearly in decay despite being renowned for the manufacture of leather objects. It is estimated to have twenty to thirty thousand inhabitants who claim to be of Tatar origin. They are Muslims with a reputation of being fanatics. There is no mention of the Armenian population, not even in the surrounding suburbs or villages. This is significant, as Muller-Simonis is a Catholic priest.

A wide tree-lined road that runs south of Khoy can be travelled for three-quarters of an hour; however, it is not maintained. Even the bridges across the canals, of which Southgate has spoken, as well as the bridge over *Qotur*, which is no longer in use, are on the brink of collapse. The preference is to ford the river (Muller-Simonis 1892, pp. 117–118). Even Develay, whom the French government sent to Persia to conduct geographic and ethnographic studies in 1890, does not provide positive feedback about the town: "une méchante ébauche en plâtre qui ne connaîtra jamais ni le bronze ni le marbre" (Develay 1898, p. 54).

The second religious person is Reverend Samuel Graham Wilson, an American Presbyterian missionary who provides a general description. He refers to the size of the city's earthen walls (eleven feet thick) and the depth of the moat (twenty feet). The city has twenty thousand inhabitants, who are all Muslims. It has five heavy iron-clad gates. The bazaar appears to be in poor shape, a sign of its reduced status in international trade. Nevertheless, it is surrounded by several caravanserais. There are many noteworthy mosques and a very popular sanctuary, the tomb of a local saint who died about forty years earlier and was considered to have performed miracles. The sixty-foot-tall *Shams-e Tabrizi* Tower is circular. Outside the city, the Armenians live in three seemingly prosperous villages. They are shopkeepers, farmers, and carpenters; however, they are not noted for their culture and spirituality. In one of the villages is *Surp Sarkis*, a six-hundred-year-old Armenian church. The interior is illuminated by small high windows. In a corner, there is a small stone carved to contain water for infant baptism. The paintings on the walls include representations of Constantine and Helen with Greek inscriptions (Figs. 41 and 42).

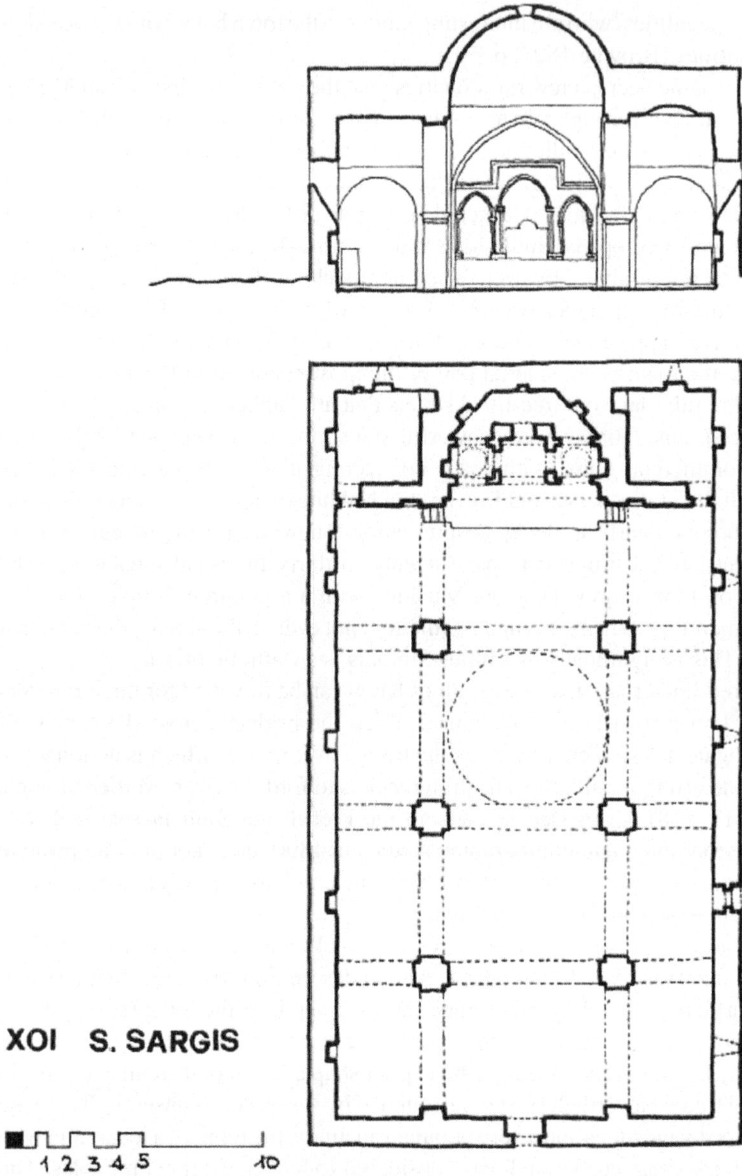

Fig. 41 The plan of the church of S. Sarkis in Khoy published by Alpago Novello et al. (1977)

Fig. 42 Wiew of the interior of the S. Sargis Church before the restorations (Alpago et al. 1977)

Another Armenian church is located in a nearby Muslim village. Sick people are brought there, and it is the site of an annual pilgrimage that involves other Armenian villages in the area. Even Muslims respect it for fear of offending the saint to whom the church is dedicated. Wilson does not mention the name of the church or village. However, it is probably the *Surp Khatch* of *Mahlazan,* a church known to have been the site of an important pilgrimage (Wilson SG, pp.83-85, (Fig. 43).

Samuel Wilson addresses the religious issues in the area in his second volume, "Persia: Western Mission". He focuses on Urmia but also provides information about Khoy and Salmas. It is a descriptive account that highlights the religious life of the Christian communities. There is no shortage of anecdotes about local religious practices. The Armenians are described as superstitious. They sacrifice animals on important religious holidays, and, above all, the clergy strongly oppose the presence of foreign missionaries. He presents their perspective: "By going to the Protestants

Fig. 43 The church of S. Sargis in Malhazam in a photo from the Qajar period (Ali Khan Vali end XIX c.). You can see the front porch which has now disappeared

you will become contaminated in two months and become lepers … They are wolves in sheep's clothing. They will steal your hearts" (Wilson 1896, p. 19).

The meeting place for the communities is the stable, which is heated by cattle in winter. It holds up to one hundred people without seats. The local population recognizes the quality of the services (schools, health care, religious education) provided by the Protestant missionaries. However, the competition that arises between the local church, which is archaic and superstitious, and the Protestant missionaries' teachings complicate the work of evangelization despite the Shah's protection of the missionaries. He devotes a section of the report to the cholera epidemic in Tabriz. According to the missionaries, the disease spreads rapidly in the region, in particular Khoy, because of poor sanitation and lack of knowledge and acceptance of good personal hygiene practices. According to Wilson, the epidemic is less severe in Urmia because hygiene suggestions are more favourably received.

The last author to visit the town in the early twentieth century is Eugène Aubin, a French diplomat. He writes several books on Oriental and, in particular, Persian themes. Visiting Khoy between 1906 and 1907, Aubin reports that the town seems to have suffered a population loss. Only six thousand of the estimated fourteen thousand residents live within the walls. This number is sufficient to defend Khoy from Kurdish attacks but not enough to maintain a modern army ("un système de defense moins efficace que pittoresque"). There are three outer districts: *Chahanagh, Imamzadé* and *Robat*. Each has its own cemetery. One of the residents is the great *Hâdjit Haïdar* Khân, owner of villages, public baths, shops, and four caravanserais. At the time, he is the owner of the *Dagh Baghi* along the road to Salmas. There are superb rows of plane trees and waterfalls that descend over rock steps. The palace is the most beautiful in the city (Aubin 1908, p. 53).

Aubin states that the walled city is inhabited by Muslims only. It is a very devout community ("un peuple de mollahs et de *seyyeds*") with as many as forty mosques. He mentions the miraculous tomb that Wilson described. Aubin refers to the site as "le plus lucratif qui soit à Khoï". He adds that the Shams-e Tabriz minaret is surrounded by trees, and the balcony and lantern appear deteriorated a lot. Finally, he describes the Armenian community: 110 families living in the suburbs. In one of these communities, there is a church, which he defines as Gregorian. It is certainly the *Surp Sarkis*. He describes it as ancient but does not name it. An interesting detail is that the lower half is stone, and the interior is shady and covered by vaults. It is a very large building because at the time of Ramerlano *[A/N: probably a mistake: Tamerlane, in Persian Timur]*, the entire Armenian community, which was larger, took refuge there. Following the Armenian massacres, a Lutheran mission was founded in Khoy. It is headquartered in Urmia (Aubin 1908, pp. 54–58).

The last traveller to report news of some significance is Colonel C. E. Stewart, who was engaged in "intelligence" operations in Persia in 1911. Before entering the city, he stops at a caravanserai located in a "very flourishing village". He has the opportunity to be hosted in the house of a middle-class family. The house is small but nice, with an orchard. The front overlooks the garden. On the ground floor are four rooms, two for men and two, separated, for women. Two are on each side of a hallway. Upstairs is a single large room enclosed by a large grill [ventilated grill]. As a sign of hospitality, he is accompanied to the public baths, which he discovers are open to Armenians. This is in keeping with Turkish customs but contrary to Persian practices. He does not provide significant observations about his visit to the governor's palace other than that the soldiers are equipped with poorly maintained antiquated weapons (Stewart 1911, p. 191).

The travel reports collected for this essay end with these last accounts. Shortly after, with the outbreak of the World War I, Khoy is attacked by Russian and Turkish forces and involved in the massacre of the Christian population. After the war, another story will begin with the Reza Khan coup.

2 Salmas in the Descriptions of Western Travellers

A European traveller's interest in the Salmas region is relatively recent. Here, I use the Persian name of the current Iranian town (the name in Armenian is *Salmast* and for some time, in the twentieth century, the town also had the name of *Shapur*).

Among the reports published in print by European travellers, if we exclude the text by Johannes de Laet, published in 1633, the first news about the town dates back to the early nineteenth century. However, the area was inhabited since ancient times, as evidenced by the numerous ruins of the Urartian period (Kleiss 1992) and Sassanid (bas-relief of *Shapur* and *Ardashir*). Local tradition traces the Christian presence in the area to even the apostle St. Bartholomew, who allegedly was martyrized in Salmas.

The town is remembered by emperor Constantine Porphyrogenitus in the first half of the tenth century ("near Chart"—ancient name of Khoy) and by geographers *Al-Istakhri* (first half of the tenth century) and *Ibn Hawkal* (second half of the tenth century): "A small Azerbaijani town, with strong walls, in a fertile region". *Al-Maqdisi* (tenth century) reports that it was the Kurds who built the city walls (Bosworth et al. 1995, pp. 998–999) (Fig. 44).

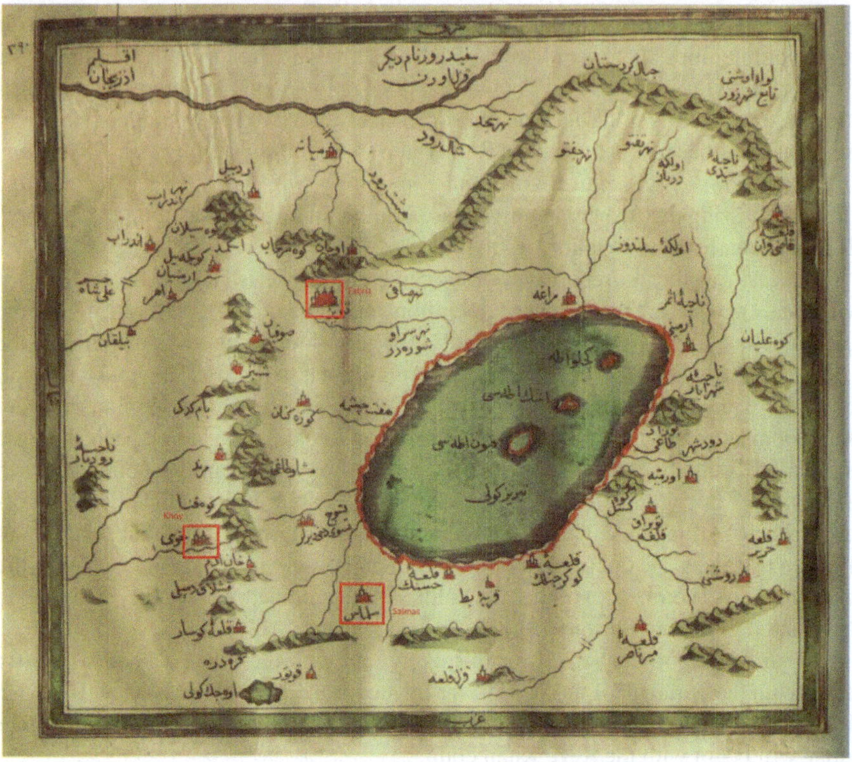

Fig. 44 Mostafa ben Abdollah Kateb Chalbi (Haji Khalife), *Map of Azarbaijan*. Original document from the eleventh century (Anvari 2015—Wikimedia Commons). The map is oriented with the east at the top: in the center, Lake Urmia, at the bottom, between the two mountain ranges, Salmas is indicated

Salams was therefore a well-known town, not only in the Armenian chronicles, but also in the Arab-Persian cartography made between the tenth and fourteenth centuries (Galichian 2007). In many maps dedicated to the sub-Caucasian area and the Persian territory south of the Araxe River, Salmas is not only present, but plays an equal role compared to other important cities such as Ardebil, Khoy and Bitlis.

> On the high road north of Urmiyah, and at some distance back from the north-western corner of the lake, is Salmas. *Mukaddasi* describes this as a fine town with good markets and a Friday Mosque built of stone; the population of the place in the 4th century [l0th century of Gregorian Calendar] was of Kurd origin. *Yakut* says that in the 7th [13th Gregorian] century Salmas lay for the most part in ruin; but the *Wazlr 'Ali Shah, Mustawfi* writes, rebuilt its walls 8000 paces in circuit during the following century, in the reign of *Ghazan Khan*, the Mongol, and the town had then regained its former importance. Its climate was cold, and a river which rose in the mountains to the west passed through it to the lake". (Le Strange 1905, p. 166)

Guy Le Strange summarizes as above, in his study on medio-oriental cartography, the descriptions handled down to us by Arabian and Persian geographers.

Fig. 45 Detail of the military map of the USA that highlights the Salmas plain with the mountains to the west and the swamp to the east on the edge of Lake Urmia (USA Army map of Iranian Azerbaijan, 1943, USA Library of Congress)

A map of 1691, preserved in Bologna (Uluhogian 2000), which describes the presence of the Armenian Church throughout the territory of Greater Armenia, indicates Salmas as a destination for pilgrimages from the whole Region, reporting the following note: "Salmast: the tomb of the Virgin *Varvare* (Barbara), come on pilgrimage from the regions of *Č'ors, Xoy* (Khoy) and Urmia. Here, the new martyr *Awag* was also martyred" (Figs. 45 and 46).

Historical information on the presence of a Christian bishop in the town dates back to 1281, and the construction of a church in the village of *Chara*, near Salmas, is documented in 1360. From the end of the thirteenth century Nestorian and Armenian bishops were present, while a first Catholic bishop (Chaldean) was mentioned in 1580; in the same year, there are news of a Christian monastery, *Mar Yohannan*, "near Salmas" (these news and the following are taken from Wilmshrust 2000). The passage of almost the entire local Christian community to Catholicism is traced back to the last quarter of the eighteenth century: a Catholic bishop has lived in *Khosrova* since 1777, as documented by his tomb, still visible in the village cemetery.

The Salmas region, which extends immediately east of the mountains that form the geographical border between Turkey and Persia, has always suffered from the centuries-old dispute among different powers: Armenian until the early eighth century, when it was conquered by the Arabs, dominated by the Mongols in 1203, by the Turkmen from 1387 (the *Kara Khoyunlu* first and the *Ak-Khoyunlu* of *Uzum Hasan* then) entered the orbit of the Persian empire at the beginning of the sixteenth century thanks to the Shah Ismail I, founder of the Safavid dynasty. It was not a

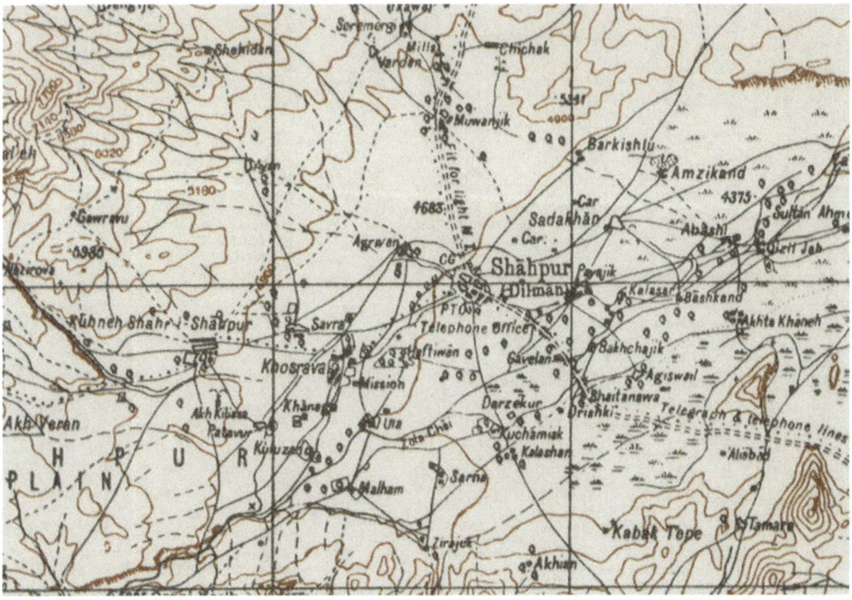

Fig. 46 Enlargement of the previous map with the detail of the Salams area (Shapur in the map) and an indication of the historic villages of the region (USA Army map of Iranian Azerbaijan, 1943, USA Library of Congress)

peaceful period: between 1514 and 1746, there were eight wars between Ottomans and Persians, about one every 30 years, not to mention the frequent raids of the Kurds who populated the mountains on the borders between the two empires.

It is known that, in order to make a scorched land for the Ottomans, Shah Abbas I ordered the deportation of the Armenian population in 1603, starting from Julfa, on the Araxe River, the most important Armenian town at the time. Destroyed the town, poisoned the wells, cut the trees, Salmas also suffered: the town became depopulated and, despite a revival in the second half of the eighteenth century, it lost its role as regional capital, to the advantage of Dilman, a new settlement at the beginning of the nineteenth century, inhabited exclusively by Muslims. About the town and region of Salmas/*Salmast/Dilman/Shapur* (the different names the town had over the years) see the recent monumental volume by Samvel Karapetian, who also researched literature in the Armenian language (Karapetian 2019, pp 483–491).

In 1828, following the Russo-Persian war, the town and the countryside saw the migration of many Christians to the Russian territory: 6000 Armenians, 1800 Nestorians and 600 Catholics according to Père Denis (O'Flynn 2016, p. 720). On the threshold of the First World War (1913), there were 10,460 Chaldeans, with 24 priests and 12 churches, of which 3700 in Khosrowa, with 7 priests and 3 churches (Wilmshrust 2000, p .326).

The two Turkish occupations (1909 and 1918) also brought to Persia the horrors of the genocide against the Armenians, who were able to return to the Salmas region, albeit to a lesser extent, only in the mid-1920s. The earthquake of 1930 and the Second World War caused a significant decrease in the Armenian population, which in the 1970s was reduced to few dozen families, mainly elderly people.

Finally, it must be said that the name of Salmas has been attributed to at least three different cities that have succeeded each other over the time: the ancient one, of which substantial ruins were still present in the nineteenth century; the one rebuilt after the destruction of the wars between the Ottomans and the Persians, which is the town the travellers of nineteenth century describe; and finally modern Salmas, built from scratch, not far from the previous one, after the devastating earthquake of 1930.

A study on historical cartography about Salmas' territory recently conveyed larger attention also on the maps published between seventeenth and nineteenth centuries. It confirmed the importance of Salmas in modern and contemporary times too, along with solving in particular the problem of the location of ancient Salmas (also called *Khone Shar*—Old Town), of Salmas as reported by the nineteenth-century travellers, and finally of Salmas as rebuilt after the 1930 earthquake. The study also precisely locates where Dilman was founded in nineteenth century and destroyed by the same earthquake (Shirazian and Tehrani 2020).

Sometimes, the name Salmas has also been given to the region surrounding the inhabited area, probably due to the large spread of villages dependent on the capital and characterized by being predominantly inhabited, until the end of the nineteenth—early twentieth century by Christians, Armenians and Chaldeans.

Two other settlements in the vicinity were the object of particular attention by travellers: *Khosrova* and *Haftvan* which, headquarters of the main Christian religious institutions of the area, were for a certain time a must for those who wanted to know the Christian communities of the area immediately north-west of Lake Urmia. In some circumstances, as we will see, the visit was also extended to the modern town of Dilman, inhabited by Muslims, which in the nineteenth century had become the administrative capital of the territory (Figs. 47 and 48).

The local importance of Salmas is evident from the way it was called in the region: *Shehir* that is "the city" (Shiel 1838, p. 55).

The first news relating to Salmas in travel reports published by European travellers dates back to 1633, by Johannes de Laet, a Dutch geographer, one of the founders (1620) and first director, until his death in 1649, of the Dutch West India Company. In his volume, *Persia seu Regni Persici Status,* written in Latin under the name of Joannes Latius, among the cities of Azerbaijan, *Choy, Ourmya, Merent* and, precisely, Salmas are mentioned, a sign that the town had a certain importance in the regional context: "Salmas, de terra inquit Armeniae: Salmas urbs est a stagno Kannudhan remota" *[A/T: Salmas begins the land of Armenia: the town is far from Lake Urmia]*. A brief but precise description of the lake is made: "stagnum vero istud salsum est caretque omni animantium et piscium generaliter … Hyberno tempore maxime excitantur in eo procellae, adeo ut cymbae naustagium patiantur" *[A/T: This lake is salty and devoid of any animal and fish in general* - In winter there are great storms, to the point that ships risk sinking, Ed]. (De Laet 1633, p. 68).

Fig. 47 View of Dilman in photography by Binder (1887)

Fig. 48 The Dilman Haftvan Gate in a sketch (Deilamqani 2006)

Almost two centuries pass before Salmas returns to the interest of Europeans: it is again a geographer who describes the town and its region, Antoine-Jean Saint-Martin. Academic of France and one of the first European orientalists, Saint-Martin dedicates a couple of pages of his *Mémoires Historiquese et géographques sur l'Armenie* (1818) to Persarmenia, a historical region whose borders are not well defined but roughly coincide with present-day Azerbaijan Western Iranian. Salmas is mentioned there as the main town in the region west of Lake Urmia: "this city is very old, it already existed in the mid-fourth century", it is located on the banks of a small river that comes from the mountains of Kurdistan and is under post to the Persian domain (Saint Martin 1818, p. 179).

The renewed interest of European travellers is justified, at the beginning of the nineteenth century, by the presence, in the vicinity, of an important bas-relief of Sassanid origin. It is precisely for this reason that Robert Ker Porter would make a detour, in order to study its appearance. The painter visited Persia between 1817 and 1820 and crossed the Salmas plain of which he did not fail to highlight the green aspect and the richness of agriculture. He does not spend a word on the town though heading directly to the small mountains that lie south of the village of Tamar. On a rocky wall overlooking the plain of one of them Ker Porter reaches the bas-relief, which he first describes extensively in a couple of pages of his book, accompanied by a relief. He attributes the work to the fourth century AD, that is to the Sassanid period, judging it severely from the point of view of material execution: "a miserable idea it gives to the then Armenian taste in such works ... all that I have seen being particularly barbarous" (Ker Porter 1822, p. 599). (Figs. 49 and 50).

Along with Joseph Wolff, visits to the region began by religious and missionaries, both Catholic and Protestant, often in competition with each other. Wolff is a German Jew who first was converted to Catholicism and then became an Anglican missionary. The information he reports in his travel journal (Wolff 1829, vol. III, pp. 131–145) concerns for the first time the structure of the Christian communities of the Salmas area, to which he adds, thanks to its origins and unique among European travellers, news on Jewish communities. The basis of his stay is in the village of Khosrowa, located few kilometres from Salmas, where most of the Chaldean Catholic community and a small part of the Armenian ones were concentrated. Wolff reports some statistical data that deserve to be evaluated, since they refer to the situation prior to 1828 (Migration of the Armenians following the Russian conquest of the Erevan region).

A small community of Jews is concentrated in the village of Shahar: there are 50 families with 3 rabbis and two synagogues. In *Sawra, Patawar, Pakajoj, Sarna* and *Melam* reside the largest communities of Armenians, respectively 400, 200, 250, 120 and 120 families and one or two churches per village. In *Khosrova,* there are

Fig. 49 Reliefs by Ker Porter (1822) relating to the Sassanid bas-relief of Salmas

Fig. 50 A photograph by Texier (1842) relating to the Sassanid bas-relief of Salmas

instead most of the Chaldean Catholics: 160 families with a church, 2 priests and 2 bishops. In *Kosrova* also resides, in exile, Alexander *Mirza*, son of the last king of Georgia then under Russian rule, whom the Shah had appointed governor of the village (Fig. 51).

The most interesting news that Wolff relates, however, concerns an argument he had with a Catholic missionary, Don Isaia, about the ease with which Catholics had operated the conversions of Nestorians, through the donation of contributions in money or with the exemption from some taxes. As we will see later, the conflict between Catholics and Protestants appears already evident at the beginning of the nineteenth century, and it will continue until the end (Wolff 1829, pp. 131–133).

Eli Smith and Harrison Gray Otis Dwight visit the region in 1831, when the effects of Armenian migration are already evident. The two American Protestant missionaries resided for some time in Khoy, in the caravanserai, due to the heat and their poor health. However, they do not fail to make an excursion to *Khosrova* and Salmas: the countryside is enchanting, the clayey earth is irrigated by canals and is mainly cultivated with wheat, which is produced here at 8 or 12 or even at 15. Each village is surrounded by a "Forest" of fruit trees: apples, pears, quinces, peaches, apricots, walnuts and sinjid [a kind of olive tree, used for medicinal purposes, Ed.]. Among these, the two missionaries visit the village of *Saoora*, on the road between *Khosrova* and Salmas (probably *Savra*): it is almost deserted since, as they are told, five or six hundred Armenian families have migrated to Georgia and very few of them remained there, together with few Muslims and a solitary family of Chaldeans, the only guardians of one of their churches, characterized by the typical oval-shaped entrance, carved out of a single stone and of dimensions for which it is difficult to pass. Nearby are three small chapels, under a single roof, dark and dirty, devoid

Fig. 51 *Map of the Turco-persian frontier made by Russian and English officiers in 1849–1855.* Detail of the valley of the Qotur river (Library of Congress—Wikipedia Commons)

of sacred texts and furnishings, if not some old stoles and some Catholic images hanging on the walls. It is celebrated only on holy feasts, and the priest comes from *Khosrova* bringing the sacred books with him.

According to Smith and Dwight, Salmas is remembered by the ancients as a town of Persarmenia; today, however, it is a modern village, occupying the site of the ancient town with poor and dilapidated scattered houses. Only three cylindrical towers remain from the past with some inscriptions in Arabic reminiscent, on a reduced scale, that of *Shamkór [A/N: Shamkir, today Ganja in present-day Azerbaijan, an image published by Dubois de Montpereux, Ed.]* (Smith and Dwight 1834, p. 357). The Armenian church appears very poor and devoid of furnishings and ornaments, probably removed during the great migration following 1828, when about 200 families left the village and only 29 remained, to which 40 were added from the neighbouring villages.

The two missionaries report that also in Salmas there is a Chaldean church, similar to that of Saoora, which however contains three or four ancient books, one of which is written on parchment in Estrangelo, the ancient Syriac alphabet. Outside the church, there are some recognizable inscriptions in characters that Eli Smith does not recognize. However, neither Chaldeans nor Nestorians are present in the village and the key to the church is kept by an old Armenian and is open only on the occasion of the great holidays, thanks to the work of priests coming from *Khosrova.* Instead, 30–35 Jewish families still live in Salmas, with a synagogue in better conditions than the Christian churches. Although Jews are the worst-treated minority in Iran, here their condition appears better than elsewhere (Smith and Dwight 1834, p. 358).

A couple of years later, Reverend Justin Perkins, an American Presbyterian missionary, visits *Khosrova* and Salmas; in *Khosrova,* a Catholic monk refuses to welcome Perkins, ostentatiously turning his back on his passage. Salmas has an ancient and venerable appearance, but, at the time of the visit, it has only three thousand inhabitants, mostly Muslims with few Armenians and Jews. Of the three towers visited by Smith, only two remain standing, which appear to be of considerable antiquity, built in bricks with writings in Arabic (Perkins 1843, p. 171).

Justin Shiel, an English military and diplomat, crossed the territory of Salmas in 1836 on his journey to Kurdistan. The starting point is Tabriz and the final destination is Van; the Salmas route is chosen as alternative to the *Júlámerik* one because, although longer, yet it appears safer. The region is inhabited by a large number of Armenians and Chaldeans Catholics, who occupy the villages. However, there is no good blood between the local Christians and the Nestorians, who inhabit the mountains of Kurdistan.

Before arriving at the old Salmas, Shiel makes a stop in Dilman, a new town few kilometres away founded by *Amir Khan Khájar* during the reign of *Fath-Ali Shah Khájar* (1797–1834) which had soon become the main administrative canter of the region. Sheil also reports that it should have 15,000 inhabitants *[A/N: We can imagine this data also include the entire countryside since, a few years later, another traveller, as we will see, gives an estimate of only 2000 inhabitants]* and describes the town as if surrounded by gardens, with clean streets; sends caravans to Van, *Júlámerik,* Tbilisi and Erzurum, yet the bazaar is poor and poorly supplied. Shiel also reports that 5–6 km west it is possible to recognize the ruins of the old Armenian Salmas, which appear to be quite vast and ancient. In the vicinity the two minarets mentioned by Perkins can be seen, which Shiel reports as attributed to a mosque built by the Osmanli at the time of their dominion over this territory. Then, at sunset, he is unable to visit the bas-relief mentioned by Ker Porter and is forced to resume his journey (Shiel 1838, p. 56).

In the same year of Shiel's visit, another British, this time Scottish, James Ballie Fraser, travelled the Salmas plain, dotted with many villages, rich and productive, especially in the part towards the mountains to the west. The plain is crossed by a river, also called Salmas, which according to the locals would be navigable for several miles from the lake but which at the time of the Scottish passage appears to be a miserable rivulet. Ballie Fraser observes that the Salmas plain, which stretched for thirty miles from the shores of Lake Urmia, must once have been part of the lake itself (Figs. 52, 53 and 54). In human memory, however, the lake has withdrawn, probably due to a minor contribution from the springs located in the mountains of Kurdistan. The emerged lands are of excellent quality, as evidenced by the abundance of crops, to the point that the local population is unable to consume all that the area could produce, both as crops and as pastures (Ballie Fraser 1840, p. 48).

Fig. 52 An image of the landscape of the shores of Lake Urmia. (*Photo* Boriani 2016)

Another British soldier, Richard Wilbraham, joins Salmas on his mission through the recently conquered Russian provinces of Transcaucasia at the expense of Persia.

The main town is Dilman, the governor's residence: in spite of that, Wilbraham camps in a miserable village near the ruins of old Salmas. The countryside is fertile and well cultivated but the gardens near the numerous villages appear to be abandoned, evidently due to the recent migration of the Armenians. Going south through the hills that divide the plain of Salmas from that of Urmia, Wilbraham enjoys a splendid view of the lake while the lands on its edge are marshy. After *Kolingi*, a village halfway between Salmas and Urmia, the countryside becomes splendid again. The traveller is admired by the numerous villages surrounded by gardens, and by the rows of trees lining the banks of the streams that intersect in every direction and, in his opinion, give the area the appearance of a typically English enclosed country (Wilbraham 1839, p. 370).

Even Horatio Southgate, an American Episcopal missionary bishop, visiting the Christian communities of Armenia, Kurdistan, Persia and Mesopotamia in the years between 1836 and 1838, realized how the villages inhabited by Armenians often appear abandoned and their churches deserted. On a stop in *Derik*, Salmas is the first Persian town he meets and the impression he gets from afar is pleasant: "the town itself broke upon us with its verdant gardens, its wide-spread plain, and, in the distance, the North-west corner of the lake of *Ourmiah*. The scene was entirely Persian, and we felt at a once that we had entered a new country" (Southgate 1840, p. 287).

Fig. 53 *Nouvelle carte_generale des Provinces_asiatiques_de_L'empire_Ottoman*, (Kiepert 1883), Detail of Urmia Lake region. You can see the marshy areas to the north and south of the lake, the mountains that delimit the area and, to the north-west, the city of Dilman (USA Library of Congress. Wikimedia Commons)

The impact with the local population is different: while in Turkey the foreigner is practically ignored by the locals and assisted by state officials for accommodation and other needs, in Salmas a crowd of onlookers flock to see and ask questions, without however move a finger to help, an evident sign of a poor habit of meeting foreign visitors. It is useless to ask for adequate accommodation: Southgate is therefore forced to make use of the local caravanserai, a clay building a dozen feet high, with the usual niches open onto the internal courtyard and devoid of any sort of furniture; he spends Sunday there, huddled in a corner, in the shade.

Fig. 54 The Tomb of Emir Arghun Agha's Daughter, Salmas, late 19th, Qajar Era (ANTOIN SEVRUGUIN—(Getty Institute—Wikimedia Commons)

In fact the local authorities did not reside in Salmas but, as we have already seen, in Dilman, where the local governor and most of the Muslim population resided while the ancient Armenian town, due to migration, was reduced to a small village along with some ruins that testify a past of greater importance. According to Southgate Dilman, on the other hand, had a population of about 2000 inhabitants, small and modest bazaar, some mosques and schools. According to the locals, the town ruled over about sixty predominantly Christian villages, both Armenian, Chaldean and Nestorian. The seat of Bishop Caldean was as we know in *Khosrova* (Southgate 1840, p.291).

It was only in 1839–40 that the visit of a Catholic missionary, Eugène Boré, French orientalist and linguist, dates back. Apart from the exaltation of the countryside ("est la terre promise de la Perse"), rich in wheat, rice, vineyards, tobacco, cotton and fruit of all kinds, the information he reports is not particularly significant, with the exception of a detailed description of how Catholicism was imported into *Khosrova* by a young man from Diarbakir (Turkey), who secretly converted a large part of the population and, subsequently, the local Nestorian bishop himself (Boré 1840, pp. 255–257).

We have to wait until 1839 to find a traveller visiting the plain of Salmas who is not a religious or a soldier. This is Charles Texier, architect and archaeologist and discoverer of the Hittite ruins of *Hattusas*, near Ankara in Turkey.

The news reported by Texier is the richest in information relating to the urban and territorial structure of the town and villages, seen through the eye of the architect. Texier is staying in a private house of Dilman, in order to avoid the annoying curiosity of the locals. The main reason for the visit, however, is given by the presence, nearby, of the Sassanid bas-relief described first by Ker Porter. Curiously, no one in Dilman is informed of its existence, which is the reason why he moves to Salmas, thinking he can find more knowledgeable people there. This allows us to have a beautiful description of the town now in decline:

> Les villes de Perse, qui de loin présentent un aspect assez agréable, offrent toujours de près le plus triste tableau. *Selmas* se présentait à nous ombragée par un rideau de hauts peupliers, ayant à droite un vaste cimetière, index d'une population nombreuse, quelques minarets, et en avant de la ville une grosse tour de briques ornée d'émaux de différentes couleurs, qui me paraissait un monument remarquable. (Texier 1842, p. 32).

The noteworthy monument, a large cylindrical tower is, according to Texier, a mausoleum, certainly prior to the time of Shah Abbas, since it is made of brick of excellent quality, a quality that Texier believed had instead failed in the time of the Abbassids. The decorations in glazed bricks are characteristic of Persia, since the marble is almost unknown there and the stone is quite rare. The building, completely isolated in the countryside, is somewhat degraded, rather by the hand of men than by time; the tower originally had two floors and a small entrance on the ground which gives access to an ancient crypt.

According to Texier, Dilman is modern, while the ruins that can be seen nearby concern the ancient Armenian town of Salmas, which dates back to no more than the third century of the Christian era, destroyed during the wars of the Mongols and Timour. The region is inhabited by a large number of Christians, and, according to Texier, the largest number of Catholics in Persia live in Salmas *[here Texier is probably wrong, not knowing Urmia].* There is no good blood among the various Christian religious confessions and the activity of Eugène Boré, who had founded Catholic schools and tried to bring order to the local clergy, has been the object of fierce criticism from other Christian denominations, criticisms that, as Texier says, seemed to be inspired more by Protestant missionaries than by local inhabitants.

Here, Christians seem to live on good terms with Muslims and this is probably due to the uniformity of customs between the two communities. Here, contrary to

Turkey, Armenians are not obliged to dress in a way that makes them recognizable, since they do not live in Persia as a conquered people but following a treaty made by their ancestors, the clauses of which are still in force. (Texier 1842, p. 34).

In Salmas, Texier obtained information on the bas-relief that had moved his initial interest in the region: he reached it through a painful march across a plain flooded by the overflowing of the many streams that run through it. The rock in which the relief is engraved is composed of a bluish limestone, a little schist and therefore difficult to work, since it is easily peeled and therefore not very suitable for being carved: for this reason, the work has suffered a lot and its execution was pretty gross. A detailed description of the scene follows, for which reference is made to the original text (Texier 1842, pp. 33–34).

A year later, it was the turn of another Englishman, William Ainsworth, a surgeon but active in the Middle East as a cartographer, with missions in Mesopotamia, Chaldea, Kurdistan and, as far as we are concerned, the area around Lake Urmia. Ainsworth is the author of several reports published in the *Journal of the Royal Geographical Society* (Fig. 55). In one of those, he describes the plain of Salmas and publishes a very detailed map of the territory between the mountains of Kurdistan and Lake Urmia. The landscape is extraordinary:

> There are moments which never slip from a traveller's memory, when, after a long journey on a heated or monotonous plain, a range of mountains with their anticipations of cool waters and refreshing breezes come into sight, or when, fatigued with mountain-toil and travel, a

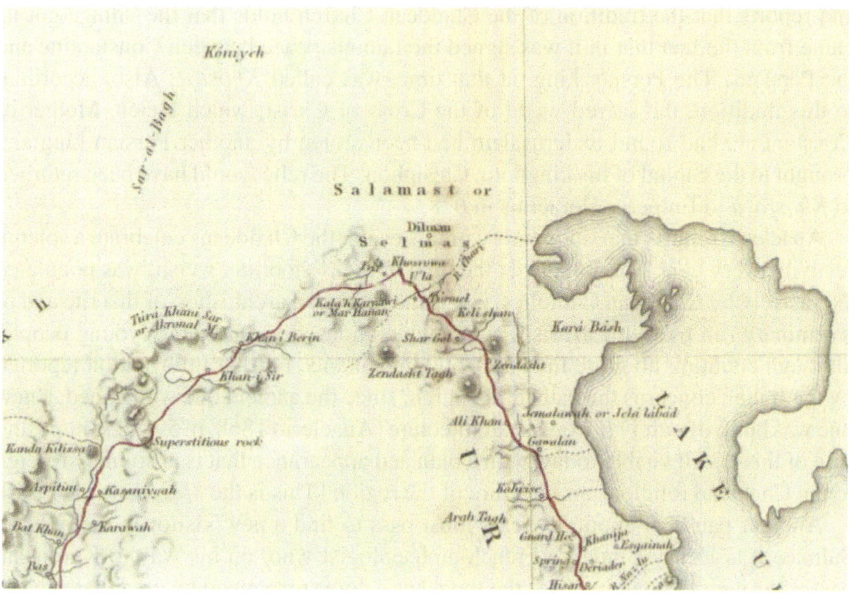

Fig. 55 Detail of the map with the itinerary followed by Ainsworth in 1841. The Salmas region is indicated with the cities of Dilman and Khosrowa (Ainsworth 1841)

plain, smiling with gardens and villages, and full of promises of delicious repose, presents itself to his delighted vision. (Ainsworth 1841, p. 59)

The numerous villages have a common market in the place where the governor resides, which is surrounded by walls, as if it were a fortress. The town is called Salamast or Dilman but is normally called *Shahr*, meaning *The City*. Bad Turkish is spoken there but the upper classes also speak Persian; here, the Christians look to the Russians as their protectors.

Just south of *Khosrova*, in the village of *U'la*, American missionaries hold a school and in the vicinity of Turmel on the rocky hills that take the name of *Kara Básh* (Black Head) there are two castles overlooking the lake. Beyond these reliefs is situated the Chaldean village of *Gawalan*, north of which is an even larger Christian village, which takes the Chaldean name of J*emalawah* and the Persian name of *Jelálábad* (Ainsworth 1841, p. 60).

For about the next 20 years, it seems that the visit to the Salmas area had lost interest both for missionaries and travellers: in fact the region was now well known, the migration of Armenians beyond the Araxe had been completed and interest in the Chaldean Catholics had moved further south to the region of Urmia. Kurdistan, above all, had attracted the interest of Westerners, for whom it was still a *terra incognita*.

The first Italian to publish a travel report on Persia is Giuseppe Anaclerio, Neapolitan, a Lazarist missionary in Persia between 1862 and 1866, sometimes confused with his brother Vincenzo who, in the same years, served as a military instructor at the Qajar court. Anaclerio spends part of his stay in Persia in Khosrova and reports that the tradition of the Chaldean Church holds that the village got its name from the fact that in it was signed the famous peace between Constantine and the Persians. The Persian king, at that time, was called *Khosroe*. Also, according to this tradition, the sacred wood of the Cross of Christ, which Helen, Mother of Constantine, had found in Jerusalem had been stolen by another Persian king and brought to the capital of his kingdom, Ctesiphon. The relic would have been returned to *Khosrova* to Emperor Heraclius in 628.

Anaclerio reports that, in memory of this event, the Chaldeans celebrate a solemn festivity every year. The village, at the time of the Neapolitan's visit, was populated exclusively by Chaldean Catholics, it was the seat of an archbishop of this rite and of a seminary run by four Lazarist Fathers which housed about twenty young people, although counting no more than 500–600 inhabitants. Further information reported by the Italian concerns the cathedral church: since the ancient one was ruined, a new one was built, of purely European architecture (Anaclerio 1868, p. 59). The imposing ruin of this is still visible today, with a plan and appearance that is absolutely foreign to the Chaldean religious architecture of the region [This is the *Mar Givargis* - Ed.].

Another pause of about 20 years must pass to find a new visitor in the parts of Salmas: it is Henry Binder, a French archaeologist who, on his way to Kurdistan, passes through Dilman in 1885: the town has a core surrounded by crenellated earth walls that collapse little by little under the action of rain and debris fill the moat in front. It is the only town within a radius of 60 km, the bazaar and streets are lively and there are over 20,000 inhabitants: Muslims, Armenians, Chaldeans and Jews.

There is a caravanserai in which the group retreats to escape the curiosity of the crowd: evidently the transit of Westerners through the town is still quite rare. The caravanserai is built in the typical Persian model: a large square building one storey high, with no windows opening onto the external walls and with a single access door. On the inner courtyard, there are some compartments raised from the ground for about one meter, two meters wide by four deep and three high; they have no furniture and all travellers bring everything they need with them. Binder reports that part of the caravanserai is rarely two-storey: in this case, the upper rooms are closed by doors. The Frenchman leaves Dilman heading south to *Guavilen* (*Gawalan*); he crosses the walls through a gate whose wooden doors appeared so worm-eaten as to be unusable (Binder 1887, pp. 82–84).

Towards the end of the century, visitors become more frequent: after Binder in 1885 it is the turn of Arthur John Maclean and William Henry Browne, two religious and orientalists, the first American and the latter English (1886–91); by Paul Muller -Simonis, French priest (1888–90); Samuel Graham Wilson, American Presbyterian missionary (1892): William Jackson, American linguist who studied Zarathustra (1903), and Aubin, a French diplomat and orientalist (1908). Evidently, the matter of the defence of the Christian populations of the region had again become worrying for the European powers, and, consequently, the competition between French Catholics and Anglo-Saxon Protestants had again become hot against Armenian Christians and Chaldean Catholics.

Maclean and Browne, Anglicans, are experts in the Syriac language, essential for being able to communicate with the Chaldeans and learn about their sacred books. They observe how the dialect spoken by the Chaldeans in the plain of Salmas is very different from that of Urmia, which is also not very distant. We can evidently deduce that these were rather segregated communities and, above all, lacking a common guide. The whole region is well irrigated by the streams that descend from the mountains of Kurdistan and which are gradually captured and diverted into artificial canals, to the point that in summer these are dry before even reaching the lake. Nevertheless the shores of the lake are marshy, and this makes travel difficult and causes malaria (Maclean and Browne 1892, p. 78).

Muller-Simonis, a Catholic, describes with some hints the old Salmas, of which he reports information already known to previous travellers, and with different details the village of *Khosrova* and the mission of the Lazarists that was there: the complex, rather vast, is distributed around the village square, near the church and is surrounded by a sad-looking *pisé* wall. Entering the enclosure, the first building you come across is the *Diwan-Khaneh*, a reception room; from this one reaches an internal courtyard around which are the common parts and, isolated, the community chapel. At the end of the courtyard is the main building of the mission, a large two-storey rectangular building, made with raw earth bricks but covered with baked bricks. The rooms are whitewashed and the furnishings are modest, but for those who have tried a Persian caravanserai, everything looks magnificent. Behind this building is a large vegetable garden and a path covered with a climbing vine and hops. Everything looks serene and restful. In the seminary 15 to 20 young people who started their training in Urmia study. Here, those with an ecclesiastical vocation can study theology: however, not all of them reach the end of the path, since there are several abandonments, to the point that, in the more than 40 years since the foundation of the mission, only about twenty priests have completed their studies, at a cost that Muller-Simonis considers difficult to bear, since all the students are housed, fed and clothed at the expense of the mission.

The seminary is flanked by a school for the boys of *Khosrova*, run by the missionaries and with professors recruited from the local population or from seminarians. Finally, the mission is completed by a set of buildings managed by the Sisters of Charity, which houses an orphanage, a kindergarten and a school. Everything is offered for free and the missionaries are usually 4 to 5, while sisters are 7. Some Chaldean priests take care of the parishes in the various villages (Muller-Simonis 1892, pp. 121–122).

Samuel Graham Wilson, a Presbyterian, visited the region in 1892 and wrote about it in two separate volumes. In the first (Wilson SG 1896), he dedicates three pages to the territory of Salmas and adds other unpublished and interesting news to what has already been reported by his predecessors. The plain is surrounded by mountains on three sides and the fourth to the east consists of the shores of Lake Urmia. There are about eighty villages and two small towns, each of about 5000 inhabitants, Dilman and *Kuhna Shahr* (literally "Old Town", i.e., old Salmas). Dilman is the administrative capital and has a courthouse, customs, post office and telegraph office and the main bazaar. All the shops open two days a week, and Sunday is the most important day, which here is called, as in Turkey, "bazaar-day". The surrounding mountains have rocks that contain many fossils of fish, snakes and other animals, and in a rocky hill is the famous Sassanid relief described by Ker Porter and Texier (Fig. 56).

Fig. 56 The rock on which the Salmas bas-relief is locatd (Flandin and Coste 1854)

The inhabitants of the plain are mostly Turks, with some Kurds, and there are however 6000 Armenians, 3000 Chaldeans and Assyrians and 500 Jews. The latter live in *Kuhne Shahr,* while the Christians live in about twenty villages. Armenians are thrifty and wealthy and have lived without particular problems until some time ago, when some disputes with Catholics and other political unrest led the Persian government to sanctions and mistreatment. In any case, the Armenian villages of the Salmas plain are the best built in Persia and both farmers and merchants have a fairly high standard of living. The chief merchant of the region even owns a house with an electric bell at the entrance and one of the sons takes delight in photography, while another takes care of telegraphy in his spare time.

However, the most important information that Wilson reports is another, concerning a church of this territory, ever meticulously described by a w Western traveller, as he did: in *Haftdewan* (*Haftvan*), there is an old church built in stone, with a dome and a large altar; as it is customary, there are no seats and the rear part of the arches is reserved for women (Figs. 57 and 58).

On the walls, there are "frescoes" and paintings representing scenes from the life of Christ and the Apostles. The canvases fall apart because of age. There is a fresco representing a Last Judgement:

> A male Angel, with black hair, holds a pair of balances. The one scale is filled with light, the other with darkness, which is over-weighted, though the devil and a group of imps are pulling with all their might on the side of darkness. On the side of light are groups sitting peaceably in Paradise; on the other is hell's gate, and behind it souls in misery. (Wilson SG 1895, p.90)

Fig. 57 The Assyrian Church of Mar Givargis in Khosrowa (Deilamqani 2006)

Fig. 58 Mohamed Zadeh Khosrowabad, Church of Salmas in Qajar Era (Wikimedia Commons)

This is the fresco recently discovered under a layer of white plaster and restored, along with others that are only partially intact (Bakhshai and Mohamadi 2015). Wilson is the only traveller who describes in such great detail the paintings that accompany an Armenian church in this region, adding another important piece of information: in the church, there is a dark well in which many sick people are lowered, especially mental, in the hope that they will be healed. Muslims themselves believe in this superstition, and Wilson encounters a procession of about thirty Muslim women, with their suffering children who came from Dilman in the hope of being able to cure them (Fig. 59).

Above the old city rises a monastery to which one goes on pilgrimage and to offer sacrifices; here there is also a pit *[a crazy-pit in Wilson's words]* for healing through prayer. The monastery had a wall that was reinforced and raised: the government had ordered to tear down this fortification, and only after a negotiation, it was allowed to keep the wall but in reduced dimensions.

Beyond the Turkish border, in the town of *Albak (Arebanes)*, there is the sanctuary with the supposed tomb of St. Bartholomew: numerous caravans of pilgrims visit the holy place every year, with a great celebration. The cost of the pilgrimage coming from Salmas is equivalent to the wages of 30,000 working days (Wilson SG 1895, p. 91).

Samuel G. Wilson left us a second volume (Wilson 1896), specifically dedicated to the missions of Westerners in Persia. The text shows that in the years preceding his visit to the Salmas region, the Protestant missions were concentrated in the Urmia region, where a large Christian community lived, especially Chaldean and, to a lesser extent, Armenian. The Armenians themselves would have invited Wilson to shift the attention of the missionaries to the Salmas region which, according to them, was in

Fig. 59 Details of the frescoes of the church of Sarkis in Haftwan described by Wilson SG in 1895 recently rediscovered and restored (Bakhshaei and Mohammadi 2015—Photo Boriani 2018)

great need of spiritual assistance and education in the "true Christian experience" *[that is, according to Wilson, that of Protestants].*

Since the environment seemed favourable, Wilson decided to found a new mission in the town of *Haftdewan* (*Haftvan*), inhabited by a good number of Armenians and in a central position in respect to a vast territory that included not only Salmas but also Khoy, *Qotur* and Maku, with a total population of 100,000 inhabitants, of which 10,000 are Armenians.

According to Wilson, 2000 Armenians lived in *Haftvan*; its streets, however winding, were sufficiently wide, lined with running water channels and rows of large willows. The houses had unusually high walls which suggested a certain prosperity of the inhabitants.

The condition apparently favourable to the establishment of the mission, however, proved to be partly wrong: since *Haftvan* was part of the villages directly owned by the Crown, the local governor raised objections, determined not to grant permission to establish the mission without an adequate and large economic contribution, and also some of the inhabitants, consequently, opposed to the arrival of the missionaries. It was necessary to turn to the Khan of Khoy, who was superior in rank to the local governor. But, even in Khoy, the resistance was apparently inflexible, although Wilson claims he was informed, indirectly, that the Khan could have settled the matter with a "contribution" of 100 pounds.

The interesting thing is that Wilson refused to give in to the blackmail and went directly to Tehran, complaining about the opposition of local rulers. Soon a reply arrived, even a telegram, which reported "Government orders house to be given, and Khan to report in person to Tehran". A new governor general of Azerbaijan harshly reprimanded the Khan, who was forced to apologize and even requested a written document stating that he had withdrawn all opposition. A sign that American diplomats evidently enjoyed a strong influence on the Shah and his government and that the Protestant missions in Persia were strongly supported by the British and US governments, even in contrast to the support that France gave to Catholic missions: a small part of the *Great Game* that was taking place in Persia by the various Western powers.

Thanks to the intervention of the central government, all opposition ceased and the mission succeeded in establishing itself in 1884. Obviously, the Armenian bishop forbade his priests to have any relationship with Protestants; however, according to Wilson, relations with the local population were good and preachers and teachers were in demand in several villages.

The missionary's story continues by describing numerous facts concerning the various missions, including some episodes of blood that involved the Protestants and the Armenian Catholics and Schismatics. Samuel G. Wilson describes in his text those and other episodes that clearly highlight the permanent conflict still brewing between the exponents of the various Christian religious confessions, competing with each other for the conquest of the local population (Christian, of course, since conversions between Christians were allowed in Persia, but any proselytizing activity against Muslim, Shiite or Sunni communities was absolutely prohibited (Wilson 1895a, b, Chaps. 7 and 8).

Today, we can keep some doubts, after more than a century and after the tragedy of the Armenian genocide, about the appropriateness of those conflicts, especially since Wilson, concluding the story of the settlement of the *Haftvan* mission, bitterly writes that:

> The work among the Armenians of Azerbaijan has been a difficult and trying one. One great difficulty which indeed manifests itself among all the races of Persia, is the mercenary idea of religion. The idea seems deep-rooted that religion is a matter of worldly policy, and is to be valued in proportion to its temporal benefits. In Islam it is the established custom to render the convert financial aid. (Wilson 1896, p. 167)

Another American from New York visited Dilman a few years later (1903): he was Abraham Valentine Williams Jackson, a pioneer of Iranian studies in the USA, a linguist who studied Sanskrit and Avestan and, above all, Zoroaster. The passage from the lake of Urmia, being Urmia considered the homeland of Zoroaster, obviously could not miss on the occasion of a trip to Persia.

Williams Jackson joins Dilman on the afternoon of Sunday, 22 March 1903; the town is the largest in the region but everything appears shabby: its post station looks like little more than a shelter, as does the "tea house" that is close to it and which later turned out to be an opium den. There is a carcass of a dead horse in the middle of the stream that runs through the city. However, the population appears to be well disposed towards foreigners, and there is a certain festive atmosphere since it is the days of *Noruz* (the Persian New Year) and Sunday is market day, the "bazaar day". The next day, leaving the town, he finds a large Armenian cemetery, where, on top of a hill, stands a large monument dedicated to the *Mamikonians*, an ancient and heroic Armenian family. Around, several tombs are marked by the characteristic ram-shaped sculpture, and there is also a burial marked by a long inscription in Syriac, with hardly legible letters. A little further on is the rocky hill with the Sassanid bas-relief about which we know. Williams Jackson tells us that the name of the hill is *Surat Daghi*, meaning "Mountain of the image" (Jackson 1906, p. 78–79).

The last traveller whose information was left to us, and seems useful to describe, is a Frenchman, Eugène Aubin, pseudonym of a diplomat and orientalist, whose real name is Léon Descos. Aubin arrives at Dilman coming, like everyone, from the north, through Diza, Khoy and the bridge over the Qotur. Overcomes the climb to Mount Kara Tepe (Black Hill) and reaches the verdant and irrigated Salmas plain. Dilman is a Muslim village of 4000 inhabitants, and surrounded by walls. The inhabitants are mostly Shiites, while the Sunnis, quite numerous, are preferably found in the surrounding villages. Near the village of *Kouné Cher [Khone Sheir],* there is an old mausoleum, as there are others in the region of Lake Urmia: these towers are built in brick and surmounted by a dome; from two opposite doors, decorated with Arabic inscriptions and blue enamelled mosaics and majolica, you enter the tomb. The dome of the *Kouné Cher* mausoleum has collapsed and the decorations are crumbling, as it is the Arabic writing, which is still legible: "Built by *Eminé Khatoun*, son of *Oghouz Aga* ... Everyone must die, except God, who is eternal" (Aubin 1908, pp. 59–60).

Kouné Cher is the only village in the region where Jews still live from the time of the captivity in Babylon, later joined by the once thriving community of Tabriz, tolerated by the Mongols, but then persecuted at the time of their conversion to Islam. Altogether are about sixty families who make a living from "usury", the sale of religious books and petty trade. Someone also owns a vineyard and, in general, no one complains, since the village is owned by the heiress princess and the taxes are not high. However, young people see no future in the village and tend to migrate to Russia.

The sixty villages of the Salmas plain are distributed along the two banks of the Zola River and are characteristic for the frequent groves of trees separated from the fields and for the watchtowers erected against Kurdish raids. Altogether 40,000 people live there, Armenians, Muslims and Chaldeans. The Armenians form about a quarter of the population and inhabit 23 villages, of which *Kelassar (Ghalasar)* constitutes their commercial center, as well as that of the Muslims is in Dilman. The real capital of Armenians, however, is *Heftevan (Haftvan)*, a large village with 446 residential houses and the seat of a Gregorian vicariate. The Armenians have redeemed the village, as well as in *Payadjuk* (350 houses) and *Mehlam (Malham*—200 houses): for this reason, Muslims only own land where Christian families live in isolation. Each village has a church and a school, while for medical care, they resort Chaldeans or Armenians who studied in the missions of American Protestants and then in the USA (Aubin 1908, pp. 61–62).

Khosrova is instead the capital of Chaldeans, and the Catholic bishop and the mission of the Lazarist Fathers reside there, and the diocese extends to the entrance to the Urmia valley and has between 5 and 6000 faithfuls. The Chaldeans are found in seven villages, of which *Khosrova* has 400 families and *Patavour* 100. Soura was once Chaldean but now 168 Armenian families, mostly Catholic, live there.

Aubin also informs us about the organizational model of Christian communities, no different from the Persian one: a "mayor" is elected annually, in charge of collecting taxes, and in the same way, a kind of Council of Elders is elected (the "White beards"). In the Lazarist mission, there is a seminary with a college where nine clerics and about fifty children study, learning French and Persian. The Sisters of Charity runs a small hospital and a school attended by 75 girls. A Chaldean doctor takes care of health. Next to the mission, the bishop's house, the church of Mar Givargis complete the religious complex, while the cemetery is located just outside the village. Many inhabitants go to seek fortune in Tehran or in the Caucasus or even in Constantinople (Aubin 1908, pp. 62–63).

During the twentieth century, interest in the town and region of Salmas is also documented by various information available in Armenian sources, which have recently been studied by Karapetian (2019). Some notes translated into English deserve to be transcribed to complete the information reported by Westerners.

A first description refers to a piece of news that contrasts with what was described by all European travellers, who had exalted the fertility of the region's soil and the richness of agricultural production that took place there: according to the writer *Arshak Atayan*, in the early 1900s the plain di Salmas appeared as

> a valley lying in the bosom of mountains which surround it from three sides. The composition of the soil and hills, and the volcanic origin of the mountains clearly show that this valley used to be the bottom of a sea: in fact, once it was the continuation of Lake Urmia. The water

level of the lake dropping year after year, water completely receded from its more shallow parts, making them fit for habitation: [in this way], the districts of Khoy and *Guney*, and certain parts of Urmia and Tabriz Districts [acquired more lands]... The soil in the valley of Salmast is mostly mixed with sand, stones, and in some parts, also lime. The areas on the shores of Lake Urmia represent vast marshlands and salt pans, due to which, the harvest is rather poor here (only 2 to 10 times as much as the amount of sown seeds). Besides, most of the lands remain unsown because of the shortage of irrigation water. As compared to the neighboring districts, only the lands of Khoy are of a quality lower than those in Salmast. (Arshak Atayan, cit. in Karapetian 2019)

If it is true that European accounts agree in the description of the marshy land at the edge of the lake, it is surprising, instead, the state of abandonment of the arable lands, which are described here as not very productive compared to the enthusiastic admiration with which they were illustrated by Westerners. This situation is probably attributable to the consistent migration of Armenian population, the instability of the political situation and the harbingers of persecution of the Christian and, above all, Armenian populations.

A second description, which dates back to 1908, tells us that

On Sunday, the center of the district, Dilman, looked like a fair and it was also the place of contact for Armenians of the villages ... The city of Dilman also resembled an exibition of nations: Persians, Kurds, Assyrians, and Jews hurried to Dilman to do shopping. (Alik 1951—p. 2 in Karapetian 2019, p. 486)

Therefore, the town and the region were active in trade and handicraft production. The news is moreover confirmed in the previous report by *Arshak Atayan* who, with regard to the Armenians who lived in the villages, indicated that they were engaged in "carpentry (one or two carpenters could be found in some villages) and blacksmithing (in some villages), also working as ploughmen, tailors (few), shoemakers (two families) and doing other petty jobs". (Karapetian, p. 67)

However, the economic situation should not have been prosperous, at least according to what another Armenian, *Melik Bahakhanian:*

There are heavy clay walls around the city wich have become dilapidated because of the ravage of time ... In the walls, in four places , are the entrances to the city: these are gates having some run-down buildings on both sides which served as watch-towers and storehouses in ancient times... The appearance of Dilman is that of a big village which lies on ruins. The walls look the same inside and outside and have some hugh gates of colored briks. All the construction are of clay, very low, and mostly one-storey. The streets are very narrow, winding and full of trash and offal of all kinds. (Melik-Bahakhanian, in Karapetian 2019—p. 486)

This story closes the information that can be gathered from travellers who visited the Salmas region during the nineteenth and early twentieth centuries. Unfortunately, history did not stop because with the Turkish conquest of Western Azerbaijan during the First World War, the persecutions of Christians began and, above all, of Armenians, which ended in the tragic genocide. Subsequently, the earthquake of 6 May 1930 gave the final blow to the Armenian community: 2500 dead out of 44,000 inhabitants and 44 villages out of 98 completely destroyed. Dilman, which had 18,000 inhabitants, was completely destroyed, as well as 31 out of 50 churches (Berberian and Tchalenko 1974).

References

From Encyclopaedia Iranica 2012

Andreeva E (2016) Russia iv. Travelers in Persia to 1917, Encyclopædia Iranica, online edition. Available at http://www.iranicaonline.org/articles/russia-iii-travelers-in-persia (Accessed on 02 Apr 2020)

Bast O (2012) Germany v. German travellers and explorers in Persia. Encyclopaedia X, Fasc. 5:555–559

Bedrosian RG (2020) Travellers' accounts journeys to the armenian highlands and neighboring lands in the 17th through early 20th centuries, in https://archive.org/details/ia_armenia_travellers/mode/2up (Accessed on 02 Apr 2020)

Hodjat M (1995) Cultural Heritage in Iran: policies for an Islamic country. Ph.D. thesis, University of York

Nasiri-Moghaddam N (2016) France viii. Travelogues of the 18th–20th centuries, Encyclopaedia Iranica, X/2, pp 146–150, Available online at http://www.iranicaonline.org/articles/france-viii-travelogues-of-the-18th-20th-centuries (Accessed on 02 Apr 2020)

Richard F (1990) Capuchins in Persia. Encyclopaedia Iranica IV, Fasc. 7:786–788

Tissot F (1994) Délégations archéologiques françaises, Encyclopaedia Iranica VII, Fasc. 3:238–242

Touzard A-M (2016) France vii. French travelers in Persia, 1600–1730, Encyclopaedia Iranica, online edition. http://www.iranicaonline.org/articles/france-vii-french-travelers-in-persia-1600-1730 (Accessed on 02 Apr 2020)

Vanzan A, Bernardini M (2012) Italy iv. travel accounts, Encyclopaedia Iranica XIV(3):250–259

Wilmshurst D (2000) The ecclesiastical organisation of the church of the east, 1318–1913. Lovanio

Wright D (2012) British travelers to persia vii, Encyclopaedia Iranica XI(Fasc. 3):246–252

Books, Papers and Bibliographies

Alemi M (2003) Giardini Reali e disegno del paesaggio ad Esfahan e nel Territorio Iraniano alla luce dei documenti inediti di Pascal Coste, in Opus: Quaderno di Storia, Architettura, Restauro 7

Alpago Novello A, Brambilla M, Uluhogian G, Vecchione N (1977) Consistenza e tipologia delle chiese armene in Iran. Rapporto della missione 1976, Ricerca sull'architettura Armena- Iran, vol 1, n. 17. Politecnico di Milano-Università Nazionale dell'Iran, Milano

Alpago Novello A, Brambilla M, Piacentini VF, Uluhogian G, Uluhogian H, Vecchione N (1978) Consistenza e tipologia delle chiese armene in Iran. Rapporto della missione 1977, Ricerca sull'architettura Armena Iran, vol 2, n.18. Politecnico di Milano-Università Nazionale dell'Iran, Milano

Anvari AH (2015) Historical atlas maps and geographic illustration of Azerbaijan. Neday-e Tarikh, Tehran

Ashtiani KH (2004) Les relations litteraires entre la France et la Perse de 1829 à 1897, Ph.D. Thesis, Université Lumiere Lyon II

Badalyan RS (2019) British travelers and the armenian question during the first half of the 19th century in Nationalities Papers 47:1, 136–148

Bakhshaei H, Mohammadi BK (2015) Peeling report and study of roles wall of the historic church of Haftavan Salmas. Peyman 19(72)

Balilan L, Jafari E (2013) Khoy's expansion from early islam to late qajar according to historical documents. In: International Journal of Architecture and Urban Development, vol 3, No. 2. Springer, Berlin, pp 21–30

Bedrosian R (ed) (2019) Historical geography of armenia and neighboring lands at internet archive, addedate 2019. https://archive.org/details/ia_armenian_hist_geography (Accessed on 21 Sept 2020)
Berberian M, Tchalenko J (1974) The Salmas (Iran) earthquake of May 6th, 1930, Annals of geophysics = Annali di geofisica June 1974
Berberian M, Tchalenko J (1976) Field study and documentation of the 1930 Salmas (Shahpur—Azerbaijan) Earthquacke. In: Geological survey Iran, n. 39, pp 271–242
Boriani M (2020) European travelers in the territory of Salmas and Khoy. In: Shirazian R (ed) Studies and documents in greater Salmas District, Tehran, Dastan, pp 1–36
Bosworth CE, van Donzel E, Lewis B, Pellat C (eds) (1995) *Encyclopaedia of Islam,* vol. VIII – NED-SAM, Leiden, Brill, pp 998–999
Brambilla MG, Tehrani F (2020) Greater Salmas District. studies and documents. Salmast Heritage Association, Tehran, Dastan. ISBN 978-600-297-166-1
Brignoli J-D (2009) Les palais royaux safavides (1501-1722): architecture et pouvoir. Ph.D. Thesis, Université Aix-Marseille I, Université de Provence, LAMM
Daman Pak M, Tehrani FD, Khakpour B, Zahedi S (2017) Atlas of the selected maps from Qajar Era. Archives of History and Diplomacy of MFA. International Institute for Central Asian Studies, Tehran
Deilamqani TM (2006) Millennial history of Salmas and West Azerbaijan, editorial Aeldar. Tabriz, 2005–2006 (in Persian)
Ebrahimi AN, Jahanzamin S (2020) Iranian Azerbaijan cities in Muslim travelers' travelogues from 7 to 13 Century (A.D.) with Focusing on Salmas. In: Brambilla MG, Tehrani F (eds) Greater Salmas District. studies and documents, Salmast Heritage Association, Tehran, Dastan, pp13–30
Galichian R (2007) Countries south of the caucasus in medieval maps. Armenia, Georgia and Azerbaijan, Art Books, Yerevan and Gomidas Institute, London
Houman I, Khan B (2015) Report of the layer removal and the study of mural paintings of historic church of Haftevan in Salams. Peiman J, N. 72, Summer 2015. (in Persian)
Jayhani H, Zali T (2015) Restore the spatial structure and the layout of dagh Baghi in Khoy. J Iranian Architect Stud 6(2015):87–105
Karapetian S (2019) History of Armenia. *Salmast*, RAA Foundation, Yerevan
Kasiri M, Taee P (2016) Investigating the development and spread of architectural photography in the qajar period. Int J Humanit Cult Stud 704–714
Kleiss (1992) Karawanenbauten in Iran.Materialien zur iranischen Archäologie, Teil. 2
Kleiss W (2016) Karavanenwege un Karavanenbauten im Nahen Osten. In: ICAANE—Proociding of the 2nd intarnational congress on the archaeology of the ancient near east, vol. II. Copenhagen, pp 93–110, 22–26 May 2000
Kroll S (1992) Survey in Ost-Azarbaidjan 1991 in Archäologische Mitteilungen aus Iran 25, pp 1–46
Marsden W, Wright T (eds) (1948) The travels of marco polo: The venetian. Doubleday, New York
Matthee R (2016) From splendour and admiration to ruin and condescension: Western travellers to Iran from the Safavids to the Qajars. J Br Inst Persian Stud 54(2016):3–22
Mehryar M and Alii (2000) Pictorial documents of iranian cities in the Qajar period. Shahid Beheshti University and Iranian Cultural Heritage Organisation, Tehran
Moradi A, Omrani B (2014) An exploration of historical recognition and decorations of Alaki Caravansary. NAZAR Q J Architect Urbanism Res Cent 11(28):71–82
Moradi A (2020) Khoi in historical texts. In: Brambilla MG, Tehrani F (eds) Greater Salmas District. Studies and documents, Salmast Heritage Association, Tehran, Dastan, pp 155–172
Mormone MR (2019) Viaggiatori Italiani in Persia. Oltre il pregiudizio: la scoperta dell'iranicità attraverso il viaggio tra il sacro e il profano nell'Iran contemporaneo. Ph.D. Thesis, Unoversité de Nantes, Università l'Orientale di Napoli
O'Flynn T (2016) The Western Christian presence in the Russias and Qājār Persia, c. 1760–c. 1870. In: Brock P (ed) Studies in Christian Mission, Boston, Brill

Ramusio GB (2015) Dei viaggi di Messer Marco Polo, Edizione critica digitale progettata e coordinata da Eugenio Burgio, Marina Buzzoni, Antonella Ghersetti, Ca' Foscari, Venezia
Reclus E (1884) Nouvelle Géographie universelle. La Terre et les Hommes—IX L'Asie Anterieure, Paris, Hachette
Reinaud MJ-T (1867) Catalogue des livres des manuscrits orientaux … composant la Bibliothèque de Feu, Paris, A. Labitte, 1867 (in particular: Géographie, pp. 115–120, Voyages en Asie, pp. 123–135)
Riahi MA (1999) The history of Khoy. Tarh-e Now publications, (1999). [in Persian]
Salzani C, Steven Tötösy de Zepetnek (2016) Bibliography for work in travel studies, version 2016.09.14 Purdue University Press. http://docs.lib.purdue.edu/clcweblibrary/travelstudiesbibliography (Acessed on 27 Mar 2020)
Shirazian R, Tehrani F (2020) Historical geography of Salmas and Khoy based upon Qajar and Pahlavi Maps. In: Brambilla MG, Tehrani F (eds) Greater Salmas District. Studies and Documents, Salmast Heritage Association, Tehran, Dastan, pp. 37–104
Steensgaard N (1974) The Asian trade revolution of the seventeenth century: the east India companies and the decline of the caravan trade. The University of Chicago Press, Chicago
Tehrani F (2019–2020) Visual documents of Iranian cities: West Azerbaijan Province, Salmas and Khoy region. Collab Asar J 1398
Uluhogian G (2000) Una antica mappa dell'Armenia. Monasteri e Santuari dal I al XVII secolo. Longo Editore, Ravenna
Vatoogian JL (1974) The image of Armenia in European travel accounts of the seventeenth century. Columbia University Microfilms
Vinson D (2000) Les Arméniens dans les récits des Voyageurs français du XIXe siècle (1796–1985). Ph.D. Thesis, Université Montpellier III, Paul Valéry
Vinson D (2009) Napoléon en Perse": la Mission Romieu. Un « lieu de mémoire » viatique au début du XIXe siècle in Astrolabe n. 26 Juillet-Août 2009. https://astrolabe.msh.uca.fr/juillet-aout-2009/dossier/napoleon-en-perse-la-mission-romieu (Accessed on 09 Mai 2020)
Waugh D, Lee A (2020) Travelers on the Silk Road, The Silk Road Foundation. http://www.silkroadfoundation.org/artl/srtravelmain.shtml (Acessed on 27 Mar 2020)
Wilmshrust D (2000) The ecclesiastical organisation of the church of the east, 1318–1913, in Aedibus Peeters
Yurdaydin H (ed) Nasuhu's Silahi (Matrakci), Beyan-I Sefer-I Irakeyn-I Sultan Suleiman Han. Turk Tarih Kurumu Basimevi, Ankara (1976) (*Chronicle of Stages of Campaign of Iraq and Persia.* translated by Rahim Rais Niya, Cultural Heritage Organization, Tehran)

Travellers in Persia

Ainsworth W (1841) An account of a visit to the Chaldeans, Inhabiting Central Kurdistán: and of an ascent of thr peak of Rowándiz in 1840. J Roy Geogr Soc London 11
Alcock T (1831) Travels_in_Russia_Persia_Turkey_and_greece in 1828–29. Clarke and Sons, London
Alik (1951) A.A. *with the eagles*, in Alik n. 196, 3 Sept 1951
Anaclerio G (1868) La Persia descritta. Relazione di un viaggio per Giuseppe Anaclerio. Napoli, Vincenzo Marchese
Anonimo (probably Francesco Romano) (1978–88) Viaggio di un Mercante che fu nella Persia, XVI century. In: Ramusio GB (ed) Navigazioni e Viaggi, Giulio Einaudi editore, collana I Millenni, a cura di Marica Milanesi, 6 voll
Armstrong TB (1831) Journal of travels in the seat of War, during the last Campaigns of Russia and Turkey, London, A.Seguin
Aubin E (1908) La Perse d'aujourd'hui. Iran-Mésopotamie. Librairie Armand Colin, Paris

Ballie Fraser J (1838) A winter's journey (Tatar) from costantinople to Tehran with travels through various parts of Persia, London, Richard Bentley

Barbaro J (1543) Viaggio dello istesso Messer Josaphat Barbaro in Persia. In: da Vinetia VF (ed) alla Tana, in Persia, in India, et in Costantinopoli, Venezia

Batuta IBN (1829) The travels of Ibn Batuta. J-L-Cox, London

Binder H (1887) Au Kurdistan, en Mésopotamie et en Perse. Maison Quantin, Paris

Bird I (1891) Journeys in Persia and Kurdistan. London 1891

Borè E (1840) Correspondance et Mémoires d'un voyageur en Orient. Tome II, Paris, Olivier-Fulgence

Browne EG (1893) A year amongst the Persians. Impressions as to life, character, and thought of the people of Persia, received during twelve months residence in that Country in the years 1887–1889, London, Adam and Charles Black

Browne EG (1910) The Persian revolution of 1905–1909.s Cambridge at University Press

Campanile G (1818) Storia della Regione del Kurdistan e delle sette di religione ivi esistenti. Fratelli Fernandes, Napoli

Celebi E (1834) Narrative of travels in Europe, Asia, and Africa in the seventeenth century by Evliya Efendi, Translated from the Turkish by the Ritter Joseph von Hammer, London, Oriental Translation Fund of Great Britain and Ireland

Chantre MB (1893) A travers l'Arménie Russe. Imprimerie Lahure, Paris

Chardin J (1686) Journal du voyage du chevalier Chardin en Perse et aux Indes orientales: par la mer Noire et par la Colchide, Première partie, Londres, Chez Moses Pitt

Clavijo RG (1928) Embassy to tamerlane—1403–1406. Translated from the Spanish by Guy Lestrange, George Routledge and Sons, London

Coste P (1867) Monuments modernes de la perse. A. Morel, Paris

Coste P (2015) Itinéraire de l'Ambassade française en Perse, sous M. le comte de Sercey, et des excursions scientifiques des deux artistes attachés à cette mission, (2 vol. text. avec dess. et 9 atl. de pl.), Fonds Pascal Coste, Bibliothèque de Marseille à vocation régionale (Publications de l'Institut national d'histoire de l'art), (Accessed 6 Oct 2015)

Curzon GN (1892) Persia and the Persian question. London, Longmans, Green & Company

Cutts EL (1878) Under the crescent in Asia. London

de Filippi F (1865) Note di un viaggio in Persia nel 1862. Milano, Daelli

de Gobineau A (1859) Trois Ans en Asie (de 1855 à 1858) Librairie de L.Hachette et C, Paris

de Hell Xavier (1855) Voyage en Turquie et en Perse. P.Bertrand, Paris

de Laet J (1633) *Persia seu Regni Persici Status. Variaquae Itinera in atque per Persiam; cum aliquot iconibus incolarum*, Secunda editio priori auctior, Lugdunum Bataviorum

de Vecchi F (2016) Giornale di carovana. Brano di un viaggio nell'Armenia, Persia, Arabia e Indostan, fatto negli anni 1841–42. Luni Editrice, Milano

Develay A (1892) Autour des lacs de Van et d'Ourmiah, Revue Scientifique Tome XLIX, jan.-jul. 1892, pp 553–557

Develay A (1898) D'Arménie en Perse à travers le Kurdistan. J des Voyages et des Aventures de Terre et de Mer n. 989, 21 juin 1898

Dieulafoy J (1887) La Perse, la Chaldée et la Susiane, Paris, Hachette

Drouville G (1819) *Voyage en perse, fait en 1812 et 1813*, Tome I, St. Petersbourg

Drouvllle G (1825) Voyage en Perse fait en 1812 et 1813. Tome II, Paris

Dubeux L (1841) La Perse. L'Univers ou Histoire et Description de tous les Peuples, Paris, Firmin Didot Frères

DUBOIS DE MONTPÉREUX (1839–43) Dubois de Montpéreux, Frédéric, Voyage autour du Caucase, chez les Tcherkesses et les Abkhases, en Colchide, en Géorgie, en Arménie, et en Crimée; avec un atlas géographique, pittoresque, archéologique, géologique... tome 1–6 and atlas. Paris, Gide, 1839–43

Dupré A (1819) Voyages en Perse fait dans les années 1807, 1808 e 1809, en traversant la Natolie et la Mésopotamie, Paris

ED'Arcy Todd (1838) Itinerary from Tabriz to Tehran, via Ahar, Miskhia, Ardabil, Talish, Gilan and Kazwin, in 1837. J Geogr Soc
Feuvrier, J-B (1900) Trois ans à la cour de Perse, Paris, F.Juven Editeur
Flandin E, Coste P (1854) Voyage en Perse de MM. Eugène Flandin, Peintre, et Pascal Coste, Architecte, attachés à l'Ambassade de France pendant les années 1840 et 1841. Paris, Gide et J.Baudry
Fowler G (1841) Three years in Persia with travelling adventures in Koordistan. Henry Colburn Publisher, London, p 1841
Fraser JB (1840) Travels in koordistan, mesopotamia etc., vol I, London, Richerd Bentley
Grey C (1873) A narrative of Italian travels in Persia in the fifteenth and sixteenth centuries. Printed for the Hakluyt Society, London
Haukal E (1800) The oriental geography of Ebn Haukal, an Arabian traveller of the tenth century, translated by Sir William Ouseley. Oriental Press, London
Jackson AVW (1906) Persia past and present. A book of travel and research. The Macmillan Company, New York
Jaubert P-A (1821) Voyage en Arménie et Perse fait dans les années 1805 et 1806. Pélicier et Nepveu, Paris
Ker Porter R (1821) Travels in Georgia, Persia, Armenia, Ancient Babylonia during the years 1817, 1818, 1819 and 1820, London, Longman, Hurst, Rees, Orme and Brown
Laurie T (1853) Dr. Grant and the mountain nestorians. Gould and Lincoln, Boston
Le Strange G (1905) The lands of the eastern caliphate. Mesopotamia, Persia and Central Asia from the Moslem conquest to the time of Timur. University Press, Cambridge
Lucas P (1704) Voyage du sieur Paul Lucas au Levant..., premier volume. Paris
Macdonald Kinnier J (1813) Geographical memoir of the Persian Empire accompanied by a Map, London, John Murray
Macdonald R (1859) Personal narrative of Military travel and adventure in Turkey and Persia, Edinburgh, Adam and Charkes Black
Maclean AJ, Browne WH (1892) The catholicos of the east and his people, eastern Syrian Christisns of Kurdistan and Northern Persia (known also as Nestorians). Society for Promoting Christian Knowledge, London
Morier J (1812) A journey through Persia, armenia, and asia minor to costantinople in the years 1808 and 1809. Longman, Hurst, Rees, Orme and Brown, London
Morier J (1818) A second journey through Persia, Armenia, and Asia minor, to constantinople, between the years 1810 and 1816. Longman, Hurst, Rees, Orme and Brown, London
Müller-Simonis P (1892) Du caucase au golfe Persique à travers l'Arménie, Le Kurdistan et la Mesopotamie, Université Catholique d'Amerique, Washington D.C.
O'Flynn TSR (1760–1870) The Western Christian presence in the Russias and Qājār Persia, c. 1760–1870. In: Brock P (ed) Studies in Christian Mission, vol 47, Leiden, Boston, Brill
Osculati G (1844) Note d'un viaggio nella Persia e nelle Indie Orientali negli anni 1841, 1842. Monza, Tipografia Di Luca Corbetta
Perkins RJ (1843) A Residence of eight years in Persia among the Nestorian Christians, with Notices of the Muhammedans, New York, Allen, Morril and Wardwell
Pichon J (1900) Journal d'une mission militaire en Perse (1839–1840) par le lieutenent Jules Pinchon, L.Gougy, Paris
Polo M (1948) In: William M (Ed) Re-Edited By Thomas Wright, the travels of marco polo: The Venetian, Published by Doubleday, New York
Price W (1825) Journal of the British Embassy to Persia, vol I. Kingsbury, Parbury and Allen, London
Purchas S (1905) Hakluytus posthumus or purches his pilgrimes. Glasgow
Ross ED, Power E (1925) Don Juan of Persia [Uruch Beg]. A Shi'ah Catholic (1560–1604), Harper and Brothers New York and London, 1925 (first edition 1603)
Saint-Martin A-J (1818) Mémoires Historiques et géographiques sur l'Arménie. De l'Imprimerie Royale, Paris

Shiel J (1838) Notes on a journey from Tabriz, through Kurdistan, via Van, Bitlis, Se'ert and Erbil, to Suleimaniyeh, in August, 1836. The London and Paris Observer, Paris, 11 Mar 1838

Shiel ML (1856) Glimpses on life and manners in Persia by Lady Shiel, London, John Murray

Sir Robert Ker Porter (1822) *Travels in Georgia, Persia, Armenia, Ancient Babylonia during the years 1817, 1818, 1819 and 1820*, London, Printed for Longman, Hurst, Rees, Orme and Brown, 1821 vol.I and 1822 vol. II.

Smith E, Dwight HGO (1834) Missionary researches in Armenia: including a journey through Asia Minor, and into Georgia and Persia, with a visit to the Nestorian and Chaldean Christians of Oormiah and Salmas, London, George Wightman

Southgate H (1840) Narrative of a tour through armenia, Kurdistan. Persia and Mesopotamia, New York

Stewart CCE (edited from his diaries by Basil Stewart) (1911) Through Persia in disguise, with reminiscences of the Indian mutiny, George Routledge and Sons, London

Stuart L-CC (1854) Journal of a residence in Northern Persia and the adiacent provinces of Turkey, London

Tancoigne JM (1820) A narrative of a Journey into Persia and Residence at Teheran: containing a descriptive Itinerary from Constantinople to the Persian Capital, William Wright, London. https://babel.hathitrust.org/cgi/pt?id=yale.39002021342671&view=1up&seq=23 (Accessed 03 Feb 2021)

Tavernier JB (1678) Les six Voyages de Jean Baptiste Tavernier, a noble Man of France now living, through Turky into Persia and the East-Indies, finished in the Year, London, R.L. and M.P., 1678 (First edition in French, 1676)

Texier C (1843) Notice sur Erzéroum, Fragment d'un journal de voyage, im Bulletin de la Société de Géographie, oct 1843, pp 213–249s

Texier C (1842) Description de l'Armenie, la Perse et la Mesopotamie, vol 1, Paris, 1842

Ussher J (1865) A journey from London to Persepolis; including Wanderings in Daghestan, Georgia, Armenia, Kurdistan, Mesopotamia and Persia. Hurst and Blackett, London

Wagner M (1856) Travels in Persia, Georgia and Koodistan; with Sketches of the Cossacks and the Caucasus. Hurst and Balchett, London

Weeks EL (1896) Persia and India. Harper and Brothers Publishers

Wilbraham CR (1839) Travels in the Transcaucasian provinces of Russia and along the Southern shore of the Lakes of Van and Urumiah, in the Autumn and Winter of 1837, John Murray, London

Wilson CW (ed) (1895) Handbook for travellers in Asia Minor, Transcaucasia, Persia, etc., J. Murray, London

Wilson SG (1895) Persian life and customs with scenes and incidents of residence and travel in the land of the Lion and the Sun. Student Missionart Campaign Library, Chicago

Wilson SG (1896) Persia western mission. Presbyterian Board of Publication, Philadelphia

Wolff J (1824–1828–1829) Missionary Journal and Memoir of the Rev. Joseph Wolff 3

Woulfe Sheil ML (1856) Glimpses of Life and Manners in Persia, by Lady Sheil. with notes on Russia, Koords, Toorkomans, Nestorians, Khiva, and Persia, John Murray, London

Zeno C (1873) Commentarii del Viaggio in Persia. In: Grey C (ed) A narrative of Italian travels in Persia in the Fifteenth and Sixteenth Centuries. Printed for the Hakluyt Society, London

Archives

Ali Khan V (2021) Album of photographs by Ali Khan Vali, Harvard Art Museums. https://harvardfineartslib.tumblr.com/post/167046736746/album-of-ali-khan-vali-born-in-tehran-in-1846-ali (Accessed on 03 Feb 2021)

Sevruguin A (2021) The getty research institute, Antoin Sevruguin photographs of Persia, http://oac.cdlib.org/findaid/ark:/13030/c85t3rn3 (Accessed on 03 Feb 2021)

Maurizio Boriani Architect, born in Milan, Italy, 1946.

Full professor in Architectural Conservation at the Politecnico di Milano (Department of Architectural Design—School of Civil Architecture)—retired in 2016.

Former Director of Department of Architectural Design at the Politecnico di Milano.

Member of the Teaching College of the Ph.D. in Architectural Conservation at the Politecnico di Milano.

Former President of the Scientific Board of Cultural Heritage Conservation Center of the Politecnico di Milano.

Research Co-ordinator on behalf of the UNESCO, World Bank and European Community, Italian Ministry of Foreign Affairs and other private and public agencies.

UNESCO Consultant for: Albania (2006–2009), Macedonia, Montenegro, Kossovo, Myanmar; expert in international cooperation on behalf of Italian Ministry of Foreign Affairs: China (2009–2012), Vietnam (2017–2019), Armenia (2019–2014), Pakistan (2011–2012, Myanmar (2012–2015).

Research Topics:

History of Architectural Conservation, Conservation and management of historic towns and archeological sites, Cultural landscape preservation and management, Gardens history and conservation, Cultural Heritage and International Cooperation. About those themes he is author or editor of about 240 papers and books.

Infrastructure for Trade Routes in Western Azerbaijan, Iran: Caravanserais, Bridges, and Fortresses

Andreh Marouti

Abstract Since 549 BC, Iranians ruled over vast territories and commanded large-scale empires. Thus, it is not surprising that Iranians have built far-reaching infrastructures for the last three millennia, ensuring safe travel and trade routes and facilitating international trade, communication, political and cultural exchanges. This phenomenon is particularly relevant in Western Azerbaijan. It was the gateway to Baghdad, capital of the Caliphate, from 632 until 1258, when it fell to the Mongols. It was also the gateway to the Mediterranean port cities, Turkey, and Russia, connecting Iran and the east to Europe. For centuries and until very recently, it was a region that was heavily populated, often governed and maintained by Armenians, who left their heritage in so many ways throughout Western Iran. Critical elements of safe travel paths require necessities to allow journeys over vast territories. It was imperative to provide a well-established and monitored travel route with strategically located stations, providing safety, lodging, food, shelter and care for the animals, storage, and a series of other services. Thus, the concept of caravanserais was developed and adopted to their particular location, climate, and topography. To protect the trade routes and provide a degree of safety in the land, building and maintaining critical points of the territory, fortresses, and garrisons were necessary. These were often not intended as major military establishments but rather as a regional garrison, protecting specific passages and regions. A further important element necessary to facilitate the movement of people and merchandise is the bridge. West Azerbaijan has few permanent, large rivers, but many seasonal rivers and water paths make traveling impossible without well-built structures over the water elements. This article intends to study, analyze, and explain the central infrastructures of the principal routes of travel in northwestern Iran, emphasizing the typologies of caravanserais, fortresses, and bridges.

Keywords Iran · Caravansarais · Bridges · Fortresses

A. Marouti (✉)
Former student (2015–2018) at PH.D program in Preservation of Architetctural Heritage, DAStU, Politecnico Di Milano, Milan, Italy

M. Boriani and M. Giambruno (eds.), *Architectural Heritage in the Western Azerbaijan Province of Iran*, Research for Development,
https://doi.org/10.1007/978-3-030-83094-6_6

1 Introduction

Due to its location on the east–west trade highway and the vastness of its land, special attention has been paid to the establishment of trade and military routes in Iran, as well as the construction of numerous bridges and the security of caravans. Based on evidence from historical documents, Iran has taken the lead in building roads and caravanserais and establishing innovative communication systems. According to researchers, the creation and development of caravanserais in Iran is one of the most important achievements of Iranian architecture, especially in the Islamic period, and the study of its various features confirms this theory (Kiani and Kleiss 1994).

The province of West Azerbaijan, in terms of proximity to the western borders of the Persian Empire, has always been one of the busiest places in the region throughout history and has been considered the gateway to Europe. Louis Vandenberg rightly refers to the province as a crossroads for nations and a center for trade (Vandenberg 1966).

On the other hand, the region once belonged to Armenia and for centuries was governed by Armenian rulers. The large number of Armenian religious monuments scattered throughout the region illustrates this fact [*A/N: For more information regarding Armenian monuments please see Karapetian (*2019*), Alpago-Novello et al. (*1977*), Alpago-Novello et al. (*1978*)*]. It seems natural, then, that trade is one of the most important inherited occupations of Armenians living in the region. Armenian merchants have long paid special attention to the education of their sons, teaching them multiple European languages and acquainting them with the customs of the eastern and western courts. Therefore, Armenian merchants have always contributed to building and repairing important commercial infrastructure [*A/N: Many bridges and caravanserais of the area have Armenian inscriptions, indicating the involvement of Armenian merchants in founding and preserving trade infrastructure*].

To have a thriving business route, the first critical requirement is safety. Iranians have long been aware of this fact and have devised special measures to ensure the security and safety of trade routes. Numerous fortifications have been built along the routes to counter bandits and repel occasional attacks by nomadic tribes.

In Iran, due to the climatic conditions and land features, there are no wide and permanent rivers. Rivers are usually seasonal. For this reason, it was possible for caravans to cross these rivers in low water seasons, but in order to have a reliable trade route that can be used in all seasons, bridges are necessary.

Another need of commercial caravans is to have a safe place to rest at the end of the day and prepare to continue the journey the next day. For this purpose, several caravanserais were built along commercial routes, many of them located approximately 40 km from each other, a safe distance for a caravan to travel in one day.

It is important to note that at different periods, for reasons such as tribal disputes, regional conflicts, or major wars between governments, trade was either halted or shifted to a different path. During the heyday of a route, commercial infrastructures such as bridges, fortresses, caravanserais, bazaars, hammams, and road guides were built along the route. They were then abandoned in times of insecurity and rebuilt and reused during booms. Nineteenth-century European travelers often described the caravanserais and bridges as being in ruins. This is due to the political and economic uncertainty and state of crisis in Persia at the time, the shift of commercial paths from land to sea through the Cape of Good Hope, moving the capital to Isfahan, Shiraz, and finally Tehran, and eventually, the lack of a powerful central government. Finally, it is important to add that the impassable mountain routes never flourished internationally and were used only in times of emergency. Accordingly, the infrastructure built along them was less important.

The lack of research related to this subject has caused a lack of awareness of the conditions of roads, transportation routes, caravanserais, and accommodations built along the way. Armenian, Persian, and European sources are completely disconnected. The same monument might have been studied by Iranian and Armenian scholars from different perspectives, and because they often use different names, even knowledgeable readers cannot recognize that it is one single monument described from different perspectives. The caravanserais identified in the province are very limited and occasional, and so far, no comprehensive research has been done in this field. The focus of this article is this secondary infrastructure.

Methodology of the Research: This is a descriptive-analytical study. The first step was to review the existing literature, done in previous years by Wolfram Kleiss, the documents created by the joint expeditions and documents created by scholars from Politecnico di Milano and the National University of Iran, Research in Armenian Studies (RAA), and the Salmast Heritage Association (SHA). The second step was to determine the trade routes and the infrastructure. Studying historical texts and related documents was the next step. The gathered information was integrated with natural and geographical features using maps, Google satellite images, and drone photography of the region in order to obtain a bigger perspective of the locations and the possible relations between the regions, which might have been lost during the time. The accessibility and availability of satellite and drone images have created a new opportunity that did not exist for previous studies. Integrating the information of individual monuments into a single GIS map gives us the perspective needed to conduct the study with a new vision that makes lost relations and connections more visible and salient.

This research addresses 12 monuments: five caravanserais (Kuzehrash, Khan/Qotur, Abbasi, Khan/Khoy, Mir Hashem), four bridges (Qaleh Juq, Panj Cheshmeh, Baran Duz, Khatun), and three fortresses (Maku, Chehriq, Berdiuk) (Fig. 1). The reason for choosing these monuments is that they are not located along the main travel route connecting Tabriz to Yerevan but on the secondary routes connecting Maku to Kars, Khoy to Van, and Salmas to Van. Fewer studies have been conducted regarding these monuments.

Caravanserais: Caravanserais are divided into two categories: urban and suburban. Urban caravanserais are often formed in combination with bazaars. Suburban caravanserais are built along the roads in the heart of the desert or the mountains. The vastness and climatic diversity of Iran, as well as the insecurity of its roads, have been serious obstacles to travel. Since the establishment of centralized and powerful governments and the presence of a favorable economic situation, the

Fig. 1 Map of the area and the studied monuments. Created by the author

prosperity of trade and commerce has always depended on the safety of the roads and the comfort of travelers. Iranians have prepared many arrangements for this since ancient times. One of the most important measures has been the construction of several caravanserais in different parts of the country and their continuous maintenance. Travelers who have been to Iran at different periods have always reported on the vital role of caravanserais in different regions and testified to how these structures have protected them from dust and sand storms, the scorching heat of the deserts, snow and mountain storms, and thieves and robbers, providing them with a safe environment to rest.

The founders of the caravanserais were governments, rulers, or benefactors. These structures were usually built next to busy roads of the period and with local materials such as mud and clay or stone and brick, or a combination. The distance between the caravanserais was often related to the distance a caravan traveled in one day. The life of the caravanserai depended on water and ease of access to water was a decisive factor in choosing a construction site. Storing water on site was also very important. Hence, water reservoirs were built in or near all caravanserais. Along with some caravanserais, there are other buildings such as bazaars and hammams that provided more comfort for travelers.

Caravanserais are simple structures. Four closed, high, untreated walls without openings but with towers at the four corners and sometimes between the longer sides make up their castle-like exteriors. The only element displayed on the outside of the building is the main entrance, which is usually higher than the total volume of the structure to display more visibility and invite caravans inside. Behind the main entrance, there is a vestibule that is the most important and significant element of the building. This space is usually independent of other elements and directs visitors to the main courtyard. Cells and guard rooms are often located on either side of the vestibule. This space usually has a high and decorated ceiling, and the surrounding spaces open to it.

The main courtyard sits in the middle of the caravanserai. The courtyard is regular and geometric in shape and often quadrangular. The courtyards are surrounded by cells, and in front of each cell, there is usually a porch considered a semi-open space dedicated to the inhabitants of each cell. In most caravanserais, there is an iwan in the middle of each of the four courtyard fronts. Behind this regular and attractive elevation of the courtyard, there are stables and cattle barns that can usually be entered from the corners of the yard. The resting place of the cattle is also at the entrance of these stables so that they can be taken care of well. In this way, the caravans rested in the cells around the yard without disturbing their cattle (Fig. 2) *[A/N: for more information regarding the caravanserais please see Haji-Qassemi (*2005*)]*.

Suburban Mountain Region Caravanserais: These caravanserais are closed on all sides and can easily protect caravans from constant winds, winter snow, and spring and autumn hail. The construction of this type of caravanserai was common in mountainous areas and in impassable passages. Closed and covered buildings, an absent central courtyard, the use of small space to trap warmth, location in the direction of the sunrise, and location on the network of communication channels are their main characteristics.

It seems that the shape and dimensions of caravanserais in the mountainous areas of northwestern Iran depended on the amount of caravan traffic and the importance of communication routes.

The caravanserais of the mountainous areas consist of rows of central domed rooms and a number of stables in the same row. Most mountain caravanserais have multiple fireplaces or wall heaters inside the building. These types of caravanserais were built in mountainous areas, especially in impassable passages, and

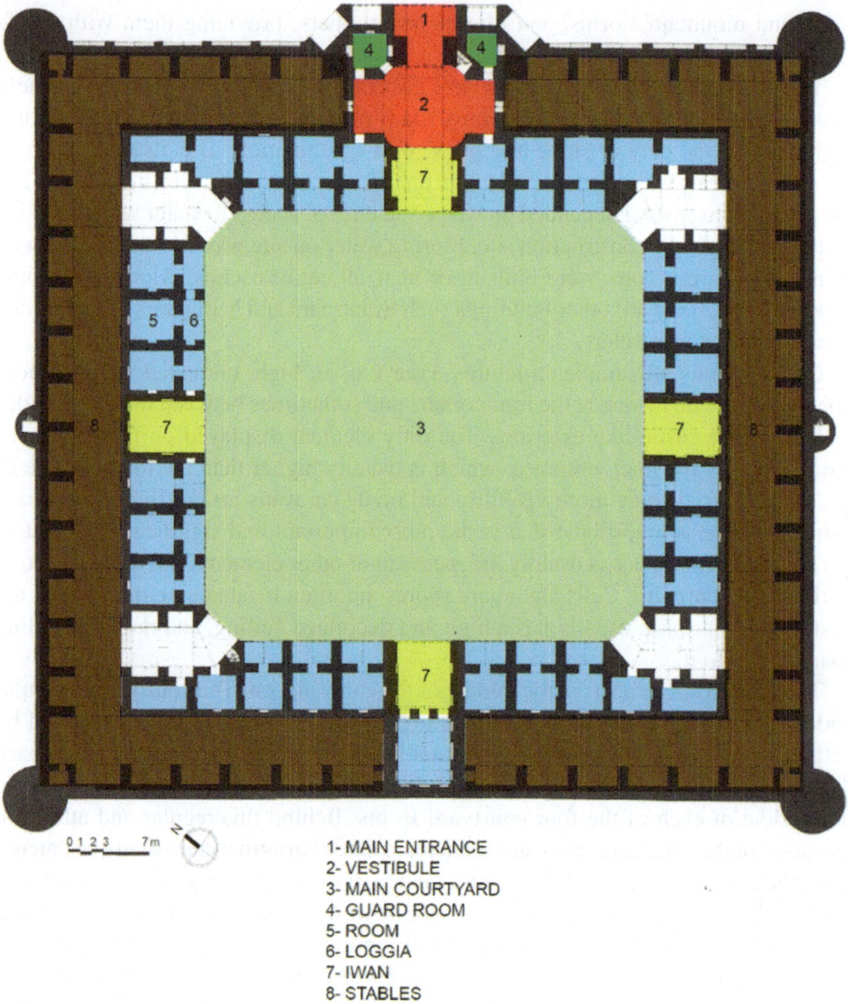

Fig. 2 Typical suburban caravanserai. Diagram created by the author

were built primarily during the Safavid era (Kiani 2000). Among these caravanserais are Shah Abbasi caravanserai in Khoy, Kuzehrash caravanserai in Salmas, and Baneh caravanserai, which belong to the Safavid period.

Maxime Siroux writes about the main reasons for the destruction of these structures: "In the nineteenth century in Iran, several buildings were erected in the form of caravanserais, but because all of them were built with raw clay and was always built in a hurry, except for the short base which was built with brick, the rest is disappeared" (Siroux 1949). Other reasons for the destruction of caravanserais along this route include heavy rainfall, neglect and abandonment, and change of use.

2 Kuzehrash Caravanserai: Geographic Coordination 38.161622, 44.466631

This caravanserai is located near the village of Kuzehrash, 40 km west of Salmas, near the Iranian-Turkish border. The building is currently in ruins and there is not much information about the date of construction. Wolfram Kleiss provides a plan of the remains of the building (Kleiss 1977). Most likely, this caravanserai was built by Armenians living in the area (Khanmohammadi et al. 2013). According to the plan created by Kleiss, the footprint of the building is rectangular, approximately 21 by 15 m. The building is built in a north–south direction, and its entrance opens to the south. It is divided into three longitudinal spaces approximately 3.6 m wide, which are covered by barrel vaults. All the walls of the building are about 1 m thick. The entrance door and the openings between spaces are approximately 1.5 m wide. The middle space (A) covers the entire length of the building, and the two side spaces are divided by a wall in the middle, forming four independent spaces with separate entrances (B, C, D, E). Most likely, space A was for travelers, and the spaces on its flanks were used as animal stables. The advantage of this division is that by placing the animals in the side spaces, the heat of the animals helped to warm the middle space. Animals were also one of the most important and expensive assets of travelers and were always at risk of theft by looters. Therefore, they were kept in spaces without direct access to the outside of the building (Fig. 3).

3 Khan or Qotur Caravanserai Near Khoy: Geographic Coordination 38.487042, 44.431364

This caravanserai is located at a distance of 37 km from Khoy and on the northern side of Khoy-Qotur road, at the foot of a mountain and next to the national railway of Iran and Turkey. This building is a square with sides of 18 by 18 m, made of stone materials with plaster and lime mortar. The main entrance looks toward the west. According to a plan created by Kleiss in 1977, the building is divided into four long corridors approximately 3.2 m wide. The exterior walls are approximately 1.5 m wide. The interior walls, dividing the space into long corridors, are approximately 1 m wide. There are steps inside the three corridors due to the slope of the ground. The entrance of the building has two large piers and opens into a preliminary room measuring 3.2 by 4 m, which was perhaps intended for thermal protection of the interior (A). This room is connected to spaces (D), (B), and (E), with an arch-like opening of approximately 1.2 m wide. Spaces (C) and (D) each have a small external window of 0.3 m wide. Most likely, these two spaces were the place where the animals were kept. There are two small niches near the corners of the western wall of the space (E) that increase the possibility of using it as a habitable space. Assuming that the western wall is buried and the animals are housed in spaces (C) and (D), the heat of the animals heated spaces (B) and (E). The niches may have been a place to safely

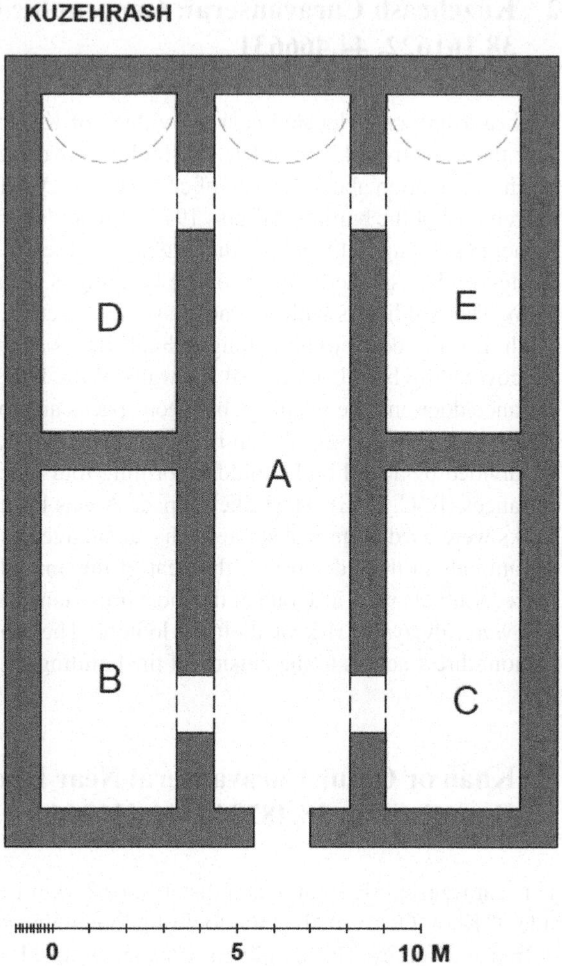

Fig. 3 Plan of Kuzehrash Caravanserai, recreated by the author using the original drawing created by Wolfram Kleiss

place fire or hot coal to heat the room. In this way, by placing the stables on one side of the living spaces, the heat of the animals' bodies was used optimally, and the animals were protected from the danger of being stolen (Fig. 4).

4 Shah Abbasi Khoy Caravanserai: Geographic Coordination 38.464229, 44.485527

This caravanserai is located at a distance of 34 km from Khoy on Khoy-Qotur road. According to the plan provided by Kleiss, this structure is built within a square of 27 by 27 m. The main entrance looks south and is divided into five corridors. The

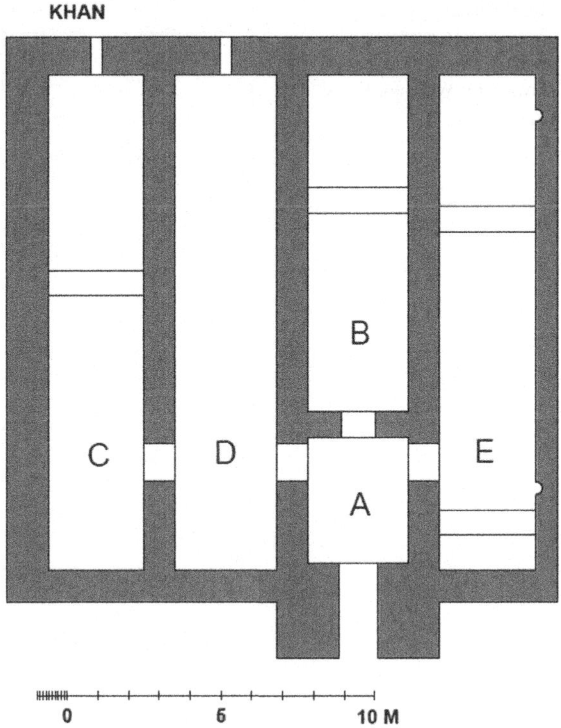

Fig. 4 Plan of Khan Caravanserai, recreated by the author using the original drawing by Wolfram Kleiss

middle corridor (A), onto which the entrance opens, is slightly narrower than the other four. The width of space (A) is 11.8 m, and the width of other spaces is 13.6 m. The external walls are approximately 1.5 m, and the internal walls are 1.2 m thick. The interior partition walls that divide spaces (B), (D), (F) and (C), (E), (G) are 1 m thick. The south and west sides of the building have several openings, all of which are approximately 0.4 m wide, except for one of the windows in space (H), which is about 1 m wide. On both sides of the entrance, there are two square rooms measuring 4 by 4 m with small windows to the outside (B) and (C). These rooms were probably places to prepare food or sell supplies to travelers or guard rooms. Apart from these two rooms, the other spaces are connected by arches approximately 3 m wide. The smaller spaces (A), (D), (E), (F), and (G) were probably habitable, and the two larger spaces (H) and (I) were stables (Fig. 5).

Urban Caravanserais in the Region: The urban caravanserais were located inside the urban bazaars and in many cases had direct access to the corridors of the bazaars (Fig. 6). This facilitated trade business. The caravans entered directly into the courtyard without passing through the narrow corridors of the bazaar, paid their taxes, unloaded their merchandise, and were ready to find customers. The two caravanserais at Khoy are not an exception. Both are connected to the main corridor of the bazaar through iwans. Both also have a secondary connection through another iwan to the adjacent bazaar corridors.

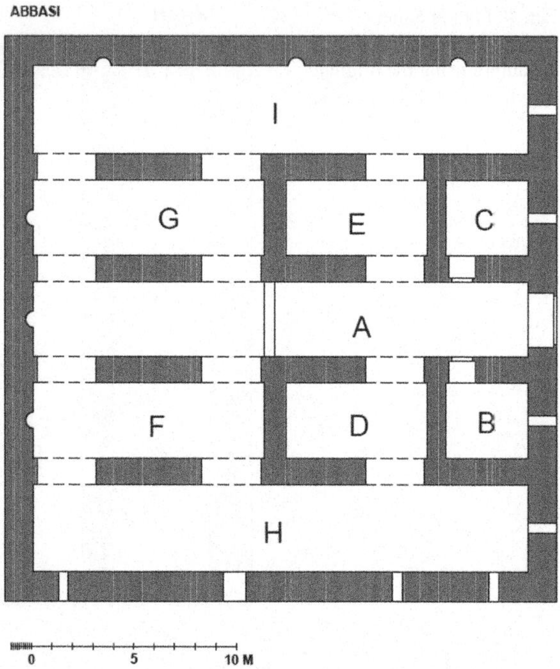

Fig. 5 Plan of Abbasi Caravanserai, recreated by the author using the original drawing by Wolfram Kleiss

5 Khan Caravanserai in Khoy: Geographic Coordination 38.5502, 44.95992

This building is made of stone and brick according to the Iranian four-iwan-style plan. The eastern iwan is directly connected to the main corridor of the bazaar. The date of construction of the building is probably the Qajar period. At present, its two north and east iwans remain relatively intact. It has a beautiful red brick lattice, and inside its arches, the brickwork decorations can be seen in the form of stars (Fig. 7).

6 Haji Mir Hashem Caravanserai in Khoy: Geographic Coordination 38.54946, 44.95995

Located next to the Khan caravanserai, Mir Hashem caravanserai is integrated into the bazaar as well. It is a standard four-iwan caravanserai but smaller than the other structure. Three of its iwans remain intact. Similar to Khan caravanserai, the caravans entered from the back of the structure and through the direct connection to the bazaar corridors; merchants could easily find local customers and do business. The structure was built by Mirza Hashem and probably belongs to the Qajar period (Sadraei 2005) (Fig. 8).

Fig. 6 Urban caravanserais in the bazaar complex. Diagram created by the author using drone photography

Fig. 7 Khan Caravanserai. Diagram created by the author using drone photography

Bridges: The vast land of Iran is a high plateau with many elevations. The existence of numerous mountain chains and hills creates a variety of climates. Thus, the rainfall in the northern and western areas is abundant and in the central part is very low. Nevertheless, Iran has a vast network of running water, both as permanent rivers and as temporary streams. Rivers were major obstacles to caravan routes. Therefore, for road construction, there was a need to build bridges *[A/N: for more information about Iranian historic bridges, see* Shalchian (1985)*, Tahini (*1990*), and* Mehdi (1977)*]*.

Prior to the emergence and expansion of bridges, the existence of natural barriers, especially large rivers in the course of various cultural, commercial, political, and economic connections, was one of the major problems of human life and hindered the progress and expansion of human civilization. It was with the construction of bridges that these problems were partially resolved.

Bridges have been popular from the very beginning because of their important role in creating, maintaining, and expanding communication. In prehistoric times, due to limited communication, simple solutions were used for transfers, such as traveling in the dry seasons by boat or building wooden bridges. The shape and materials of the original bridges were much simpler. Most of these bridges were wood. But with the formation of governments and the expansion of commercial and cultural communications, military invasions, and the advancement of architectural technology, bridges became a special way for architects and their builders to perform.

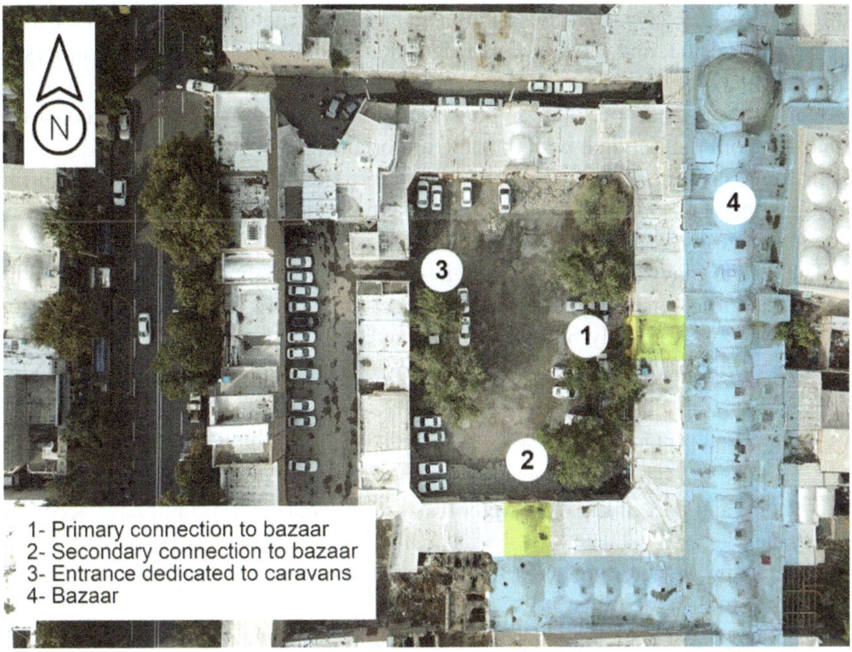

Fig. 8 Mir Hashem Caravanserai. Diagram created by the author using drone photography

In modern-day Iran, the oldest surviving bridge is one built by the Urartians on the Aras River. Although the bridge was destroyed around the eighth century BC, there are still enough traces of it to understand the existence of the bridge and its design and size.

From the Achaemenid period, bridges built in the fourth and fifth centuries BC can be found in the area of the palaces and gardens of Pasargadae. No Parthian bridges have been found, but the Silk Road, which ran through the heart of the empire, could not have existed without bridges. In future archeological excavations, traces of bridges of this period may be found.

Political and economic conditions of the Sassanid period demanded that bridge construction be considered on par with other architectural activities. Therefore, the bridges built in this period are considered masterpieces of their kind. During the Seljuk period, according to Ibn Balkhi, the rulers and governors attached great importance to the construction and repair of the bridges.

During the Ilkhanid period, the construction of bridges and roads continued. The Timurid period is one of the most important periods in Iranian architecture. From this period, an extraordinary example (PolDokhtar in Miane) remains, which shows the extensive knowledge and skill involved in the field of bridge construction (Razani and Mohammadi Sefidkhani 2017).

In the Safavid period, due to the expansion of the network of roads and the great interest in architecture, bridge construction enjoyed significant progress. In no period of Iranian history have road construction and related structures spread as much as in the Safavid period, and most of the bridges that can be seen today in the corners of this vast land are valuable works of this period or attributed to it.

During the Afshari period, not much attention was paid to road bridge construction. But from the Zandieh period, some important works have survived.

In contrast to the advanced architecture of bridges in the Safavid period, during the Qajar period, some old bridges were restored, and a relatively large number of simple bridges were built only for traffic facilities.

The bridge plays an important role in establishing, maintaining, and expanding communications, enabling humans to connect with more lands. Therefore, in the past, bridges were mainly built along caravan routes and on rivers and canals. The function of the bridge was not only in terms of communication but also included other issues:

A. Since the Achaemenid period, many bridges have had the dual function of bridges and dams, storing water for dry periods. The dams also divided the water, and part of the river water was sent to areas that did not have access to water.
B. Water mills were built next to and under many bridges and used the flow of water to turn the millstone.
C. Bridges have been a good place to control roads and caravans, and there were often guard stations on the bridges, in addition to tolls and trade taxes that were collected on the bridges.
D. Inside cities, bridges were a place of gathering people for entertainment and shopping.
E. In addition to economic, political, and social issues, these bridges also had a strategic and decisive role in military issues. Bridges have played an important role in communication between the front line of the war and operations behind the front in order to send equipment and food, and the conquest of bridges has played an important role in conquering territories. For this reason, bridges also functioned as defensive fortresses.

This article will explore five bridges in the area [*A/N: I recreated the plans and 3D models of the next five bridges based on the Reza Meshkini Asl dissertation (Meshkini Asl* 2001*)*].

7 Qaleh Juq Bridge in Maku: Geographic Coordination 39.281824, 44.473308

This bridge was built in the south of the city of Maku, near the village of Qaleh Juq, on Zangmar River, and is a Safavid period construction. In the past, this bridge provided the connection of the residents of Qaleh Juq with Maku. This east–west bridge is 10 m long, 4.5 m wide and 9 m high and has a large span and a small span. Due to the rocky nature of the bed, this bridge was built without foundations directly

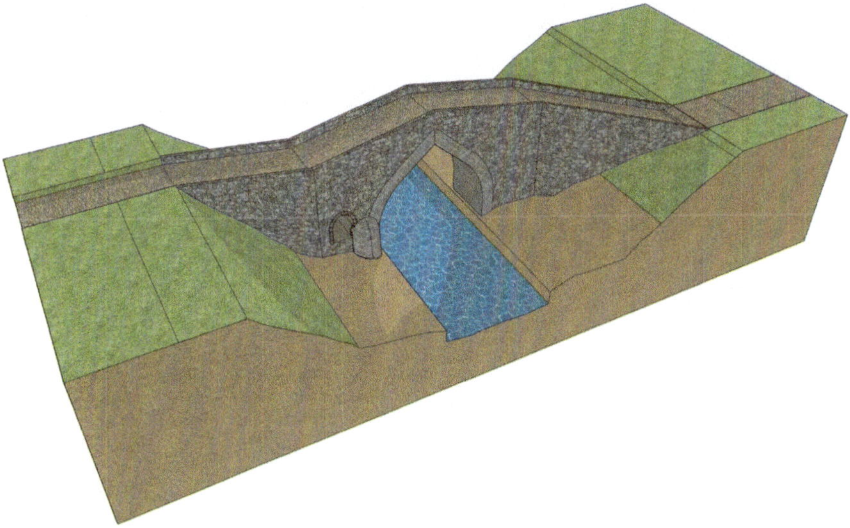

Fig. 9 Axonometric view of the Qaleh Juq Bridge. 3D created by the author

on the natural rocks of the river. It has a base of 4.5 by 2 m. This base in the opposite direction of the river has a semicircular water breaker with a diameter of 2.10 m, which is made in a conical shape with rubble stones and cut stones. The large span of the bridge is 9.10 m long, and its height is about 7 m above the water level of the river, which is built with a pointed arch with a sharp ridge with carved stone materials and mortar. The small span is covered by a Roman arch that is 1.7 m long and 2.1 m high. The body of the bridge is raised to the level of the bridge crossing with white and black rubble stones and lime mortar. On both sides of the passage, 60-cm-high parapets have been built with rubble stones. Because the road is curved at the beginning and end of the bridge, it is not aligned with the condition of the road (Figs. 9 and 10).

8 Panj Chesmeh Bridge in Maku: Geographic Coordination 39.29442, 44.51316

Panj Cheshmeh means "five span" or "five spring" in Farsi. The bridge is located 3 km east of Maku on the south side of the asphalt road and on the Zangmar River. The bridge is 43 m long, 3.2 m wide, and 5.5 m high. The bridge has four springs or spans and probably had five springs in the past. Apparently, after the destruction of its two arches, a large arch replaced them. The size of the first span, the largest one, is 5.7 m long and 3 m high, with a circular or Roman arch, the edge of which is decorated with white stones. The second span, with a pointed arch, is 3.9 m long

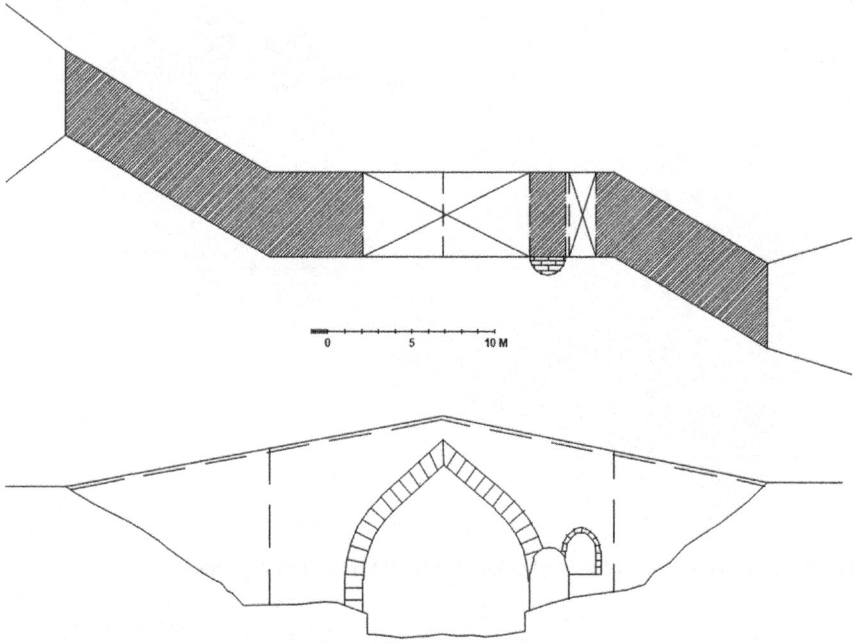

Fig. 10 Plan and elevation of the Qaleh Juq Bridge. Recreated by the author using the original drawing by Meshkani Asl (2001)

and 2.25 m high. The third span is made with a pointed arch and is 3.7 m long and 2.3 m high. The fourth span is 3.6 m long and 2.3 m high. There were four Armenian inscriptions on the eastern body of the largest arch, of which no trace remains today. The history of the bridge dates back to the early era of Shah Abbas Safavid (reigned 1588–1629), when religious minorities had a lot of freedom. With the construction of this bridge on the Zangmar River, Armenians could easily cross the river for the pilgrimage to Saint Thaddeus (Figs. 11 and 12).

9 Baran Duz Bridge: Geographic Coordination 37.41495, 45.14962

This bridge, which is about 15 km from the Oshnavieh-Urmia road on the Baran Duz River, is a Qajar period construction, and a significant part of it has been destroyed. The bridge is 40 m long, 5 m wide, and 4.8 m high. It has six spans or openings. The spans are equal in pairs. The middle spans, the third and fourth ones, are the largest, at 3.2 m long and 4.4 m high. The first and sixth spans are 3.2 m long and 3.4 m high. The second and fifth spans are 2.7 m long and 3.8 m high. All of these spans have been erected with a pointed arch by bricks measuring 25 by 15 by 5 and

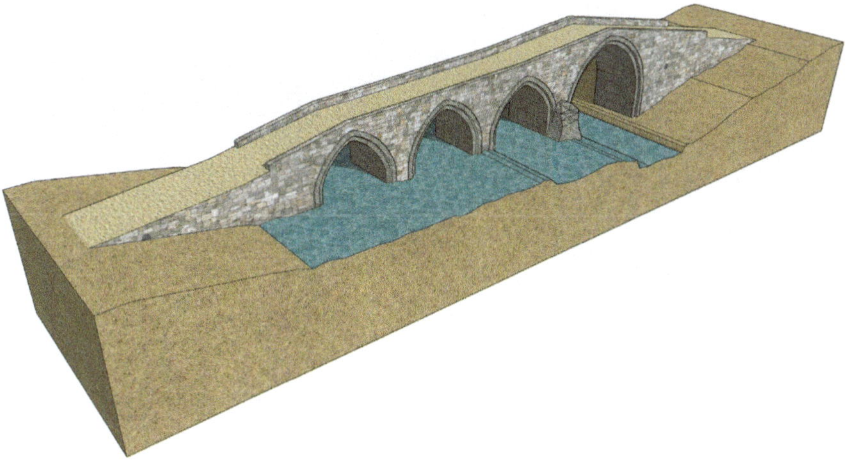

Fig. 11 Axonometric view of the Panj Cheshmeh Bridge. 3D created by the author

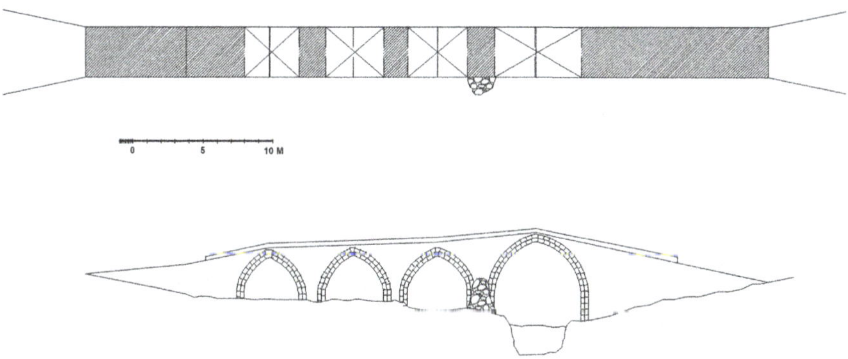

Fig. 12 Plan and elevation of the Panj Cheshmeh Bridge. Recreated by the author using the original drawing by Meshkani Asl (2001)

20 by 20 by 5 cm with lime mortar. In addition to the large spans, there were small arched openings on five pillars, of which only one remains intact. The size of this small span is 1.2 m long and 1.2 m high with a Roman arch.

The Baran Duz Bridge has five pillars made of large cubic boulders. These bases have triangular water breakers in the direction of the water flow. In the opposite direction of the water flow, the condition of the breakers is different, so that the first base and the fifth base have triangular water breakers, but the rest of the bases are flat. This is in addition to the blind, arched, decorative half-columns or guide mills are installed in the body of the bridge. These half-columns are made in the front of the bridge in a semicircle with a radius of 90 cm. In the studies carried out by Reza Meshkini Asl (2001), an irregular boulder was found at the base of the bridge, on

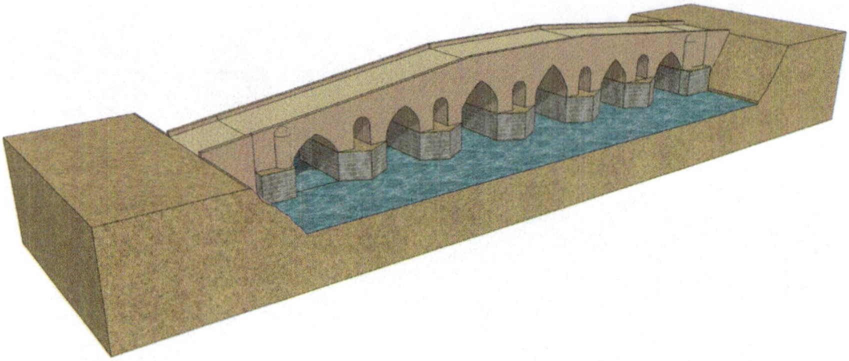

Fig. 13 Axonometric view of the Baran Duz Bridge. 3D created by the author

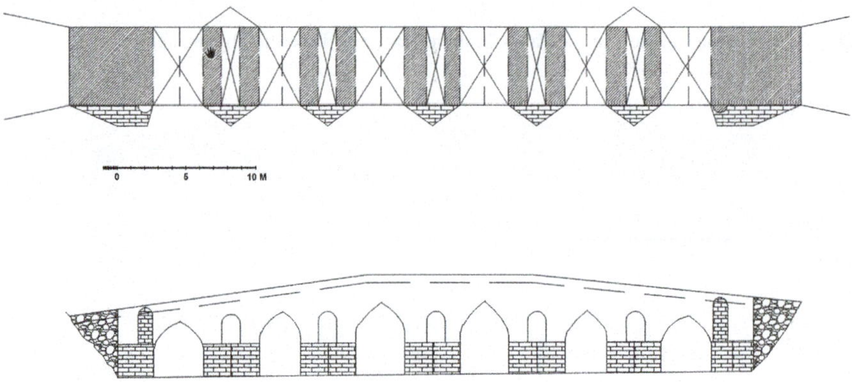

Fig. 14 Plan and elevation of the Baran Duz. Recreated by the author using the original drawing by Meshkani Asl (2001)

which the signs carved on it are in the form of two crosses, and it is probable that this bridge was built by Armenians in the late Qajar period (Figs. 13 and 14).

10 Khatun Bridge: Geographic Coordination 38.51199, 44.95725

This bridge is located 2 km southeast of Khoy, on one of the branches of the Qotur River, 50 m to the left of the Khoy-Salmas road. This bridge is one of the works of Ahmad Khan Danbali, one of the great men of Khoy, and was built between 1757 and 1786. The length of the bridge is 67.5 m, and the width and height of the bridge are 7 m each. The direction of the bridge is north–south. Most of Khatun Bridge is sedimentary and sandy; so, in recent years, due to the prevention of erosion and the collapse of foundations and arches, the province has strengthened all the lower levels of the bridge with concrete materials. The bridge has six stone foundations, which have triangular breakers in both directions of water flow to facilitate water flow.

Khatun Bridge has seven large openings and seven small arched springs. The dimensions of the first and seventh large springs are 4 m long and 4.5 m high. They are made of bricks measuring 25 by 25 by 5 and 18 by 18 by 5 cm with lime and gypsum mortar. The second and sixth springs are 4.5 m long and 5 m high. The third and fifth springs are 5 m long and 5 m high. The fourth spring (or middle spring) is larger than other springs or openings, and its length is 5.5 m with a height of 5.2 m. In Khatun Bridge, the sizes of small arched springs are different from each other, so that the third and fourth arches have dimensions of 2.2 m in length and 2.75 m in height, but the rest of the arched springs are 2 m long and 2.3 m high. The main materials of the building were irregular and regular white carved stones, rubble stones, and bricks with lime and gypsum mortar (Figs. 15 and 16).

Fortresses: From ancient times, trade and economic prosperity have been components of the permanence and stability of governments and states. Transportation of

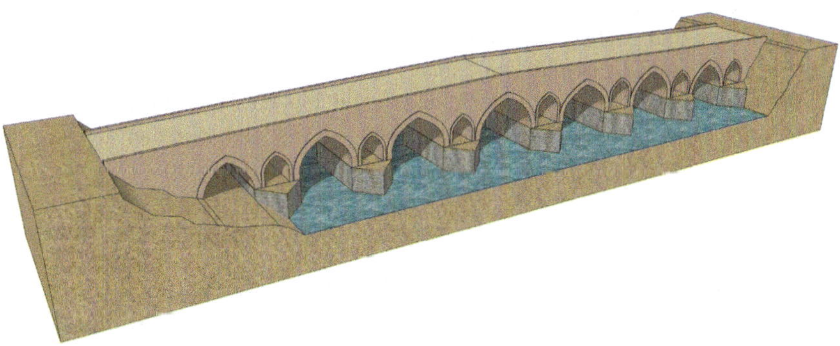

Fig. 15 Axonometric view of the Khatun Bridge. 3D created by the author

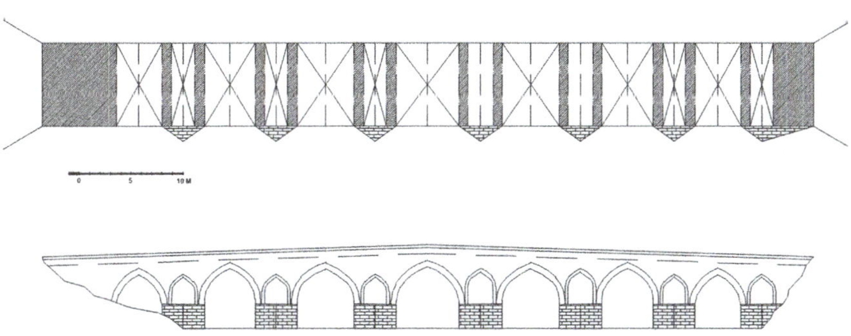

Fig. 16 Plan and elevation of the Khatun Bridge. Recreated by the author using the original drawing by Meshkani Asl (2001)

goods, services, and passengers requires important infrastructure such as accommodation centers along the route, proper access roads, and, most importantly, security. Hence, the construction of fortresses, bridges, caravanserais, and road signs has always been considered important by rulers.

During the first centuries of the first millennium BC, northwestern and Western Iran were widely exposed to repeated invasions by the powerful Assyrian Empire and the kingdom of Urartu, as well as the Scythian nomadic tribes. Such an insecure situation and instability forced the tribes living in the area to build strong fortresses. On the other hand, the invading governments, in order to protect their interests, built strong forts in the region, especially Urartu, which, by building many strong forts in northwestern Iran, popularized a certain style of military architecture in Iran. Today, the remains of many of these fortresses are left. The most important of these remains is the ancient site of Bastam in the northwest near Lake Urmia, which was formed in the seventh century BC.

In addition to the military architecture of Urartu, the tribes living in the region (Manichaeans, Medes, and other ethnic groups), in order to counter the aggression of regional and supra-regional powers, built forts and fortifications, which are described in reliefs. Assyrian and Urartian inscriptions have been found, and their remains have been identified and studied. The most important of these fortifications are located in the ancient sites of Hasanlu, Suleiman Prison, Jan Agha (in West Azerbaijan), Ziviyeh (Kurdistan), Noshijan Tappeh, and Godin Tappeh (Hamadan). Forts in this period were often built of stone materials and on hills and mountains, with high fences and congressional defensive towers. Ordinary people's houses were located outside the fortresses because their livelihoods were based on agriculture and animal husbandry, and their inhabitants took refuge in the fortresses only when there was danger.

During the Achaemenid period, due to the increased power and wealth of the central government and the expansion of the territory, which included large areas of civilization such as Mesopotamia, Asia Minor, Greece, and Egypt, military and defense architecture, like other artistic and technical branches, flourished.

In the Parthian period, according to historical texts and archeological data, many enclosed cities were built along with forts and military and defensive fortifications. The Parthians had many conflicts with the Seleucids and later with the Romans during this period, which justified the construction of fortresses and many military fortifications. Among the important fortresses of this period, which are located within the current borders of Iran, are Yazdgerd Fortress, Sarpol-e Zahab Fortress, and Zahak Hashtrood Fortress.

The Sassanid period witnessed the development and expansion of military and defense architecture, the prominent examples of which are the throne of Suleiman Takab and the fortress of Dokhtar Firozabad. In the military and fortification architecture of the Sassanid period, arch and dome coverings, carved stone materials, rubble and rubble stone, gypsum and lime mortar, as well as clay and brick were used. Proper use of circular defense towers that have better strength and durability is another feature of military architecture of this period.

With the fall of the Sassanid dynasty and the victory of the Muslim Arabs, although many of the fortifications of the Sassanid period were destroyed, many more were still used by the conquerors, local rulers, and especially the emirs who did not surrender to the Arabs and continued their resistance.

With the formation of national and regional governments, the construction of strong and defensive buildings flourished again. Castle and fortification building continued during the Ilkhanid and Timurid periods. The important building of Dokhtar Fortress in Miyaneh, which protected the bridge and the important communication road, belongs to this historical period.

During the Safavid period, along with other parts of architecture, the construction of defense and military buildings flourished, especially those buildings with the task of preserving and protecting important trade routes, caravanserais, straits, and important communication crossings, as well as border fortresses that have the task of confrontation upon the invasion of desert tribes or the Ottomans. The Miankaleh Fortress in Gorgan and Berdiuk Fortress in West Azerbaijan belong to this group.

During this period, due to the presence of European colonizers in the Persian Gulf, we see the construction of strong fortresses on the Iranian islands of Hormoz, Khark, and Qeshm by the colonizers.

Defensive architecture continued during the Afshari and Zandi periods. The military forts and fortifications of Kalat Naderi and the citadel of Karim Khan of Shiraz are among the constructions of this period.

In the Qajar period, we see the construction of many fortresses, especially fortresses that were built by local rulers and leaders and were used as abodes. During this period, due to the weakness of the central government, a number of fortresses and military fortifications became bases and shelters for insurgents and rebels. During this period, a number of cities were surrounded by fortifications, and government citadels were built with their own architecture and gates, some of which have survived.

In the late Qajar and early Pahlavi periods, with the development of various types of military weapons, increased security and other technological advances, especially in the field of communications, fortifications and the construction of other defense structures lost relevance and gradually faded.

The fortresses of Iran can be divided into two general types: plain fortresses and mountain fortresses. Plain fortresses were usually built according to a pre-designed plan and geometric shape. Their main defensive element was formed by ditches and fortifications. In mountain fortresses, there are no pre-designed and geometric plans. Because their plan was formed in proportion to the surface of the rocks and peaks, just as these rocks did not have a specific shape and design, neither did the built fortresses. This diversity is also reflected in the interior of the fortresses. In a number of fortresses, due to the natural condition of the mountains and rocks, there was no need to build additional towers and fortifications, because the rock walls and deep precipices played the role of fortifications. In such cases, only the entrance and possible crevices of the mountain were blocked and protected by strong walls.

Mountain fortresses also show great diversity in terms of construction materials. While a number of architectural spaces have been excavated in the rock, other spaces were built freely with common materials. The materials used in mountain fortresses, due to the abundance of stone materials in the area, are mainly rubble stone, rubble, and in some cases cut stone that was worked with lime, mud, or gypsum mortar, or, in some cases, dry stone walls. Bricks, clay, and mud were also used in these fortresses. The coverings used in these fortresses are flat, arched, or domed.

In mountain fortresses, defensive towers were built at irregular intervals and only where necessary. These towers from the Islamic period generally have a circular plan with various diameters. Small-diameter towers were mostly built in solid form with larger towers with empty and functional spaces in the middle. In the towers of the last category, spaces were installed on several floors along with staircases to climb to higher floors. Holes were made in the top of the tower and fortifications so that the guards, while watching, could shoot at the enemy during the attack and drop objects and materials on them.

The main gate of a fortress was usually located at the end of the lane leading to it, and two strong defensive towers were built on its sides. These gates had arched doorways with strong wooden doors. To prevent the doors from being set on fire by the enemy, they were usually covered with metal cladding. There were also various types of metal and stone doors. Fortresses usually had side gates and secret passages that were used in emergencies.

Inside the fortress gate, there were rooms for guards, which usually had openings to the outside. The architecture of the interior of the mountain fortresses did not have specific plans, and it was the elevation of the mountain or rock that determined the location of these facilities. The interior spaces were also diverse in terms of number, size, and construction and usually included the location of soldiers and fighters, the preparation spaces, and the headquarters of the commander or ruler of the fortress. In important and large fortresses, there have been numerous other buildings with different functions such as mosques, pavilions, production workshops, and prisons. The fortress commander's headquarters were built at the highest and most inaccessible point, often creating a strong castle in the middle of the main fortress. The architecture of these spaces was also diverse and was subject to different conditions and factors, especially the importance and size of the fortress.

One of the most important issues in the construction of mountain fortresses was the issue of water supply. Given the remains of the surviving fortresses, this issue was resolved in various ways. In a number of fortresses, including Alamut Fortress, water was brought from springs and streams flowing at higher altitudes to the surface of the fortress by creating canals and underground waterways with the help of interlocking pipes made of clay "tanboushe." A number of fortresses had running springs inside or nearby. In others, such as Lambsar Fortress, by creating an underground corridor from the fortress site to the river, the fortress dwellers could use the river water even in times of danger. In a number of other fortresses, such as Falak al-Aflak, Bam Citadel, and Dokhtar Fortress in Kerman, deep wells were dug. In other fortresses without such facilities (and in some cases in addition to such facilities), rainwater storage ponds or reservoirs were created. These ponds were built in certain parts of the rock or mountain where rainwater or snow flowed, and in some cases, by creating streams and grooves around them, they directed the flowing water into these reservoirs. The construction of these reservoirs was also completely different. Some of them were dug in the heart of the rock and some of them were made of common materials and mortar, while others consisted of small walls or dams that were placed in suitable locations with running water collected behind them. Some reservoirs were roofed with flat covers or with the help of arches and domes, some had temporary coverings, and some had no cover at all *[A/N: for more information regarding the fortresses in Iran, see Mollazadeh and Mohammadi* (2006)*]*.

11 Maku Fortress: Geographic Coordination 39.30045, 44.51261

The remains of this fortress are located on the northern front of the city of Maku and at the entrance of Maku Strait from the Bazargan border and are considered to be Safavid period monuments, probably built in the site of an older building. To reach Maku Fortress, one has to go north inside the city. Remains of the fortress tower and castle can be seen from the side of the Blue Mosque. The surviving ruins indicate the greatness of the building during its heyday. These include high walls, various corridors, nested inner fences, carved stone terraces, and watchtowers made of carved stone materials. The southern front of Maku Fortress was completely destroyed during the construction of later periods. In addition to carved stone pieces, clay, mud, and lime were also used in its construction. At the top of the fortress, there is a spring that boils from inside the mountain and that in the past could have provided the water needed by the fortress dwellers. In this fortress, there is also an inscription measuring 60 by 80 cm, an important part of which has become illegible today. However, according to readings that were made in the past, these are poems describing the fortress's destruction by the order of Shah Abbas in 1642. This destruction was due to the rebels and thieves sheltering inside the fortress (Sadraei 1994) (Fig. 17).

12 Chehriq Fortress: Geographic Coordination 38.07955, 44.5881

Chehriq, also known as Chahriq, Chiriq, or Charik, is a citadel located near the Turkish border. The remains of this fortress are located on a hill near the village of Chehriq Olya, 23 km southwest of Salmas, along the Zola River. Until 1828, Chehriq was an entirely Armenian-inhabited village. The original structure of the fortress may have been built before Islam. According to Samvel Karapetian, it is probable that the fortress served its purpose from at least the times of the Van Kingdom (ninth to sixth centuries BC) (Karapetian 2019, pp. 530–535). However, its main importance is related to the Islamic and especially the Qajar periods.

Fig. 17 Maku Fortress. Image from Wikimedia Commons

According to historical texts, Chehriq Fortress was the center of the Shakak tribe, located on the border of Azerbaijan with Turkey. Chehriq Fortress was famous from the Afshari and Zandieh periods onwards. One of the reasons for this fame was the imprisonment of Seyyed Ali Mohammad Bab, the Muqtada of the Babi tribe, in this fort during the Qajar period. After being imprisoned in this fort for some time, he was killed in Tabriz at the beginning of the reign of Nasser al-Din Shah. In 1911, during the war between the constitutionalists against the Russian invaders, this fort was used by the constitutionalists for some time.

There is not much left of Chehriq Fortress today, but, fortunately, a map of this fortress is available, prepared in 1863 by Colonel Alijah Abdul Ali Beg, a military figure of the Nasserite period. According to this plan, the height of the fortress was more than 80 m, and its area was more than 12,000 m^2. It had twelve main towers and four towers inside the mansion of Yahya Khan Chehriqi. According to this historical plan, the mansions in the fortress were divided into two parts, internal and external (Karimian Sardashti 2003).

Samvel Karapetian has prepared a general plan of the fortress and surrounding structures. In his drawing, the remains of two Armenian churches are visible (Figs. 18, 19 and 20).

Fig. 18 Remains of Chehriq Fortress. Image from Wikimedia Commons

Fig. 19 Chehriq Fortress in the early twentieth century. Photo courtesy of Dr. Farhad Tehrani from Golestan Palace Archives

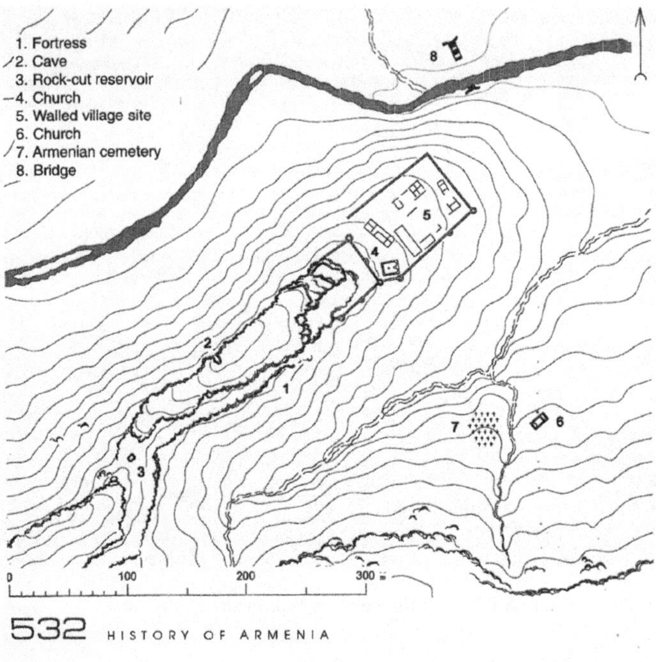

Fig. 20 General plan of Chehriq Fortress. Created by Samvel Karapetian (p. 532)

13 Berdiuk Fortress: Geographic Coordination 37.8294, 44.50206

The remains of this fortress are located above the heights of Kushan, 6 km west of the village of Berdiuk, at the strategic heights and 85 km northwest of the city of Urmia. In this place, there is a gorge called Hassani (Kotal), which is one of the channels used to enter Iran from the west.

It is useful to point out the possible Armenian root for the name "Berdiuk" ("berd" in Armenian means castle, and "juk" means small). Currently, the mule track that goes up to Godal constitutes a secondary passage through the Kurdistan chain, reaching the Turkish territory near the well-known Armenian monastic complex of St. Bartolomeo *[A/N: it is worth mentioning the hypothesis that in the past the path might have had much more significance]* (Alpago-Novello et al. 1978, p. 171).

Berdiuk Fortress was built during the Safavid period on top of a mountain overlooking a gorge. The architectural remains left in this place include the fort and its internal facilities, water reservoirs, a three-story building, and a watchtower. The materials used in the walls are rubble stone and sarooj (a traditional, water-resistant mortar used in Iranian architecture for the construction of bridges and water reservoirs; it is made of clay and limestone mixed in a six-to-four ratio). This wall is reinforced at various locations by semicircular supports. The main gate is located on the east side, which is made of limestone and covered with a pointed arch. In this gate, volcanic carved stones have been used for decoration, and there is a marble inscription. According to this inscription, the structure was built in 1668 by Taqi Sultan Ibn Nazar Beyk Ibn Ghazi Beyk. From the entrance of the fortress to the middle of it, there is a narrow passage created with a gentle slope in the rock. At the end of this passage is a three-story-high building with a rectangular plan, parts of which remain today in front of the entrance gate. The materials used in this building are rubble stone and sarooj mortar.

In the central part of the fortress, the remains of several buildings can be seen, one of which is more durable. This building is made with a square-shaped plan and materials of rubble stone and mortar and has an arched cover. It seems that this building was used for water storage.

On the northern front of the fortress and on the highest peak of the mountain, the remains of a large building can be seen, which was probably the watchtower of the fortress. This building or tower is built with a circular design and has an unobstructed view over the surrounding areas. There is a spring next to this tower.

On the eastern slope of the mountain, 500 m from the castle, there are the remains of a beautiful mosque, which, according to the existing historical inscription, was built two years after the castle was built (Figs. 21, 22 and 23).

Stacked Rocks as Road Guides: In the mountains of the region, pyramid-shaped rock masses are seen. These guides are created with dry stones techniques. They are usually more than 2 m high and about 50–100 cm in diameter. At the heights

Fig. 21 Berdiuk Fortress. Photo from Adriano Alpago-Novello's archive

Fig. 22 General plan of the Berdiuk Fortress and the adjacent structures. Plan from Alpago-Novello et al. (1978, p. 178)

overlooking the main and secondary roads, this type of rock mass can be seen. Their use is unknown, and they have their roots in the distant past of the region. The elderly believe that these were route-finding guides for caravans and travelers; the fallen rocks were put back in place during travels so that these structures were always higher than about 2 m and were visible from a certain distance.

Fig. 23 Mosque entrance. Photo from Adriano Alpago-Novello's archive

References

Alpago-Novello A et al (1977) Ricerca Sull'Architettura Armena Iran, vol I. Politecnico di Milano, Milan

Alpago-Novello A et al (1978) Ricerca Sull' Architettura Armena Iran, vol II. Politecnico di Milano, Milan

Haji-Qassemi K (2005) Ganjnameh, vol 17. Shahid Beheshti University, Caravanserais, Tehran

Karapetian S (2019) History of Armenia: Salmast. RAA, Yerevan

Karimian Sardashti N (2003) History and historical map of Chehriq Castle. Asar J 24(35):280–295

Kiani MY, Kleiss W (1994) Iranian Caravanserais. Mirase Farhangi, Tehran

Khanmohammadi B, Sadrai A (2013) Survey and introduction of Urmia caravanserai and the silk road route in West Azerbaijan. Asar J 34(60):40–54

Kiani MY (2000) Iranian architecture: Islamic period. Samt, Tehran

Kleiss W (1977) AlteWege in West-Iran. AMI 10:137–151

Mehdi F (1977) History of engineering in Iran (from the beginning to the present century). Pahlavi University Publications, Tehran

Meshkini Asl R (2001) Study and identification of the bridges in East Azerbaijan in the Safavid and Qajar periods, Tehran University

Mollazadeh K, Mohammadi M (2006) Fortresses and military fortifications. Surah Mehr, Tehran

Razani M, Mohammadi Sefidkhani R (2017) Pol-Dokhtar in Mianeh: history, documents and repairs. Asar J 38(78):19–32

Sadraei A (1994) Investigation of Safavid period buildings in West Azerbaijan, Dissertation, Tarbiat Modares University

Sadraei A (1996) Investigation of Safavid period buildings in west Azerbaijan from archaeological perspective. Asar J 17(26):186–203

Sadraei A (2005) Monuments of the Safavid Era in West Azerbaijan. Ganjineh Honar, Tehran
Siroux M (1949) Caravansérails d'Iran et Petit Constructions Routières, l'Institut français d'archéologie orientale
Shalchian H (1985) Bridge construction, vol I. Dehkhoda, Tehran
Tahini S (1990) Bridge design. Tehran University Publications, Tehran
Vandenberg L (1966) Archeology in ancient iran with an introduction by Roman Ghrishman, translated by Issa Behnam. Tehran University Publications, Tehran

Andreh Marouti, Iranian Armenian, is a practicing architect with Ph.D. in Preservation of Architectural Heritages from Politecnico di Milano. In his Ph.D. dissertation, he reconstructs the image of Yerevan and Gyumri during the nineteenth century through the lens of European travelers and highlights the preservation status of their major monuments. He participated in numerous symposia and gave lectures in Europe and the US.

Christian Churches in Salmas, Khoy, and Urmia Counties. Conservation Problems and Enhancement Perspectives in the Cultural and Territorial Context of Western Azerbaijan

Maurizio Boriani and Mariacristina Giambruno

Abstract The territory of Salmas, Khoy and Urmia Counties still preserves traces of the strong presence of the Armenian community and the churches they built to practise their faith. Damaged by the strong earthquake that shook the area in 1930, and no longer used for worship because the Armenian communities are no more present today, they are now often in a state of ruin. During three successive missions (2016, 2017, 2018), 22 villages and 29 churches were visited. The paper describes their current state of preservation, highlighting the problems of degradation and instability that afflict them, with the aim of outlining possible solutions and defining strategies for their new use in line with the needs of the communities that now inhabit the area.

Keywords Iran · Khoy · Salmas · Armenian churches · Preservation

1 In the Footsteps of Alpago Forty Years Later

Adriano Alpago Novello, together with a small group of young scholars, made two successive study missions in Iran to rediscover the traces of the Armenian presence and the architectural heritage it had been left in that country.

The occasion for these studies arose, as Alpago himself reports (Alpago Novello 1977, p. 1) from the presentation in Teheran of the exhibition *Architettura Medievale armena dal IV al XVII sec*, opened in Autumn 1975. The exhibition presented the study material on the architecture of historical Armenia collected by a group of researchers of the Institute of Humanities of the Faculty of Architecture of the Politecnico di Milano, Centro Studi e Documentazione della Cultura Armena, over the course of about ten missions. The limited and not organic researches carried

M. Boriani
DAStU, Politecnico di Milano (retired 2016), Milan, Italy

M. Giambruno (✉)
DAStU, Politecnico di Milano, Milan, Italy
e-mail: mariacristina.giambruno@polimi.it

M. Boriani and M. Giambruno (eds.), *Architectural Heritage in the Western Azerbaijan Province of Iran*, Research for Development,
https://doi.org/10.1007/978-3-030-83094-6_7

out until then in Iranian territory convinced Alpago to deepen the studies in order to discover and catalog not only "major" examples of Armenian architecture, that is, the most famous and well-known monasteries in Iran, but also a conspicuous number of small religious buildings realized in the remote villages and, in the seventies, challenging to reach. Marco Brambilla—part of the group that participated in these missions and author of the research reports with Alpago Novello, Nice Vecchione, and Gabriella Ulohugian (Alpago Novello 1977, 1978)—gives a good account of the "pioneering" spirit of these missions, in his essay *Adriano Alpago Novello and research on Armenian architecture in Iran*, published in this volume.

The first mission took place in Autumn 1976, in the areas of Tabriz, Khoy (Xoy), Salmas (Shapur), and Urmia (Rezayeh).

In Khoy, Salmas and Urmia Counties, Alpago identified 27 Armenian and 4 Assyrian churches from written or oral sources and documented 23 Armenian and 4 Assyrian churches, describing them with photographic images and/or with plans in scale 1:100 and documenting not only the main architectural features (domes, apses, windows, entrances, etc.) but also the consistency of the buildings, pointing out the collapsed parts on the plan (Alpago Novello 1977, pp. 130–137 and Alpago Novello 1978, pp. 30–36).

Already then, Alpago recorded "the severe and progressive deterioration of most of the buildings (also as a consequence of the abandonment of the villages by the Armenian community)" (Alpago Novello 1977, p. 2).

In Autumn 1977, the second mission took place at first retracing and completing the surveys in the territories investigated in the first mission and then moving northward to Julfa.

In this second mission, 16 churches were described and surveyed, even if the volume that collects the studies carried out (Alpago Novello 1978) reports the plans of about fifty cases, including also the buildings analyzed during the first mission.

Also in this case, Alpago notes "the precarious state of conservation of the buildings, which in many cases are total abandonment" (Alpago Novello 1978, p. 6), consequently stressing the urgency of the realization of a catalog as complete as possible of the architectural heritage, not only Armenian, of the region.

Forty years after the missions of the group of scholars led by Adriano Alpago Novello, three of our missions (2016, 2017, 2018) conducted in part of the territories then investigated have retraced the tracks in situations certainly easier than then, especially concerning the road links, but not without difficulties related in particular to the finding of some buildings of which only few traces are now preserved, or there is little memory today.

The missions organized by us—in which participated Farhad Tehrani, Majid Rassekhi (archival documentation, organization of surveys and consultations), China Yavali (drone shooting), and Panthea Karimi (control measurements and translations)—follow those carried out in 2011, 2012, and 2017 by Research on Armenian Architecture (RAA) and were solicited and supported by Salmast Heritage Association (SHA).

The primary purpose of the survey was to verify the current state of preservation of the buildings—which also changed in the three-year duration of the mission, to document the phenomena of degradation (of materials and structures). And to verify the problems related to their protection in order to outline possible strategies of valorization and new uses.

The investigated territories mainly concern the counties of Salmas and Khoy, with a brief detour in part of the County of Urmia, for the Assyrian villages of Jamalabad, Rehanabat, and Nachcevan Tepe.

In all, 27 churches were visited in 22 towns or villages. This volume describes (see the chapter *Directory of the Armenian churches of the Salmas and Khoy Counties*) only 23 churches present in 19 cities or villages of the three counties, chosen among those that are still recognizable in their consistency, although often now in a state of ruin (Figs. 1 and 2).

In addition to the Armenian churches, Assyrian buildings were also documented, both because of the common Christian roots of the two peoples and because of the close relationship (not always peaceful) that historically existed between the two communities.

During the missions, no architectural surveys were carried out, for which reference is made to those of the Alpago Novello group, but a photographic catalog was produced (addressed, among other things, to document the various forms of degradation and disruption). In 2018, a survey with a drone was also carried out to verify the state of conservation of the roofs of the different churches and, even more, to capture the relationship between these buildings and their territorial context.

Far from being an exhaustive catalog, this reconnaissance, and the consequent Directory, wants to take stock of what the state of this heritage is today in order to promote, through knowledge, its protection and preservation, and to verify what its

Fig. 1 Location of the villages of the Counties of Salmas and Urmia where the Armenian and Assyrian churches are set. Their consistency might allow useful conservation and reuse interventions. 1. Kohneshahr, 2. Savra, 3. Khosrova, 4. Haftvan, 5. Ghalasar, 6. Ghezelja, 7. Akhtekhaneh, 8. Drishk, 9. Sarna, 10. Malham, 11. Gharabagh, 12. Jamalabad, 13. Payajuk (USA Army map of Iranian Azerbaijan, 1943, USA Library of Congress)

Fig. 2 Location of the villages of the County of Khoy. 1. Gris, 2. Khoy, 3.Mahlazan, 4. Fanai, 5. Qoruq. USA Army map of Iranian Azerbaijan, 1943 (USA Library of Congress)

role could be, if properly enhanced, in this part of Iran (Fig. 3). A role linked to the possible development of religious and cultural tourism and, above all, of support to the collective memory of the plurality of thought and religions that historically characterized the area, as well as at the service of the communities that now live there.

2 For a Knowledge of Materials and Construction Techniques

While the well-known monasteries of St. Thaddeus and St. Stephen, protected by UNESCO as World Heritage Sites, feature dimensions, architectural and material characteristics comparable to the complexes located in the territory of the present-day Republic of Armenia, the churches of Salmas and Khoy counties differ profoundly in size and construction techniques.

Except for a few "major" examples (S. Sarkis in Khoy and S. Khateh in Mahlazan, to name but two), these are almost always buildings of modest dimensions, constructed for the "daily" religious practices of the Armenian communities once

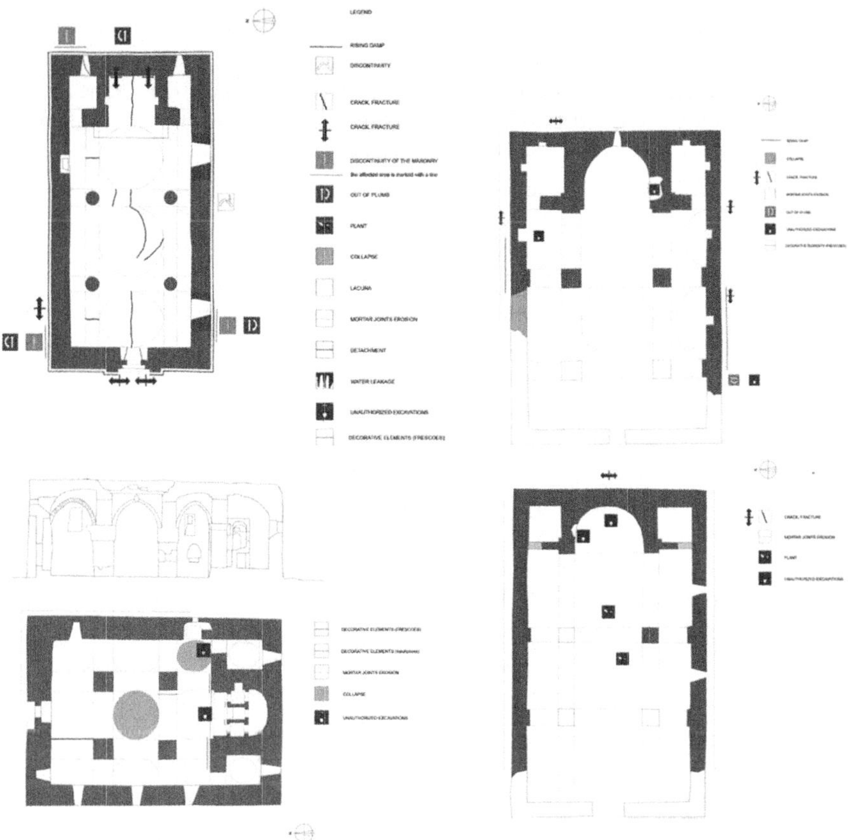

Fig. 3 Qualitative mapping of decay status drawn up following the survey (drawing by B. Silva, F. Vigotti). S. Astvatsatsin in Akhtekhaneh, S. Sargis Zoravar in Drishik, S. Sarkis in Fanai, S. Hovhannes in Sarnaq

numerous in these villages. Not noble and wealthy patrons wanted to embody their wealth and power in the churches, but small communities built these constructions, often more than one per village, to express and cultivate their religious faith.

Therefore, the construction materials of these buildings were available locally and easily procured without traveling great distances, as were the construction techniques employed those known and in everyday use by local communities and constructors.

> Basically, there are two construction choices adopted: the use of stone and clay (*khagil* or fired brick). Stone is used in the mountain regions, particularly in the Salmast, Giulfa, and Tabriz areas. In general, it is rarely the perfect masonry found in medieval Armenian architecture (cf. e.g. St. Thaddeus and St. Stephen's) with a double face of tufa filled with conglomerate, but rather more or less worked stone. (...) The use of unbaked clay (*khagil*) is typical of lowland areas: it is also found in the region of Salmast, but especially in that of Rezayeh and in the province of Peria (Alpago Novello 1977, pp. 29–30).

The description made by Adriano Alpago Novello summarizes very well what we have been able to find out during the inspections carried out.

Many of the churches of the area use masonry made of small-sized stone ashlars, generally roughly hewn and arranged in irregular courses, bound with large quantities of clay or gypsum mortar (Fig. 4). This technique is primarily found in the buildings of the mountain areas, as Alpago noted, even if it is found in some cases the coexistence in the same place of churches made of stone and churches made of raw earth, as in Mahlam or Khosrova, probably a sign of a different economic capacity of the community that built them or of different construction periods.

The angles and the access portals, sometimes even the openings, are often made of larger, regularly squared ashlars. It is the case, for example, at St. Astvatsatsin in Akhtekhaneh, where the masonry consists of small rough-hewn limestone ashlars, while the angles show large basaltic blocks as does the entrance portal.

In some cases, the main front is well set, while the side ones, as mentioned, have an irregular texture. It is visible, due to the collapses, in Mar Givargis of Khosrova. The masonry is sack-like, with a front of squared basalt ashlars and the filling of limestone flakes and mortar, a technique that recalls, as Alpago Novello recalled, the one used in Armenian medieval architecture (it should be remembered, however, that this is an Assyrian church, built at the end of the nineteenth century when there was already a Lazarite Catholic mission in the village, which certainly must

Fig. 4 Masonry in small-sized stone ashlars arranged in irregular courses. Note the lack of box behaviour due to corner collapse (S. Hovhannes church, Ghezelja-photo M. Boriani 2017)

have inspired the appearance of the European church noted by the Italian Giuseppe Anaclerio) (Anaclerio 1868, p. 59) (Fig. 5).

The masonry, when irregular, is covered in the interior and exterior fronts by a thick layer of plaster bound with clay.

Fig. 5 The masonry sack-like, with a front of squared basalt ashlars and the filling of limestone flakes and mortar of Mar Givargis church (Khosrova-photo M. Boriani 2018)

The plan is a basilica, divided into three naves, usually of equal size, marked by massive pillars with a square base.

The roofs are vaulted in stone, usually lowered, supported by the pillars of the aisles (Fig. 6). In correspondence with the transept, one of the vaults can assume the aspect of a dome or even of a *tiburium* (in Khoy, Mahlazan, Haftvan, and Akhtekhane). In some cases, as in the S. Astvatsatsin at Akhtekhaneh or at Fanai, the domes and the columns on which they are set are made of brick.

The frequency in finding domes and vaults indicates that specialists were probably employed, and there was certainly no lack of them in Persian territory. Several repairs

Fig. 6 The stone vaults of church of S. Astvatsatsin (Gris-photo M. Boriani 2017)

or reconstructions, carried out in the nineteenth century, both to the masonry and the vaults, with baked bricks, even present evident Persian characters (at Akhtekhane, for example), suggesting cultural permeability.

Access to the churches is through small doors of reduced height, often framed by a trilithic structure. Generally, the entrance is placed at the center of the west elevation, but accesses along the south wall are not excluded. An apse, usually semicircular, opens to the east, in correspondence with the nave, completed on the sides by two small rooms that serve as sacristy. The apse rarely emerges from the perimeter wall.

As already mentioned, in the villages of the plain, the presence of churches with walls in raw earth is frequent, according to the technique of pisé (*khagil* in Farsi language, that is clay mixed with straw) at the most with a basement of coarse stone material (Fig. 7). The plaster that covers the walls is also made of earth with straw (plaster of mortar-straw). The coverage is flat, supported by slender and close wooden beams, resting on thin wooden pillars with small and straightforward capitals (Fig. 8). In some cases, the wooden pillar, sometimes covered by a thick layer of plaster, can be supported by a stone base and worked, as in St. Gevorg in Malham (Fig. 9).

Sometimes, brick courses were added in the upper parts to the raw earth masonry, evident works of subsequent repair imagined (with not many reasons) to be helpful to better resist rainwater runoff.

The access, always in small dimensions, occurs from the west side in most of the visited buildings: also, in this case, a trilithic structure is used, generally recovered

Fig. 7 The stone basement covered by earth plaster of S. Sargis church in Khosrova (photo M. Boriani 2018)

Fig. 8 The roof supported by slender and close wooden beams, resting on thin wooden pillars of S. Sargis church in Khosrova (photo M. Boriani 2018)

from more ancient buildings or tombstones. The interior can be divided by pillars into three or two naves, except in smaller churches or chapels, which have only one.

3 The Current State of Conservation. A Reading of Decay and Instability Phenomena

On May 6, 1930, the area of Salmas was hit by a very violent earthquake that caused the destruction of sixty villages and the loss of about 2500 human lives (Tchalenco Berberian 1974). Many of the churches present at the time were destroyed, totally or in part, or at least reported the severe damage that can still be seen today.

The degradation brought in by that event has continued over time because of several causes and still makes the buildings remain since then in the state of ruins.

Surprisingly enough, if one compares the schematic surveys and photographs published by Tchalenko and Berberian in 1974 with the present situation, one observes relative stability of the conditions after more than forty years, in many cases without any maintenance.

Fig. 9 Detail of the wooden column covered with plaster, resting on a stone plinth in the church of S. Gevorg (Malham-photo M. Boriani 2017)

The almost total disappearance of the Armenian population from these territories is undoubtedly the first and most important cause, from which many others cascade, of the current state of these buildings. In fact, on the one hand, there is no way to watch over these buildings and, on the other hand, there is no interest in carrying out any maintenance work because the churches are not and cannot be used by the local population in any way, except illegally and improperly. Only a new use would allow to make convenient and durable the interventions necessary for the survival of these architectures. There are, however, many factors that strongly hinder this process, which will be mentioned later, together with the issue of reuse compatible with the architectural features and the state of preservation of these churches (and the areas under their jurisdiction), which today seems crucial to offer these buildings, even those now reduced to the state of ruins, a real chance of survival.

Many churches in Iranian West Azerbaijan are protected by law (29 for the counties of Khoy, Salmas, and Urmia, of which 5 in Khoy, 11 in Salmas, and 13 in Urmia—see the essay by Panteha Karimi published in this same book).

During our inspections in only five churches (between Khoy and Salmas), we found that recent restorations had been carried out (Haftvan, Akhtekhane, Gris, Malhazan, Garabagh, some still in progress at the 2018 inspection). Much of this heritage is unprotected and abandoned to itself, sometimes with attempts at misappropriation and often the subject of clandestine excavation activities to try finding hypothetical treasures that remained buried due to the earthquake. Intentional vandalism and stripping of materials to be used in new constructions are not excluded.

The lack of interest on the part of the population that currently inhabits these territories means that there is no daily care that could prevent the worsening of many decay phenomena; the diocese of Tabriz, on which formally these churches, for the Armenian part, depend, has no resources to maintain them and the Armenian communities that still live in Iran have moved to large cities and are gradually losing their original territorial roots; in addition to this, the even if sporadic restoration works underway at the initiative of local Heritage Offices are often interrupted for lack of adequate funding, leaving even urgent interventions unfinished.

The picture of the phenomena of degradation and instability that affect these buildings is severe in many cases.

Almost all of the buildings present, in fact, a worrying cracks pattern (Fig. 10), indeed originated by the earthquake of 1930 and accentuated by the causes described above and some subsidence of the foundations that are recognizable due to the presence of lesions which have a 45° trend. It is not rare to find cases of masonry discontinuities passing through the corners of the buildings, which determine the loss of

Fig. 10 Deep crack in the masonry of S. Sargis Zoravar church (Drishik-photo M. Boriani 2017)

Fig. 11 Cracks pattern and detachment of angle in the church of S. Sargis (Ghalasar-photo M. Boriani 2017)

the masonry grip, very dangerous because it causes the loss of the box-like condition of the building, which becomes in this way even more vulnerable to seismic events (Fig. 11).

Injuries and collapses of the top parts of the masonry or in the roofs imply that the walls, now disconnected from each other, can also present substantial out-of-plumb or overturning.

All these instabilities should be monitored over time to see if they are still progressing or if the static structure of the churches has reached a sort of new equilibrium. However, it is evident that a new earthquake, even less violent than the one of 1930, could destroy many of the surviving churches and the ruins of the others. In this sense, interventions of structural consolidation or, at least, the realization of appropriate provisional works appear urgent, given the high seismicity of the area and the more than 90 years that have passed since the last earthquake of high intensity.

To exacerbate the structural pattern outlined are many decay phenomena, the nature and extent of which varies according to the characteristics of the material and the building techniques used. Familiar to many of the buildings, in stone masonry or unfired earth, is the rising damp from the ground that manifests itself through erosion, more or less deep, of the bottom parts of the buildings, Further aggravating the situation are the phenomena of runoff, which erode the masonry base and are due to the lack or inefficiency of the drainage systems of rainwater, which falls during the short but intense rainy season (Fig. 12).

Fig. 12 Collapse caused by rainwater in S. Gevorg church (Mahlam-photo M. Boriani 2018)

In buildings made of stone material, a deep erosion is often detected of the mortar joints between ashlars or of the mortar binding the irregular stones that make up the masonry, especially in the parts most exposed to rain or affected by the presence of infesting vegetation.

This phenomenon decreases the resistance of the masonry and causes gaps in the same, dangerous especially during seismic events, even of modest entity, to which the area, as we said, is very subject.

The decay of the mortars protecting the roofs, due to lack of maintenance and, above all, to the infesting vegetation, facilitates the percolation of the meteoric waters and the erosion of the joints in the vaults causing their partial or total collapse.

Other gaps or lacks are due to the numerous clandestine excavations carried out to look for improbable "treasures" or pure vandalism. These interventions often involve the foundations of the buildings or concern removing many stone ashlars in correspondence with the baptismal font and niches in the apsidal area (Fig. 13). There have also been some cases—as mentioned above—of removing the best quality stone ashlars to reuse them in new constructions.

These actions damage the connection of the masonry with the respective foundations and/or reduce the resistant section.

Fig. 13 Clandestine excavations in the area of the foundations in S. Sargis church (Ghalasar-photo M. Boriani 2017)

In the case of buildings made of unfired earth, the most severe phenomenon is the erosion of the building material, often no longer protected by plaster. In many cases, there are gaps and cracks through which rainwater penetrates the masonry apparatus giving rise to the disgregation of the same, masonry discontinuities, and consequent failure and collapse (Fig. 14). In Mahlan, in S. Gevorg, the comparison of the state of the building in 2016 with that of 2018 highlighted severe damage of the apsidal part, caused by the washout produced by rain.

As for the oldest plasters placed inside the buildings, which in some cases still present traces of decorations or fragments of paintings, it should be noted that these generally show detachments from the substrate, discoloration, erosion, loss (sometimes affecting a large part of the surface where the plaster is now present only in fragments), and surface deposits of various kinds (Figs. 15 and 16).

4 Practices in "Restoration." Interventions from Local Preservation Bodies and Repair Works of the Population

There is no doubt that the situation of Armenian and Syriac churches in Western Azerbaijan and those found in the counties of Salmas and Khoy investigated in the

Fig. 14 Deep masonry erosion caused by rainwater (S. Nišan church, Jalabad-photo M. Boriani 2018)

recent study campaigns is critical. The earthquake of 1930, the abandonment of the cult caused by the relocation of Armenian communities, and the consequent lack of maintenance led many cases to their ruderization or, at least, to a state of preservation very close to it.

In such a situation, the theme of restoration work, of the theoretical principles to which it must adhere, of the relationship between reconstruction and authenticity, become particularly compelling, as does the need to plan and carry out urgent works that shall avoid further collapses or the worsening of instabilities.

In Iran, there is undoubtedly a solid tradition in the field of protection and restoration, born from the collaboration with experts from various European nations (see in this volume the essays by Brambilla-Ghadiri and Karimi), with substantial adherence to the principles of the Venice Charter, and therefore to the refusal of complete reconstructions, in respect of the authentic text. Despite this, if we analyze the interventions carried out in some of the most known churches of the area (see in this volume the appendix to Karimi's essay), for example, in the cases of St. Thaddeus monastery or St. Dzourdzour chapel in Maku, we can notice that they do not renounce to integral reconstruction of the lost parts or of the parts that are no longer preservable. It is certainly still an open matter, that of the ancient-new relationship and of the recognizability of the intervention, which is common to many cases, not only in emerging countries but also in contexts where the theoretical debate is much more

Fig. 15 Detachment, erosion and deposits in painting part of S. Sarkis church (Fanai-photo M. Boriani 2018)

consolidated, especially when the aim is the touristic "enhancement" of an asset or when there is the problem of new use that requires the completion of a building.

Let us consider the current consistency of the artifacts object of the recent survey, as it was often said in the state of ruins. A reflection in this sense before proceeding to an intervention should be conducted since it is clear that their documentary interest lies today mainly in being a testimony of deep religious faith, construction techniques, materials, and craftsmanship knowledge of a population. Handing these testimonies to the future, preserving their authenticity, could be challenging with an even more symbolic and practical meaning than the intervention of restoration or integral reconstruction.

What appears, however, from the interventions we could observe in the visited churches, recently conducted by the local authorities of protection, does not reveal these issues yet. If anything, it shows a bit of "backwardness" in the conservation techniques used. Perhaps because the buildings are considered less critical, perhaps because the isolation that Iran has suffered for years has prevented its officials in the peripheral areas from keeping up to date on these issues, the techniques used do not always respond to criteria of non-invasiveness and durability as proper conservation of the architectural heritage would require.

Fig. 16 Lacunae in the pictorial layer in S. Gevorg church (Haftavan-photo M. Boriani 2017)

It is possible to detect the use of cement mortars to fill in loss or restore joints between stone or brick quoins (Fig. 17), the complete remaking of plasters even when not strictly necessary, the failure to control rainwater that leads to deposits on surfaces that have been recently cleaned. The subject of meteoric waters, in particular, often does not seem fully resolved in the recent interventions carried out, neither in the lower part of the buildings nor in the upper part, so much that descending and ascending humidity are often present even after the works have been completed (Fig. 18). Therefore, particular attention must be paid to the damming of rising water, developing effective interventions, and falling water, with guttering or downpipes that are both efficient and compatible with the architecture of the buildings, unlike many of those that have been implemented so far. In the latter case, the usually practiced intervention consists of coating the flat roofs with aluminum foil and tar and the realization of metal downspouts to protrude, not fully effective and significantly impacting the architecture of the building.

Some significant cracks were repaired using the so-called undoing-doing technique, especially where the continuity of the masonry of the building was lost in correspondence with the windows of the sacristies (and therefore at the corners of the buildings). If this technique is the only one that can be used in many cases, it must be said that it is an extremely delicate operation that requires a specialized

Fig. 17 Joints filling with cement mortar and insertion of flooring incongruous with the character in the church of S. Khach Kristosi (Mahlazan-photo M. Boriani 2018)

workforce to be effective and avoid demolitions and reconstructions greater than those necessary.

Not dissimilar issues arise from small works carried out by the population, perhaps for the legitimate desire to open the building to worship. It is these interventions, even if minute, that profoundly modify the consistency of these buildings with materials borrowed from the construction site of new buildings, for example, new flooring, plaster, or heating systems.

If a renewed interest in the care of these buildings on the part of local communities is desirable, it would be necessary and appropriate at least to have a general "direction" by competent and trained personnel who would suggest to the population how to intervene and with what materials, even more so in cases where the need arises to rebuild parts in order to be able to perform rituals in an enclosed space or reuse the buildings for other purposes. The latter is a very delicate operation, as we said, that requires sensitivity and skills both in architectural design and in the project of preservation of the existing and, therefore, the involvement of experts in the field.

Although it is customary in Iran to carry out studies and surveys before interventions on historical buildings, it would seem appropriate to suggest for the churches that are the object of this research an accurate survey campaign, the study of existing

Fig. 18 Waterproofing with aluminium foil, a technique frequently used in churches in the area (S. Hovhannes church, Sarnaq-photo M. Boriani 2017)

materials, and the decay phenomena in order to establish the causes and plan efficient and lasting interventions, also considering the limited funds available that, as mentioned, often lead to leave the works unfinished.

Waiting for a desirable organic campaign of studies and surveys by the local authorities of protection, maybe supported by the technical faculties of the area, it would be opportune, however, to set up some interventions that are considered of the utmost urgency.

In the ruined parts, that is, for the masonry no longer protected by the covers, it would be opportune to avoid the worsening of the phenomena of degradation and instability caused by the infiltration of the descending meteoric waters, realizing adequate protections with mortars of the masonry crests (capping). This intervention would also allow controlling the infesting vegetation the roots of which can aggravate the instability.

The removal of the rubble, the infesting vegetation, and the constant cleaning of the surroundings and of the building itself would guarantee a certain decorum and, maybe, the possibility of public use, even if still occasional. Removing debris or accumulated material is also essential to prevent soil and vegetation from causing further degradation problems. In the same way, with simple nets or cloths, it would be appropriate to close the openings now without doors and windows to avoid the entrance of birds (an intervention often performed but rarely maintained).

Another crucial issue that should be addressed is implementing provisional safety works that could prevent new collapses while waiting for decisive and definitive intervention.

We could think of closing, with sheets fixed on wooden frames, those portions of the roofs or walls now open to the weather because of collapses. This simple intervention would allow preventing, at least partially, the entrance of rainwater and birds and dust and soil, of infesting vegetation and guano, which accelerate the phenomena of degradation, especially in the presence of painted or decorated surfaces.

The latter could be safeguarded while waiting for a definitive conservation intervention by protecting them with Japanese paper veils or at least sealing the edges with plaster mortar to prevent water and humidity from worsening the detachments currently underway.

Particular attention should be paid to the realization of provisional solutions that make the building safe while waiting for studies to verify if the crack pattern is in progression or the result of previous seismic events.

Wooden scaffolding could support damaged arches and vaults, and the box-like behavior of the buildings, which were now absent due to cracks and gaps, could be temporarily reconstituted, with tapes that act as "belts." These works, easy to put in place, are essential—even in case of earthquakes, as we know they are frequently given the high seismicity of the area—to try minimizing further collapses; they can be carried out by local labor but must be designed by engineers or experts in the field, to be effective and not, on the contrary, prove harmful.

5 Prospects in Conservation, Enhancement and Reuse

As pointed out during our surveys on site, many Armenian churches in Khoy and Salmas districts are no longer in use: Armenian people once living there have mostly migrated abroad (USA and Armenian Republic in particular), while those who stayed in Iran have moved to big towns (Tehran, Tabriz, and Isfahan in particular).

So, the abandonment of those buildings is the main problem to be solved in order to make sure about the effectiveness, in the long run, of restoration interventions either already performed or still to be made.

In fact it should be noted that maintenance interventions made in the last few years by local people, and mostly the restorations made by Iranian Conservation Bodies, very seldom could face, for various reasons, the issue of reusing those buildings.

A first step should be made toward conservation and awareness in using this heritage, by preparing an inventory of the current owners of the buildings and testing their availability in cooperating at their safeguard and reuse. In lack of direct initiative from the owners, specific policies might be defined for stimulating the assignment in use of the churches to interested public bodies.

A further problem refers to unauthorized excavation and vandalism performed by alleged "treasures searchers": a fact happened in almost all the churches of the region, very dangerous for the static conditions of the buildings.

Same thing happened in Armenian cemeteries, subject to looting and devastation which, in addition to damaging the remains of Armenian presence in those territories, did not even respect the poor remains of the dead, leaving their bones uncovered.

Obviously, such a diffused phenomenon means that local people does not keep things under control, because of fear, indifference, or even complicity. And, it is equally obvious that any form of repressive control has no hope of being successful in such decentralized areas. The only possibly successful action would be for local people to start feeling those churches as "of their own," thanks to their usefulness for the community. Fortunately, during our surveys, there was no lack of signs of interest and attention for that heritage on the part of those who accompanied us.

Thence, a plan for the reuse of those sites and buildings is even more necessary right for the aforementioned reasons, that is for people—even though belonging to different religion and cultural traditions—to make them "of their own" becoming their caretakers.

Some of the numerous churches now reduced to the state of ruins have an exclusively "archeological" interest, being impossible to think of their use for the local communities. In addition to safeguard works aimed at avoiding further collapses, the only suggestible policy would be protecting the remained ruins, preventing from abusive recovery of building material and vandalism, along with appropriate excavation campaigns. Thus, it would be possible to save whatever meaningful might be still buried after the devastating earthquake of 1930.

Other ruins instead have kept a monumental character of their own which would deserve to be enhanced. Though impossible would be a recovery of these buildings for new uses, yet their pertaining areas might be appropriately reused instead of being sometimes, as today, turned into deposits for agricultural products and materials or more frequently left abandoned among trash of any kind.

In fact, those areas can be easily identified and outlined since the churches were almost always surrounded by a wall (generally built with raw bricks), of which evident marks are still visible. In such cases, in addition to securing the ruins with appropriate consolidation works and enhancing any finds of historical interest, it would be well requalifying the surrounding area with light interventions (such as trees and shrubs, seats, water outlets, etc. ...) and allocating it as public space, obviously except for the cemeterial areas (Fig. 19).

Some ruins (like Mar Givaris in Khosrova, or Surb Sargis in Dirshk) have constructive features that might be of interest for the visitors, mainly with communication tools like illustration panels and/or small guide-maps. The usability of the areas as public space would help better control on the part of local people that is a kind of an indirect safeguard of the remains of the old churches.

Fortunately, a fairly good number of whole churches is left, albeit some of them in bad conditions. A fair part of those has already been the intervention subject from their pertaining Safeguard Bodies, while others (mostly the raw brick ones) were the subject of modest interventions by the local population. These churches,

Fig. 19 Haftvan, S. Gevorg yard in the village middle (photo from drone by Sina Yavarinia 2018)

to whom attention and care was already paid, shall be first of all secured in their components in danger of collapse and furtherly made the subject of a specific reuse and enhancement policy, without which no maintenance or restoration intervention would result, as already said, scarcely sustainable in the long term.

At this point, we ought to wonder what the vocation of use of these buildings could be for these buildings, and what criteria shall be adopted in order to outline new destinations, within the boundaries of maximum respect for their original historic and religious characters.

What criteria should be followed in view of those interventions?

We suggest some general principles, kind of a check list to be usefully followed:

1. Keeping in any case the worship destination and forecasting reuse activities consistent to the space, constructive, and material nature of the buildings.
2. Respecting the architectural and decorative characters and their historic-artistic and cultural values.
3. Providing a destination that will be of utility to the local people.
4. Enhancing the cultural potential of the buildings (even those in ruins), first of all to keep memory of history and then to support the identity of the different villages, including the possibility of their involvement in view of tourism—both religious and cultural—which might be of interest, in the future, also in this area of Iran.

5. Conserving also the structures in ruin, or wholly collapsed, on order to preserve interest in ethnographic and archeological research.
6. Involving both civil and religious authorities responsible for competence, as well as the local people and the Armenian Diaspora in the management and surveillance of the different sites. An environmental alarm system might protect the churches from vandals.
7. Devoting to use of collective interest the areas and buildings best fitting to the purpose.
8. Planning visit itineraries of the most important monuments and churches in the area, thanks to the collaboration of local people and in particular of university students.
9. Providing a map of the archeological and architectural heritage of the area, both in paper and in electronic format.

Such a refreshed interest by the Safeguard Bodies encourages the perspective of the reuse of those monuments, possibly supported by local authorities that might find there suggestive places, also fit for collective services inside villages totally or highly lacking them. The precincts of the churches—as said—are sometimes full of waste material, other times destined to orchards. They might be turned into useful green areas to serve the local population.

As said, the reuse of all these buildings would be possible only if property problems were solved, which is sometimes obscure today, but anyhow asserted by the Armenian religious authorities and communities still living in the big Iranian towns.

It would be good for the National or Regional Government to promote the definition of specific agreements involving the Regional Safeguard Bodies, the religious Armenian authorities, the civil organizations of Iranian Armenians, the Armenian Diaspora abroad and, of course, the villages and the local population, for a strategic plan of safeguard and enhancement of that heritage where right and duties of the different parts would be well defined. The plan might be prepared by a commission composed of representatives of the various stakeholders and integrated by technicians from local, national, or international universities (Fig. 20).

6 For a Tourism Development Perspective

Talking about perspectives of Iranian tourism in times of pandemia and economy sanctions might look off topic. Yet, in the hope, these big problems could find their solutions very soon, it seems appropriate to make some reflections. It is not about magnifying, as often happens, the economy advantages resulting from the enhancement of Cultural Heritage for tourism purposes rather than assessing, instead, whether the involvement of Armenian and Assirian Churches of our areas in tourism policies could help preserve that Heritage.

The Urmia lake region has great development potential for cultural and environmental tourism. It is located at few car hours from Tabriz town, one of the largest

Fig. 20 Akhtekhaneh, S. Astvatstatsin Church yard at village edge (photo from drone by Sina Yavarinia)

Iranians, where a small but active Irano-Armenian population lives. Moreover, Tabriz has an international airport, while two more national airports serve the towns of Khoy and Urmia (where a consistent Assirian community lives, as well as a small group of Armenian Iranians).

Two routes give land access from Turkey (via Kars-Dogubayazit-Maku-Khoy-Salmas) and from Nakichevan (a detour from the main road Marand-Tabriz). A third access, at Julfa on the Araxe, also allows entry to the Armenian Republic. Other minor road access are possible from East (Van-Khoy road in particular).

A railway is also active from Van to Tabriz running across the Qotur valley—where a spectacular arched reticular bridge is located—stopping also in Khoy. The railway travel passionates can go through the Istanbul-Ankara-Tatvan line and reach Van by ferry through the homonymous lake (a 6 h journey on the lake where on a small island stands the Aghtamar Monastery, one of the most important all over Armenia). Moreover, a Russia-Armenia-Iran rail connection, once solved some politic-economic problems, is in advanced realization.

The two most important Iranian-Armenian monasteries, St. Tadeus (Kara Kilise) and St. Stephen, were declared UNESCO Heritage in 2008; they are still destinations of important pilgrimages and can be reached with short detours from the roads coming from North. Thus, the Khoy area has great potential both for cultural and religious tourism, which could involve not only the Armenians living in Iran and in the neighboring countries, but also all the Diaspora ones, that already visit the Armenian Republic in large numbers.

The Urmia Salt Lake, the Zagros mountains, and the fertile plains of Khoy, Salmas, and Urmia with their orchards, besides, offer different opportunities for natural-environmental tourism.

Khoy town has important historic monuments (Bazar, Hammam, Ilkhanide Mosque, Shams Tabrizi tomb and minaret), and many Qajar palaces. In the surroundings, important remains of the Urartu period can be seen (fortifications, dams, canals) and of the caravan system built in Abbassid era (caravanserais, masonry bridges). Not far from the Urmia area are Zoroastrian monuments (Takht-e Soleyman—UNESCO Heritage since 2003) (Fig. 21).

Finally, the Armenian churches of notable appeal are at least three: Saint Khathé at Malhazan, Saint Sarkis at Khoy and Saint Gevorg at Haftvan (Salmas). Other important Assirian Christian religious buildings (Khosrova, Mar Church Sar Giz, etc.) fulfil the cultural resources of the area.

Apart from the two monastic complexes of Saint Thadeus and Saint Stephen, which are much promoted, specific initiative for the churches in the Khoy and Salmas County is lacking.

A decisive action could be the extension of the UNESCO safeguard today related to St. Tadeus and St. Stephen monasteries and their chapels, also involving the main churches of the region. The international appeal of the two monasteries might be the

Fig. 21 The region on the border between Turkey and Iran. In evidence the main territorial poles and the two monasteries of S. Taddeo and S. Stefano (UNESCO World Heritage) (from Google Maps)

steering wheel of the enhancement of the two churches which, although seeming of lesser importance, yet witness a millenary presence of the Armenian and Assirian communities in these territories. In that way, the two famous monasteries will be demonstrated as being part of a quite larger system of very old Armenian religious centers.

The scientific and administrative management of West Azerbaijan Christian sites might be entrusted to Tabriz University and Urmia University, also involving the territorial jurisdictions of civil and religious Authorities, as well as the Armenian and Assirian cultural associations that are still present and active both in Iran and in the Diaspora.

The study presented here actually aims at being a first contribution to the mentioned direction, though the potential for each site and the economy and management resources needed are to be assessed case by case, along with possible synergies with the local administrations and with the Iranian Tourism Organizations. An important contribution might also come from the involvement of UNESCO as well as of the different agencies dealing with the international cooperation in the field of Cultural Heritage, that would certainly participate enthusiastically, but definitely after the solution of the political problems that today make nearly impossible any actual collaboration.

References

Alpago Novello A, Brambilla M, Uluhogian G, Vecchione N (1977) Consistenza e tipologia delle chiese armene in Iran. Rapporto della missione 1976, Ricerca sull'architettura Armena-Iran, vol 1, n 17, Milano: Politecnico di Milano-Università Nazionale dell'Iran

Alpago Novello A, Brambilla M, Piacentini VF, Uluhogian G, Uluhogian H, Vecchione N (1978) Consistenza e tipologia delle chiese armene in Iran. Rapporto della missione 1977, Ricerca sull'architettura Armena Iran, vol 2, n 18. Politecnico di Milano-Università Nazionale dell'Iran, Milano

Anaclerio G (1868) La Persia descritta. Relazione di un viaggio per Giuseppe Anaclerio, Napoli, Vincenzo Marchese

Berberian M, Tchalenko J (1974) The Salmas (Iran) earthquake of May 6th, 1930, Annals of geophysics = Annali di geofisica

Cultural Heritage, Handicrafts and Tourism Organisation of Iran (2007) The Armenian monastic ensembles in Iranian Azerbayjan. Unique emblems from the fringe of its Heartland, UNESCO, Tehran

Haghnazarian—Armen Haghnazarian, The Monasteries of St. Thaddeus the Apostle and St. Stepannos Nakhavka, RSA, s.d.

Iranian cultural heritage organization I.C.H.O (2003) Takht-e Soleyman—proposal for the nomination of Takht-e Soleyman on the World Heritage List as a World Cultural Property, UNESCO

Karapetian S (2019) History of Armenia. Salmast, RAA Foundation, Yerevan

Maurizio Boriani Architect, born in Milan, Italy, 1946.

Full professor in Architectural Conservation at the Politecnico di Milano (Department of Architectural Design—School of Civil Architecture)—retired in 2016.

Former Director of Department of Architectural Design at the Politecnico di Milano.

Member of the Teaching College of the Ph.D. in Architectural Conservation at the Politecnico di Milano.

Former President of the Scientific Board of Cultural Heritage Conservation Center of the Politecnico di Milano.

Research Co-ordinator on behalf of the UNESCO, World Bank and European Community, Italian Ministry of Foreign Affairs and other private and public agencies.

UNESCO Consultant for: Albania (2006–2009), Macedonia, Montenegro, Kossovo, Myanmar; expert in international cooperation on behalf of Italian Ministry of Foreign Affairs (China (2009–2012), Vietnam(2017–2019), Armenia (2019–2014), Pakistan (2011–2012, Myanmar (2012–2015).

Research Topics:

History of Architectural Conservation, Conservation and management of historic towns and archeological sites, Cultural landscape preservation and management, Gardens history and conservation, Cultural Heritage and International Cooperation. About those themes he is author or editor of about 240 papers and books.

Mariacristina Giambruno Architect, Ph.D. in Preservation of Architectural Heritage and graduated in Restoration of Monuments, she teaches at the School of Architecture, Urban planning, Construction engineering, and Graduate School in Architectural and Landscape Heritage (Politecnico di Milano).

Full professor in Architectural Conservation at the Department of Architecture and Urban Studies (Politecnico di Milano).

Member of Faculty Board of the Ph.D. program in Architecture, Urban Design, Conservation of Housing and Landscape until 2012, from 2019 Head of the Ph.D. program in Preservation of Architectural Heritage (Politecnico di Milano).

The teaching and research activity is oriented on the history and theories of Architectural Heritage, Conservation of Historical Heritage, Historical Centers Conservation and Management, and guidelines and management plans to preserve and promote Cultural Heritage.

In recent years, she has focused on emerging and transitioning countries, working and directing international projects in collaboration with organizations such as UNESCO, World Bank, AICS in Albania, Iran, Armenia, China, Pakistan and Vietnam.

In these countries, she investigates the role of Cultural Heritage as a "trigger" for sustainable development.

As a speaker at conferences in Italy and abroad, she held courses and workshops in Armenia, Vietnam, Portugal, where she is part of a research unit coordinated by the Lusìada University of Lisbon. She authored about 150 publications on the preservation of Historical Centers and Architectural Heritage.

Directory of Armenian and Assyrian Churches in the Villages of the Salmas and Khoy Counties

Maurizio Boriani and Mariacristina Giambruno

Abstract The directory gives a detailed account of 23 churches surveyed in the counties of Salmas, Khoy and Urmia, chosen from the 29 visited from 2016 to 2018 because they still retain a recognizable architectural conformation that characterized them, although many of them are now in a state of ruin. For each of them, a sheet has been drawn up detailing the period of construction, the planimetric layout, the description and the state of conservation, defining the main problems but also the possibilities for reuse and enhancement. Where possible, the current state is compared with that found in the 1970s during the missions of Alpago Novello et al. (1976, Nice Vecchione, Consistenza e tipologia delle chiese armene in Iran. Rapporto della missione 1976, Ricerca sull'architettura Armena–Iran. Milano: Politecnico di Milano-Università Nazionale dell'Iran (1977)).

Keywords Iran · Churches · Khoy · Salmas

The directory gathers and systematizes the data collected during the surveys carried out in 2016, 2017 and 2018, comparing the current situation with what was found by Alpago Novello in 1976–77 and by Berberian and Tchalenko in (1974). In order to complete the framework of information, when necessary, the work carried out by Research on Armenian Architecture (RAA) (Karapetian 2019) and some Western and Persian bibliographic sources were also used Al-Jeloo (2010), Berberian and Tchalenko (1976), Bosworth (1995), Brambilla and Tehrani (2020), Mauroy (1972), Deilamqani (2006), Flemming 1986), Kleiss and Kroll (1992), Kleiss (1973), Moradi (2020), Flynn (2016).

Present Address:
M. Boriani (✉)
DAStU, Politecnico di Milano (retired 2016), Milano, Italy
e-mail: maurizio.boriani@polimi.it

M. Giambruno
DAStU, Politecnico di Milano, Milano, Italy

M. Boriani and M. Giambruno (eds.), *Architectural Heritage in the Western Azerbaijan Province of Iran*, Research for Development,
https://doi.org/10.1007/978-3-030-83094-6_8

The buildings presented in the directory are exclusively those that still retain architectural consistency and interest in their preservation, albeit in the state of ruins, while those of which only a few remnants remain at present were excluded.

In order to make the data collected comparable and to favour a synoptic reading of the various churches, the repertory has been organized in cards. The denomination gives both the current Persian name and the name used in the missions directed by Alpago Novello.

The cards report:

- Location, i.e., the position of the church about the nearest town;
- Date of construction, the primary date of construction of the building, mainly deduced from the writings of the Alpago Novello group, compared, where existing, with the dates reported by Tchalenko and Berberian, Kleiss or those deduced from other more recent studies;
- Planimetric layout;
- Consistency and state of the building in the 1970s, where significant parts of the descriptions of the buildings made by Alpago Novello and by Tchalenko and Berberian are better compared with the current situation;
- Description and state of preservation, with particular emphasis on the currently present phenomena of degradation and instability; in case the churches be no longer used for worship, some reflections are also reported on their potential for reuse and enhancement.

Some general or detailed photographs of their current state are added to the cards, compared, where available, with historical images in order to better document the modifications made to them over the last fifty years.

1 Salmas County

AKHTEKHANEH (AKHTAKHANA)

It is situated 10.4 km east of Salmas in a plain lying at an average altitude of 1318 m. Until a century ago, the village was surrounded by malarial swamps. Recently established, there are no records earlier than the mid-nineteenth century (in 1896, it had 130 houses, 45 of which were inhabited by Armenians, with 336 inhabitants). In 2011, the village had seen only a modest expansion since the end of the nineteenth century, with 144 houses and 500 inhabitants, all of whom were ethnic Turks. (Karapetian 2019, pp. 99–100).

S. Astvatsatsin/Asdvadzadzin

Location: on the northern edge of the settlement.

Period of construction: 1342, reconfiguration in 1892 (Alpago Novello et al. 1977).

Planimetric layout: basilica with three domes.

Consistency and condition of the building in the 1970s: "The church of S. Asdvadzadzin in Akhtekhaneh survived entirely even though it was fissured. It is a fourteenth century construction with the traditional double wall (kiln-brick inside, rough stone masonry outside), and the interior N-S arches developed cracks near their keystone" (Berberian and Tchalenco 1974) (Figs. 1 and 2).

Description and State of Preservation

The area where the building is located is currently a farmyard, cluttered with waste materials and rubbish. There are only few remains of the earthen fence that must have surrounded the church.

The building is thought to date from 1342, and, according to the plaque above the entrance door, it was "restored", essentially rebuilt, in 1892.

An analysis of the masonry might suggest that the remaining portion of the fourteenth-century church is made of stone ashlars, while the brick parts might date from the so-called nineteenth-century restoration, showing clear influences from the Qajar period. According to this hypothesis, the north, south, east and partly west perimeter walls date back to the fourteenth century; the four columns, the vaults and the inserts corresponding to the two windows on the south and the entrance portal

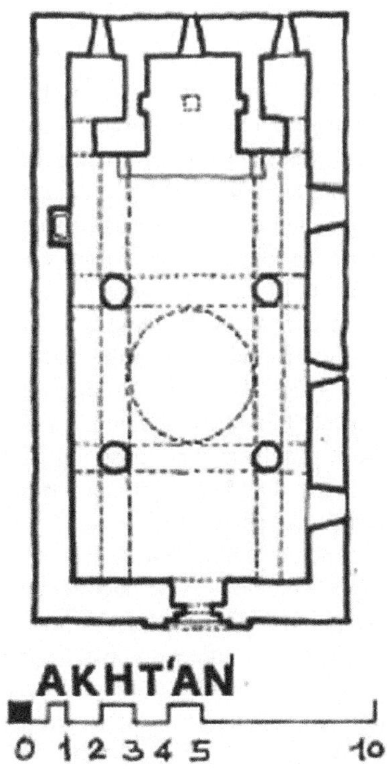

Fig. 1 Akhtekhanéh, S. Astvatstatsin: plan (Alpago Novello et al. 1977)

Fig. 2 Akhtekhanéh, S. Astvatstatsin: inner view towards the apse area (photo M. Boriani 2017)

on the west, all made of brick, could be ascribed to the end of the nineteenth century. Evidence of this hypothesis could be the size of the open windows in the south wall, much larger than typically found in the oldest Armenian churches in the Salmas area; the brick portal, clearly influenced by Iran; the two-star vaults and the large windowed dome, also attributable to Iranian artisans. There must have been a *portico* in front of the entrance to the church, as evidenced by the presence of a remnant of an arch shutter.

Therefore, the building is a rare example of a stylistic mixture, probably due to a desire for modernity on the part of the Armenian patrons and/or the use of workers who were also active in Iranian religious and civil buildings. The church has unusual monumental characteristics for having been built in Islamic context; it is much brighter than those previously built. Because of this exceptionality, the building deserves great attention and requires interventions appropriate to its historical consistency.

Another exciting feature of the building is the arches that form the naves and support the domes: the central ones are rounded, and those in the first and third bays are pointed. This aspect also makes the church unique in the local area.

There are three domes: the central one has a cylindrical drum—somewhat irregular—in which there are four large windows, surmounted by a hemispherical dome. The other two domes are supported by four pendentives, which in turn carry triangular sails. The central dome is thus higher than the two side domes and appears so from the outside.

The west façade has a round stone arch engraved with two rather crude figures of angels and, in the centre, the 1892 dated inscription mentioned above. Above this arch, a second one, also round, is made of bricks, arranged to form an X-shaped decoration. Two lateral pilasters framed a brickwork with bricks arranged to form a decorative motif, which has now largely disappeared. The whole structure must have been slender and at least as high as the first dome, connected by a gabled wall, a sort of tympanum, which joined the north and south walls. The façade must therefore have had an unusual monumentality.

From some regular holes in the brickwork, probably a sign of beams that no longer exist, one can assume the presence of a front porch protecting the entrance and supporting the bell tower, all probably made of wood, since no significant wall traces are found on the ground. As the use of bells was generally prohibited in Persia, they were likely replaced by a wooden board (*simandro*).

The inspection carried out in 2017 had shown that the building was in a severe state of disrepair: the masonry arch of the west façade was damaged at the kidneys, the left part of the tympanum had collapsed and revealed the first and second cupolas, which in turn had severe cracks suggesting an opening of the first dome (Figs. 3 and 4).

Fig. 3 Akhtekhanéh, S. Astvatstatsin: west front in 2017 earlier than the restoration works (photo M. Boriani 2017)

Fig. 4 Akhtekhanéh, S. Astvatstatsin: west front in 2018, during the restoration works (photo M. Boriani 2018)

The entire left side of the crowning of the façade had collapsed and what remained was in a severe state of decay. The perimeter masonry showed deep erosion of the mortar joints for about five courses of stone in the lower part and the same number in the upper part, quite severe in the south wall, which also showed small instabilities, probably due to the presence of infesting vegetation above the roof. The angles, consisting of well-made ashlars superimposed on courses of hewn stones, had heavily eroded joints.

As for the north and south perimeter walls, some through cracks were also visible, producing a wall discontinuity between the west façade of the church and the side walls. The east wall also showed a serious pass-through lesion at the north sacristy (Figs. 5 and 6).

Inside the church, the keystone of the two transversal arches supporting the central dome was damaged and a portion of the arch was lowered as a result; there were also small wall discontinuities at the keystones of the pointed arches. Finally, there was an evident total detachment of the back apse wall from the body of the church.

The south sacristy was also somewhat damaged, while the north sacristy was in better condition. In addition to the lesion already visible from the outside, the south wall had a second lesion at the point of connection between the brick and stone masonry of the first window. Damage from rising damp was visible both inside and

Fig. 5 Akhtekhanéh, S. Astvatstatsin: north front in 2017 earlier than restoration works (photo M. Boriani 2017)

outside: large portions of the plasterwork had fallen off, and the remains of these small collapses remained on the floor.

A visit in 2018 showed that the state of the building had changed considerably, due to significant, though not definitive, restoration work: the interior was re-plastered and whitewashed, the joints of the brick pillars were stencilled, and a new terracotta floor was laid, a few centimetres higher than the existing one. The altar, now totally destroyed, was also rebuilt. The two domes were restored, and the roof was protected by a bituminous sheath covered with aluminium foil (a type of protection against rainwater infiltration that can be seen everywhere, even in rural buildings in the various villages). On the outside, the wall of the east façade was repaired with sew–unsew interventions, while the north wall still had a rather severe lesion which local sources report should be repaired in the same way. Since these lesions were undoubtedly caused by the 1930 earthquake, which caused a detachment with the consequent translation of the north façade and the rotation of the west façade, it would seem appropriate to also carry out seismic improvement work by inserting some chains to restore an adequate box-like behaviour to the building.

Moreover, on the west façade, the sizeable rough stone placed at the south-west corner of the basement has been removed but not yet replaced with another. That creates a potential danger to the structure, although the thickness of the existing masonry is such that urgent intervention is not required.

Fig. 6 Akhtekhanéh, S. Astvatstatsin: the north-east corner in 2018 after restoration (photo M. Boriani 2018)

The current restorations have repaired the upper part of the stone masonry and only part of the brick insert on the west front.

The restorations were carried out under the direction of the Superintendency of Western Azerbaijan and with a contractor from the city of Maku. On the whole, they do not reveal any particular criticalities, although it would be advisable not to undertake complete reconstructions, for example, of the brickwork portions, which should be left in their current state, safeguarding the wall crests with a capping intervention to prevent the ingress of rainwater. The work had not been completed at the date of the last inspection.

Since the work still to be done is low cost and the necessary skills are available—except for the installation of the chains, for which further training of the local workforce would be required—it would be desirable for it to be completed in a short time and for control systems to be installed to prevent birds from entering through missing windows and openings in the domes.

Crucial, however, is a decision on the future destination of the building, which does not appear to be used as a place of worship: it could be a good attraction for visitors and thus constitute a small resource for the community. According to the local population, the church should be converted into a village library and the enclosed area into a public space, open to the community. This is an important project, not only for the local context but also because it could set a precedent to be exported to other villages with unused religious buildings, subject to vandalism, and at risk of physical deterioration.

DRISHIK (DIRESK/DIRSHK)

(Persian: درشك; also known as Derīshk).

It is a village located 6.5 km SE of Salmas, on the edge of an agricultural area. Until 1928, the population was entirely Armenian, and there were more than 300 houses. Reduced to ruins by the 1930 earthquake, "Village completely destroyed. Population 1930: 390, 2 casualties, most slept out. Present population: Turk" (Berberian and Tchalenko 1974). A community of Turkish immigrants occupied the abandoned houses and land.

At the 2006 census, its population was 508, in 116 families of Iranian Turks.

S. Sargis Zoravar/Sarkis Zoravar.

Location: at the northern edge of the settlement.

Period of construction: 1400 (Alpago Novello et al. 1977).

Planimetric layout: apsidal basilica.

Consistency and condition of the building in the 1970s: (Figs. 7 and 8).

Description and State of Preservation

The church is located at the northern edge of the village in an area not protected by fences that appears to be used for various activities related to agriculture and breeding.

The existence of a church in this place seems to be documented as early as the fifteenth century, but the present ruin can be dated between the end of the seventeenth and the beginning of the eighteenth century. The building is mentioned in documents from the second half of the nineteenth century. A description from 1931 suggests that the church had withstood the previous year's earthquake: "a spacious church that looks like that of a monastery" (Karapetian 2019, p. 130). In 1968 it was described as "… built of stone and lime … it has arches remaining on four pillars" (Karapetian 2019, p.130).

Visited by Kleiss in 1969 and by Alpago Novello in 1976, it was already ruined. A description from 2013 reports that " … the roof collapsed because of the Salmas earthquake of 1930 because it was made of wood … judging by the remains of the

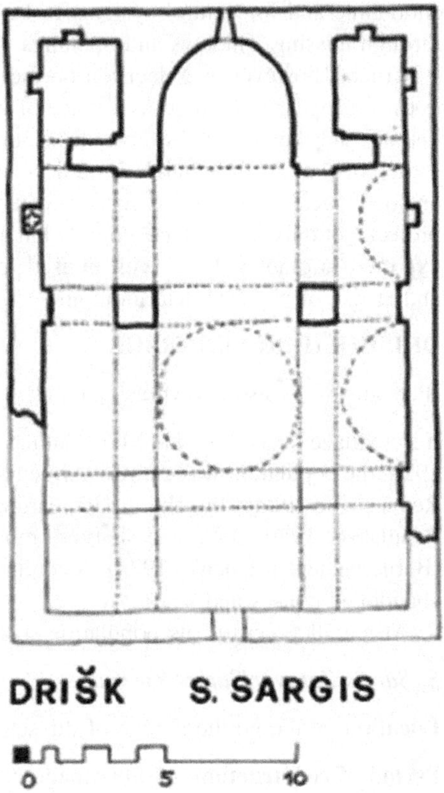

Fig. 7 Dirsk, S.Sargis: plan (Alpago Novello et al. 1977)

foundations of 4 pillars that were in the centre of the church, it probably had a dome" (Karapetian 2019, p. 131).

According to RAA, the church had a nave and two side aisles with a vaulted roof and probably two false domes in the centre. It was built with partially hewn limestone, connected by low-quality mortar. On one pillar is a stone with some inscriptions, probably what remains of the previous church (Karapetian 2019, pp. 131–132). Overall, although only a few portions are preserved, it is an exciting building that would need urgent consolidation and protection of the walls and increased protection against vandalism.

The building is in a state of ruin, and its condition must have worsened from the 1970s to nowadays if we compare its current state with the findings of Alpago Novello in 1976: the masonry on the east side with the apse and the two sacristies, about half of the north wall and almost all of the south wall survive without the roof vaults. The west wall of the entrance is lost. At present, two of the four pillars also remain, up to the height of the arches shutters and the two sacristies on either side of the apse, with the stairs leading to the upper rooms.

The north wall has lost about 3 m of its length, and the phenomenon is still in progress, despite some spontaneous interventions such as the rebuilding of the

Fig. 8 Dirsk, S.Sargis: the apse area in 1977 (Alpago Novello et al. 1977)

joints between the base stones (for 2 or 3 courses) in the western portion. Up to about half of the original church, the north wall is in good condition; the remaining part, reduced to a sort of spur, is losing the joints between the wall ashlars.

The south wall is almost intact but has an out-of-plumb of about 25 cm, decreasing as one reaches the west corner. There is a general lack of mortar joints and several cracks, including two through ones. The most serious is in the central area, where part of the top masonry has been lost. The corner with the east wall, made of large square stones at the base, is in good condition except for the usual disintegration of the mortar in the joints of the two lower courses and the top ones.

The east wall presents the same problems, accentuated by a severe lesion at the north sacristy, affecting two finely worked stone windows. The central window is engraved with solar symbols and a small cross (Figs. 9 and 10).

The apse is relatively well-preserved, but a small lesion in the bowl and a hole, with a considerable loss of material, made by intruders on the right side to enter the south sacristy, which was closed at the time of this vandalism. The right side of the

Fig. 9 Dirsk, S. Sargis, details of inner south wall (photo M. Boriani, 2017)

apse contains a massive block of black basalt with an Armenian inscription, which appears to be reused material from an earlier building.

The left side of the apse, corresponding to the north sacristy, has, in addition to the severe pass-through damage mentioned above, gaps and disintegrating mortar. A clandestine excavation has weakened the masonry in the basement area.

Nevertheless, the south sacristy is in a good state of preservation, and it would be sufficient to remove the stones lying on the ground. A small staircase leading to the upper rooms, which are not accessible today, is recognizable. The north sacristy is in a worse state of repair, partly due to an illegal excavation in front of the small staircase. The two remaining pillars appear to be in a reasonably good state of conservation and retain substantial traces of plaster.

It would be advisable to secure the ruin which, as mentioned, is of some architectural interest. Some measures should be taken to this end, including sealing the wall crests to prevent the penetration of rainwater and the effects of erosion and the related freeze/thaw cycle; mending the walls damaged by vandals to restore continuity to the surviving walls; repairing the cracks and setting up temporary works to prevent further collapses.

Finally, the stones scattered around the perimeter of the church and the soil from the clandestine excavations should be removed and the holes filled in to prevent rainwater from pooling near the walls.

Fig. 10 Dirsk, S. Sargis, the outer east front with a typical crack in correspondence of two small windows (photo M. Boriani, 2017)

The fencing of the site and the general cleaning of the area would allow for greater decorum of the place and its possible use by the community as a public open space.

GHALASAR (QALASAR)

Qaleh Sar (Persian: قلعه سر, also Romanized as Qal'eh Sar, Ghalasar, Kuilasar and Kuilavar).

It is village 6.7 km east of Salmas on a plain extending east of the Zola River, at an average altitude of 1336 m. It is documented as an Armenian village from at least the seventeenth century, and an important *Scriptorium* developed there, the most active in the Salmas area judging by the number of manuscripts still preserved today. Ghalasar prospered until the end of the nineteenth century when it had an Armenian school and library, a theatre and a public bath and many fine private houses. It was considered one of the most prosperous and best-built villages in the region.

From 1894 onwards and especially during the Turkish occupation of 1914/15, the village suffered continuous war, persecution and devastation. The population, which in 1905 amounted to 684 inhabitants distributed in 121 houses, was reduced to 79 due to the destruction caused by the earthquake of 1930, which also caused the partial collapse of the church. In the 1970s, the population was predominantly Turkish (Berberian and Tchalenko 1974).

S. Sargis/Sarkis.

Location: outside the settlement, isolated in the countryside.

Period of construction: eighteenth century or earlier (Alpago Novello et al. 1977).

Planimetric layout: basilica with three naves (today hardly recognizable).

Consistency and condition of the building in the 1970s: "In Ghalasar, the roof of S. Sarkis collapsed partly, and the northern wall fell to the north. This collapse to the north, also observed for the small parapet wall surrounding the roof of the church, is exceptional and may be compared to the dominant northward displacement of the tombstones in Khosrova (…)" (Berberian and Tchalenko 1974).

"… located on a modest relief of the ground (…), it is a basilica-type construction, with three vaulted naves, and three small domes on the lowered axis. The roof has almost completely collapsed (earthquake 1930), but it is still clearly legible from the remains of the south side" (…) The vaults appear to have been abundantly lightened by the insertion of amphorae: on the outside, the roofing system was concealed by the raised field masonry, which formed a sort of parapet. The masonry apparatus consists of rough stones, with corner ashlars of squared black basalt and wooden tie rods inserted in the walls, placed horizontally at about 70 cm from the floor. The windows are splayed on the inside, and the entrance portal has a lowered arch with a superimposed pointed arch lunette. (…) The church was much better preserved at the time of Kleiss' visit (1969)—now the northern side has completely collapsed" (Alpago Novello et al. 1977, p. 51) (Figs. 11 and 12).

Description and State of Preservation

There is no information about the foundation year of the church, which certainly dates back to earlier than 1655, the year of publication of a Gospel that is said to have been realized in the building.

It was most likely rebuilt around 1720, and traces of the earlier church would have been recognizable in two carved stones, which have now disappeared and were reused as building material in the new church (Karapetian 2019, p. 143). A cornerstone commemorating the donation of a door to the church dated 1723 would confirm the date of reconstruction of the building. "The church is built of undressed, half-dressed, and finely-dressed stones which are bound by mortar. Finely-dressed stones are used in the section adjacent to the entrance, in the corners, pilasters, pillars and vault-bearing arches" (Karapetian 2019, p. 148).

In 1915, during the Turkish occupation, it suffered several thefts of furniture and damage to the floor and interior decorations. Following the 1930 earthquake, it was reduced to the state of ruins remaining today.

Of the ancient building, which must have had a basilica layout with a nave and two aisles enclosed by an apse and two sacristies, all that remains are the north wall and a corner of the south wall. Three domes once topped the roof.

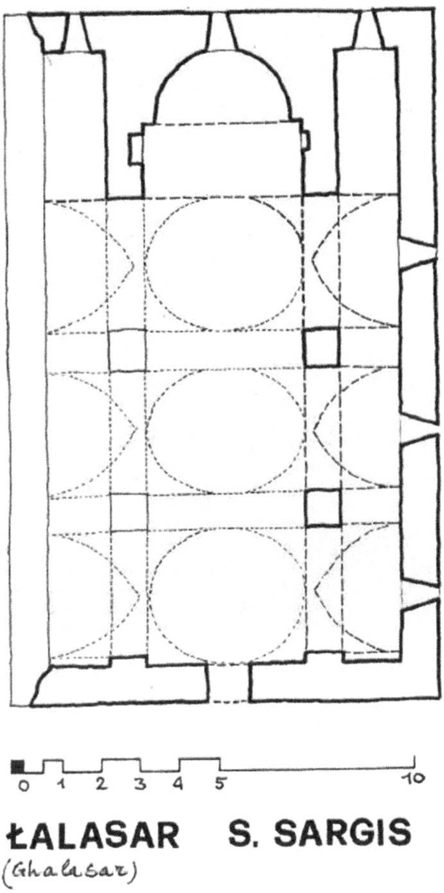

Fig. 11 Ghalasar, S. Sargis: plan (Alpago Novello et al. 1977)

The studies carried out by Kleiss and Alpago Novello, compared with the current consistency, testify to a progressive deterioration of which the rubble of the collapses following the 1930 earthquake remains on site. Even the two pillars documented in the Alpago Novello survey have now disappeared (Figs. 13 and 14).

The ruins of the building show masonry made of roughly squared stones, except for the arches and several corner elements, composed of carefully squared basalt stones.

Around the church, the agricultural landscape is characterized by fields of sunflowers, clover and other fodder or vegetables; nearby is an apricot orchard. Double rows of poplars line the road that winds through the countryside, flanked by irrigation channels with their diversions and masonry outlets with iron gates. Not far from the road are the ruins of what must have once been a mill, now demolished but of which there are still considerable traces of the underground parts.

The characteristics of the agrarian landscape, of some interest, and the presence of the mill would suggest an enhancement of the system and the documentation of

Fig. 12 Ghalasar, S. Sargis: south front (Alpago Novello et al. 1977)

its history and constructional features through appropriate displays to be placed near the ruins of the church.

GHEZELJA (GHEZELDJA)

It is a village located 11.3 km east of Salmas, on the right bank of the Zola River. Its origins probably date back to the late seventeenth/early eighteenth century. In the nineteenth century, it was in poor economic condition due to Kurdish raids and very high taxation by the local lord, plundered and destroyed during the First World War. The population was initially all Armenian until 1828, the year of the great migration to the Russian territories. In 1860, it had 234 Armenian inhabitants in 35 houses, rising to 262 in 47 houses in 1905. After 1914, the Armenian population disappeared. In 1930, at the time of the destruction of the village by the earthquake, there were 420 Turks present (Karapetian 2019, p. 154–155).

S. Hovhannes/Hovhanes.

Location: on the edge of the settlement, in a cultivated countryside.

Period of construction: eleventh century (Berberian and Tchalenko 1974); 1685 (Karapetian 2019, p. 155).

Fig. 13 Ghalasar, S. Sargis: west front (photo M. Boriani 2017)

Planimetric layout: Trapezoidal basilica plan with a wide nave and two side aisles; semicircular apse inscribed in the outer perimeter. It has two sacristies on two floors.

Consistency and condition of the building in the 1970s: "(…) In Ghezelja, S. Hovanes (…) survived, even though with more damage than S. Asdvadzadzin [in Akhtekhaneh]. This church, which we dated provisionally 1007 A.D. by an inscription on the doorway, is a rough stone masonry construction with four pillars supporting a central dome. All four walls were fissured diagonally, and an important central crack crossed the whole roof from E to W" (Berberian and Tchalenko 1974).

"… stands on the edge of the settlement, characterized, as usual, by the typical courtyard-gardens enclosed by high *khagil* walls and rows of poplars (surrounding the dwellings) (…) The interior, completely plastered, offers the usually vaulted basilica typology, but without the focal element of a dome: the vaults of the main nave are essentially sailed (a rather unusual motif). There is no mention of pointed arches: on the contrary, the small transversal arches of the smaller ships seem to be horseshoe-shaped" (…) The construction is very compact in volume, without any indication outside the apsidal area (semicircular on the inside, flanked by two small rooms) or the vaulted roof. About 5 m high, a sort of cornice marks the essentially terraced extrados of the roof, concealed by the masonry band above (…) The

Fig. 14 Ghalasar, S. Sargis: remains of the inner south-west corner after the collapse of the apse still existing in 1976 (photo M. Boriani 2017)

building is made of roughly cut stone, with extensive use of clay mortar, partially covered by plaster, with various reconstructions in later periods. The earthquake badly damaged it: in particular, two large cracks are visible running through it in a north–south direction" (Alpago Novello et al. 1977) (Figs. 15 and 16).

Description and State of Preservation

The building may date back to 1007, but "it does not seem possible to find in situ confirmation of the date proposed for the church by Tchalenko …, although some of the motifs noted and a certain anomaly of the building compared to the current typology could suggest a certain archaism" (Alpago Novello et al. 1977, p. 62). According to Karapetian, its origins are much more recent, having been built in 1685 and mentioned in 1890 as "an arched church built of stone and mortar" (Karapetian 2019, p. 155).

Situated in the middle of the countryside, near a farm, among fields of vines and maize, the church appears as a massive building without domes. Internally four pillars, supporting three ribbed vaults, divide it into three naves leading to the apse, next to which are the two sacristies. The structure is in stone bound with mortar. With a raised altar, the interior must have been covered with plaster, as tested by the significant traces still existing.

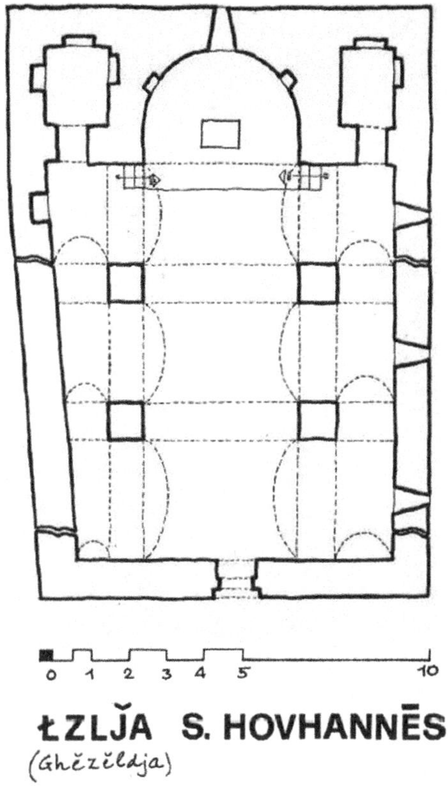

Fig. 15 Ghezelja, S. Hovannes: plan (Alpago Novello et al. 1977)

The state of conservation of the building, which was severely damaged by the 1930 earthquake, has deteriorated in recent years if one compares not only the surveys conducted in 1976 by Alpago Novello, but also the most recent photographs of the SHA (SALMAS HERITAGE ASSOCIATION—2011, 2017) study campaigns with the current situation (Figs. 17 and 18).

The cracks pattern is indeed very worrying: several through cracks, out of plumb and rotation of an upper portion of the south wall, almost total collapse of the north-west corner (which occurred, according to the comparison of images, between 1976 and 2011) and damage to the keystone of the entrance. Internally, the situation is just as serious: the apse is deeply cracked, the masonry of the first north bay has a lack of reinforcement (with the collapse of a small part of the roof vault), and the second bay has a through crack. The south bays are in no better condition: the keystone of the arch of the third bay is damaged, the second bay has a masonry discontinuity, and the first bay is badly damaged. The vaults of the nave have partially collapsed, with other ashlars no longer bound as they fall. Illegal excavations complete the framework of the deterioration of the building, although its architectural features are still fully legible (Figs. 19 and 20).

Fig. 16 Ghezelja, S. Hovannes: south front, severly cracked (photo M. Boriani, 2017)

In the absence of consolidation work, even if temporary, the building will soon be in ruins, and access to it should be prohibited for safety reasons. Work should be preceded by a rigorous study of the phenomena, including a necessary check of the ground conditions that could have led to the collapse of the foundations, as probable cause of the severe cracks that can be seen today.

In addition to the necessary and urgent temporary works, the church deserves overall restoration, taking into account—as Alpago Novello notes, the rarity of other examples of vaulted roofs in the area (Alpago Novello et al. 1977, p.26).

A few kilometres away from the church, there is an Armenian cemetery: the poverty of the burials is impressive, most of them simple, barely rough-hewn stones, always without any identifying marks. Other burials present a vertical portrait with some poor signs, usually crude crosses. There are very few graves with square stones and inscriptions or decorations.

HAFTAVAN (HAFT'WAN).

It is currently the largest village in the county of Salmas, located 3 km south of the city. There are records of the village since the sixteenth century. It enjoyed prosperity, especially in the seventeenth and eighteenth centuries, when it could be considered the Armenian capital of Salmas County. In 1828, a large part of the Armenian population emigrated to Russia. Nevertheless, in the 1830s, Haftavan

Fig. 17 Ghezelja, S. Hovannes: the west front in 1976 (Alpago Novello et al. 1977)

became the seat of an Armenian Prelature, i.e. the seat of a Vicariate (Karapetian 2019, p. 171).

The village experienced a certain revival at the end of the nineteenth century when feudalism was abolished, and the local population was enabled to buy the village's land (Karapetian 2019, p. 167). In the early 1900s, Haftavan had Armenian schools, a theatre, and a library.

The village was then involved in the persecution and destruction that characterized the early years of the twentieth century, which only ended in the 1920s/30 s when part of the Armenian population that had emigrated to the Caucasus returned (Karapetian 2019, p. 171).

In 1860, the census reported 858 Armenians in 154 dwellings, which increased to 2314 in 423 dwellings in 1905. Persecution in the first 20 years of the twentieth century and the 1930 earthquake reduced the Armenian population to 700 in 1934 and only 87 in 1969. By 2011, Armenians were absent from the village, which had

Fig. 18 Ghezelja, S. Hovannes: the west front in 2017 (photo M. Boriani 2017). The collapse of north-west corner is evident

2200 houses, 2000 inhabited by Kurds and 200 by Iranian Turks (Karapetian 2019, p. 172).

It suffered heavy damage in the 1930 earthquake, with the destruction of three of the four churches present at the time. During the reconnaissance mission carried out in 1976 by Alpago Novello, the presence of a chapel dedicated to San Plos was reported, with a straightforward architecture and made of raw earth, partially in ruins and used as enclosure for animals; of another church, San Tadeos, only the external enclosure, 50 m per side, made of clay, remained even then.

S. Gevorg.

Location: inside the built-up area, near a school.

Period of construction: seventeenth century (Alpago Novello et al. 1977).

Planimetric layout: single-nave hall, corner chambers and gavit with cupola.

Consistency and condition of the building in the 1970s: "In Haftavan, two of three churches were destroyed, the survivor being the 13th-century church of S. Gevork. The central dome of this massive construction collapsed, and the eastern façade developed oblique fissures, but otherwise, the building was intact (…) The

Fig. 19 Ghezelja, S. Hovhannes: the apse area with the altar today disappeared (Alpago Novello et al. 1977)

nearby school and assembly hall, built-in klinsbrick with a stone foundation, collapsed" (Berberian and Tchalenko 1974).

"The sacred building consists of two parts: basically, a two-bay single-nave hall with an apse (in wall thickness), flanked by two corner chambers and a kind of 'gavit' with a dome on four free pillars. The volumes are blocked, and massive, almost two cubes placed side by side, from which only the dome surmounted by a pyramidal spire and the small bell tower emerge. The interior, entirely plastered, is characterized by vaults with a pointed arch profile: traces of frescoes can be seen under the various layers of mortar (…) the dome has a wooden structure (probably rebuilt after the earthquake). In the *bema*, the *iconostasis* in plastered masonry has the form of a church façade. Above the entrance, traces of a tribune supported by wooden corbels" (Alpago Novello et al. 1977) (Figs. 21 and 22).

Fig. 20 Ghezelja, S. Hovhannes: the severe collapse of the vaults in the north portion of the church (photo M. Boriani 2017). This building, very interesting in its structure and features, is probably the most at risk of all those we detected in our surveys

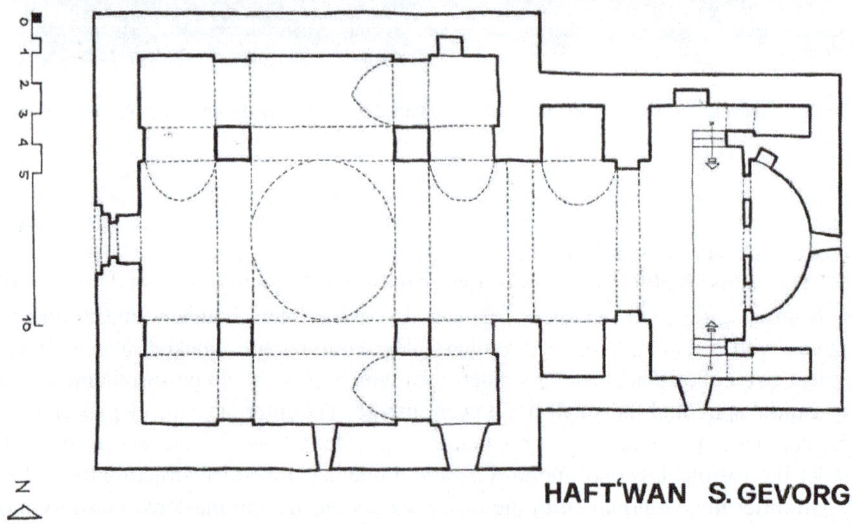

Fig. 21 Haftvan, S. Gevorg: plan (Alpago Novello et al. 1977)

Fig. 22 Haftvan, S. Gevorg: aerial view of the pertaining area of the church; on the north-east corner, the ruins of the theatre

Description and State of Preservation

Oral tradition attributes considerable antiquity to the church (at least as a foundation), perhaps the thirteenth century (Berberian and Tchalenko 1974). The original church seems to have been a Scriptorium, housing several manuscripts, the oldest of which dates back to 1538. It also seems that the relic of the right hand of the Apostle St Bartholomew was kept there (Karapetian 2019, p. 192). Alpago Novello dates the present building, more likely, to 1652, as some epigraphic evidence, albeit fragmentary, would attest.

There is news of repair work on the building financed in 1913, followed by damage and looting in 1915. The church was renovated in the 1920s, and, although damaged, the building did not collapse in the 1930 earthquake.

The church is accessed by a modern fence in fired brick on the street side and in unfired earth on the north side of the plot (mixed with the west side).

The building appears as a compact parallelepiped. The entrance, central to the front, is made of regularly shaped basalt ashlars emerging from the façade in light-coloured simply hewn stones. The interior of the church is divided into three naves by large painted pillars. In the centre is the dome with a theory of small openings that allow light to filter through. A photograph from the beginning of the 1900s shows the existence of a two-storey bell tower made of stone ashlars that

Fig. 23 Haftvan, S. Gevorg: west front (photo M. Boriani 2018)

surmounted the narthex roof to the west. This was probably damaged in the 1930 earthquake and replaced by the lighter wooden structure now visible (Figs. 23 and 24).

The church proper is a small, ancient, square-plan building (9.59 × 9.52 m), with a single nave covered by an ogival vault and a semicircular apse flanked by two sacristies. A rectangular plan narthex (*gavit*) (15.25 × 13.31) with four pilasters supporting a dodecahedron drum covered by a pyramid roof was probably added to this older part to the west. The narthex would then have been incorporated into the body of the church, constituting the space specifically for the worshippers, while the old church was probably reserved for the *chancel* and the *bema* (the space reserved for the clergy, including the apse and the altar) (Figs. 25 and 26).

There was a cemetery around the church: several graves dated between 1782 and 1927 are still preserved. Another cemetery, located 450 m SE of the church, has over 500 graves, the oldest of which date back to the fifteenth–seventeenth centuries (Karapetian 2019, p. 210 ff).

Also in the church precinct is the ruin of a building from the late nineteenth century, once used as a theatre and public assembly hall.

The church underwent recent restoration work by the local directorate of monuments at the time of the last visit (2018), which was almost completed: the cracks on the west front were repaired, and a new interior floor was laid (so far in concrete). On the other hand, no work has been carried out to repair the large gaps in

Fig. 24 Haftvan, S. Gevorg: interior looking to the apse area (photo M. Boriani 2018)

the interior plasterwork or to preserve the rather damaged paintings on the basement band.

Recent restoration work has revealed and recovered substantial traces of wall paintings, which, according to an inscription in the church, date back to 1819, hidden under various layers of paint. These, which are still being restored in 2018, show a very significant picture of subjects, including Adam and Eve, The Virgin and Child Jesus, St Gevorg, floral motifs and above all an exciting and evocative Last Judgement (Bakhshaei and Mohammadi 2015).

This cycle of paintings is unique in the entire Khoy and Salmas area and testifies to the wealth of the Armenian community in Haftvan (Figs. 27 and 28).

Lastly, a gas plant has been built to heat the church, which is to be converted for tourist and cultural use.

Fig. 25 Haftvan, S. Gevorg: view of the interior of the church, still in use in1976 (Alpago Novello et al. 1977)

Outside, the weed vegetation has been completely removed, making now accessible the ruins of the theatre, which were inaccessible in the previous inspection in 2016. In front of the church façade, a two-level "churchyard" has been created, and the access to it from the entrance gate has been paved in stone. All around the north, east and south sides of the church, a concrete pavement covered with stones was built, under which a cavity seems to have been created to minimize the effects of rising damp, the effectiveness of which will have to be verified over time.

The tombs alongside the north and south sides of the church have been cleared of vegetation and, in part, roughly reassembled.

Contrary to Alpago Novello's findings in 1976, the building is no longer used for worship, as there is no longer any Armenian community in the village.

Since the religious function is no longer present, considering the recent interventions—much appreciated by the village population, as well as the interest of the architectural complex, church and theatre ruins, it could become a documentation centre for enhancing the Armenian cultural heritage of the Armenian Salmas area.

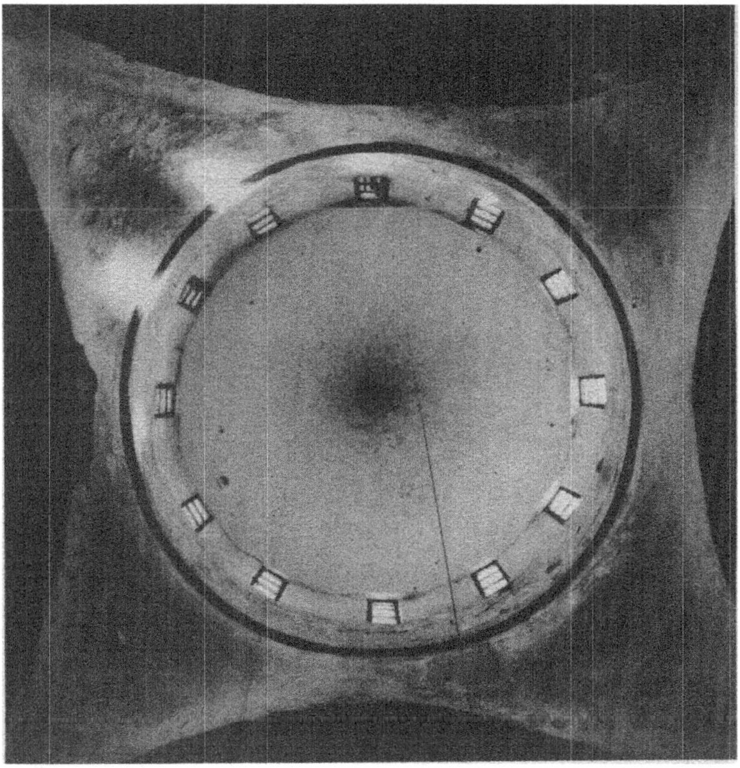

Fig. 26 Haftvan, S. Gevorg: view of the dome in 1976 (Alpago Novello et al. 1977)

In particular, the old theatre, after conservation of the historical remains and the adjacent construction of a new volume harmonizing with it but in contemporary architectural language, could house not only meeting/exhibition hall, but also a small visitor centre documenting the Armenian presence in the area and the numerous religious buildings in the Salmas district. A lapidary could document the most exciting and ancient tombstones present in the cemeteries of the area and collect finds recovered from the churches of the area destroyed by the 1930 earthquake.

The area enclosed by the fence of the complex could become a green space for use of the local community, which would become its watchful guardian to some extent.

KOHNESHAHR (KOHNEH SAHR/QYOHNASHEIR).

The village is about ten kilometres from Salmas. The ancient Zarehavan took the name Salmas in the Middle Ages and became known as "the Old City" when a new Salmas was rebuilt nearby. Historical evidence suggests that it has existed since the sixth century. The Arab historian Ibn Haikal describes it in the tenth century as

Fig. 27 Haftvan, S. Gevorg: one of the painted pillars in 2017 earlier than the restoration (photo M. Boriani 2017)

"isn't a big city either, but it is prosperous, populous, and rich in fruits … and it has strong city walls of stone" (Karapetian 2019, p. 460). In the Middle Ages, it was the most populous Armenian city in the area and an important centre of literary production. In the nineteenth century, several travellers made mention of its still visible ruins.

The local cemetery shows that in ancient times, at least from the thirteenth century, the population was composed only of Armenians. However, after the Turkish–Persian wars of the sixteenth–seventeenth centuries, Kurds, Chaldeans (Assyrians) and (in the nineteenth century) Jews also began to settle here. In 1896,

Fig. 28 Haftvan, S. Gevorg: the same painted pillar after its 2018 restoration (photo M. Boriani 2018)

there were 683 Iranian Turks, 179 Jews and 411 Armenians. The latter was reduced to a few dozen after the persecutions of 1914–15. At the time of the 1930 earthquake, there were only 49, and they had all but disappeared by 1959 (Karapetian 2019, p. 464). There were three Armenian churches that, with the emigration of 1828, were practically abandoned and reduced to ruins that were no longer recovered after the 1930 earthquake. Already in 1976, Alpago Novello testified that there were no longer any traces of the largest of the three churches, in the locality of Tazeshahr, which must have been a hall with a lowered dome resting on four pillars, and that the chapel of St. Varvar was unrecognizable, reduced to a heap of stones that made it doubtful that it could be attributed to a building (Alpago Novello et al. 1977).

Mar Hakob/Hagop.

Location: about 10 km west of Salmas, isolated in the countryside.

Period of construction: unknown. The date 1952 engraved on the lintel of the entrance may date back to restoration work (Alpago Novello et al. 1977).

Planimetric layout: single hall with a barrel vault.

Consistency and state of the building in the 1970s: (Figs. 29 and 30).

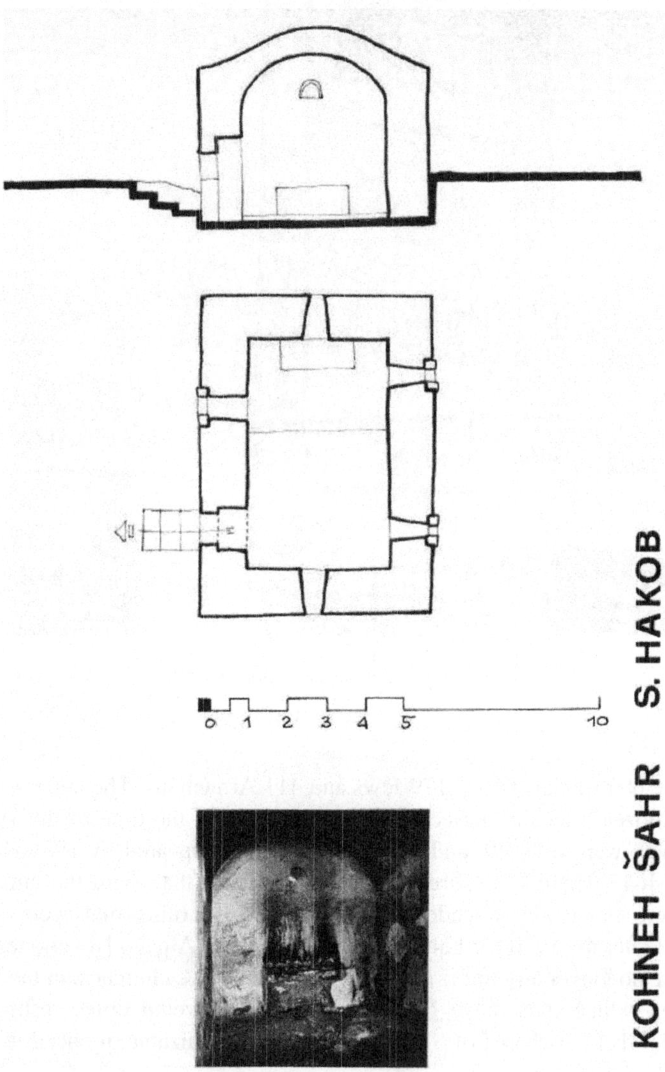

Fig. 29 Kohnehshahr, S. Hakob: plan, section and inner view (Alpago Novello et al. 1977)

Fig. 30 Kohnehshahr, S. Hakob: the altar after the recent restauration (photo M. Boriani 2018)

Description and State of Preservation

This is a small building for Assyrian Catholic worship, mentioned by Raffi in the mid-nineteenth century, then used by the few local Armenian Catholics. In 1905, it was unused, and in 1952, it was repaired (Karapetian 2019, p. 468). There is no definite information about the construction time, although the lowered height of the chapel floor compared to the current ground level would suggest an early foundation.

The small building is located within a large orchard area, surrounded by a recent stone fence.

Externally, the chapel has rough-hewn stone walls, five small windows (one on the north, east and west sides and two on the south side) illuminate the interior, which is accessed through a low door, marked by a basalt ashlar on which an inscription and the date of a past intervention (1952) are engraved, and by climbing four high steps (Fig. 31).

The interior is very simple, white-plastered, with a barrel-vaulted roof, the shape of which is also visible on the outside.

The building has no particular conservation problems; it is maintained and is still in active use, albeit occasionally.

The maintenance work, evidently recent, must not have been carried out by skilled labour, as evidenced by the filling of the joints between the stones with

Fig. 31 Khonehshar, S. Hakob: aerial view of the landscape around the chapel

heavy cement mortar and the waterproofing of the roof with bituminous sheets covered with aluminium foil.

KHOSROVA (KHOSROWABAD/XOSROVA)

It is located 4.4 km SO of Salmas in a plain not far from Haftvan. At the 2006 census, its population was 158, in 44 families.

An Armenian population has been documented since at least the sixteenth century. There was also a Scriptorium, but only one manuscript, dated 1783, is documented. Like all villages in the county, it suffered devastation and massacres in 1915–17 and fell into ruin.

The population appears to have been essentially Armenian until 1828, later replaced by a large community of Assyrian Catholics, who were often in sharp contrast to the few remaining Armenians. At the end of the nineteenth/beginning of the twentieth century, 400 houses were occupied by Assyrians, 36 by Armenians and ten by Kurds.

At the time of the 1930 earthquake, the population was utterly Assyrian (Karapetian 2019, p. 297): 34 people died out of the 280 living there, and the seven churches that existed there at the time were destroyed (Berberian and Tchalenko 1974).

The only surviving historical Armenian church is dedicated to St. Sargis, while only a few remains of Mar Givargis remain.

A new church, dedicated to Mar Givargis like the older one, was built after the 1930 earthquake: it is maintained in good condition and regularly used by the Assyrian Catholic community.

According to the village head, there are still some Armenian families, about 20 people in total.

***Mar Givargis*.**

Location: at the edge of the built-up area.

Period of construction: 1845 (Alpago Novello 1977).

Planimetric layout: rectangular with three naves and triapsidal.

Consistency and condition of the building in the 1970s: "Mar Givargis, a large stone masonry structure with walls built in the traditional style, i.e. about 1.25 m. thick, with an outer layer of hewn stone and an inner layer of rough stone and good cement mortar. The inner pillars and the vaulted roof collapsed, but most of the walls excepting the corners of the building remained standing" (Berberian and Tchalenko 1974).

"The church…is a large three-nave basilica with three semicircular apses, encapsulated in the rectangular perimeter of the church. The apses are also semicircular outside, with a narrow passageway separating them from the E side. (…) The walls are very thick, and the masonry is not very neat, except for the façade, which is made of very carefully squared black basalt, with ashlars of considerable size. Inside, the cornices, both in the apse and in the naves, appear to be much later additions (…) The Syriac inscription above the entrance (there is a second one on the inside of the architrave) recalls the date of foundation (1825). According to oral tradition, the church was destroyed by earthquakes and was eventually spoliated by the occupying Russian troops" (1940–45) (Alpago Novello et al. 1977) (Figs. 32 and 33).

Description and State of Preservation

Giuseppe Anaclerio, who visited Khosrova in 1862/65, writes: "The cathedral was ruined when the missionaries founded the mission: but thanks to the care and attention of the Christians, a new cathedral has been built according to European architecture" (Anaclerio 1868).

Although now reduced to a state of ruin, a building with an interesting architectural layout enclosed within a large enclosure of more recent construction, the south wall of which has bricks laid at 45 degrees. The church has kept the two north and south walls relatively intact; the three apses are more degraded, while the façade is no longer attached to the side walls (the two corner parts of the walls, which collapsed following the 1930 earthquake, are missing). From the plaque above the lintel of the west entrance, it appears to be a church of Assyrian worship.

Fig. 32 Khosrova, Mar Givargis: plan (Alpago Novello et al. 1977)

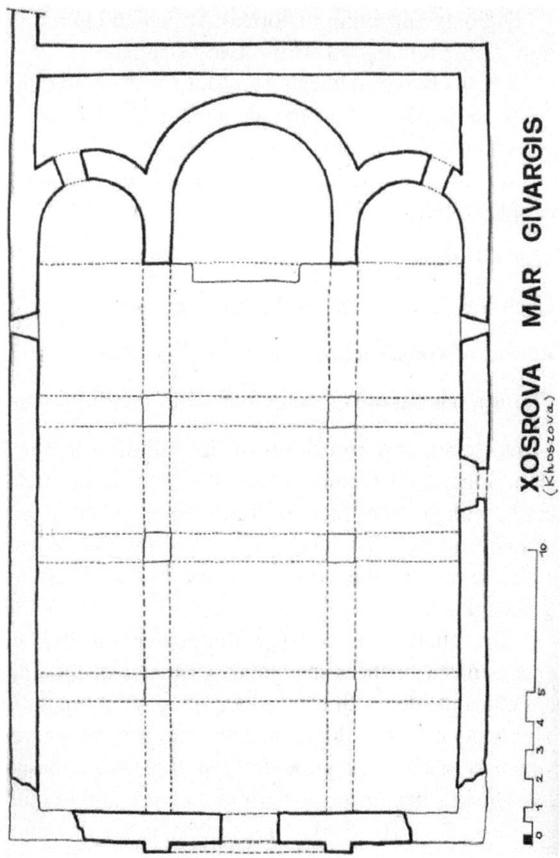

The masonry is mainly made of rough-hewn stone and appears to be bound by large amounts of mortar, while the principal front, of good workmanship, is covered with squared basalt ashlars. The apsidal part consists of three apses, hidden on the outside by an open cavity, visible through a beautifully crafted window on the south side of the church.

At the end of the south side, a small door seems to give access to an underground part, which is not accessible today.

A comparison with the images taken during the Alpago Novello study campaign in 1976 shows that the building conditions have not substantially deteriorated. However, the lack of reinforcement of the walls endangers the stability of the façade, which is very interesting for several traces of graffiti and engraved figures. Above the lintel, on the exterior side, is a plaque with an Assyrian inscription (published, translated into Italian, by Alpago Novello). The inner side of the lintel, which is fractured, bears an undeciphered inscription. The figures engraved on the outer facade deserve in-depth study to suggest an interpretation. The fractured lintel

Fig. 33 Khosrova, Mar Givargis: zenith view of the remains of the church

is now supported by a pillar made of stones and bricks and covered with an earthen plaster.

The apse on the north side is well recognizable, although seriously compromised by a lesion with large detachment of material at the point of connection with the north wall. The apse on the south side has been filled in by non-recent masonry and, from what little can be seen from the window mentioned above, might be seriously degraded. Numerous fragments of plaster are still present, although detached from the support.

The small chapel recently built in the central apse of the church ruin, already surveyed in 1976 and undergoing a complete renovation in 2017, was extended and finished in 2018. The only older trace is a stone placed on the altar, painted blue and engraved with a red-painted cross. Overall, the intervention is not of excellent quality, but it testifies to the vitality of a religious community that is still present and active (Figs. 34 and 35).

There is an urgent need to make the building safe by installing at least some temporary structures and filling in the wall crests with hydraulic lime mortar or made hydraulic by *cocciopesto/pozzolan* to prevent the infiltration of rainwater and the settlement of infesting vegetation. It would also be helpful to remove the rubble still present on the outside of the three apses and reopen access to the small space between the apses themselves and the external fence, making accessible some pencil inscriptions in Arabic characters that could be of some interest. Finally, the

Fig. 34 Khosrova, Mar Givargis: assyrian rite church. Remains of the western façade (photo M. Boriani 2018). A safety intervention of the whole building is urgent

access in the south wall to the underground crypt, still to be investigated, could be reopened and made accessible.

Since it is reported that, during religious services held in the chapel built in the central apse, the faithful also gather outside, it would be desirable to properly

Fig. 35 Khosrova, Mar Givargis: view of the apse area with the small chapel recently rebuilt. (photo M. Boriani 2018). The area being still in use for religious functions needs an arrangement of the internal space of the ruin

arrange the space of the current ruin after carrying out an archaeological survey that would be useful to do, since the church was in use at the time of the earthquake, and, under the rubble, there may still be preserved furnishings or other traces of the original building (flooring, decorations, etc.). The new intervention, on the model of what has been carried out, for example, in Coventry in England, would make it possible to create a public square that would enhance the ruins of the church. The greening of the entire enclosed area (including an orchard) could also enrich the quality of the settlement of the village (Figs. 36 and 37).

S. Sargis/Sarkis.

Location: in the village, along a secondary road, enclosed by a bare earth wall.

Period of construction: 1717 (Alpago Novello et al. 1977) (Berberian and Tchalenko 1974).

Planimetric layout: hall with two naves.

Consistency and state of the building in the 1970': "The church is located in the center of the settlement, along an inner street, within a large enclosure (…) The building seems to have been reduced on the N side, where traces of the foundation

Fig. 36 Khosrova, Mar Givargis: south façade (photo M. Boriani 2018)

remain. This also explains the misalignment of the entrance door and the corresponding small window of the *bema*, in which case the ancient building must have had three naves (…) Note the bench-grading that runs along the entire south side, which is still preserved: it is to be assumed that the motif was repeated symmetrically (…) *Khagil* walls, flat wooden roof. The altar is of the isolated type, all plastered (…) The church is still used and in a relatively good state of preservation. On the façade, two inscriptions (1832 and 1900) and a plaque with the date of restoration 1955. On the ground, lintel or inscribed tomb piece. In the box tombs of the adjoining cemetery, various inscriptions, all from the twentieth century" (Alpago Novello et al. 1977) (Figs. 38 and 39).

Description and State of Preservation

The building may date back to 1717, although sources do not agree on the date of construction. A plaque walled at the entrance of the present church dates the construction to 1832, and a second one describes repairs carried out in 1953. It is known that the church was one of the few allowed to ring a bell, which is why a bell tower was added in 1890. It is reported that around 1892 all the furnishings and decorations were stolen (Karapetian 2019, p. 300). Hayk Atjemian writes in 2016, "Probably the Armenian church of St.Sargis is about 300 years old, [but] it remains obscure how many times it underwent renovation" (Karapetian 2019, p. 299).

Fig. 37 Khosrova, Mar Givargis: south façade with a detail of the openings located at the right end of the wall (photo M. Boriani 2018)

It is a rectangular, two-nave building, accessed through a modestly sized entrance surmounted by two stone ashlars with inscriptions, located off-centre from the front.

The building, in Armenian rite, is made of unbaked earth, built on stone foundations; the walls taper upwards to form a trapezoidal section. The simple interior has exposed wooden beams supporting the flat roof. Two slender wooden pillars on which the ridge beam rests divide the rectangular plan into two naves. The altar is raised above the floor level of the church. The interior fittings, the drape on the altar and the carpets covering the floor show that the place is still used for worship (Figs. 40 and 41).

On the whole, the structure does not show severe signs of disruption and deterioration, although there are detachments and gaps in the unbaked earth plaster

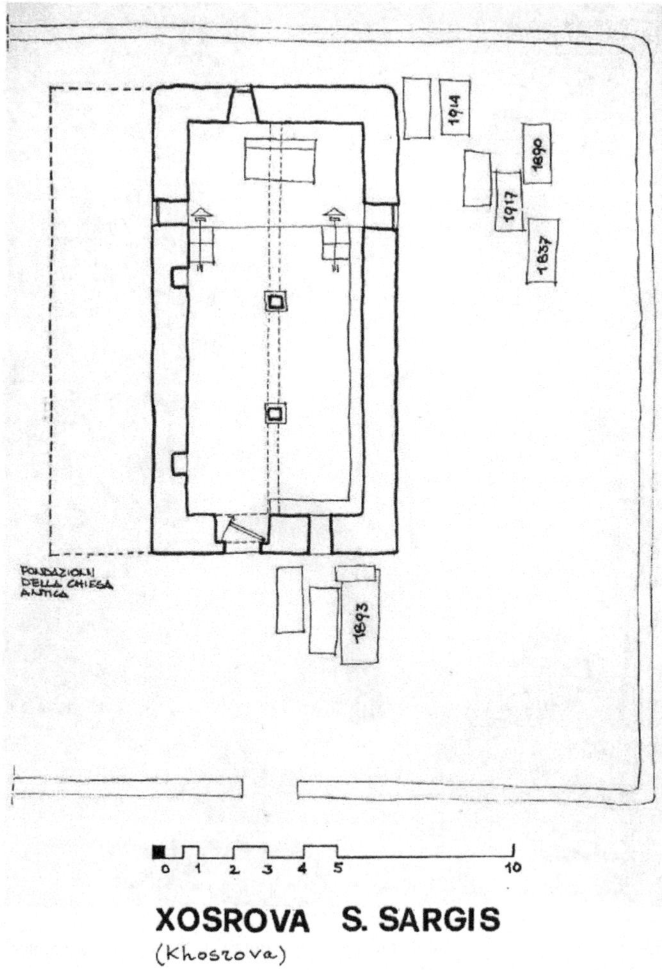

Fig. 38 Khosrova, S. Sargis: plan (Alpago Novello et al. 1977)

and, in 2017, a severe damage on the north wall about 1/3 from the entrance, which was repaired in 2018, at least on the outside of the masonry. A further, less significant lesion affects the south side. Evident signs of rising damp damage are visible in several parts in the basement area. In the vicinity to the church, there are about ten graves with dates around the end of the nineteenth century.

The church is surrounded by an earthen wall, which in 2017 appeared to be seriously damaged (props, gaps, etc.) and was repaired the following year. On the side of the entrance, outside the wall, runs a canal lined with walnut trees. The area inside the walls has several fruit trees left abandoned.

If compared to the images resulting from the missions in the 1970s, the state of preservation of the building appears to have deteriorated in recent decades, while

Fig. 39 Khosrova, S. Sargis: west façade (photo M. Boriani 2018)

the "step-bench" described by Alpago Novello on the south side is no longer visible (Fig. 42).

MALHAM.

It is a village situated at 7.5 km south-west of Salmas and 1.3 km from the right bank of Zola River, near the Karnearegh mountains. In the cemetery, to the south-east of the village on a gentle slope, some graves date back to the Middle Ages, and a church exists as early as 1641. In the second half of the nineteenth century, it had a school, a library and a theatre hall.

The Armenian population amounted to 444 in 70 houses in 1860, rising to 1048 in 185 houses in 1905. Most of the population emigrated during the years of the Turkish occupation to escape persecution, so that after the Turkish occupation, there were only 219 Armenians in 1927, falling to 110 in 1964 and 46 in 1969 (out of a total population of 800, the vast majority of whom were Kurds) (Karapetian 2019, p. 330–331).

A last emigration of Armenians took place in 1946 when they decided to move to the Soviet Union, to the then Socialist Republic of Armenia. In the 1970s, the remaining Armenian families numbered 2–3 (Alpago Novello 976).

Destroyed by the 1930s earthquake, which caused 48 deaths out of a then population of 327 (Berberian and Tchalenko 1974), today the village is inhabited by about 4000 people.

Fig. 40 Khosrova, S. Sargis: interior view in 1976 (Alpago Novello et al. 1977)

The agrarian landscape surrounding the village is characterized by double rows of cypress poplars placed at the edge of roads and/or canals, fruit trees arranged in rows along land worked in "rooms" to facilitate irrigation, small canals for water distribution through diverters/partitions (in masonry and cement) and sluice gates (in iron).

S. Gevorg.

Location: inside the village, on a small rise.

Period of construction: nineteenth century (Alpago Novello et al. 1977).

Planimetric layout: rectangular, divided into three naves.

Consistency and state of the building in the 1970s: "The building is made of *khagil*, cubic in shape. Single entrance on the N side, shallow with a stone lintel and

Fig. 41 Khosrova, S. Sargis: interior view in 2018 (photo M. Boriani). The church appears still in use

an inscription with the date 1882 (...) "Very narrow interior, with a grit tile floor, horizontal wooden roof, three pairs of wooden pillars on stone bases, with a wooden shelf, one of which is carved. Painted wooden '*iconostasis*' with popular features" (...) The church is functional, in it is a *Šarakan* (hymnal) printed in 1888 in Vagharšapat. The building dates back to the end of the last century and became the village church after the 1930 earthquake. In the space in front of the entrance door, is a small cemetery from the nineteenth-twentieth century on the N side, where there are family graves (e.g. the Mouradians ...)". (Alpago Novello et al. 1977, p. 67) (Figs. 43 and 44).

Description and State of Conservation.

The building, dating back to the end of the nineteenth century, was turned into a stable by the Turks in 1905; in 1930, the church was damaged by the earthquake but continued to be used. Repaired in 1950, it is still in a good state of repair during studies by Kleiss (1969) and Alpago (1976).

The building stands on a small relief within a modern enclosure: to the right of the entrance is a series of tombs from the late nineteenth and early twentieth centuries. The rectangular apse occupies the entire width of the church and is divided from the naves by a sort of *iconostasis* with three arches.

The perimeter walls of the building are made of unbaked earth; inside, the supporting pillars, made of wood covered with plaster on a stone base and with a

Fig. 42 Khosrova, S. Sargis: zenith view; in the middle, the courtyard of the church

crude wooden capital, divide the church into three naves, with an exciting solution of three arches at the altar.

A large lesion severely damages the east side, now transformed into a gap, which seems to be due to the lack of a downpipe and the consequent erosion caused by the flow of rainwater, but perhaps also to the intervention of men to enter the building illegally (the church is currently the subject of clandestine excavations, perhaps in search of precious material). The angles of the building show cracks that are not yet visible.

The interior has two large excavation holes; the altar appears to have been destroyed, probably to search for a hypothetical treasure. The roof is badly damaged and partly collapsed (east side). Next to the altar, traces of some paintings emerge under the plaster.

Provisional works should be carried out as a matter of urgency, filling in the holes caused by clandestine excavations, removing the rubble and reinforcing the damaged roof beams.

The building, no longer an object of worship, could be put to public use again for the local community to be guarded and maintained once more.

***S. Sargis Zoravar/S. Sarkis Zoravar*.**

Location: in the centre of the village, along a secondary road.

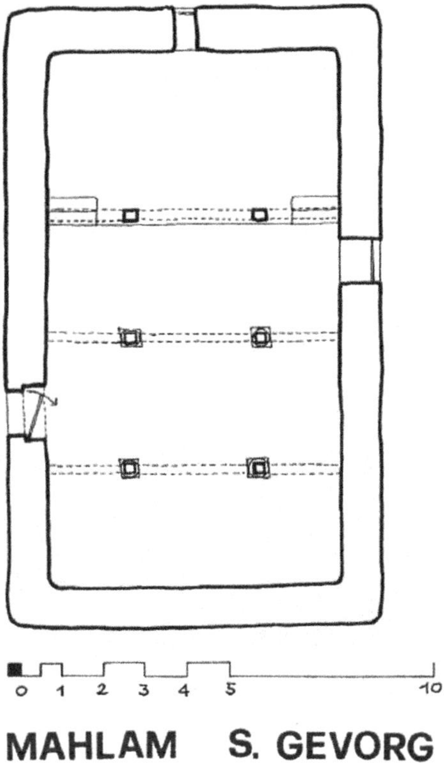

Fig. 43 Mahlam, S. Gevorg: plan (Alpago Novello et al. 1977)

Period of construction: 1641 (Alpago Novello et al. 1977).

Planimetric layout: rectangular, triapsidal (currently hardly recognizable).

Consistency and condition of the building in the 1970s: "…the southern and eastern walls of S. Zoravar—a construction similar to Mar Givargis—remained partly standing, but the rest of the church collapsed" (Berberian and Tchalenko 1974).

"The building almost completely collapsed in the earthquake of 1930, but the S side remains standing (…) The masonry is generally in stone; the basement and two pointed arches of the south side are in brick. Imposing ruins, never excavated, remain towards the east end. Traces of one of the four free pillars, overturned, but still towards the north, where there is also a niche, perhaps the baptismal font (…) The enclosure surrounding the area above the church and cemetery has an entrance door, of modest size, but surmounted by a carved and modelled stone lintel. (…) Some tombstones on the north side. To the south of the church is a small cemetery with many graves of the victims of the earthquake of May 6, 1930: further to the east of the same cemetery are older graves, dated 1800–1870–1822, with *khatch-kars*. (…) According to oral tradition, the church had three domes on an axis, emerging from the roof, of which the central one was taller; three altars and a

Fig. 44 Mahlam, S. Gevorg: east front (photo M. Boriani 2017). Evident collapse of the wall in rough earth, due to rainwater flow

'matroneum' tribune at the entrance; vaulted side aisles. Tradition still has it that the building was over six centuries old. It had no valuable furnishings, which in any case were lost after the earthquake" (Alpago Novello et al. 1977) (Figs. 45 and 46).

Description and State of Preservation

The ruin of the building is surrounded by a masonry enclosure, probably of more recent construction, which is accessed through a basalt stone portal probably carved to depict the birth of Eve from Adam's rib and an angel holding scales (perhaps a synthetic representation of the Last Judgement).

Although not easy to access, it is possible to see what remains of the building, namely part of the south wall characterized by an ogival arch. Other traces of the former masonry of the church, albeit of a lesser height, can still be seen beneath the weedy vegetation. The collapses result from the violent earthquake of 1930, as already noted by Alpago Novello and Tchalenko, Berberian in the 1970s.

A clean-up of the area and the ruins might be appropriate to allow the start of an archaeological excavation campaign to be carried out under expert architects and/or archaeologists to avoid the destruction of evidence of any importance.

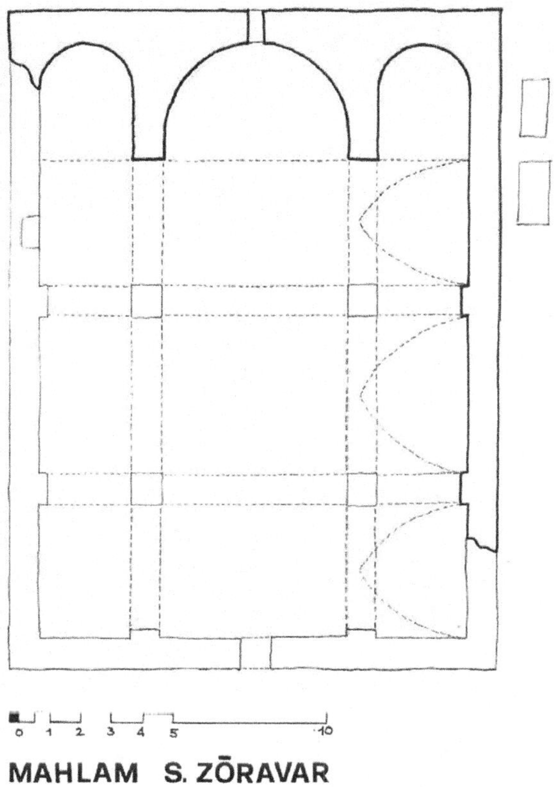

Fig. 45 Mahlam, S. Zoravar: plan (Alpago Novello et al. 1977)

Pakajik (Persian: پکاجیک, also Romanized as Pakājīk; also known as Bīā Jīk, Pagājek, Pagā Jīk, Parch, Pārchī, Payadzhik, Paya Jīk, Pega Jik, Pīkāchīk, and Yakāchīk) is a village in Gowharan Rural District, in the Central District of Khoy County, West Azerbaijan Province, Iran. At the 2006 census, its population was 1,216, in 298 families.

It is a village 3 km east of Salmas, documented as a writing centre as early as 1557. Until 1828, it was inhabited exclusively by Armenians, who remained in the village until the Turkish persecutions during the First World War. The Armenian population amounted to 657 in 1860, which increased to 1,478 in 1905, in 294 houses. After the First World War and the persecution that followed, the Armenians were drastically reduced: 137 in 1930 and only 10 in 1969 (Karapetian 2019, p. 408). At the end of the nineteenth century, the village "looks like a small town: the streets are clean, the houses have neat gardens and are surrounded by high walls" (Karapetian 2019, p. 409).

Fig. 46 Mahlam, S. Zoravar: door in the enclosure wall with two roughly sculpted angels

Paiadjuk is the birthplace of the famous Armenian writer Raffi (Hakob Melik-Hakobian), born there in 1835. At the end of the nineteenth century, a school, library and theatre were located here.

Raffi's birthplace, which was a fine example of a middle-class house, was destroyed in the 1930 earthquake (Karapetian 2019, p. 413).

Near the church of St. Gevorg, on a hill to the south of the village, stands the Armenian cemetery with some older graves at the top and others, generally more modest, dating from the late nineteenth—early twentieth century, scattered along the slopes. The more affluent tombs (regularly shaped tombstones with inscriptions and rare decorations) have unfortunately been vandalized by treasure hunters, who have left deep holes and sometimes exposed bone remains.

There is an urgent need for works to fill in the excavations and bury the remains to restore decorum and safe accessibility to the site.

S. Gevorg.

Location: the church is on the southern edge of the settlement, isolated and surrounded by a recently built wall.

Period of construction: eighteenth century (Alpago Novello et al. 1977); before 1557 (Karapetian 2019).

Planimetric layout: basilica plan with three naves and a single apse flanked by traditional service rooms. According to Alpago, it was vaulted and not domed as Kleiss claimed.

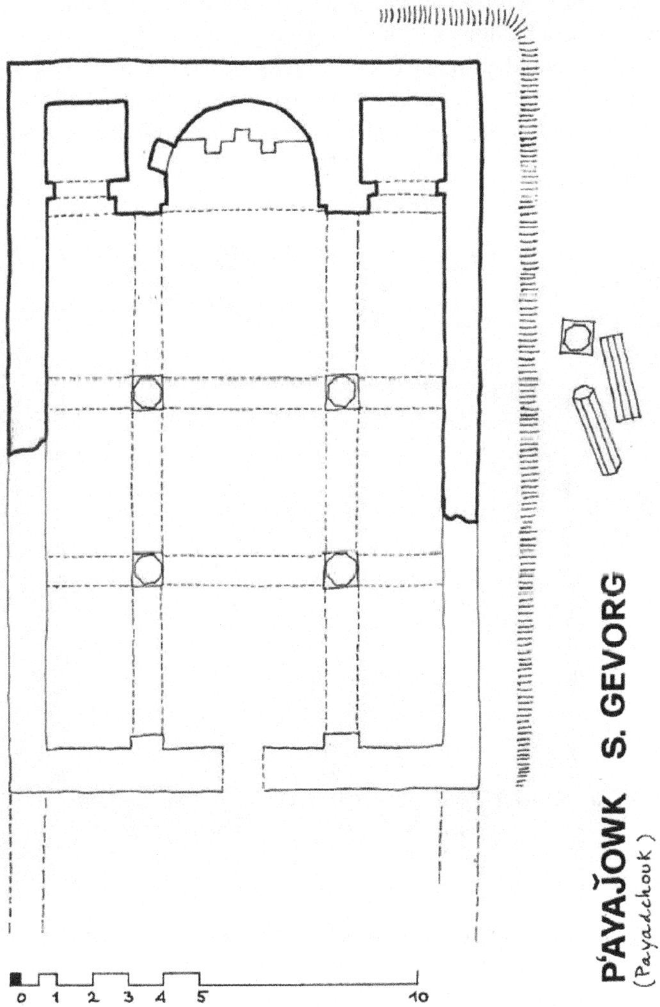

Fig. 47 Payadjuk, S. Gevorg: plan (Alpago Novello et al. 1977)

Consistency and condition of the building in the 1970s: (Figs. 47 and 48).

Description and State of Preservation

In a state of ruin after the destruction caused by the 1930 earthquake, the building is thought to date back to the sixteenth century, or more likely to the eighteenth, as Alpago Novello reports, also thanks to the discovery of a votive plaque dated 1806.

As early as the 1970s, the apse, the small chamber to the side of the apse on the south side, and one of the four monolithic basalt pillars mentioned by Alpago Novello and some finely worked stones and traces, perhaps, of a *gavit*, were preserved.

Thanks to the modern fence and gate that still protect the site from vandalism, much of what Alpago documents is still in place, except for the apsidal vaults that have now largely collapsed.

Due to the presence of architectural and decorative finds in the church, which indicate that this is a building of some importance, an archaeological excavation is suggested, which could also uncover the religious furnishings that were in the church at the time of the earthquake and the recovery of the most exciting finds in a lapidary. An orchard arrangement, also thanks to the presence of a spring, could

Fig. 48 Payadjuk, S. Gevorg: zenith view of the enclosure of the church. The area keeps its archaeological interest, but might be partially arranged as collective green space

give new decorum to the site and contribute to the preservation of the church ruins through control if these were used as a green space usable by the population.

SARNAQ (SARNA/SARNAY).

It is a village 6.7 km south of Salmas, 1.5 km from the right bank of Zola River, in a well-cultivated area. At the 2006 census, its population was 914, in 223 families (defined as Azeri).

Of the modest size, it is located along the main road that runs through it. Documented only from the eighteenth century, it probably has older origins. In 1914–15, it suffered the fate of the other villages in the county. At the beginning of the twentieth century, the population amounted to 543 souls, almost all Armenians; these were reduced to 86 in 1927 (Karapetian 2019, p. 497).

The 1930 earthquake destroyed a large part of the village, causing 18 deaths among the 180 inhabitants there. St. Asdvadzadzin, St. Hovhannes and Mar Khinah were severely damaged, and today only St. Hovannes, a *khagil* building, was severely damaged; the ruins of St. Asdvadzadzin remain.

In the 1970s, the population was predominantly of Turkish origin (TCHALENKO BERBERIAN 1974).

S. Hovhannes/Hovhanes (Mar Yukhanna Marikhna).

Location: at the western edge of the village.

Period of construction: 1896 (Karapetian 2019).

Planimetric layout: a hall with trapezoidal plan and two side sacristies. It has a flat wooden roof, supported by six wooden pillars. Two side sacristies.

Consistency and state of the building in the 1970s: (Fig. 49).

Description and State of Preservation

A place of pilgrimage for both Armenians and Assyrians, the building was repaired in 1984 and surrounded by a brickwork fence. The Assyrian church is still used today for religious services.

Fig. 49 Sarnaq, Mar Hovhannes: plan (Alpago Novello et al. 1978) and two recent images (photo M. Boriani 2017)

Externally, it appears as a massive and straightforward parallelepiped: the masonry is made of unfired earth, covered by plaster of the same material. At the top, brick courses were evidently added during the 1984 intervention, when the roof was probably rebuilt.

The interior is in the form of a hall marked by four wooden pillars. On the north side, there is a small niche with probably a baptismal font. The altar was recently rebuilt.

Although not regularly maintained, the church does not show any significant deterioration, although signs of rising damp are visible in the basement area.

On the other hand, the exterior cladding shows detachments, gaps and profound erosion of the joints between the bricks in the upper part and widespread cracks in the plaster. The roof and the rainwater conveyance system are not able to protect the masonry from the washing away effects of the rain.

There is a large cemetery outside the church enclosure, with relatively recent graves and many probably older ones, carved from simple boulders without any finishing work or symbols.

S. Avatzatzin/Asdvadzadzin.

Location: situated on the north-eastern edge of the present village.

Period of construction: 1625 (Alpago Novello et al. 1977; Berberian and Tchalenko 1974).

Planimetric layout: basilica with three naves.

Consistency and condition of the building in the 1970s: "S. Asdvadzadzin survived partly, even though its roof collapsed and 3 out of 4 central pillars and the western façade fell to the west" (Berberian and Tchalenko 1974) (Figs. 50 and 51).

Description and State of Preservation

There is written evidence of the church's existence in 1864 ("a church of stone and lime"). In that year, there were apparently still tombstones dating the building to 1625, which disappeared (Karapetian 2019, p. 499).

The construction is made of rough-hewn stone connected by mortar. The window surrounds and arches are made of basalt. Originally, it was supposed to be plastered inside, while there are no traces of plaster on the external stone face.

Destroyed by the 1930 earthquake, it is now a large ruin of which only part of the stonework survives.

The north and south walls are badly damaged in the upper part due to the penetration of rainwater that has broken up the mortar joints.

The church initially had a single apse, flanked by two sacristies, whose access is now walled up, unlike they were in the 1970s when Alpago Novello carried out his surveys. The apse, which does not protrude to the outside, has a pass-through crack and is affected by holes caused by clandestine excavations.

Fig. 50 Sarnaq, S. Astvatzatzin: plan (Alpago Novello et al. 1977)

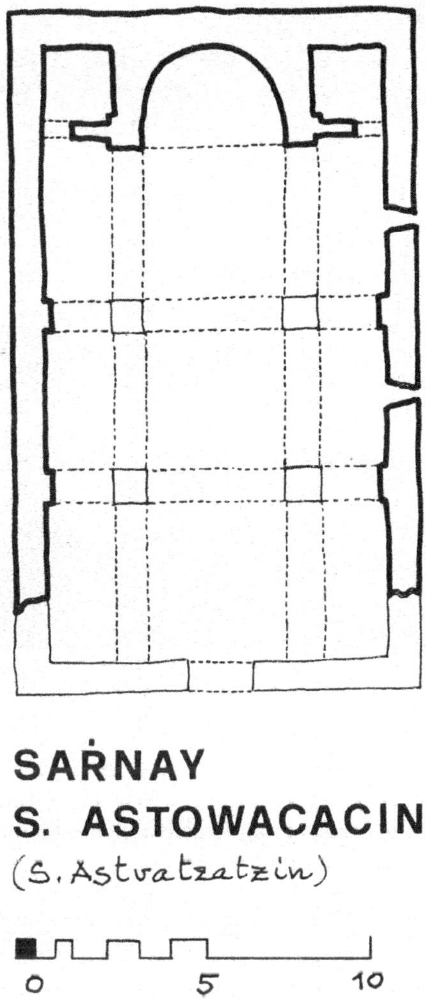

Of the four masonry pillars that should have divided the interior space and probably supported a dome, only one remains (SE side), which appears to be surmounted by the remains of a lowered arch.

A mound of earth from the excavations mentioned above prevents from a straightforward reading of the dimensions that the two arches must have once had. There are no traces of the west façade: its stones were probably used to build a fence that started from the north wall of the church and continued obliquely for about twenty metres. Today, only the foundations can be seen. Next to the north side of the church, outside, is a large millstone of uncertain origin. There are also some fairly recent graves (nineteenth/twentieth century).

Fig. 51 Sarnaq, S. Astvatzatzin: detail of the interior part of the south wall (photo M. Boriani 2017)

Two large shrubs have grown inside the nave and should be removed to prevent further damage to the ruins of the building, just as it would be essential to have greater control of the accesses, now free, to fill in the holes caused by clandestine excavations, to clean the area of vegetation and to carry out provisional works to secure what remains of the ancient building.

The building can only be used as an archaeological find, but it is an imposing one and deserves to be preserved. The area around the ruins would be turned into a public green space, which would ensure greater and better access and control.

2 Khoy County

FANAI

The village is part of Khoy County and is located about 4 km north of the capital. In 2006, it had a population of 652 in 134 households.

Dedication unknown, (**S. Sarkis/Sargis?**—Alpago Novello et al. 1978).

Location: inside the village, in the enclosure of a private property.

Period of construction: ("a fairly old building with interventions in the apsidal area of the sixteenth–seventeenth century", Alpago Novello et al. 1978).

Planimetric layout: inscribed cross.

Consistency and state of the building in the 1970s: "On the outside, the apse is flat with three tiny windows framed by large, roughly defined stone elements. The north side is made of cyclopean stone masonry: it is worth noting the brickwork placed in the central part to create a side window and a kind of passage to the roof. On the western side, the tiny entrance portal is framed by massive stone elements, surmounted by a kind of lunette, vaguely horseshoe-shaped, also monolithic. The inner and outer walls are strongly tapered, and the windows are splayed. On the inside, the entire apsidal part appears to be covered with a 20 cm layer of *terracotta*. This is also evidenced by the small, partially walled-in *khatchkar* at the bottom of the modest niche on the right side of the apsidal area. (…) The inner perimeter wall and pillars are of coarse stone, partially plastered in *khagil*, with a thick layer of white lime overlying it. The dome, resting on four pillars, with huge cube-shaped capitals, is now broken through. The arches are pointed and made of bricks laid in a knife shape. The vaults are flattened and made of very irregular stones and do not emerge from the roof (except for the central one). The brick vault of the apse area appears to be of later date. The basin of the apse and the triumphal arch are also made of brick, with the insertion of glazed ceramic bowls, probably from the Safavid period (…) On some of the corner joints one notices the presence of large *amphorae*, perhaps for lightening the load or, more likely, for acoustic purposes. There are traces of frescoes: on the second central arch and on the front of the triumphal arch, floral decorations in the colours brown, yellow and green. On the left, in the upper part of the triumphal arch are figures of winged knights. Underneath, traces of other frescoes: a standing figure in perspective with a halo. (those two seem superimposed layers; reference can be made to the pictorial decoration of Mahlazan). A very characteristic element is the Persian-style altar-*iconostasis* (see the example of Khoy) with two stalactite arches and decorations of palmettes and twisted columns in green and brown. The small domes of the passages in the *iconostasis* are minutely decorated with stucco. A brickwork revival can be seen in the upper part. The jambs are probably reused tombstones, although the inscriptions are illegible. The baptismal font, removed from its original location, is now lying on the ground in the south aisle. The walls are very thick, more than a metre, and made of granite and schist blocks, sometimes of very considerable size: this would suggest for the building to be quite old, with subsequent interventions in the apsidal area in the sixteenth-seventeenth centuries. In any case, the local environment, still characterised by persistent archaic features, must be taken into account. Certain similarities with the church of Gharabagh are particularly noteworthy, both in terms of typology and construction, and in certain details such as the entrance portal and the amphorae in the corner joints. But in Fanai the dome was undoubtedly emergent, probably without a drum, of the type of that at Khoy." (Alpago Novello et al. 1978) (Figs. 52 and 53).

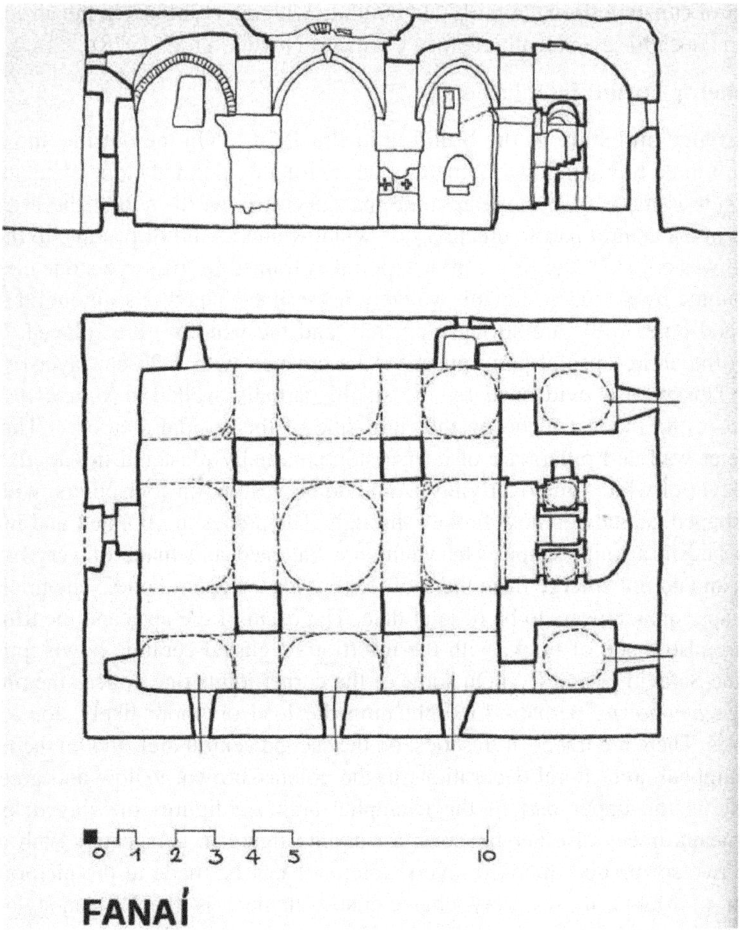

Fig. 52 Fanai Church: plan and longitudinal section (Alpago Novello et al. 1977)

Description and State of Preservation

The church is now within a private property, and two boundary walls lean against it, one of which hinders the full view of the west façade.

The masonry of the building appears to be made of coarse stone blocks: cyclopean those of the basement. The walls, which are about 120 cm thick at the base, taper considerably towards the top. On both the exterior and interior parts of the basement, deep erosions of the joints between the stones are recognizable, due to the action of rainwater. These erosions are particularly severe in the upper parts, where the masonry appears wholly decayed. In contrast to the interior, which has elements manufactured with care, the exterior appearance is massive and straightforward. It is accessed from the west through a shallow doorway: the interior has

Fig. 53 Fanai Church: general view from drone. The church is clearly inserted in the village context

Fig. 54 Fanai Church: West façade in 1976 (Alpago Novello et al. 1977) and 2018 (photo M. Boriani 2018). Evident erosion of joints in clay mortar

three naves, marked by large, squat pillars of about a metre on each side, supporting vaguely pointed brick arches which, in turn, supported a dome in the centre, also made of brick, now largely collapsed and, for the remaining part, unsafe. The church walls have superficial cracks, except the wall of the third north bay, partially destroyed by the collapse of the vault (Fig. 54).

Fig. 55 Fanai Church: the absid iconostasis in 1976 (Alpago Novello et al. 1977) and in 2018 (Foto M. Boriani 2018)

Beautiful is the altar part with frescoes on the side pilasters, vaults with painted stucco stalactites and an altar, which has unfortunately been vandalized. The apse does not show any severe deterioration. The vault of the apsidal area has amphorae at the four corners, perhaps for acoustic purposes.

The traces of frescoes with winged knights described by Alpago have disappeared, as has the baptismal font.

The NE pillar is partly made by reusing tombstones with ancient decorations. A break, not serious, is at the SO corner. The two sacristies are in good condition, except for a deep excavation at the entrance to the north sacristy (Fig. 55).

The building could soon suffer further collapses to its roofs if urgent safety measures will not be taken. It is absolutely essential to consolidate the few remaining interior plasters, restore the margins and preserve the *iconostasis* paintings, which are quite rare for the area and reminiscent of those of the S. Sarkis of Khoy. The two collapses in the roof allow birds access, and the whole interior is littered with excrement, feathers and animal corpses. If there is no adequate funding to rebuild the roofs, it would be helpful to at least close off the openings with nets to prevent birds from entering. External access is almost impeded by the placement of various materials (tractor wheel tyres, manure, straw bales, concrete block walls): all should be removed and the access to the church protected by a suitable door. The north-west corner, at the base, has evidently been damaged by the passage of heavy vehicles. Although the exterior walls of the church are massive, it is still appropriate to repair the deficiencies and avoid further anthropogenic damage. It is also important to consolidate the weathered wall ridges and remove the weeds that damage them while rebuilding the mortar joints in the upper stone courses, which are also exposed to rainwater (Fig. 56).

The site could be the subject of a general redevelopment plan which, in agreement with the private property on whose land it is located, would grant access and redevelop the surrounding open space.

It has not been possible to verify the information given by oral sources. However, it seems that negotiations are underway to achieve protection of the

Fig. 56 Fanai Church: the central nave looking the absid (Photo M. Boriani 2018)

building, better accessibility by the public and a use more compatible with the historical character of the church. In the opinion of the current owner, there have also been recent attempts by treasure hunters to gain access: a solid door to close off the access would be very useful.

GRIS (QIRIS)

Qeris (Persian: قریس, also Romanized as Qerīs, Ghris, Grnis) is a village in Khoy County, West Azerbaijan Province, Iran. At the 2006 census, its population was 625, in 118 families.

The village is located about 30 km NW of Khoy. It can be reached by driving along the valley of the Gudik stream, in the bottom and on the edges of which there are oases rich in vegetation, finally climbing a slope that leads to a more secluded valley.

The settlement is composed of traditional buildings made of *khagi*l (unbaked earth, in this case, red). The historic buildings are still largely in use, which has led to various extensions or replacements of materials and finishes that have often damaged the authenticity of the place. However, several traditional parts are still recognizable, often in a reasonably good state of preservation: *tandoori* (tonir in Armenian) ovens, grain stores, barns and stables, flat roofs with holes for venting the fireplaces below, etc.

The relative proximity to Khoy and the beauty of the landscape on the way to the village, perhaps visiting some of the more exciting oases on the valley floor, would also allow an exciting occasion for tourism valourization.

S. Astvatsatsin/Asdvadzadzin/Unknown dedication (Alpago Novello et al. 1978).

Location: in the northern part of the village.

Period of construction: seventeenth century ("date can be deduced from several inscribed and dated crosses", Alpago Novello et al. 1978).

Planimetric layout: inscribed cross.

Consistency and state of the building in the 1970s: "The church of Gris presents an original typology in the form of an inscribed cross and, above all, in the presence of pillar-columns supporting the dome, a solution that is not at all common (except for the example of Payadchouk, which is now in a state of ruin). The two-storey side chambers are also an unusual motif, never found in the area.

(…) A flat apse with three small windows framed by a block of pale stone. At the height of the windows, a sort of band of stone blocks, much larger than the rest, runs along the outer face. On the outside of the wall of the apse, a large lesion is visible (according to oral tradition, a consequence of the 1930 earthquake). Also on the outer wall of the apse are a few simple engraved crosses without inscriptions. (…) the floor plan is basically cruciform, with a central dome, on four monolithic octagonal-section reredoses, before the capital. The base of the pillars is considerably sunken (…) The *bema* is fairly raised, with three side steps. There is no trace of an altar; in the apse, two niches open on both sides. The baptismal font is in its usual position, surmounted by an arch made of large ashlars of the usual light-coloured stone. On either side of the apse are two accessory rooms, with a singular two-storey layout. The lower floor is below the level of the nave (down two steps); the entrance is architraved, reinforced inside by a wooden beam; the room to the north is covered with a false vault, the one to the south with an almost flat vault. The upper floor is accessed from the apse by two stone staircases, each with 6 steps. The two small rooms have a small niche in light-coloured stone on the outer wall, with shaped moulding. The side aisles are vaulted, with a kind of lowered sail. Between the pillars in the centre, the bulge caused by the debris of the dome is clearly visible. (…) The outer face is made of large square blocks in the lower part, then continues in coarser, roughly squared stone. The N wall, as well as about half of the W wall, are now deeply buried (the earth has been removed—2017). The roof, the dome of which has collapsed (no one remembers seeing it erected), has a slightly rounded profile. The entrance portal, on the W side, which is considerably buried, has a sizeable monolithic lunette in light-coloured stone, with a moulded lower profile. The jambs are formed by two blocks each. (…) A very simple *khatchkar*, evidently of rubble, is inserted transversally into the S perimeter wall. There is an opening on this side, but it does not appear to have been an entrance. Several stones on this wall bear engraved crosses with some inscriptions … In the interior some stones bear engraved crosses and some inscriptions: the transverse

Fig. 57 Gris S. Astvatstatsin: plan (Alpago Novello et al. 1978)

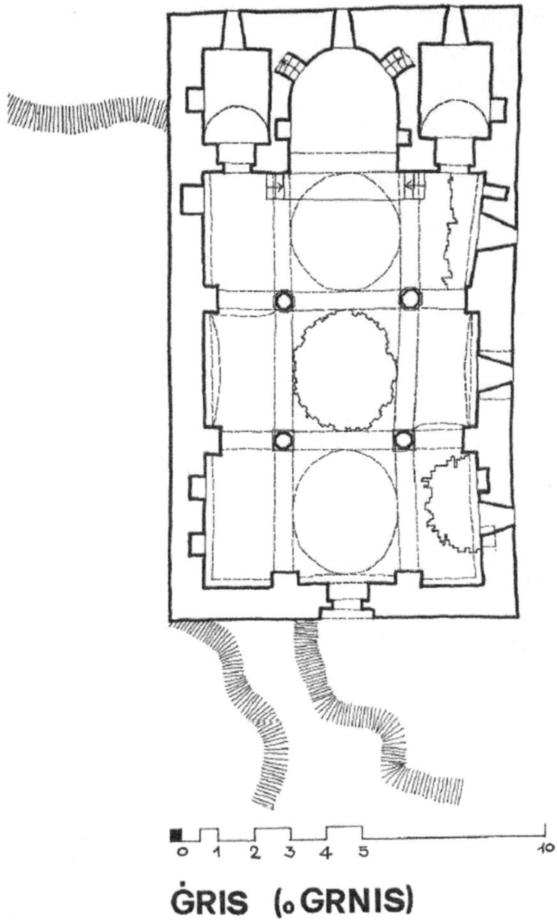

position of these indicates that they are rubble stones. To the SW of the village, on the hill, stood the cemetery … now almost completely destroyed, there are still a few red stone tombstones … of which only one bears an inscribed cross. It is singular that none of the inhabitants remember the Armenian presence there" (Alpago Novello et al. 1978) (Figs. 57 and 58).

Description and State of Preservation

The church overlooks a square with a children's playground and is preceded by a multi-level stone pavement, recently built. It is a massive building with a rectangular plan, made of larger rough-hewn stones in the basement. It is accessed through a low entrance made of squared and worked stone blocks.

The interior is divided into three naves by four columns, and the central nave is surmounted by three domes.

Fig. 58 Gris S. Astvatstatsin: south façade (photo M. Boriani 2017)

Fig. 59 Gris S. Astvatstatsin: south west corner in 1977 (Alpago Novello et al. 1978)

The building appears recently restored on the outside (note the incorrect use of cement mortar for the joints and the probable replacement of part of the stone ashlars in the upper part of the masonry), while inside the central dome has collapsed. Birds nesting in the church can enter through it and the narrow windows not protected by netting.

The plan drawn up for the 1977 mission shows indications of damage to the roof of the south aisle, which does not appear today. The baptismal font and *kachkhar* mentioned by Alpago Novello have disappeared, the mounds of earth that buried part of the pillars have been removed, and the damage to the apse caused by the 1930 earthquake is no longer present (Fig. 59).

Fig. 60 Gris S. Astvatstatsin: the small left nave with a two story sacresty (Alpago Novello et al. 1978)

The two two-storey sacristies are attractive, an architectural solution that is quite rare in this area. On the whole, probably thanks to the recent work carried out in keeping with the historicity of the building, the state of conservation of it does not present serious deterioration problems.

Involved at least in the urban layout with the new arrangement of the public space, a new and compatible use for the building does not seem to be envisaged, perhaps for reasons related to property rights (Figs. 60 and 61).

However, a new use would be desirable, as the one linked to worship, seems to be unthinkable at the moment, involving the resident population in order to guarantee its protection and maintenance.

The character of the village, as already mentioned, could attract discrete, albeit modest, flows of cultural tourism once the current pandemic phase is over and international travel is resumed. The church could be included as an exciting element in this circuit, collecting and setting up within it documentation on the building itself but also on the construction techniques of the village and local customs and traditions.

KHOY (XOY)

Khoy, the capital of Khoy County, is located north of the province capital and largest city Urmia. The regional economy is based on agriculture, mainly fruit,

Fig. 61 Gris S. Astvatstatsin: the dome collapsed and now without adequate protection (photo M. Boriani 2017)

grain and timber production. At the 2006 census, the city had a population of 178,708.

Khoy is now populated by both Azerbaijanis and Kurds, with the majority being Azerbaijanis. Occupied since Median times, it was for centuries a contested border town between Ottomans and Persians. As a result, it was one of the few cities in Iranian Azerbaijan to be provided with a ring of walls, which were reinforced in the European style in the early nineteenth century.

Khoy shares a long history as an important Christian centre: until 1828, it had many Armenians; however, many emigrated when the Treaty of Turkmenchay (1828) gave the Russians the right to encourage Armenians to immigrate into Russian Empire. Nevertheless, a small Armenian population remained living in Khoy.

The Armenian community lived mainly in a suburb called Imamzadeh, about 3 km north-west of the walled city. According to Hommaire de Hell (1855, pp. 526–528) in the mid-nineteenth century, there were 2,800 houses inside the walls and 3,700 in the outer suburbs, and only about fifty Armenian families remained in the town.

S. Sargis/Sarkis.

Location: within the modern city, in the old Imamzadeh district, now semi-central.

Period of construction: 1120 (?), interventions in seventeenth century (Alpago Novello et al. 1977).

Planimetric layout: domed basilica.

Consistency and state of the building in the 1970s: "This is a large domed basilica, with three naves and four bays, with cruciform pillars and a polygonal apse, flanked by small corner chambers, all non visible externally (…) The volume is elementary, barely animated by the low emergence of the dome, which has a zenithal hole for lighting. The interior, made of cut stone, which is quite accurate, then plastered and now being restored, is very dark. In the *bema*, is a sort of *iconostasis-ciborium*, with a triple arch opening towards the worshippers, all frescoed (probably in the nineteenth century). The pendentives of the dome have a sort of stalactite decoration. (…) The masonry on the outside consists of a plinth of perfectly squared blocks about 1.60 m high. Baked bricks are then used (stone is used for the squares of the thin slot-windows) (…) A certain oral tradition (reported by Vardjavand) would have the building built on an earlier pagan temple. The foundation date of 1120 (mentioned by the same author) cannot be verified at present. The church has undergone a series of restorations, in particular in the eighteenth century. The only decorative elements on the portals show clear Islamic influences. The numerous epigraphs, not all of which are legible due to the ongoing restoration work, do not generally give chronological references, except for the small *khatchkar* on the W front (1657). The fact that, for example, for the jambs of the entrances, tombstones have been reused leads to think of radical reconstructions

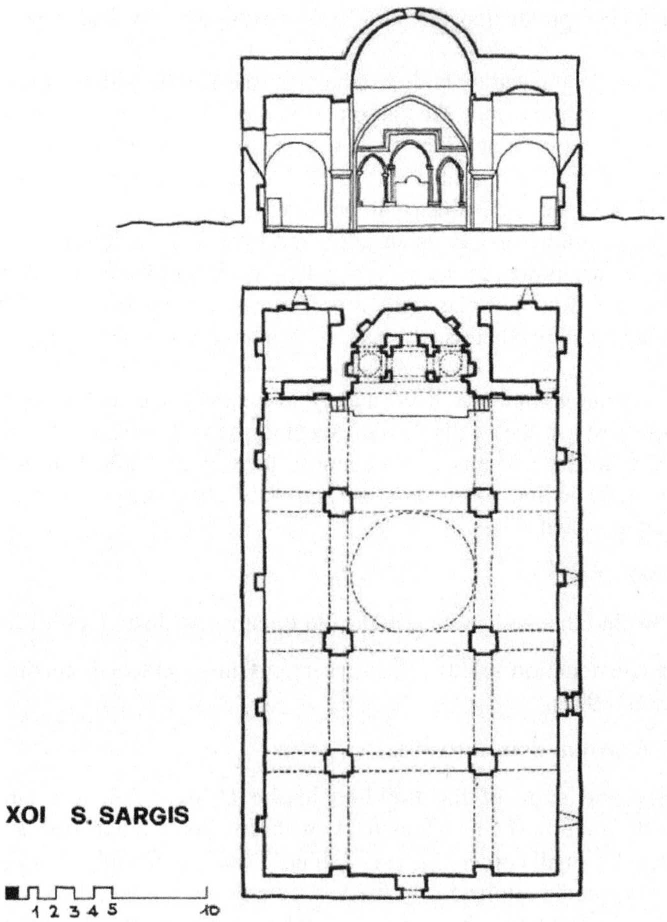

Fig. 62 Khoy, S. Sarkis: plan and cross section (Alpago Novello et al. 1978)

of the structure, which, in any case, makes it one of the most significant in the region" (Alpago Novello et al. 1977) (Figs. 62 and 63).

Description and State of Preservation

The church is located within the inhabited settlement, surrounded by a paved open space. It is (along with Haftvan) the best-known Armenian church in the area and the one in the best state of preservation.

It probably once had a structure entirely made of stone, which today survives for only eight courses, visible in the upper parts of the building only in a few places. Today, the upper masonry consists of bricks made in several phases, as shown by the visible repair of gaps and lesions.

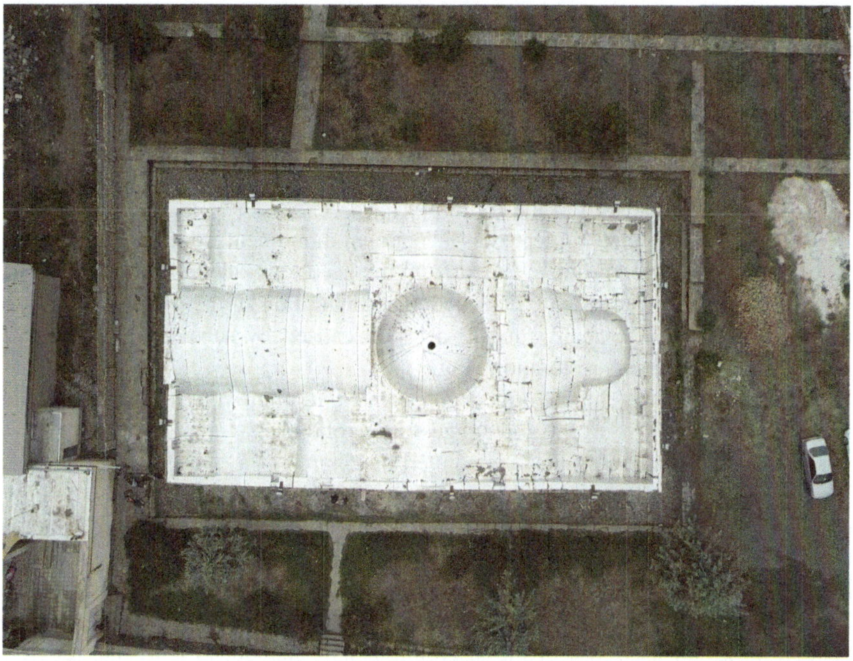

Fig. 63 Khoy, S. Sarkis: zenith view of the roof recently restored

The church has undergone several restoration works in recent times: the first documented studies date back to the years between 1975 and 1978; according to the archives of the local monument protection office, a series of subsequent interventions took place throughout the 1980s and were completed in 1995 (ICHHTO Archive no. 4424–1995).

Fig. 64 Khoy, S. Sarkis: the iconostasis in 1977 (Alpago Novello et al. 1978) and 2018 (photo M. Boriani 2018)

It is now in a good state of preservation, well maintained and valued as a town monument.

With a nave and two side aisles, the interior is also in good condition: the apsidal part is beautiful, with an iconostasis, restored in the 1980s, which still bears substantial traces of decoration. The paintings of Helena and Constantine mentioned by Samuel G. Wilson (Wilson 1895, p. 85) have almost disappeared. From the north sacristy, a ladder leads up to the roof.

The building, which is protected as a "national monument", does not present any particular conservation problems apart from the need to re-adhere and seal the parts of painted plaster that are now detached. The disruption in the polylobate arch of the west entrance is probably due to a lesion caused by subsidence, which was resolved during the last restoration (Figs. 64).

The location of the church, now deconsecrated, in a fairly large town with other monuments worth visiting, would suggest its public use.

Oral sources report a proposal to convert the church as a "museum of religions". That would seem to be opposed by the Iranian Armenian community, which claims exclusive building use. While the proposal for a museum of this kind should not be ruled out a priori (obviously everything would depend on its arrangement and layout, as well as respect for the morphological and material characteristics of the building), it is essential that a decision on the matter be taken as part of a more general policy of use/reuse of the large number of abandoned Christian religious buildings in the area. At the moment, a less complicated proposal to get approved could use the building for public meetings for cultural or social use, compatible with the original religious function. Since it is located in a relatively central area of Khoy town, the public hall could be flanked by a centre for the documentation of the buildings and culture of the local Christians, supporting both the Christian communities that still survive, and visitors, who could find information here and visit the monuments of the area (Figs. 65 and 66).

MAHLAZAN

The farming village of Mahlzan, eight kilometres north-west of Khoy, is located in a densely cultivated area, 300 m east of the main road linking Khoy to the crossroads leading to Maku to the north Julfa to the west. At the 2006 census, its population was 462, in 127 families.

S. Khach Kristosi.

Location: at the northern edge of the settlement, immediately west of the site of archaeological interest of Mahlazan (Tepe) Hill, which is registered in the list of Iranian national monuments.

Construction period: seventeenth-eighteenth century (Alpago Novello et al. 1977). Mongol-Safavid according to the registration report of the church in the catalogue of national monuments (Bakhshaei and Mohammadi 2015).

Planimetric layout: three-nave basilica with polylobate apse and two side sacristies.

Fig. 65 Khoy, S. Sarkis: central nave (photo M. Boriani 2018)

Fig. 66 Khoy, S. Sarkis: south façade (photo M. Boriani 2018)

Consistency and state of the building in the 1970s: "The church is one of the most important monuments in the region, both in terms of its size and its architecture: typologically, it is a vaulted and domed basilica, with a large central dome emerging from the roof with a drum: polygonal and with a pyramidal cusp in terracotta (…) Despite its elongated layout, the internal space is inspired by examples with a central plan, with the cupola supported by four pillars, as *fulcrum* of the composition. The surfaces, completely plastered, were probably partly frescoed: there are considerable traces in the apse area and the entrance (…) The interior has, as usual, an elementary volumetry: the masonry is in irregular stone, with extensive use of mortar, and the arches and dome are in brick. In the wall structure wooden tie rods are arranged horizontally at the height of the keystones (visible from the outside). A large portion of the north side has now collapsed. The *gavit* has also disappeared (since 6/7 years according to oral tradition: the church was seen in one piece by Kleiss in 1969) (…) There are no longer any Armenian families on the site, which must once have been the seat of an important community: rather than a church, it is more appropriate to speak of a 'sanctuary', an ascertained destination for pilgrims who also came from distant regions (Giulfa–Georgia), as can be seen from the simple engravings left graffitied on the plaster (from the second half of the seventeenth century). It is worth noting the apsidal basin that is plastically modeled with a series of undulations (or small flattened niches). (…). The arrangement of the *bema* is unusual: it is not so high and communicates with the nave by means of a flight of steps extending the entire width of the apse (while the Armenian liturgy provides only two side access steps). On the façade to the right of the entrance, where there is now a rectangular hole, there was a black stone with an inscription (perhaps the dedication) that was removed 20/25 years ago, perhaps together with the stones that formed the doorpost and were used for the construction of a bridge over the Navayi stream, which later collapsed due to a flood" (Alpago Novello et al. 1977) (Figs. 67 and 68).

Description and State of Preservation

The church is located on the edge of the settlement, with a small wooded area to the south and arable fields to the north. To the east is the above-mentioned relief (*tepe*) from which substantial traces of ancient buildings emerge; on its top, it is possible to observe the building from an elevated position.

The building appears as a massive rectangular construction, surmounted by the drum of a large central dome. Horizontal wooden beams are inserted into the stone masonry at the height of about 2 m, probably for anti-seismic purposes.

The interior is divided into three naves by six massive pillars and ends in the apse, not visible from the outside, flanked by the two sacristies. The roof is vaulted: two vaults strongly lowered to the east and west and a large vault with a central drum. This drum is unique in the Khoy-Salmas area and is reminiscent of the traditional roofing system of St. Thaddeus and St. Stephen's central vaults, which are derived from models in Armenia proper.

On the west elevation, on either side of the entrance door, there are traces of shutters supporting arches, which correspond to two truncated pillars: as already

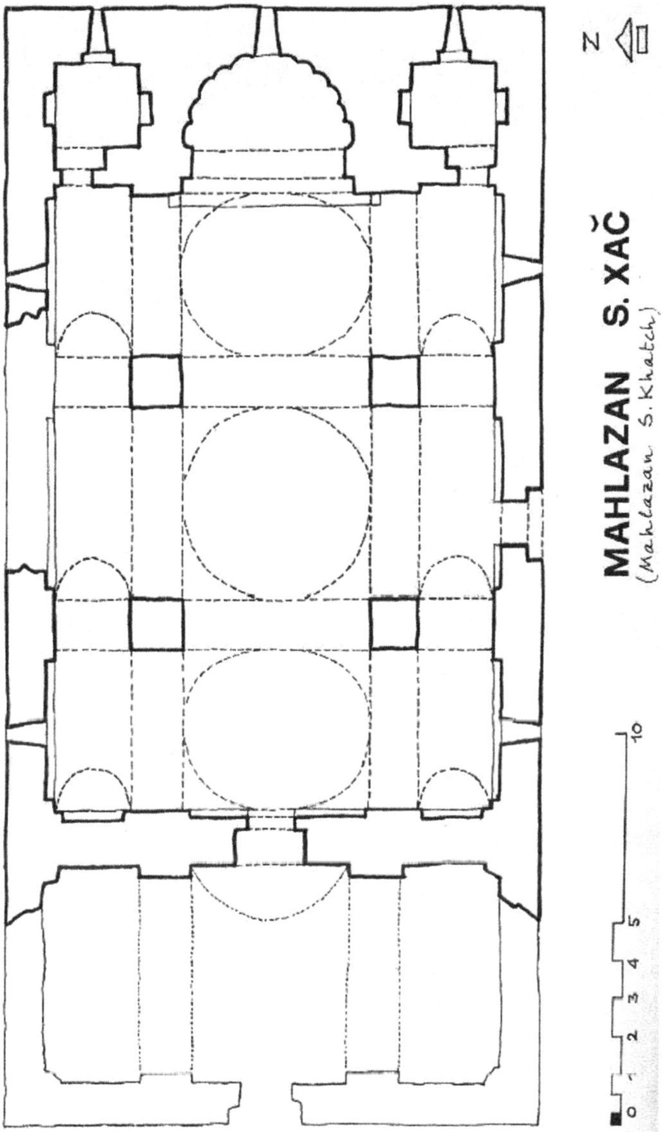

Fig. 67 Mahlazan S. Kateh: plan (Alpago Novello et al. 1977)

noted by Kleiss and Alpago, there was a vaulted structure that had already disappeared in 1977. Alpago defines it as a *gavit* but, due to its small size, it seems to have been more of a simple portico, a relatively rare construction in the area of Khoy, Salmas and Urmia. A pre-1930 photograph documents the portico was still intact.

Fig. 68 Mahlazan S. Kateh: dome (photo M. Boriani 2017)

In the 1977 inspection, the eastern portion of the building showed a significant collapse, which now appears to have been repaired. The reconstructed part used the collapsed materials as far as possible and, for the missing parts, a local stone of a greenish colour, which differs from the materials used initially. This differentiation method does not seem to have been dictated by a precise design intention, but it is appreciable in any case, as makes it possible to recognize the two construction phases, ancient and recent, of the church.

It would have been preferable using more compatible materials, a sample of which was easily available by analysing the old joints still in place (probably in plaster and clay). The upper portions of the north, west and south walls also appear to be of recent reconstruction, and here too the use of cement for the joints between the stones is evident.

Photographs, taken during the 1977 study campaign and others before 2011, show that a portion of the north wall and its vault had also collapsed. The same information can be drawn from the 1977 plan, which locates the gap at the north central bay. By carefully observing the external state of the north wall, it is possible to identify the portion of masonry recently rebuilt, also using the greenschist stones used in the east wall (Fig. 69).

Recent interventions have also included protecting the roof with the usual waterproofing panels and the channelling of rainwater. Traces of numerous drips on the external surfaces present at the 2017 inspection have now been partially removed.

Fig. 69 Mahlazan S. Kateh: the church before (Alpago 1976) and after restoration (photo M. Boriani 2017)

In order to combat rising damp, a drainage system was installed inside the building, which runs around the entire perimeter. This is unusual, as it is typically carried out on the outside of the masonry to counteract rising damp and prevent rainwater from collecting.

Fig. 70 Mahlazan, S. Kateh: west façade (photo M. Boriani 2017)

Photographs from 1977 documented a fairly good state of conservation of the west façade. As of 2017, the condition appeared to have deteriorated, mainly due to erosion of the joints between the stones of the upper portion of the masonry: recent interventions have also completely reconstructed the missing parts. As for the west entrance, the stone pillars and architrave, on which several crosses were carved, documented in 1977, appear to have disappeared in 2011, and an iron door is now in their place, already documented in 2011 (Figs. 70 and 71).

The drum and the pyramidal roof of the central dome, made of exposed bricks, have undergone extensive restoration and are now in perfect condition. Twelve large windows pierce the drum and some damage, not yet severe, can be seen near the keys of the supporting pointed arches.

Several missing bricks can still be seen, but these do not seem to endanger the statics of the building for the time being.

In the interior, the paintings found by Alpago have almost disappeared, while there are still considerable traces of ancient inscriptions mixed with many recent ones of vandalism.

Here, recent interventions, in this case highly critical, have entailed the complete removal of the pre-existing plasterwork, which has been replaced by new, polished plasterwork, in addition to the pseudo travertine flooring finished to a high gloss and the insertion of a stone plinth running along all the walls, which seems rather

Fig. 71 Mahlazan, S. Kateh: south façade (photo M. Boriani 2017)

Fig. 72 Mahlazan, S. Kateh: central nave looking west (Alpago Novello et al. 1977)

unsuitable: all of which compromises the historical substance of the building (Figs. 72 and 73).

Numerous traces of bird traffic suggest that the windows of the dome should be closed off with netting or, at the very least, bollards, to prevent birds from roosting.

In the vicinity of the church, there is much scattered material, presumably intended for external paving, which, if carried out, would further damage the appearance of a rural sanctuary that the site must have had initially been.

The hypothesis of the existence of a monastery annexed to the church (Bakhshaei and Mohammadi 2015)—the site was a place of pilgrimage even from distant regions—would suggest carrying out archaeological surveys to verify the permanence of traces of this and other annexes of the sacred building.

Fig. 73 Mahlazan, S. Kateh: central nave plasters remaking (photo M. Boriani 2018)

QORUQ (GHOROUGH)

Qoruq (Persian: قروق, also Romanized as Qorūq; also known as Bāzār-e Kargārūd and Qoroq) is a village in Saheli-ye Jokandan Rural District, in the Central District of Talesh County, Gilan Province, Iran.

The village is located about 10 km south of Khoy, on the edge of the road to Salmas, in an area of flourishing agriculture. At the 2006 census, its population was 1,160, in 274 families.

Dedication unknown.

Location: the church stands in the open countryside a few hundred metres from the village.

Period of construction: not known, although it should be quite old.

Planimetric layout: basilica layout.

Consistency and state of the building in 1970s: (Figs. 74 and 75).

Description and State of Preservation

It is a building with three naves and three bays, with an apse flanked by the two standard small service rooms. According to oral tradition, four pillars (two still

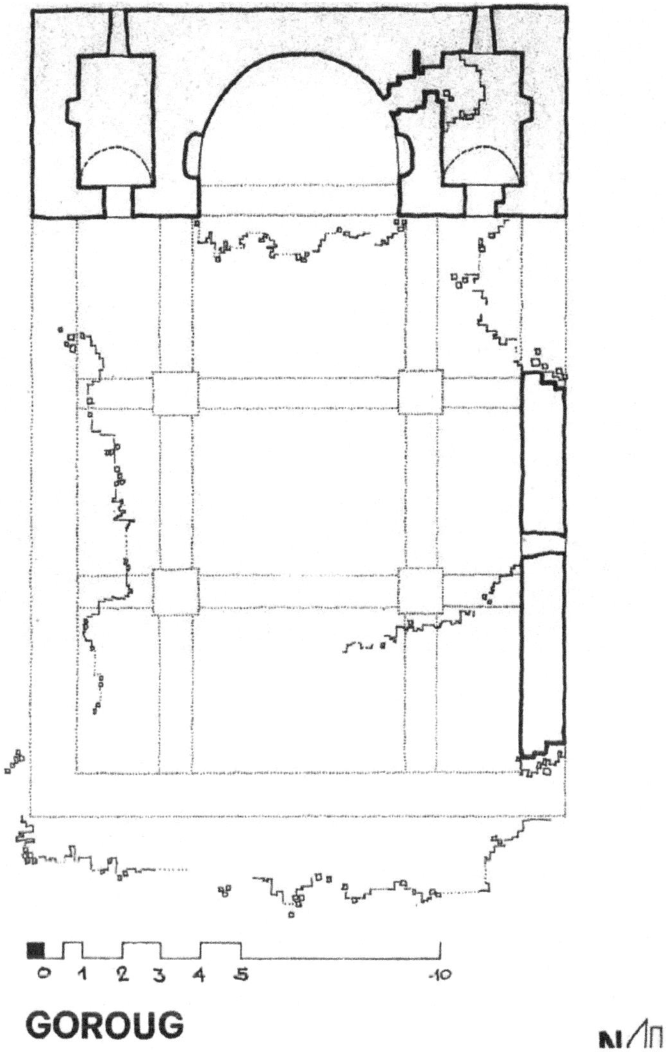

Fig. 74 Qoruq Church: plan (Alpago Novello et al. 1978)

visible in Alpago's time) supported a dome, but no one remembers when it collapsed.

The building is now in a state of ruins, although the features of the layout are still clearly recognizable. The structure of the apsidal part, facing east, is made of well-crafted stone, even if the outer covering of the masonry, made of regular ashlars, has mainly been removed (only a portion remains on the north wall), leaving the conglomerate sack of flakes and pebbles visible. No severe deterioration was noted with respect to the geometric and photographic survey carried out during

Fig. 75 Qoruq Church: zenith view

Alpago's mission, except for a thickening of the infesting vegetation, which occupied the western side of the construction. Alpago also refers to small niches with terracotta arches in the apse and in the two side rooms, which are now invisible due to the infesting vegetation and the impracticability of the area due to the danger of collapse of the vaults. The large stone with three inscribed crosses reported by Alpago has disappeared.

It is unthinkable to reuse the building, which is now a find of archaeological and landscape interest. However, it is suggested that maintenance be carried out to control weeds, protect the remaining wall crests, prohibit the removal of other materials and consolidate the apsidal vaults that still exist but are structurally significantly damaged and in danger of collapse.

3 Urmia County

GHAEABAGH

The village of Gharabagh is located about 75 km north-east of Urmia and 35 km south-east of Salmas, 1480 m above sea level and about five kilometres from the shores of Lake Urmia. It is one of the Anzal County (Urmia Region) main centres and is inhabited by some 500 mainly Turkish families, who engage in agriculture and livestock breeding. The few Armenians who used to live in the village have migrated to Urmia and Salmas. The Gharabagh settlement is made of adobe and features some exciting buildings with windows looking directly onto the public street, which is rare in rural Iran. Remains of a defensive wall with two semicircular towers on the western edge of the settlement.

S. Petros (S. Gevorg).

Location: on the western edge of the settlement, overlooking the fertile valley that descends towards Lake Urmia.

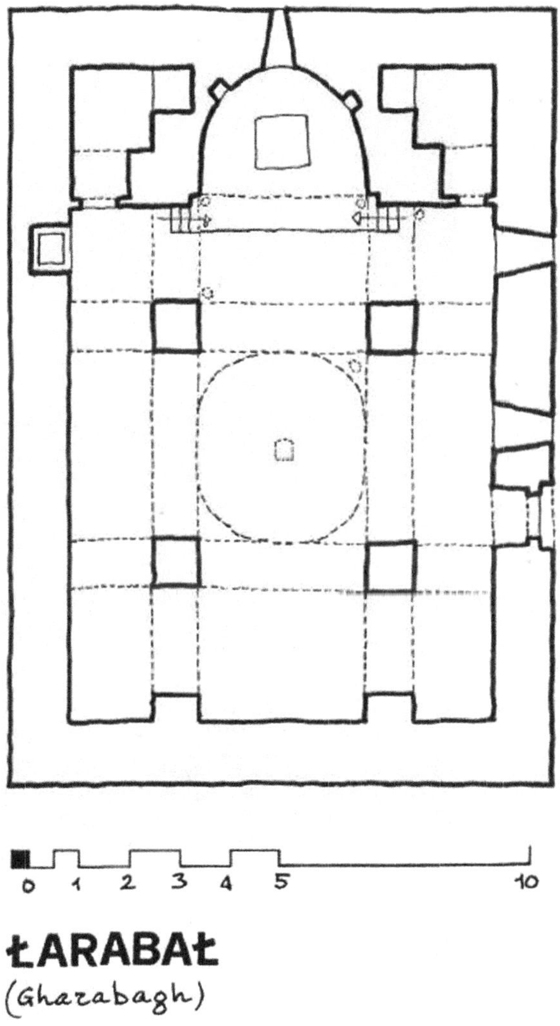

Fig. 76 Gharabagh, S. Petros: plan (Alpago 2017)

Period of construction: seventeenth century according to Alpago Novello and local tradition, 1650 according to MIRAS.

Planimetric layout: rectangular outside and Greek cross inside.

Consistency and condition of the building in the 1970s: (Figs. 76 and 77).

Description and State of Conservation

St. Petros, whose dedication would ascribe the church to a very ancient period, or St. Gevorg (Alpago Novello et al. 1977), could date back to the seventeenth century, restored or reconfigured in 1784 (or 1785) as evidenced by the date engraved on the entrance lintel (written in Armenian, also bearing the names of the benefactors who financed the intervention).

Fig. 77 Gharabagh, S. Petros: detail of the apse area (photo M. Boriani 2018)

The church is attached to a small group of residential and rural buildings. The keys are kept by an older woman who lives in one of these buildings and has provided some information on the recent history of the building. The church is listed as historical monument in Iran.

Externally, it appears as a simple rectangular volume with masonry of scope covering the roofline, marked by a frame of small regular ashlars. The face is made of various types of hewn stone (granite and volcanic stones) with ashlars of cyclopean dimensions in the basement area (Fig. 78).

The state of preservation documented in 1976 during the mission of Alpago Novello showed evident failures of some large ashlars in the lower part, a deep erosion of the mortar joints between the stones and considerable deficiencies in the masonry of the area.

The entrance to the interior is from the south side, through a lowered doorway, whose shoulders and lintel are made of regularly squared stones. A round arch emphasizes the entrance (Fig. 79).

The interior, which is considerably lower than the ground level, has walls, arches and a vault of exposed stone, again roughly hewn. Some traces of plaster are in the apse. In 1976, Alpago Novello documented the presence of an altar and *amphorae*, particularly interesting, inserted in the corners of the masonry, perhaps for acoustic or aeration reasons, which have now disappeared following, in all probability, the restoration work of the 1990s and the most recent (2017), suspended without being completed.

In fact, at the time of the visit, traces were still clearly visible of a restoration work in progress (beginning 2017, work already documented by SHA 2017) but interrupted, according to local witnesses, after three months of work, due to lack of funding.

The partly flat and partly vaulted roofs have been covered with mortar, and rainwater drainage is via newly installed gargoyles (two on the west side and two on the east side). According to a MIRAS technician, the original stone roofing, part of the boundary wall and the amphorae mentioned by Alpago have been removed to

Fig. 78 Gharabagh, S. Petros: view of the south front (Alpago Novello et al. 1977) and south west corner after restoration (photo M. Boriani 2018)

Fig. 79 Gharabagh, S. Petros: the very small original entrance (Alpago Novello et al. 1977), and its current arrangement after restoration (photo M. Boriani 2018)

reduce the load on the masonry. The part of the ambit wall is now much smaller than it was in 1976.

Inside, the masonry is still without plaster, although it is unclear whether this will be the final arrangement, and there are still work tools. The baptismal font has been destroyed, and the altar has disappeared.

The church was in use until the Khomeinist revolution and was abandoned following the emigration of the Armenian community.

It is hoped that any resumption of work will not lead to further changes to the church and that the interior work will take into account the historical nature of the building.

In order to ensure unquestionable durability of the works carried out, it would be appropriate, considering that the religious function is no longer present, to propose a new use for the building, which could become a public hall (meetings, library, exhibition space, etc.) for the local community.

JAMALABAD (DJAMALABAD)

Jamalabad (Persian: جمال اباد, also Romanized as Jamālābād; also known as Jamlav and Sūlīng) is an Assyrian village in Anzal-e Shomali Rural District, Anzal District, Urmia County, West Azerbaijan Province, Iran. At the 2006 census, its population was 209, in 67 families. According to local sources, only two Armenian families now remain in the village.

Mar Khné (S. Hovannhes?).

Location: near a poplar grove and uncultivated land.

Period of construction: undefined; the church seems to date from the eighteenth century but appears to have been greatly altered in recent years.

Planimetric layout: rectangular hall.

Consistency and state of the building in 1970s: (Figs. 80 and 81).

Description and State of Preservation

A church of Assyrian rite shows a plan with a single rectangular hall to which a body used as a sacristy or for other accessory functions is annexed to the south. The church has recently been joined to a different body to the south, also serving as a narthex.

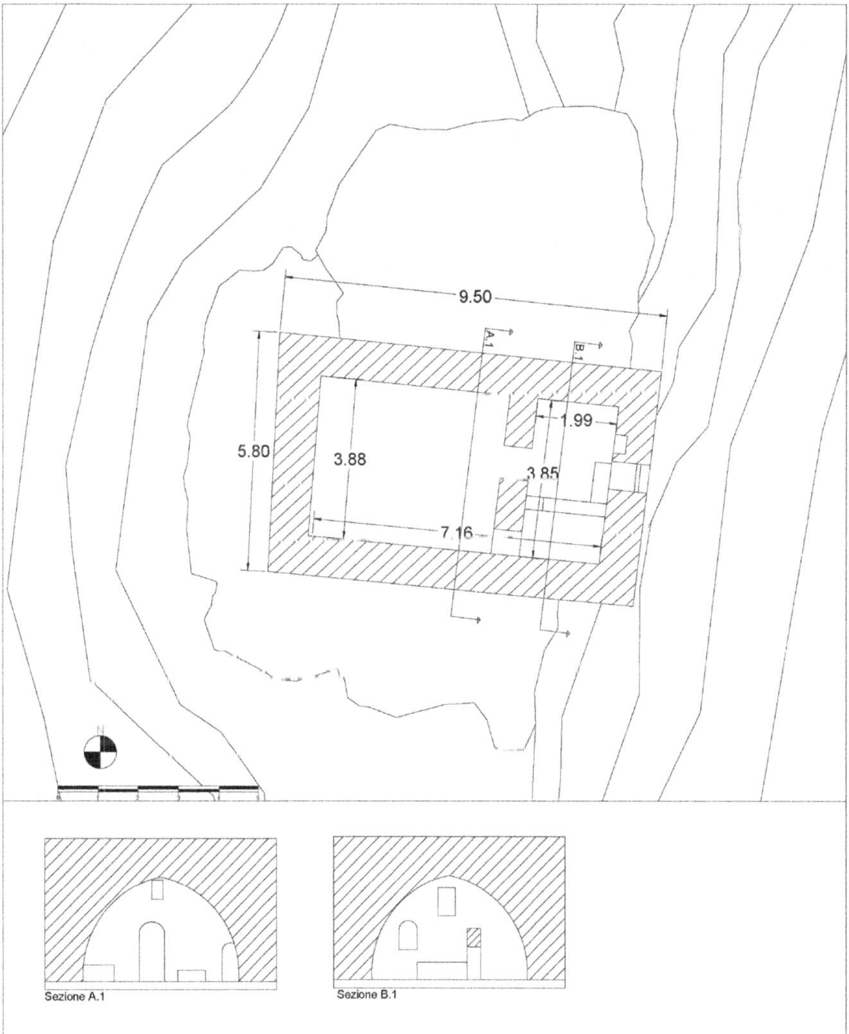

Fig. 80 Jamalabad Mar Khné: plan and sections (P. Karimi 2018)

Fig. 81 Jamalabad Mar Khné: east façade (photo M. Boriani 2018)

The altar is divided from the presbytery by a wall with a central opening, almost like an *iconostasis*. A service room is on the right side of the altar and is recognizable what could be an immersion baptismal font.

The roof is vaulted with a polycentric section. The older structure is made of stone masonry in regular courses, while the newer body is made partly of bricks and partly of cement blocks. The complex also had some outbuildings, which are difficult to interpret today (perhaps a rustic or a chapel). The whole is surrounded by a recently built wall whose entrance is closed by an iron door. The interior floor has been raised three steps above the original level, in order to preserve the church from frequent flooding due to heavy rain. Today, the building, which is still in use due to the presence of four Assyrian families, retains few historical features but is interesting for the articulation of the rooms mentioned above, which would confirm the attribution to the Assyrian rite (Figs. 82 and 83).

***S. Nišan/Nšan*.**

Location: isolated, in the middle of now uncultivated land. Near the Assyrian church of Mar Knhè.

Period of construction: eighteenth century.

Planimetric layout: basilica plan with three naves and four bays. Central apse flanked by two service rooms.

Fig. 82 Jamalabad Mar Khné: south façade (photo M. Boriani 2018)

Consistency and condition of the building in the 1970s: (Figs. 84 and 85).

Description and State of Preservation

The building is mentioned in the list in the volume of the Alpago mission, but it does not appear to have been visited at the time of the 1977 mission.

The church, made mostly of roughly hewn stone blocks, with the exclusion of the apsidal part, made of bricks, is now in ruins: most of the perimeter walls and the apsidal area remain. The latter has a rare convex ogival arch.

Severe erosion at the top and at the base of the outer walls (south and west) is probably due to un-channelled rainwater run-off.

A recent flood, caused by the overflow of the nearby river, which also affected the nearby church of Mar Knhè, may have accentuated the visible damage to the basement part of this building.

Without an enclosure, the church has been plundered by treasure hunters, which further compromises its stability.

Fig. 83 Jamalabad Mar Khné: internal view (Sina Yavarinia 2018)

Fig. 84 Jamalabad S. Nishan: zenith view

Fig. 85 Jamalabad S. Nishan: remains of the south wall of the church. The image shows the use of a mixed structure: stone for basement and vaults of the apse and rough earth for the walls (photo M. Boriani 2018)

Bibliography

Al-Jeloo N (2010) Evidence in stone and wood: the Assyrian/Syriac history and heritage of the Urmia region in Iran, as reconstructed from epigraphic evidence. "Parole de l'Orient" (35):1–15

Alpago Novello A, Brambilla M, Uluhogian G, et al (1977) Nice Vecchione, Consistenza e tipologia delle chiese armene in Iran. Rapporto della missione 1976, Ricerca sull'architettura Armena–Iran. Milano: Politecnico di Milano-Università Nazionale dell'Iran 1(17)

Alpago Novello A, Brambilla M, Piacentini VF, Uluhogian G, Uluhogian H (1978) Nice Vecchione, Consistenza e tipologia delle chiese armene in Iran. Rapporto della missione 1977, Ricerca sull'architettura Armena Iran. Milano: Politecnico di Milano-Università Nazionale dell'Iran 2(18)

Anaclerio G (1868) La Persia descritta. Relazione di un viaggio per Giuseppe Anaclerio, Napoli, Vincenzo Marchese

Bakhshaei H, Mohammadi BK (2015) Peeling report and study of roles Wall of the historic church of Haftavan Salmas. Peyman 19(72)

Berberian M, Tchalenko J (June 1974) The Salmas (Iran) earthquake of May 6th, 1930, Annals of geophysics = Annali di geofisica

Berberian M, Tchalenko J (1976) Field study and documentation of the 1930 Salmas (Shahpur–Azerbaijan) Earthquake. Geological Survey Iran (39):271–242
Bosworth (1995) Salmas. In Bosworth CE, van Donzel E, Lewis B, Pellat Ch (eds) Encyclopaedia of Islam, vol VIII—NED-SAM. Leiden, Brill, pp 998–999
Brambilla MG, Tehrani F (2020) Greater Salmas district. Studies and documents, Salmast heritage association, Tehran, Dastan, ISBN 978-600-297-166-1
de Mauroy H (1972) Lieux de culte, anciens et actuels, des églises syriennes orientales dans le diocèse d'Ourmiah-Salmas en Iran, azerbaïdjan occidental. "Parole de l'Orient" 3(2):313–351
Deilamqani TM (2006) Millennial history of Salmas and West Azerbaijan, editorial Aeldar, Tabriz, 2005–2006 (in Persian)
Flemming (1986/1995) Khoi. In: Bosworth CE, van Donzel BL, Pellat Ch (eds) Encyclopaedia of Islam, vol V—KHE-MAHI. Leiden, Brill, pp 28–29
Hommaire de hell X (1855) Voyage en Turquie et en Perse. Paris, P. Bertrand
Karapetian S (2019) History of Armenia. Salmast, RAA Foundation, Yerevan
Kleiss, Kroll S (1992) Survey in Ost-Azarbaidjan 1991 in Archäologische Mitteilungen aus Iran vol 25. pp 1–46
Kleiss W (1973) Bericht über Erkundungsfahrten in iran in Jare 1972. In: Archives mitteilungen aus Iran, N.F.B. vol 6. Berlin, pp 17–46
Moradi A (2020) Khoi in historical texts. In: Brambilla MG, Tehrani F (eds) Greater Salmas district. Studies and documents. Salmast Heritage Association, Tehran, Dastan, pp 155–172
O Flynn (2016) Thomas O Flynn, The western christian presence in the Russias and Qājār Persia, c. 1760–c. 1870. In: Brock P (ed) Studies in christian mission. Boston, Brill
Riahi MA (1999) The history of Khoy. Tarh-e Now publications. [in Persian]

Maurizio Boriani Architect, born in Milan, Italy, 1946. Full professor in Architectural Conservation at the Politecnico di Milano (Department of Architectural Design—School of Civil Architecture)—retired in 2016. Former Director of Department of Architectural Design at the Politecnico di Milano. Member of the Teaching College of the PhD in Architectural Conservation at the Politecnico di Milano. Former President of the Scientific Board of Cultural Heritage Conservation Center of the Politecnico di Milano. Research Co-ordinator on behalf of the UNESCO, World Bank and European Community, Italian Ministry of Foreign Affairs and other private and public agencies. fUNESCO Consultant for: Albania (2006–2009), Macedonia, Montenegro, Kossovo, Myanmar; expert in international cooperation on behalf of Italian Ministry of Foreign Affairs (China (2009–2012), Vietnam (2017–2019), Armenia (2019–2014), Pakistan (2011–2012), Myanmar (2012–2015). Research Topics: History of Architectural Conservation, Conservation and management of historic towns and archeological sites, Cultural landscape preservation and management, Gardens history and conservation, Cultural Heritage and International Cooperation. About those themes he is author or editor of about 240 papers and books.

Mariacristina Giambruno Architect, Ph.D. in Preservation of Architectural Heritage and graduated in Restoration of Monuments, she teaches at the School of Architecture, Urban planning, Construction engineering, and Graduate School in Architectural and Landscape Heritage (Politecnico di Milano). Full professor in Architectural Conservation. Member of Faculty Board of the Ph.D. program in Architecture, Urban Design, Conservation of Housing and Landscape until 2012, from 2019 Head of the Ph.D. program in Preservation of Architectural Heritage (Politecnico di Milano). The teaching and research activity is oriented on the history and theories of

Architectural Heritage, Conservation of Historical Heritage, Historical Centers Conservation and Management, and guidelines and management plans to preserve and promote Cultural Heritage. In recent years, she has focused on emerging and transitioning countries, working and directing international projects in collaboration with organizations such as UNESCO, World Bank, AICS in Albania, Iran, Armenia, China, Pakistan and Vietnam. In these countries, she investigates the role of Cultural Heritage as a "trigger" for sustainable development. As a speaker at conferences in Italy and abroad, she held courses and workshops in Armenia, Vietnam, Portugal, where she is part of a research unit coordinated by the Lusìada University of Lisbon. She authored about 150 publications on the preservation of Historical Centers and Architectural Heritage.

Studies in Armenian Architecture in Iran/Western Azerbaijan

Elnaz Yousefisalekdeh

Abstract Armenians living in the northwestern part of Iran, from the Aras River to the bottom of Lake Urmia, which is now part of the East and West Azerbaijan provinces of Iran, are the native inhabitants of this region. It means that this geographical area has always been the Armenian people's residence throughout history, despite the numerous displacements of political borders. The study of the dispersion of Armenian churches in Western Azerbaijan shows that many of them were built in the Northern part of the region. Unfortunately, most of those were gradually abandoned and destroyed for various reasons, most notably the forced migration of Armenians to Russia in 1828, The Armenian Genocide, the 1930 Salmas earthquake, and Iran's 1979 Islamic Revolution The paper documents the architectural features of these churches, highlighting their construction specificities.

Keywords Iran · Churches · Khoy · Salmas · Urmia

West Azerbaijan Province is one of the 31 provinces of Iran. It is in the northwest of the country, bordering Turkey, Iraq, and Azerbaijan's Nakhchivan Autonomous Republic, as well as the provinces of East Azerbaijan, Zanjan, and Kurdistan. The study of the dispersion of Armenian churches in Western Azerbaijan shows that many of them were built in the northern part of the region. In total, 111 historical Armenian churches have been identified in this region so far (Khanmohammadi 2005, p. 86). Unfortunately, most of those were gradually abandoned and destroyed for various reasons, most notably the forced migration of Armenians to Russia in 1828, The Armenian Genocide, the 1930 Salmas earthquake, and Iran's 1979 Islamic Revolution caused the damage.

Since most of the churches in the province are collapsed or half-destroyed, the Administration of Cultural Heritage in West Azerbaijan has only registered and documented 39 churches; the architectural features of which can still be examined (Shojadel 2005, p .54). The lowest number is built in cities south of the western

E. Yousefisalekdeh (✉)
Sapienza-University of Rome, Rome, Italy
e-mail: elnaz.yousefisalekdeh@uniroma1.it

M. Boriani and M. Giambruno (eds.), *Architectural Heritage in the Western Azerbaijan Province of Iran*, Research for Development,
https://doi.org/10.1007/978-3-030-83094-6_9

Azerbaijan region, such as Naqadeh with four churches, Miandoab with three, and Mahabad. Based on this, it could be said that the range of influence of Armenians in southern cities was more limited, and fewer Armenians lived there between the seventeenth and twentieth centuries. Churches in the northern part of the province, the St. Thaddeus Monastery in Chaldoran and the Chapel of Dzordzor in Maku, along with Saint Stepanos Monastery in East Azerbaijan Province, were inscribed as cultural heritages in the thirty-second session of the World Heritage Committee on July 8, 2008, under the UNESCO's World Heritage list. Since many studies have been done related to these buildings, in this article, we attempt to examine other churches in the province, especially those for Khoy, Salmas, and Urmia.

Although Armenians have a long history of interaction and settlement with Iran and within the modern-day borders of the nation, Iran's Armenian community emerged under the Safavids in the sixteenth century. At the dawn of the sixteenth century, Iran was unified under a new dynasty, the Safavids (1501–1732). Following the seizure of power in Iran by the Safavid dynasty, Armenia became the battlefield in the wars between the Safavid and Ottoman armies. The devastation of the Armenian regions by the armies of Safavid Shah Abbas I was accompanied, from 1604, by a campaign of forced emigration of the Armenian people from their homelands to northern regions and other parts of Iran, particularly to Isfahan. After this great migration, many churches were built in cities and villages of Iran where Armenians lived, and in the northwestern regions of Iran where Armenians already lived, their number increased.

By studying some Armenian churches in the West Azerbaijan region, we find that these buildings have undergone many changes over time and have moved away from their original form, especially in the dome. The reduction of the dimensions of the central dome, the replacement of the single ruling dome with several small domes, the change of materials, and the dome construction methods are the manifestations of these changes.

The Armenian churches in this region of Iran are still following in volume and type of plan the ones in today Armenia. They choose a more straightforward form in the exterior and plan, from among the rich treasures of Armenian architecture, never referring to other types and examples. Some buildings have (or rather had) a portico (Mahlazan, Afatekahneh) or even a narthex (Haftvan), but generally today disappeared or incorporated into the body of the building.

The churches do not have a prominent appearance in their surroundings and may be distinguished from the adjacent buildings by some signs such as the dome. That is why, the West Azerbaijan churches are often integrated into the surrounding urban texture, with simple shapes and small dimensions. Their facades are seldom decorated, with few exceptions that generally concern the entrance (carefully cut stone, inscriptions, or small figures). The apse is always incorporated into the east wall, almost as if to be hidden, to reduce the appearance of a church to a minimum. The windows are usually small in size, often simple loopholes. The only emerging element is a dome which, in addition to providing the necessary lighting, obviously performs a symbolic function.

All these features contrast with the characters of the most important Northern Armenian churches in that the ones built there dominate the surrounding environment and play a prominent role in the rural and urban landscape.

Azerbaijan Province was located between the two Islamic states, Persia and Ottoman Turkey, and most of the churches in this region were built during the Islamic domination. As a result, although the settlement of Armenians in this region has a historical background, Armenians should have to be careful not to hurt the religious sensibility of those Islamic states. This is probably why the churches in this area are so simple. In addition to economic reasons, the Armenians of the region built them only based on need and minimal decorations not to compete with Islamic religious buildings.

In contrast to the exteriors, the interiors are much more refined and architecturally complex (if we exclude the chapels and the small rural churches built-in raw earth). Most churches in the Azerbaijan region of Iran have a basilica plan, and the interior space is divided by pillars into three areas (in which the central nave is the largest one and corresponds to the apse, which constitutes its visual and ritual terminal).

These buildings have common characteristics, and concerning the general architecture, vaulted or domed roofs, with pointed, round, or lower arches. Central dome/lantern is in correspondence with the transept, in which the windows at the base or oculi at the top create a well of light that stands out clearly in the penumbra of the rest of the buildings. The altar is constantly raised by three to five steps to separate the clergy from the rest of the church, and at the same time, this makes the liturgy more visible from the nave. In the most important buildings, the altar has an architectural structure (Khoy, Fanai, Haftvan) that plays the role of an iconostasis as well as a monumental scene in the space of the public liturgical celebrations, usually with three pointed or roman arches with stucco moldings, *muqarnas*, painted geometric, or floral decorations. On the sides of the apse, two small rooms served as a sacristy. Primary access is from the west side (and sometimes, a minor access is from the south side). The internal decorative apparatus anyhow does not present the magnificence of the Armenian or Georgian examples of the north (Fig. 1).

The interiors were generally plastered, and cases of paintings with figures of saints on the pillars of the naves or the sides of the apse were not rare. Unfortunately, these figures, being the most fragile and badly seen by Islamic culture, have been preserved only in rare cases (the most important in Haftvan, but wall paintings today are documented also in Mahlazan, Khoy, Malham).

In the northern regions of Iranian Azerbaijan, near Aras River, there are also cruciform churches such as the Chapel of Chupan (East Azerbaijan) and the Chapel of Dzordzor (Western Azerbaijan). The reason why we see very few examples of the cruciform plan in this region may be because this type of plan belongs to an older period of the construction of Armenian churches, or perhaps because, this type of plan was attributed to isolated small prayer halls, not used as churches for the people of the neighborhood.

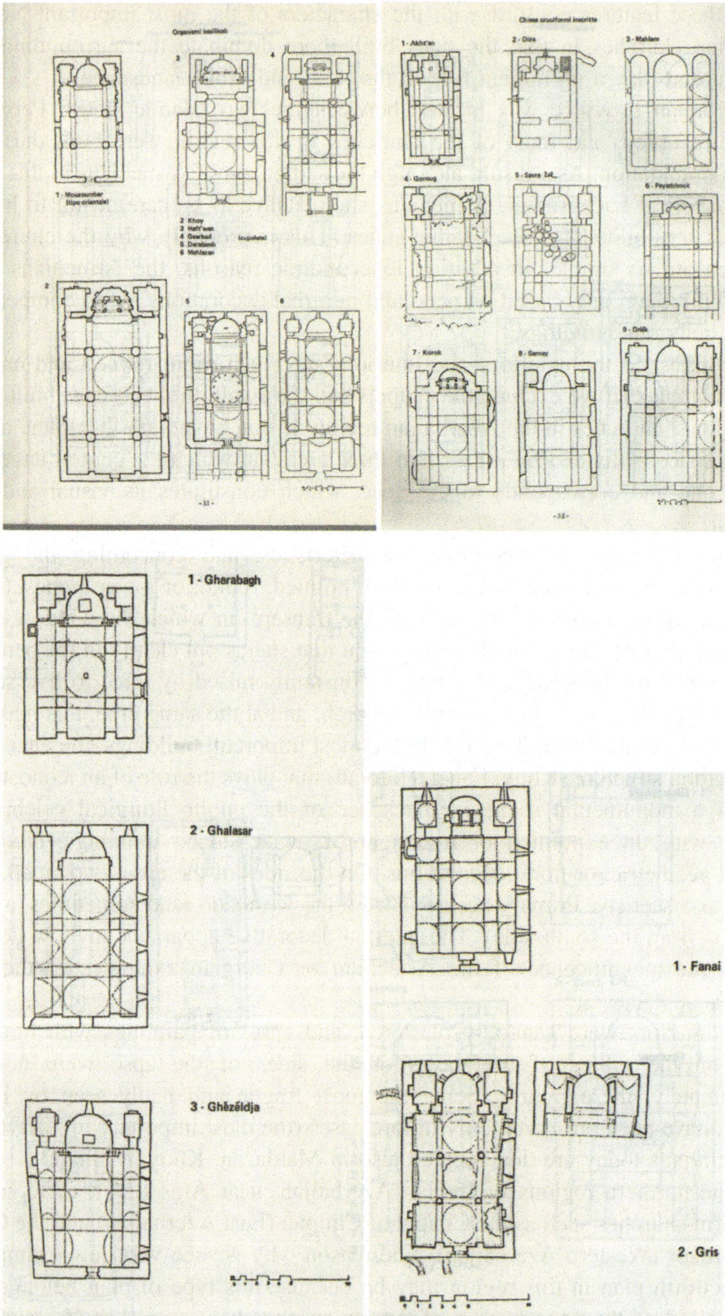

Fig. 1 Some reliefs of Armenian churches with the stone structure of the counties of Khoy and Salmas (Alpago Novello 1978, pp. 31, 32, 33, 34) [Bring the reliefs to the same scale]

The primary materials used in the construction of Iranian Azerbaijan churches include adobe, brick (rarely), and in high and mountainous areas, large rubble stones and sometimes square stones, and secondary materials, including plaster and wood. The climatic conditions of the region have influenced the primary use of adobe and stone in constructing here (Fig. 2).

In the Khoy/Salmas area, instead, roughly hewn stones have been use for the entire wall section (the case of S. Sargis di Khoy is different), except for some corner pieces or for the adjacent wall parts to the west entrance, where the stones appear worked in regular shapes. A justification can probably be found in that there

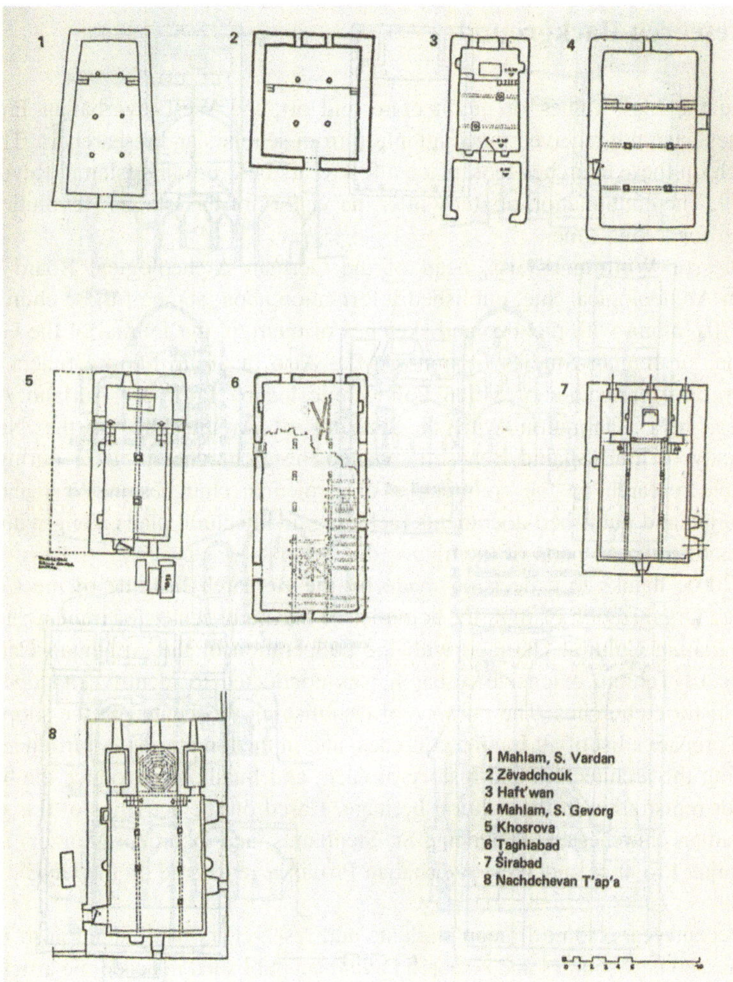

Fig. 2 Some reliefs of Armenian churches with the earthen structure of the counties of Khoy and Salmas and Urmia (Alpago Novello 1978, p. 35) [Bring the reliefs to the same scale]

is abundance of tuffs (black or reddish) in the northern historical Armenian area, easy to work and sculpt, while in the area of Lake Urmia, schists, and in part, granites and basalts are more frequent. In recent times, the influence of Persian construction culture is also evident, which preferred fire-clay bricks, especially in reconstructions, as is evident in Khoy or Afatekahneh, or arched or vaulted roof structures (such as a Mahlazan or Fanai). The use of stones worked in traditional forms in Khoy's S. Sargis could be justified because it is undoubtedly one of the most important churches in the region, located in an entirely Armenian district, being part of a large and populous city.

1 Research Background

Despite the high values of architecture and art, the West Azerbaijan Province churches have not received much attention from scholars and researchers. The first research on these churches took place a few years before Iran's Islamic Revolution in 1979. The continuation of study after the victory of the Islamic Revolution was delayed for a long time.

Professor Wolfram Kleiss, head of the German Archeological Board at the Bastam Archeological Site, published information about some of these churches in the 1970s, along with pictures and sketches of them, in the Journal of the German Institute for Iranian Studies in Iran (AMI). Also, a group of researchers at the Faculty of Architecture of Milan Polytechnic led by Professor Adriano Alpago Novello, in collaboration with the Faculty of Architecture of the National University of Iran (Shahid Beheshti), visited these churches in 1977 during their extensive research on the architecture of Armenian churches in West and East Azerbaijan and published documents including architectural plans along with some pictures of different parts of them.

In 2005, thanks to a decision made by the Research Institute of the Cultural Heritage Organization of Iran, the activities of the study center for Iranian churches at the Iranian Cultural Heritage with the cooperation of the Armenian Caliphate Council of Tehran, extensive research was conducted to identify and document Iran's historic churches. The purpose of establishing this center, in the short term, was to prepare a list of all Iranian churches, and in the long term, to introduce them, including the architecture and history of each, and finally to introduce the historic ones for registration in the cultural heritage. Based on the activities of this center, The Iranian churches St. Thaddeus, St. Stephanus, and Dzordzor (Zorzor) in East Azerbaijan Province and West Azerbaijan Province registered on the UNESCO list in 2008.

In recent years, some Iranian students and researchers paid attention to Iranian churches, so new sources and research have been published in academic articles and dissertations. One of the reliable Persian sources currently available to researchers in this field is Peyman Quarterly, which has been published in the spring of 1996 under the supervision of the Hoor Translation and Research Institute and through the collective efforts of Armenian and Iranian researchers interested in Armenian culture.

2 Armenian Churches in Khoy County

Khoy is one of the major cities of West Azerbaijan Province and is bounded on the north by Mako, on the south by Salmas, on the west by Turkey, and the east by East Azerbaijan Province. Thirteen historic Armenian churches have been identified in the city and its suburbs, four of which are listed as national monuments. The registration of some others in relatively good condition is also in the middle of being implemented (Khanmohammadi 2017, p. 202.). Unfortunately, most of these churches are at risk to collapse over time.

2.1 Surp Khatch Kristosi Church of Mahlezan

Over time and the emptying of Armenian village led to so much damage to the southern, northern, and western walls of the church that it was long abandoned as a half-ruined building. After the formation of the Cultural Heritage Organization in the post revolutionary years and with the efforts of experts from the General Directorate of Cultural Heritage of West Azerbaijan, this church was examined and documented. In the registration report, due to architectural style, the materials used, and the comparison of its plan with the contemporary buildings, they have attributed its antiquity to the Ilkhanid or Safavid period (Figs. 3 and 4).

The Church of Mahlezan consists of a rectangular chapel measuring 26 × 13 m. This building has an east–west axis, and its architectural elements are arranged according to the direction of the axis. The plan is symmetrical for this axis. The building has two doors on the south and west sides. The level of the church is slightly lower than the surrounding ground. According to Kleiss's plan, it seems that a rectangular portico with two columns was attached to the main building in the post-construction period. The church is based on a basilica plan, divided into a nave and two aisles by four square columns, which are covered with a thick layer of plaster on their surfaces and are placed in two parallel rows. The central dome is made of bricks and is conical, and the drum of the dome is divided into twelve arches. The semicircular apse of the church is placed at the eastern end and has the same width as the middle nave, and on both its sides, there are two niches. On the north and south sides of the altar, there are side sacristies. The facade is made of rubble stones bonded with gypsum mortar and other compounds. In the recent renovations of the Cultural Heritage Organization, cement mortar has been used in the joints as well as dark stones in the upper parts of the walls.

Fig. 3 Surp Khatch Kristosi church of Mahlezan, detail of the southern side (Alpago Novello 1977, p. 77)

2.2 *Surp Asdvadzadzin Church of Qeris (Goris)*

This historical Armenian Church with an Armenian cemetery is located at the entrance to the village. Italian researchers during 1977 and 1978 (Alpago Novello 1978, p. 83.), in a comprehensive study on Armenian churches typology in Iran, published maps, pictures, and complete information about, dating it at sixteenth and seventeenth centuries. The primary materials used in the church are rubble stone and, in some, igneous rock. The plan is rectangular with a length of 16.30 and a width of 9.70 m and has a central east–west axis. The church has a main entrance

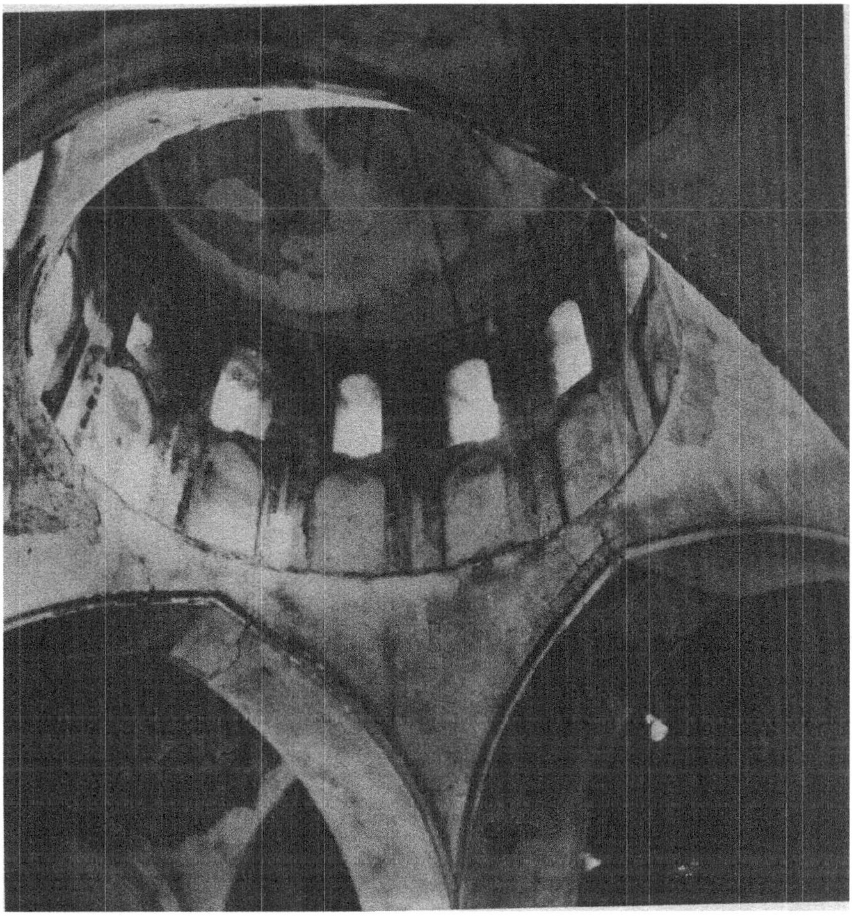

Fig. 4 Surp Khatch Kristosi church of Mahlezan, glimpse of the Dome (Alpago Novello 1977, p. 78)

on the west side, right in front of the altar. The interior is lower than the surrounding ground, and the plan, like most churches in the northwestern region, is built in a basilica type. Access to the upper floors of two-story side rooms is possible through two narrow staircases; they are embedded in the altar, a method that is rarely seen in the churches of the region of Iranian Azerbaijan. The roof in the middle nave is covered with shallow stone domes. Unfortunately, the upper part of the central dome is collapsed over time due to damage caused by rain, and in April 2008, it was covered with wood and metal sheets (Figs. 5 and 6).

Fig. 5 Surp Asdvadzadzin church of Qeris Goris), South-east corner (Alpago Novello 1978, p. 85)

2.3 *Surp Sarkis Church of Fanai*

The historic Church of Fanai is located inside the village, among the houses of the villagers. The town has probably been abandoned since the early twentieth century, and Armenians have migrated from it. The following residents, regardless of the privacy and importance of this building, have built their houses around so that it is difficult to be seen and recognized from near the buildings. There are different opinions about the date of construction of the church, and each of which probably refers to different periods of construction and restoration inside of it. Italian researchers (Alpago Novello 1978, p. 67) mention the construction date between the sixteenth and eighteenth centuries. Professor Kleiss also visited and studied it during his archeological excavations in the northwest and dates it to the nineteenth century. In a registration report prepared in 2000 for the Cultural Heritage Administration of the province, according to the architectural features and comparison of the plan and materials used with the Surp Sarkis Church of Khoy, the construction of this building is attributed to the thirteenth and fourteenth centuries (Khanmohammadi 2017, p. 217) (Figs. 7 and 8).

The primary materials used in the church are rubble stone and, in some parts of it, igneous rock. The historical Church of Fanai consists of a rectangular prayer hall with a length of 14.70 and a width of 11.10 m and has the central east–west axis. The main entrance is on the west side. The inside is about half a meter lower than

Fig. 6 Surp Asdvadzadzin church of Qeris (Gris), Interior of the church (Alpago Novello 1978, p. 86)

the surrounding ground. Like most churches in Iran's northwestern region, its interior is built in a basilica type. On both sides of the altar, like most churches in the area, there are two rooms. The altar is located on the highest level and consists of three pointed arches that make the middle arch taller and more extensive than the lateral ones. Above the truncated arches, plasterwork decorations and colorful paintings with floral and plant motifs had been skillfully created. Unfortunately, most of these valuable paintings and decorations were destroyed and disappeared over time due to negligence. The roof of the church, in the central aisle, is covered with brick domes. Unfortunately, the upper part of the main dome ruined over time due to rain damage.

Fig. 7 Surp Sarkis church of Fanai. Entrance side (Alpago Novello 1978, p. 69)

2.4 *Surp Sarkis Church of Khoy*

This church is located in the northwest of Khoy in the old neighborhood of Imamzadeh. According to the researches done so far about the antiquity of this church, the time of its construction is attributed to two different periods, namely the fourth and ninth centuries, and its later repairs, which include the current form of the building, belong to the Safavid period, seventeenth century. The plan is a rectangle measuring 18 by 23 m. Inside, there are six stone pillars. The inner floor of the church is about eighty centimeters deeper than the outer base. There are two entrances from the south and west sides. The primary materials used are stone and brick. This building is simple in terms of decorations. Only, the arched stones at the entrance have designs and carvings, and in some parts, the shape of the cross is engraved on the stone. Moreover, the apse has the remains of religious paintings with bright colors on the plaster. Approximately, two-fifths below the walls of the building are stone, and three-fifths above it are brick that clearly belong to a recent phase of restoration. The rocks are of dark limestone type, cut into rectangular cubes, and carefully placed on top of each other, but not enough attention has been paid to the brick part. The church was restored in recent years (Figs. 9 and 10).

Fig. 8 Surp Sarkis church of Fanai. Detail of the apse area (Alpago Novello 1978, p. 70)

3 Armenian Churches in Salmas County

Salmas is one of the relatively densely populated West Azerbaijan cities, located in the north of Urmia and south of Khoy. It is bounded on the west by Turkey and the east by Lake Urmia. More than 39 Armenian churches have been identified in the Salmas country so far, mostly built between the thirteenth and nineteenth centuries (Khanmohammadi 2014, p. 263). Unfortunately, most of them have been destroyed so much over time that no trace of them remains today.

Fig. 9 Surp Sarkis church of Khoy, interior and apse area (Alpago Novello 1977, p. 109)

3.1 Surp Asdvadzadzin Church of Afatrkahneh

The village of Aftarkhaneh, also referred to in some sources as Akhte Khan, has an archeological site, an Armenian historical church and an Armenian historical cemetery. In the registered report prepared, the church has been attributed to the Safavid period according to the architectural features, the comparison of the plan, and its materials (Khanmohammadi 2014, p. 263). According to studies by the Italian group (Alpago Novello 1977, p. 121), the church was rebuilt by the Armenians in 1891. The interior of the building is covered with plaster, and the primary materials used in the construction are rubble stone, brick, and lime mortar. The church floor is slightly lower than the surrounding ground, and its interior, like most churches in the northwestern region, is basilica plan. The goal is divided into a nave and two aisles separated by two rows of two columns. The northern facade has

Fig. 10 Surp Sarkis Church of Khoy, North west side (Alpago Novello 1977, p. 109)

no windows, but on the south side, there are three, one of which was the entrance in the past, and in later restorations, it was blocked and turned into a window. The roof of the church in the middle nave is covered with three brick domes. At the drum of the central one and in the four main directions, four skylights increase the light of the built-in interior space. For reasons unknown to us, another conical dome made of metal and wood has been created above this semicircular dome, which gave a strange and asymmetrical shape to the central dome. This method of constructing, which we do not know to what date goes back, is unique and not seen in other churches in the northwestern region (Figs. 11 and 12).

3.2 *Surp Gevorg Church of Haftvan*

The history of this church is mentioned in various sources in different periods. At the building entrance, several inscriptions in Armenia have been placed on the wall. The date of its original construction dates back to the thirteenth and fourteenth centuries. Moreover, the church was rebuilt during the Safavid period. Tawhid Malekzadeh Dilamghani, author of the book of *Tarikhe dah hezar saleye Salmas (Ten thousand years history of Salmas)*, has stated: "This church was built by order of Shah Abbas and was destroyed in the earthquake of 1930 and was restored a few years later" (Figs. 13 and 14).

Fig. 11 Surp Asdvadzadzin church of Afatrkahneh or Akht'an church. West facade (Photo Boriani 2018)

The general plan is in the shape of two adjacent rectangles. The external dimensions of the larger rectangle, which includes the main hall of the church, are 13.3 × 15 m. According to the studies of Italian researchers (Alpago Novello 1977, p. 35), this part is of the church is a kind of "gavit" with a dome on four free pillars. The smaller rectangle, which consists of the altar and the surrounding spaces, is a square with 9.5 × 9.7 m. In the central part of the plan and on four pillars, a conical dome had been erected. When the dome was rebuilt after the earthquake, its drum was made of wood, and the dome was covered very lightly. The other parts of the building are covered with arches and ribbed vaults. The interior surfaces of the church walls were once covered with plaster. This coating was poured in parts, and in the restoration, operations carried out in recent years, due to the signs of colored patterns, parts of the walls were peeled off. At the end of the works, colorful patterns with human, animal, plant, and geometric designs appeared on the columns and some walls (Bakhshayi and Khanmohammadi 2015, p. 226).

Fig. 12 Surp Asdvadzadzin church of Afatrkahneh or Akht'an church. Glimpse of the dome (Photo Borlani 2017)

4 Armenian Churches in Urmia

Urmia or Orumieh, the capital of West Azerbaijan Province, is located south of Salmas and north of the town of Naqadeh and is bounded on the east by Lake Urmia and on the west by Turkey. More than forty historic Armenian churches have been identified in the city and its suburbs (Khanmohammadi 2011, p. 178), most of them built in from seventeenth to nineteenth centuries. Unfortunately, most of the churches in this city have been destroyed over time due to lack of attention, and no trace of them has been left. Among the surviving, ones are six with adobe (Nakhjavan Tappeh, Jamalabad, Qaraqiz, Issa Lou, Spurghan, and the half-ruined church Digaleh), two with stone materials (Qarabagh, Surmanabad), and two with a combination of materials (brick and stone at Gardabad, Reyhanabad).

4.1 Surp Petros Church of Qarahbagh

The exact date of the original construction of the church is unknown. According to the inscription on the western wall, the current building was built in 1655 at the

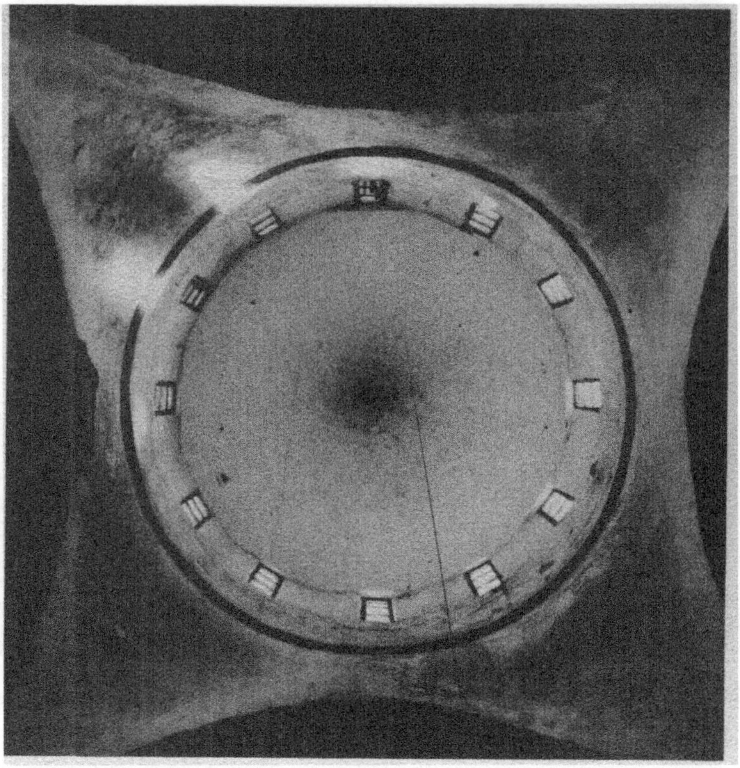

Fig. 13 Surp Gevorg church of Haftvan. Glimpse of the dome (Alpago Novello 1977, p. 37)

same time as the reign of Shah Abbas II Safavid. It consists of a single rectangular shape integrated with the adjacent residential buildings from the east and north. The church is made of plaster and rubble stones. The level is same as the surrounding ground, and there is a main entrance on the south side. Like most churches in Azerbaijan, this one is built in a rectangular plan with a nave and two aisles. Four prominent arches between four free-standing pillars play a significant role in supporting the dome. Besides, the semicircular apse of the church is located at the top of the eastern part, and it is possible to enter it with three steps, which are installed symmetrically on both sides. The shallow central dome of the church consists of small and large rubble stone with gypsum mortar, which has been masterfully placed next to each other. Above the central dome, a square stone skylight is installed to add to its interior light (Figs. 15 and 16).

Fig. 14 Surp Gevorg church of Haftvan. General view of the interior of the church (Alpago Novello 1977, p. 37)

4.2 Surp Asdvadzadzin Church of Reyhanabad

The church does not have any inscription indicating the date of its construction or its name. There is no documented evidence about the date of construction; scholars date it to the seventeenth century. An inscription in the Armenian script at the church entrance mentions the date of its restoration as 1892. The building is in good condition today and is regularly used by Armenians in religious ceremonies. The church consists of a rectangular hall on the western edge of the rectangular courtyard. And next to this, there is another church belonging to the Assyrians. It is made of stone, brick, and clay as the principal materials and plaster and wood as the secondary materials (Figs. 17 and 18).

Fig. 15 Surp Petros church of Qarahbagh. Outside, towards the apse (Alpago Novello 1977, p. 57)

The building has a main entrance on the west side; its surface is level with the surrounding ground and, like most churches in Azerbaijan, is divided into three aisles by six wooden columns and two stone pillars. The roof is covered in two ways. The first part, including the roof of the altar, is executed in a dome shape. In this section, two small irregular conical domes have been built, comparable to the ones of Surp Gork Church in Gardabad village: A group of Armenian immigrants built the two churches. The pointed arches transfer the weight of the domes to the stone pillars installed for this purpose. The second part is flat, and six wooden pillars support the weight of the roof. The main importance of this church is due to the use of stone pillars and wooden columns together to support the roof, which is a unique feature not seen in other Armenian churches in the region.

4.3 Surp Gevorg Church of Gerdabad

The church is located in a large courtyard with an irregular geometric shape. Italian scholars (Alpago Novello 1978, p. 73) who have studied the church date it to the

Fig. 16 Surp Petros church of Qarahbagh (Alpago Novello 1977, p. 58). Note the extremely small entrance door

seventeenth century. Over time, much damage has been done to the south, north, and west walls. The damages have sometimes been repaired by the villagers with unprincipled and non-technical repairs that have destroyed the originality of the significant parts of the church (Figs. 19 and 20).

The materials used in the construction of this building include adobe and plaster, but bricks and stones were also used as secondary materials. The church has a main entrance on the west side, and access is possible with few small steps. Like most churches in Azerbaijan, this is built in a basilica type, and its interior is divided into a nave and two aisles separated by two rows of three pillars. The structure has four columns along with two pillars connected by Roman arches. The roof in the middle nave is covered with two irregular conical and three shallow domes. Other parts of the church (side aisles) are covered with small shallow domes (Fig. 21).

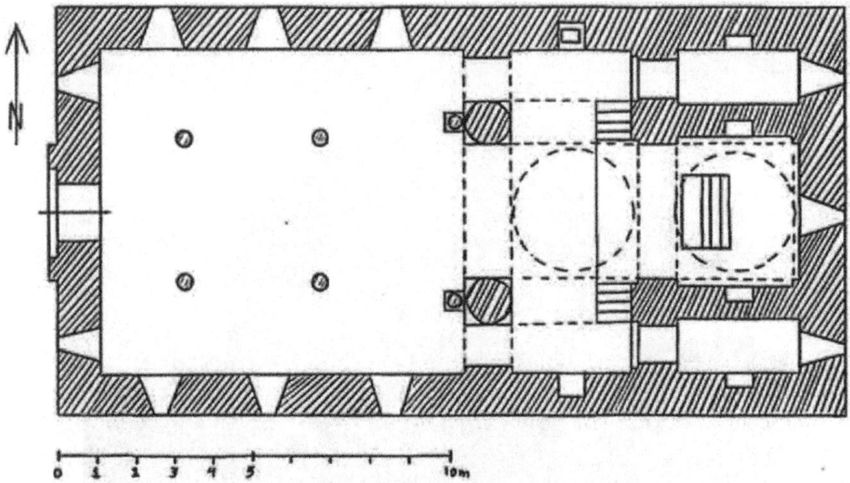

Fig. 17 Surp Asdvadzadzin church of Reyhanabad (Salmast Heritage Association)

Fig. 18 Surp Asdvadzadzin church of Reyhanabad (Photo Boriani)

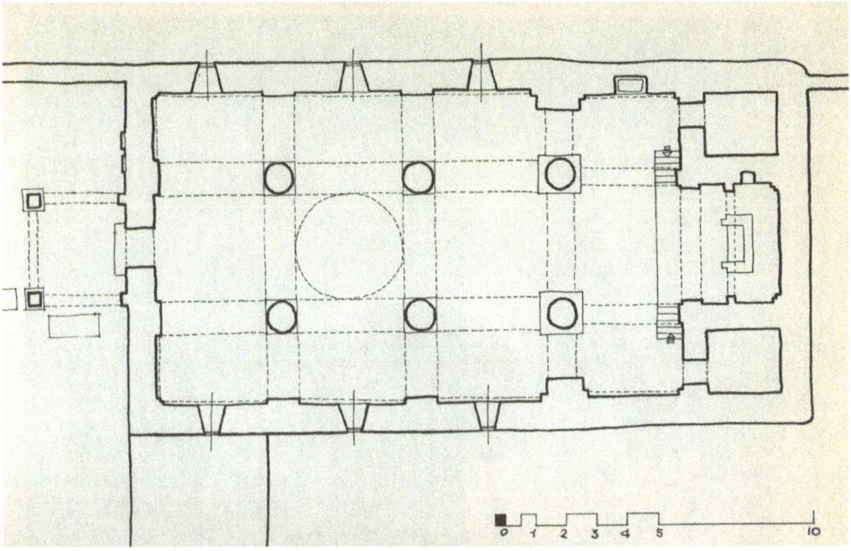

Fig. 19 Surp Gevorg church of gerdabad (Alpago Novello 1978, p. 72)

Fig. 20 Surp Gevorg church of gerdabad. Details of the dome (Alpago Novello 1978, p. 74)

Fig. 21 Surp Gevorg church of gerdabad. Interior of the church (Alpago Novello 1978, p. 75)

4.4 Surp Asdvadzadzin Church of Nakhichevan Tepe

The group of Italian scholars (Alpago Novello 1977, p. 81), who have studied the church in detail, date its construction to the seventeenth century and at the same time as the Safavid period in Iran, but it is also possible for the church to be older than this time. The current building was renovated in 1803. This date is based on an inscription in Armenian in three lines located at the entrance. In 1957, the Armenians repaired the church to be in relatively good condition today, and it is used for religious ceremonies. It is made of stone, adobe, plaster, and wood (Fig. 22).

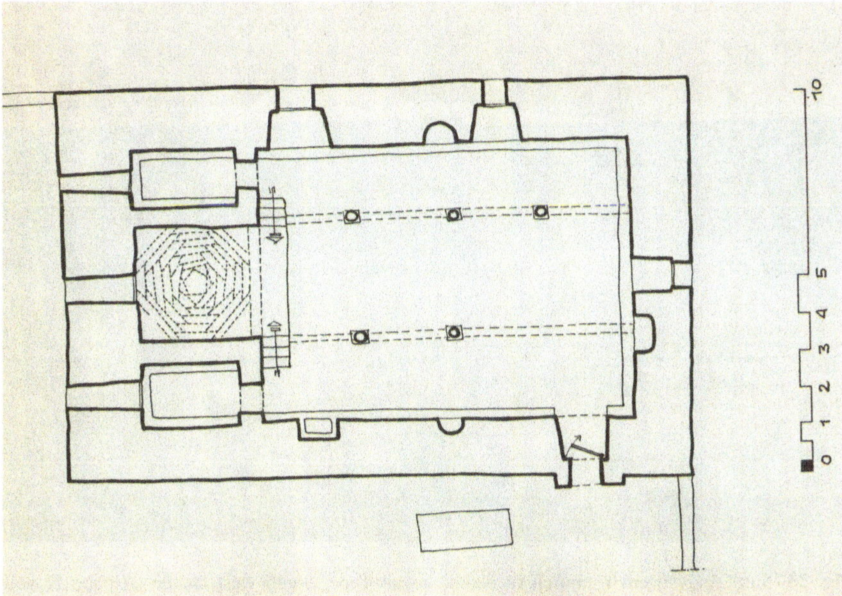

Fig. 22 Surp Asdvadzadzin church of Nakhichevan Tepe (Alpago Novello 1977, p. 80)

The level of the church is same as the surrounding ground and has a main entrance in the north wall, which has two half-columns on both sides. The building is in a basilica type, and in its interior, five wooden pillars are linearly arranged in two rows according to the direction of the altar, thus bearing the roof's weight. These columns are cylindrical and have no base, being directly located on the floor. Each is about 4.5 m high, and all have wooden capitals. The rectangular altar of the church is created at the top of the eastern part and on its highest level; it is possible to enter with four steps symmetrically installed on its sides. Like most churches in Azerbaijan, this has two side rooms on both sides of the altar. The roof is covered in two ways: flat roof and dome in *Hazarashan* style [*A/N: Armenian term hazarashen means «the one thousand beams». The original building system is devised to cover a square room with wooden beams shorter than the span of the room to be covered. Four pillars are erected at a distance from the walls in the middle of the room, and four thick wooden beams are placed on them in a strong horizontal coil. After placing four beams on the pillars to form a square, four other beams for each layer overlap gradually, rotated by 45° on the underlying elements, thus forming a so-called false dome. The beams are placed layer by layer till the remaining uncovered space is a smaller and rotated square*].

This conical wooden dome, essential in Armenian churches, was built on top of the altar (Figs. 23 and 24).

Fig. 23 Surp Asdvadzadzin church of Nakhichevan Tepe. South west façade (Alpago Novello 1977, p. 83)

5 Kahgel Churches of Urmia

The remnants of the adobe (*kahgel)* churches will be soon collapsed due to neglect, abandonment, rainfall, and sometimes human destruction; so, we decided to examine them in this article.

5.1 *Surp Nshan Church of Jamalabad*

Jamalabad village is located fifty kilometers northeast of Urmia city. Surp Nshan Church neither has any inscription about the date of its construction or its founder's name, nor documented evidence exist about that. Armenian researcher Andranik Hovian (Hovian 2002, p. 156) dates it to the eighteenth century. On the east side of the building, another church is located in the fenced courtyard and belongs to the Assyrian Christians (Nestorians). The Surp Nshan Church roof has been wrecked, and parts of the side walls are collapsed. It has a main entrance in the south wall, which has ruined over time, leaving only traces of its foundations; besides, no trace is found of the stairs that most likely gave access to the building. The plan is basilica type, and although collapsed, the effect of the six columns on the floor remains, indicating the structure feature for the roof support. In this church, like most in Iran, two small rooms are placed on both sides of the altar, with the difference that here the principle of symmetry is not observed, and the two side

Fig. 24 Surp Asdvadzadzin church of Nakhichevan Tepe, dome in Hazarashen (Alpago Novello 1977, p. 85)

rooms have different interior spaces. The Armenian Church of Nishan and the Assyrian Church of Mar Khaneh in the village of Jamalabad show that this village was one of the leading Christian areas of Urmia.

5.2 *Surp Hovhannes Church of Qaraqiz*

About 35 km northeast of Urmia is the small village of Qaraqiz. Italian scholars (Alpago Novello 1977, p. 126) who have studied the church date it to the seventeenth. Over time, Armenians' neglect and migration left the church abandoned and made it gradually collapse; so today, parts of the north wall and of the roof have gone destroyed. The altar is located at the end of the eastern side and above other places; entrance is possible through two steps, symmetrically placed on both sides of the altar. In this church, two methods have been used to cover the roof. The cover of most of the church is flat and wooden. In these sections, the weight of the roof is transferred to the four wooden pillars and the north and south walls. Due to the demolition of part of the south wall, which supports horizontal roof beams, another pillar has been added to strengthen this part. The covering of the other part of the church, which includes the roof of the altar, is made in the shape of a dome, built with adobe 10 × 10 × 20 cm bricks supported in the corners by four wooden pegs, thus turning the square base into a circle.

5.3 *Surp Asdvadzadzin Church of Isalu*

Isalu village is located twelve kilometers northeast of Urmia. The probable date of construction of the church is considered to be the seventeenth century. The materials used, the method of construction, and the similarity of the plan with the churches of the region such as Surp Asdvadzadzin Church of Qaraqiz and Surp Asdvadzadzin Church of Nakhichevan Tappeh, and also the use of wooden capitals, which are the main feature in the seventeenth and eighteenth centuries, strengthen this theory. Time pass and negligence have caused so much destruction in the church. Most of the north wall, the entrance, and the roof have collapsed. The building, about one meter above the surrounding ground, has a main entrance on the north side, which unfortunately collapsed along with most of the north wall, and only a small part of it remains. The church is built in a basilica type and has two pillars. Like most churches in Iran, this one has two rectangular side rooms on both sides of the altar with narrow and low entrances. The building had wooden columns placed along with each other. Unfortunately, due to the collapse of the roof, these columns also fell into interior.

5.4 *Surp Sarkis Church of Sepurghan*

The village of Sepurghan is located about forty kilometers northeast of Urmia and about four from Lake Urmia. The church is located in the middle of the village, in a large courtyard. In addition to the church of the Surp Sarkis, two other churches belong to Assyrian (Nestorian) Christians in this courtyard. Italian scholars (Alpago Novello 1977, p. 126) who have studied the church dated it to the seventeenth century. In recent years, the church was restored by the Armenians of the village. During that, parts of the upper adobe walls of the building have been restored with bricks. Of course, these non-technical restorations have damaged the original composition of the building. The church has a main entrance on the south wall and, like most of them, is built in a basilica type and has four pillars inside supporting the weight of the flat wooden roof. The altar is located in the eastern part and is distinguished by a wall in the tradition of other regions. In fact, unlike most churches in Azerbaijan, Surp Sarkis Church does not have side rooms, and it is possible that during the renovation works, the walls of the side rooms were removed, and a retaining wall was created instead. The roof is flat and wooden, and its weight are transferred into the columns inside the building and the side walls.

Examining the churches of this region, we conclude that the Armenian churches of West Azerbaijan, in all their spatial and temporal extent, belong to Armenian culture and are always in the same plan and interior as their contemporary examples in Armenia. In terms of external volume, despite enduring changes over time, mainly in the dome, it is faithful to this architectural culture.

6 Hazarashen, the Continuation of the Ancient Tradition of the Armenian Dome

A house of "*Hazarashen*" type is a traditional construction on a single level, covered by a roof composed of wooden logs laid upon each other, thence the name "*a thousand logs*" or "*Hazarashen,*" with an opening for smoke evacuation in the center. This roof structure can be supported by four posts or none. This type of construction is normally half-buried on flat ground, or its rear part is buried when the flat ground slopes. This method involves closing a square by placing wooden beams across the angles of the square, touching adjacent sides, making squares one on top of the other, alternately at 45 degrees to each other. The Armenian conical development of the domes is inspired by the tops of Armenian peasant houses following the *Hazarashen* technique. Due to its permanently open roof (with a small window called "*yerdik*"), *Hazarashen* style came to suggest a spiritual opening, which could mean a connection with divine spirits, the universe, and later in Christianity, with God. These mound buildings, partially buried with a wooden roof, remained in the local settlement tradition until a few decades ago, called Armenian *Hazarashen* and Georgian *darbazi*. The conical domes of Armenian churches have also significantly impacted the domes of buildings in Northern Iran and near the Caspian Sea. In this area, architecture, especially in sanctuaries, shows pyramid domes. Examples of wooden covers can also be seen in Northwestern Iran. For instance, we can mention the church of Nakhjavan Tappeh village, Ushtibin village, and Hematabad village houses, and an example of four arches in Azarshahr.

6.1 Wooden Domes of Ushtibin Village Houses

Ushtibin is a border village located at the foot of a steep mountain in Jolfa city of East Azerbaijan Province of Iran. This village is registered in the list of national monuments, and the documents obtained from the inscriptions found there show the dates 1440 and 1569. Buneh Baghi is one of the unique architectural elements of this village, which can also be called a hiding place because of its hidden location. Except for the opening in the roof, there is no else opening for lighting and ventilation. The construction method was *Hazarashen* as illustrated above.

6.2 Hemian Abad (Hemmat Abad)

It is a village located in Salmas district of West Azerbaijan Province of Iran. Part of the village is situated on a rocky bed. An inscription carved on the part of this rock shows that the village is old, but it is not possible to comment on the exact text due to its illegibility. One of the critical features of the houses in this village is the use of wooden domes to cover one of the architectural spaces, which is usually the family living space. Generally, the area with a wooden dome is square or rectangular, and the walls of the site are made of brick or adobe, with only one entrance. The light is provided only from the opening above the wooden dome.

6.3 Chaharsogh of Azarshahr

Azarshahr is one of the towns of East Azerbaijan Province of Iran. Chaharsogh thermal spring water is located in a small square in the neighborhood, which is the intersection of four narrow alleys. An arch with eight wooden columns casts a shadow on the spring, dating back to the early Qajar period. The plan is octagonal, and the dome-shaped cover is entirely wooden. First, a wooden beam was placed on the columns according to the octagonal plan; then, in the second step, the sticks are placed with a rotation of 60 degrees compared to the last beams and a little deeper, and in the third step, like the first phase, the sticks are arranged and connected again. In this way, the dome cover is gathered in the form of a basket woven from wood.

References

Adriano Novello A, et al (1977) *Ricerca sull'architettura armena 17*, vol 1, Milano

Adriano Novello A, et al (1978) *Ricerca sull'architettura armena 18*, vol 2 , Milano

Arce I (2014) *Qasr Kharana Revisited*, 9th ICAANE, vol 2

Bakhshayi H, Khanmohammadi B (2015) Report on the restoration and study of wall paintings of Haftavan Salmas historical church (گزارش لایه برداری و مطالعه نقش های دیواری کلیسای تاریخی هفتوان سلماس), Payman 72, Tehran

Hovian (2002) Armenian Iranians (ایرانیان ارمنی), Tehran

Kamali FP, Omrani B (2013) Structure of wooden domes in the north west of Iran, Adv Environ Biol

Khanmohammadi B (2005) Armenian historical churches in West Azerbaijan (پراکندگی کلیساهای تاریخی ارمنیان در آذربایجان غربی), Payman31, Tehran

Khanmohammadi B (2011) An Analysis of the Architecture of Armenian Historical Churches in Urmia (تحلیلی بر معماری کلیسا های تاریخی ارمنی در ارومیه), Payman 56, Tehran

Khanmohammadi B (2014) Surp Asdvadzadzin Church of Afatrkahneh (کلیسای مریم مقدس روستای آفتارخانه سلماس), Payman68, Tehran

Khanmohammadi B (2017) Two historic churches of Khoy city, Qeris Church and Fanaei Church (نگاهی به دو کلیسای تاریخی شهرستان خوی،کلیسای قریس و کلیسای فنایی), Payman79. Tehran

Kleiss W (1979) *Armenische Kirchen in Azerbaidjan*, Archäologische Mitteilungen aus Iran, 12

Shojadel N (2005) An overview of the Armenian churches in West Azerbaijan Province (مروری بر کلیساهای ارامنه استان آذربایجان غربی), Payman 33, Tehran

Simoni P (2016) Architecture of Armenian churches in Azerbaijan (معماری کلیساهای ارامنه در آذربایجان), Kharazmi. Institute of Higher Science and Technology, Shiraz

Elnaz Yousefisalekdeh graduated in Architecture in 2008 from the University of Art and Architecture of Tehran (Iran). In 2014 obtained a master's degree in Architecture-Restoration of architecture at the Faculty of Architecture, Sapienza University of Rome. After completing her studies, she returned to Iran. She collaborated in the role of architect and researcher with the Organization of Ministry for cultural heritage and activities in a restoration project of the Iranian sculptor museum "Ali Akbar Sanati" in Tehran. In 2016, she began her Ph.D. in History of Architecture. Her dissertation is titled "interaction between Armenia and Persia in architectural culture in Isfahan". She yielded her Ph.D. degree in 2020.

Western Travellers in Iranian West Azerbaijan Anthology

Maurizio Boriani

Abstract Many travel descriptions of the territories of Salmas and Khoy counties deserve to be published in full. The reader will find information ranging from agriculture to the construction techniques of village dwellings, from the ways in which people travelled to the social and economic conditions of local populations. Particular attention is paid to the activities of Western missionaries, to the presence in the territory of Armenian and Assyrian Christians and their living conditions. A chapter is also dedicated to the description of the territory of the County of Urmia, which is outside the research field of this volume but was in any case an important destination for those wishing to get to know other important Christian communities.

Keywords Persia · Iran (West Azerbaijan) · Travelogues · Khoy · Salmas

1 Descriptions of the Ancient City of Julfa and the Araxe River in the Persian Western Travelogues

Two Voyages of Master John Newberie, One in the Holy Land; The other to Balsara, Ormus, Persia, and backe thorow Turkie, 1579, **in S.Purchas, *Hakluytus Posthumus or Purches His Pilgrimes*, Glasgow 1905**, vol. VIII, p. 468.

The fourth day from *Merent*, and the same day at Julfa: and there is a bridge of wood upon Boates, there was a stone bridge but it is broken. And *Aras* is the name of the River which runneth before the Towne, and the Towne standeth under a Mountanye. There are in this Towne three thousand Houses, and they have seven Churches, and the richest pay every house twentie Merchel a yeere, and the rest according to their abilitie. The fifth day from Julfa and the same day we lodged at a good Towne called *Naxnan*, which hath great plenty of victuals, and especially Angour, and also hath Silke. And in the middle way betweene *Jolfa* and this Citie, is a very high round Rocke, much like unto a Castle.

M. Boriani (✉)
DAStU, Politecnico di Milano (retired 2016), Milan, Italy

M. Boriani and M. Giambruno (eds.), *Architectural Heritage in the Western Azerbaijan Province of Iran*, Research for Development,
https://doi.org/10.1007/978-3-030-83094-6_10

***Observation of Master John Cartwright in his Voyage from Aleppo to Hispaan and backe again …*, 1603, in S.Purchas, *Hakluytus Posthumus or Purches His Pilgrimes …*,vol. VIII, Glasgow 1905**, pp. 496–497.

At length our Caravan ferried over the foresaid Rivcr, and so we arrived at *Chiulfal*, a Towne situated on the frontiers betweene the Armenians and the *Atropatians*, and yet within Armenia, inhabited by Christians, partly Armenians, partly Georgians: a People rather given to the traffic of Silkes, and other sorts of wares, whereby it waxeth *[grows]* rich and full of money, then instructed in weapons and matters of warre. This Towne consisted of two thousand houses, and ten thousand soules, being built at the foot of a great rockie Mountayne in so barren a soyle, that they are constrained to fetch most of their provision, onely Wine excepted, fom the Citie *Nassivan*, halfe a dayes journey off, which some thinke to be *Artaxata*, in the confines of Media, and Armenia. The buildings of *Chiulfal* are very fair, all of hard quarrie stone: and the Inhabitants very courteous and affable, great drinkers of Wine, but no brawlers in that drunken humour, and when they are most in drinke, they powre out their prayers, especially to the Virgin Mary, as the absolute commander of her Sonne Jesus Christ, and to other Saints as lntercessors. It is subject and tributarie to the Scepter of Persia, and contrariwise, both by nature and affection great enemies to the Turke. This Towne was much indangered in the warres betwixt *Amurat* the great Turke, and *Mahomet Codibanda* the Persian King, ready to be swallowed up of both.

John Baptista Tavernier, *Les six Voyages of John Baptista Tavernier, a noble Man of France now living, through Turky into Persia and the East-Indies, finished in the Year 1670*, London, R.L. and M.P., 1678 (First edition in French, 1676), pp. 19–20 (Fig. 1).

From this Bridge to *Zufa* is but one day journey: which Town being altogether ruin'd, shews the reason wherefore the *Caravan* lodges upon the Bank of the River five hundred Paces on this side.

Zulfa, the ancient Habitation of the *Armenians* which *Sha-Abas* carried into *Persia,* is a Town squeez'd between two Mountains, through which the *Aras* runs leaving but little Land on either side. It carries no Boats till about two Leagues below, (for upward it will hardly bear a piece of Timbcr) and in regard the Country grows low and extends itself into Plains, there is no fear of Rocks, the course of the Stream being very quiet. There was a fair Stone-Bridge, which *Sha-Abas* caus'd to be broken down; when he wholly destroy'd the Town, that it might be no harbour for the *Turks.* Neither by the Ruines, nor by its Situation, doth it appears to have been a City of any ancient Beauty; the Stones were clapt unskilfully together without Morter, so that the Houses were more like Caves than Houses. The North-West side was most inhabited, there being nothing on the other side worth taking notice of. The Lands about *Zulfa* being very fertile, there are certain *Armenian* Families return'd, who live very quietly. *Cogia Nazar,* one of the chief *Armenians* that went out of *Zulfa,* growing rich by Trade, and being in great reputation with *Sha-Abas,* and *Sha-Sefi* his Successor, who made him *Kelonter,* or Chief Judge of the *Armenian* Nation, built two great Inns, for the

Fig. 1 Jean-Baptiste_Tavernier (1605–1689) (Wikimedia Commons)

Honour of his Country, in *Zulfa,* upon each side of the River one. He spent above an hundred thousand Crowns but dying left two remarkable Pieces of Work unfinish'd.

Half a Legue on this side *Zulfa,* before you cross a Torrent that falls into *Aras,* you may take your choice of two ways to go to *Tauris.* The one upon the right hand, leading to the South-East, which is the ordinary Road; the other upon the left hand, toward the North-East, which we took eight or ten in company together on horseback, the last time I went to *Ispahan.*

……

We travel'd upon a descent for two or three hours, till we came to a great Village neatly situated, where there grows excellent Fruit. There we repos'd for an hour or two; and from thence we came to a great Stone-Bridge on a River where there is no Water but when the Rains fall. It falls into the Lake *Roumi*; but the water of the River is so sowr and ill-tasted, especially when it is low, that it is not to be drunk. About a quarter of a League from the Bridge are three long Stones set in the ground like Pillars: the Natives say that they were plac'd there for a Monument, in the same place where *Darius* the Son of *Histaspes* was elected King by the cunning of the Gentleman of his Horse: from whence to *Tauris* is but half a League. The Mountains of the *Medes* which we cross'd in this Road, and those which run along toward the ancient *Parthians*, are the most fertile in all Persia. They bear Corn and Fruit in abundance; for upon the high Mountains there are fair Plains sow'd with Wheat, which are extraordinary fertile. The Springs which rise there, and rains which fall give a fresher beauty and a higher taste to what grows there, more than in any other part of Persia that wants Water; and the products of those Fields are of a higher price.

Chardin, Jean, *Journal du voyage du chevalier Chardin en Perse et aux Indes orientales: par la mer Noire et par la Colchide*, Première partie, Londres, Chez Moses Pitt, 1686, p. 181 (Fig. 2).

Ce fleuve *[l'Araxe]* est grand et fort rapide. Il s'enfle, durant son cours, de plusieurs petites fleuves qui n'ont point de nom, et de beaucoup de torrens. On a bâti diverses fois des ponts dessus à Julfa, et en d'autres endroits; mais quelque forts et massifs qu'il fussent, comme il paroît à des arches, qui sont encore entières, il n'ont pû tenir contre l'effort du fleuve. Il est si furieux, lors que le dégel le grossit des neiges fonduës des monts voisins, qu'il n'y a ni digue, ni autre bâtiment qu'il n'emporte; et à la verité, le bruit de ses eaux, et la rapidité de son cours, étonnent les gens. Nous le passâmes dans un grand bateau, fait pour passer vingt chevaux et trente personnes à la fois … Le courant l'emportoit avec une indicible impetuosité, et lui fit faire cinq cens pas en un instant. Voilà comme les bateliers de l'*Araxe* le traversent: Ils mettent plus de deux heures à aller et venir: à cause des efforts qu'il leur faut faire pour le remonter. L'hiver, que les eaux sont basses, on passe le fleuve sur des chameaux. Le gué est à demi lieuë de Julfa, en un endroit où son lit étant fort large, il court beaucoup plus à l'aise.

Chardin, Jean, *Journal du voyage du chevalier Chardin en Perse et aux Indes orientales: par la mer Noire et par la Colchide*, Première partie, Londres, Chez Moses Pitt, 1686, Tome I, pp. 180–181.

Le 13 nous partîmes de *Nacchivan*, et fîmes sept lieues, à la première lieue nous passâmes sur un fort grand pont, un fleuve, à qui les gens du pays ne donnent point d'autre nom que celui de fleuve de *Nacchivan*. Le pays que nous traversâmes est sec et stérile, l'on n'y voit que des côteaux pierreux. Nous couchâmes sur le bord du fleuve *Araxe*, que les Orientaux nomment *Aras* et *Ares*. On le passe à *Esquijulfa*, ou *Julfa la vieille*, ville ruinée, que quelques Auteurs croient être celle que les Anciens appelloient *Arriammene*. On l'appelle vieille, pour la distinguer d'une ville de *Julfa*, qui est bâtie vis à vis d'*Isphan*. On a véritablement raison d'appeler celle-ci vieille, car elle est toute ruinée et abbatue. On n'y connoît plus rien, excepté la grandeur qu'elle avoît. Elle étoit située sur la pente d'une montage le long du fleuve, et sur ses bords. Les avenues, qui sont naturellement difficiles et fortes, étaient gardées par plusieurs Forts. La ville avoit quatre mille maisons, à ce que disent les Arméniens, cependant à en juger par les ruines, il n'y en pouvoit pas avoir la moitié, encore n'étoit-ce la plupart que des trous, et des cavernes, faites dans la montagne, plus propres à retirer des troupeaux, qu'à loger des hommes. Je ne pense pas qu'il y ait au monde un endroit plus stérile et plus hideux, que celui de *Julfa la vieille*; on n'y'voit ni arbre, ni herbe. A la verité il y a dans le voisinage des endroits plus heureux et

Fig. 2 Jean_Chardin (1643–1713) (Wikimedia Commons)

plus fertiles, mais toujours est-il vrai qu'il ne se peut voir de ville située en un lieu plus sec, et plus pierreux. La figure en étoit belle en récompense, ressemblant à un long Amphithéâtre. Il n'y a présentement qu'environ trente familles qui sont toutes Arméniennes.

Ce fut *Abas* le grand qui ruina *Julfa*, et tout ce que l'art avoit contribué à la fortifier. Il le fit par la même raison qu'il ruina *Natchivan*, et les autres places d'Arménie, qui étoient sur la même ligne, afin d'ōter les vivres à l'armée Turquesque. Ce fin Politique, et grand Capitaine, voyant ses forces inégales à celles de son ennemi, et songeant aux moyens de l'empêcher de revenir tous les ans en Perse, d'y faire des conquêtes, et de les conserver, résolut de faire un désert des pays qui étaient entre *Erzerum* et *Tauris*, sur la ligne d'*lrivan* et de *Nacchivan*; qui était la route que les Turcs tenoient d'ordinaire, et où ils se fortifioient, parce qu'ils y trouvoient des vivres suffisamment pour faire subsister leur armée. Il en transporta donc les habitants et le bétail, il ruina toute sorte d'édifices, il mit le feu par toutes les campagnes, et aux arbre, il empoisonna même plusieurs fontaines, à ce que l'histoire reporte, et ceux qui l'ont lue savent que cela lui réussit tout-à-fait bien.

Pour retourner à notre gite, l'Araxe est ce fameux fleuve qui sépare l'Arménie de la Medie. Il a sa source dans le mont, où l'on tient que s'arrêta l'Arche de *Noé*; et c'est peut-être de ce mont célèbre d'Ararat qu'il tire son nom. Il se rend de là dans la mer Caspienne. Ce fleuve est grand et fort rapide. Il s'enfle, durant son cours, de plusieurs petits fleuves qui n'ont point de nom, et de beaucoup de torrents. On a bâti diverses fois des ponts dessus à *Julfa*, et en d'autres endroits, mais quelque forts et massifs qu'ils fussent, comme il paroit à des arches, qui font encore entières, ils n'ont pu tenir contre l'effort du fleuve. Il est si furieux, lors que le dégel le grossit des neiges fondues des monts voisins, qu'il n'y a ni digue, ni autre bâtiment qu'il n'emporte; et à la vérité, le bruit de ses eaux, et la rapidité de son cours, étonnent les gens. Nous le passâmes dans un grand bateau, fait pour passer vingt chevaux et trente personnes à la fois.

Paul Lucas, *Voyage du sieur Paul Lucas au Levant, seconde volume*, Paris 1704, pp. 33–35.

Le trentième nous partîmes à cinq heures au matin et nous entrâmes dans une Vallée toûjours entre deux montagnes qui nous conduisit à Julfa le vieux sur les deux heures après midi. C'est une Ville tout ruinée, dont les habitants ont été emmenéz à *Hispaham* où le Roy leur fit bâtir le fauxbourg du même nom. L'ancienne Julfa étoit bâtie en amphithéâtre au pied d'une montagne si escarpée, qu'il auroit été impossible d'y approcher de ce côté-là. L'Araxe passe à son pied et il y a sur le haut de la montagne une grande colonne de pierre blanche quarrée avec une inscription qui marque que c'est le grand *Scha Abbas* qui a détruit la Ville, et qui en a emmené tous les habitants.

Nous voulions passer cette même journée l'Araxe, mais les passeurs ne s'accordèrent point de prix avec notre *Moucre*. Le trente-unième les Bateliers se relâchèrent un peu, et nous passâmes dans un grand bateau, où il pouvoit bien tenir 20 chevaux ensemble. Enfin toute la Caravane étant passée sur les dix heures du matin, nous continuâmes nôtre route, et après avoir marchés heures nous campâmes dans une plaine où il y a une quantité prodigieuse de sepulcres. Les pierres qui les couvrent sont d'un très bel albâtre transparent, dont quelques montagnes voisines sont toutes pleinse. Les eaux de ces lieux ne valent rien, et sont toutes un peu salées.

Tous les Marchands se résolurent en cet endroit de quitter les Mules, et de charger leur marchandises sur des chameaux pour ne pas payer à *Marante* tant de Doüane. En effet, on n'y paye pas davantage de la charge d'un chameau que de celle d'un mulet; et le chameau porte le double, c'est puorquoi il y a du profit. Il sémble que la terre soit là semée de sel par tout, tant elle en produit. On me dit que tous ces

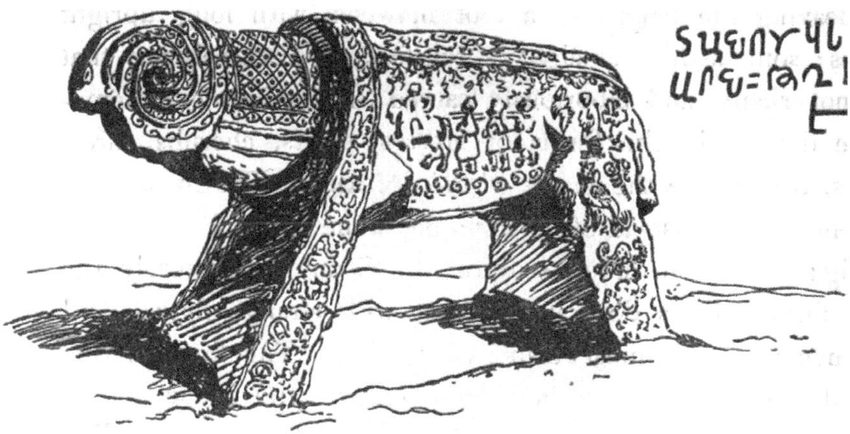

Fig. 3 Robert Ker Porter, An ancient ram-shaped gravestone in Julfa cemetery (Ker Porter 1821)

tombeaux étoient de ceux qui furent tuez dans une bataille qui se donna dans cette plaine entre Turcs et les Perses.

Sir Robert Ker Porter, *Travels in Georgia, Persia, Armenia, Ancient Babylonia during the years 1817, 1818, 1819 and 1820*, London, Printed for Longman, Hurst, Rees, Orme and Brown, London, 1821, Vol. I,, pp. 214–215, 215–216 (Figs. 3 and 4).

We rode for nearly a mile, and then came forth on a small plan, which appeared to be completely surrounded by mountains. Through an enormous chasm to the west, as if the opening scene of this great theatre, I had a distinct view of the magnificent windings of the Araxes, with the ruins of Eski *Julpha* on its banks. That city was not, like that of *Nackshivan*, merely sacked and dismantled, by the victorious arms of *Shah Abbas*, but absolutely battered into ruin; and its surviving inhabitants transported to *Ispahan*, the victor's capital, where they formed a suburb to that city, which still bears the name of *Julpha*. Extensive remains of fortifications, on various points, close to the town, and on each side of the river, still show what must have been its former consequence. The piers of a bridge may also be traced, which, probably, was the very same that Augustus ordered to be erected, on some part of the *Araxes*, in this neighborhood.

This noble river flows in a bending direction, almost in the shape of a sickle, from its rise in the west, to the point of its junction with the *Kur*, in the east. Its source, is said to be near *Hassan Kala*, about eight *agatches* (or eight hours' journey) east from *Arzerum*; and thence it flows onward, in a waving course, till, in traversing the plains of Ararat, it takes a deep curve southward; which, according to its convex sweep, embracing the provinces of *Erivan*, *Nackshivan*, and *Kara-Bagh*, finishes its point in the north-east, near the castle of *Kalagan*; where it meets the *Kur*, or Cyrus; and immerses its own celebrated name, in that of the more famous flood to which it has united its waters. A very short ride brought us from the mountain-plain, to the verge

Fig. 4 Robert Ker Porter, The so-called Tomb of the Virgin in Julfa (Ker Porter 1821—II)

of the river. We were to cross it, at almost the most southern stretch of its winding banks, where their sickle form, having made its curve, turns eastward. We found a raft-like boat, of a lozenge shape, that was to convey us to the opposite shore; and, by that translation, land us in *Azerbijan*, the government of *Abbas Mirza*, Prince Royal of Persia, and what was part of ancient Media. The *Araxes* was not more than fifty yards wide at the place of embarkation; but the waters were rapid, and occasioned no small difficulty in getting our animals afloat. When we were all on board, the boat was pushed off, and rowed with apparent ease by the ferrymen, till it got into the full flood of the current; and then we were carried down the stream to a considerable distance.

……

At *Julpha*, and for a great extent above those ruins, the *Araxes* (otherwise called the *Aras*) flows through a precipitous valley; and, while in that course, it is largely augmented by the numerous mountain-torrents which fall into its stream. At times of thaw, or in the rainy season, these accessions of water are very formidable; and the valleys of the *Araxes* often suffer by their too abundant supply. However, at only a very few miles below the ferry, the river is almost always! fordable, from the width of its bed at that part allowing the waters to become shallow by expansion. In the winter, it is sometimes frozen so hard, as to admit whole caravans to travel over its surface.

Having crossed the river, and a plain on its border, of about three miles in extent; we arrived at the foot of a steep bank, which we ascended; and passing through the ruins of a place called *Sooja*, travelled on four miles further, where we began another ascent, which we surmounted speedily, being spurred forward by the increasing cold of the evening; and gladly saw several large villages scattered over its summit. We halted at *Gurgur*, one of the nearest to us; and the night being so severe, were happy to find any shelter, though in a poor little hut.

Robert Ker Porter, *Travels in Georgia, Persia, Armenia, Ancient Babylonia during the years 1817, 1818, 1819 and 1820*, London, Printed for Longman, Hurst, Rees, Orme and Brown, London, 1822, Vol. II, pp. 610–615.

September 21st. Started this morning at half-past seven, and in two hours reached the banks of the *Aras*, or *Araxes.* I found the stream very low, but extremely rapid; the latter striking feature being occasioned by the visible long descent of land over which it rolls to the eastward. After wandering like troubled ghosts for near an hour on its southern margin, we at last got ferried over in one of the lozenge-shaped boats I described in my former passage two years before. The bridge in the vicinity having for nearly two centuries been impassable, the government of *Azerbijan* established the present ferry at its own expence, for the accommodation of merchants and others. Having passed, I ordered my baggage to proceed, and went off the road myself across some barren hills, which form the bank of the river in a western direction for about four miles, and then suddenly terminate in a vast naked flat expanse, once constituting the fertile plain of *Eski Julfa* (or *Julpha*), marked out into fruitful fields, and blooming gardens. At this point the *Aras* makes a bend; and here the ruins of the bridge just mentioned, form a picturesque and melancholy object. Its remains appear a few hundred yards from the spot which once supported the great arch of the eastern gate of the city. The masonry of the bridge is of the first order, finely hewn, and the stones admirably joined. Only two of its immense arches have been required to stand immediately over the stream; and their lofty and massy bulwarks yet defy the utmost fury of the current. Two others advance on the land, one on each side. The span of each arch may have embraced about ninety feet; and from the firm and noble structure of the whole, I should be inclined to say, that had not the contests of man's ambition been engaged in its overthrow, "the indignant waters" of this proud stream, with all its mountain-torrents, might have poured through it for ages, without starting one stone of its mighty fabric. At the present comparatively dry season, the water flowed through its northern arch only; so that this sometimes formidable river, is not now so wide here as at the ferry; and at a mile lower down, they tell me it is fordable. The first ruin of the bridge is attributed to *Shah Abbas*, who ordered its destruction when he transferred the inhabitants of the city to the great suburb of *Ispahan*, which afterwards adopted the Armenian name. About fifty yards above the bridge, the river rushes through a narrow and rocky valley, or rather vast chasm in the side of the mountain. On its northern acclivity, remains of houses, churches, walls, and other mouldering masses of considerable extent, mingle with the rugged cliffs. The town anciently touched the verge of the river, which formed its defence on that side; while the high mountains, over whose slope the buildings extended, seemed an impregnable bulwark on the other. The two extremities then were all that needed protection from the art of man; and we find the remains of strong walls, in the wide and massy places of entrance still standing. Beyond that to the eastward, a very beautiful relic presents itself, of an ancient chapel and tomb. One side of the later shows a finely wrought cross cut in stone, and embosomed in the most intricate fretwork, intermingled with mazy interlacings, bands, tendrils, and a variety of climbing, interwoven leaves, the whole of which is enclosed by a Saracenic arch; exhibiting altogether a rich specimen of this style of Armenian architecture bearing date, probably, about the middle of the eleventh or twelfth century. An old peasant, our guide, (who, on our chancing to meet him at the bridge, had offered to show us the ruins), said, this was the burial-place of a female saint, whose name was no longer known; but tradition always called it-the Tomb of the Virgin.

Armenia had been long trampled underfoot by the Turks, when *Shah Abbas* wrested the wealth of Julfa, and the industry of its opulent inhabitants also, from the treasure-house of his Ottoman rival.

It was about the year 1603 when this transaction took place; leaving the city utterly deserted. Since. which time the whole has become a mass of vacant ruins; echoing the hooting of the birds of night alone, or now and then the startling voices of a few travellers like myself, who turn aside to read, again and again, in such mouldering monuments, the transitory nature of a human greatness.

After this place had lain about a hundred years in an almost forgotten solitude, seven poor Armenian families wandered towards the spot, and peaceably established themselves amidst its desolated walls. Having abided there ever since, at least in their descendants, they have increased to thirty-five houses; and being quiet and industrious, live in unmolested content. The father of one of the present resident families was the good man who acted as our guide; and who, from the separated holy spot that contained the tomb of the saint, led us across the site of the city to beyond its western gate, where we found the great cemetery of the ancient inhabitants, a vast, elevated, and thickly marked tract of ground. It consists of three hills of considerable magnitude; all of which are covered as closely as they can be set, leaving the length of a foot between, with long upright stones; some as high as eight or ten feet; and scarcely any that are not richly, and laboriously carved with various commemorative devices in the forms of crosses, saints, cherubs, birds, beasts, etc. besides the names of the deceased. The most magnificent graves, instead of having a flat stone at the feet, present the figure of a ram rudely sculptured. Some have merely the plain form; others decorate its coat with strange figures and ornaments in the most elaborate carving. The form of this animal appears to have been a favorite type in sepulture, throughout Armenia after the introduction of Christianity.

But having observed it first in some old burying grounds near *Ouroomia*, and in others beyond *Tabreez*, I conceived that it must be over *Mahomedan* graves, and probably marked the last beds of some heroes of the Black and White Ram factions; but finding it thus at Julfa, I can have no hesitation in supposing it an emblem connected with the Christian faith. Subjoined is a sketch I made from one of the largest, and most ornamented; which may be sufficient to give an idea of the uncouthness of shape, but indefatigable labour of ornament, exhibited in these strange memorials of the dead. On one of the sides, the legend of St. George slaying the dragon, is most curiously carved; and on the other, some second champion followed by three walking personages bearing a long pole.

The tail also is honoured with the figure of an equestrian saint, bound round with a scroll stampt with Armenian letters. A sort of network covers the neck of the animal.

I shall not exceed the fact, when I say that thousands of gravestones marked this awful depositary of the old Armenian race. Indeed the various sorts of recollections which present themselves to the mind in this particular quarter of the East, continually suggest the idea of treading over some vast tomb. And it is so: for here the first fathers of all the families of the earth were buried; here immense empires rose, and crumbled into dust; here we find the remains of cities, whose founders died in the infancy of the world; and the monuments of people, sunk so long into the depths of time, that the name of their nation is no more remembered. The cemetery of the Armenians at Julfa was not yet in such a case, though the individuals which filled it were mingled in the common mass; but the time may come, when the ram on its hills, may be as little assignable to them there, as I found it on the tombs beyond *Tabreez*; or as the rude sculptures dug up in the endless tumuli north-west of the Caucasus, are now explainable of the people whose vast remains they cover.

From this melancholy spot my conductor led me to one well calculated to disperse its gloomy apparitions; his own little abode, clean, cool, and cheerful; and, how he had managed it I cannot tell, but I found a nice repast prepared for my entrance, consisting of fish, eggs, bread, butter, and acidulated milk.

Frédéric Du Bois de Montperreux, *Voyage autour du Caucase, chez les Tscherkesses et les Abkhases, en Colchide, en Géorgie, en Arménie et en Crimée*, Librairie de Gide, Paris, 1839–1849, Tome IV, pp. 21, 23–25, 25–26, 26–27, 27–28 (Fig. 5).

L'*Araxe* est ici … encaissé, l'espace des plusieurs verst, par ces rochers dont les parois, bizarrement rongées en forme de pics et d'aiguilles ruinées, couronnent ses rives en y reflétant leur teintes sanguinolentes. C'est là qu'est *Djoulfa*. Les anciennes histoires arméniennes en font déjà mention dans les siècles les plus reculés, et la citent dans le nombre des villes et des bourgs que *Tigrane I*er*(Dikran)* assigna pour patrimoine aux parents d'*Ajdahag (Astyages)* … *Djoulfa* à la portée d'un des gués les plus commodes de l'*Araxe*, s'augmenta considérablement.

L'industrie arménienne s'y concentra, et le commerce l'enrichit plus qu'aucune ville de l'Arménie, jusqu'à ce qu'en 1605 elle fut tout à coup rayée du nombre des demeures des hommes. Cependant une ville de 40.000 habitants devait avoir laissé quelques traces de son existence. Effectivement, il en reste assez pour qu'en voyant tant de débris d'une riche industrie et partout le silence de la mort, le cœur se serre à la vue de ces étranges révolutions que le caprice d'un despote peut opérer.

On arrive au bord de l'*Araxe* gonflé et troublé par la fonte des neiges. Vous ne passez plus sur le pont dont les voûtes des arches se sont écroulées dans le fleuve.

Fig. 5 Frederic Dubois de Montpereux_(1798–1850) (http://www.men.ch/de/histoires/portraits/dubois-de-montperreux-frederic/—Wikimedia Commons)

La garde ne vous arrête pas au pied de ces tours carrées qui en commandent l'abord des deux côtés.

Lorsqu'on veut entrer dans la ville, on remonte la vallée de l'*Araxe*; les rochers sanguinolents qui le bordent commencent ici …: chaque saillie, chaque assise du rocher tant soit peu accessible est hérissée des murailles d'un château qui défendait l'entrée de la vallée. Un mur qui part du fort et qui s'étend jusqu'à l'*Araxe* la ferme encore plus étroitement … Devant la porte même, un pic isolé de rocher s'était trouvé assez large pour établie à son sommet un monastère en l'honneur de la sainte Vierge: l'église est encore là; les morts dont les riches tombes couvrent les pentes du rocher y sont aussi: mais les vivants, où sont-ils?

Nous entrons dans la ville: entre la muraille à pic de rochers et l'*Araxe*, il y a un espace de quelques centaines de pas où gisent çà et là d'énormes blocs qui se sont écroulés des pentes voisines. Y reconnaissez-vous *Djoulfa*? Tenez, voilà son bazar au bord de l'*Araxe* et quelques-unes de ses églises au pied du rocher. Celle qui est là sur un tertre était, dit-on, catholique, et c'est devant le seuil de sa porte qu'on a enseveli, sous una belle tombe de marbre blanc, le riche *Khatchababa*, le plus puissant des habitantes de *Djoulfa*, sous le règne de *Chah Khodabendeh*, qui, dit-on, le fit mettre à mort.

……

Sur le deux rives de l'*Araxe* s'élèvent, en face l'un de l'autre, deux vastes caravansérais, bâtis en pierre de taille, et qui n'ont jamais été terminés; une dizaine de familles qui composent toute la population actuelle de *Djoulfa*, se sont établies dans celui de la rive gauche.

Tout autour, jusqu'à un verst de distance sur le bord uni du fleuve, dans les ravins, sur les rochers pittoresques et contre leurs parois escarpées, sont disséminées, les habitations, à demi-lavées par les pluies, à demi-entraînées par les torrents.

……

Cependant, à ma grande surprise, je ne trouvai aucun des édifices de *Djoulfa* très-remarquable, soit par la richesse, soit par la majesté de son architecture; l'incendie qui causa leur ruine les a trop maltraités. Toutes les maisons étaient en pierres de grès bigarré, liées avec de l'argile rouge. Les églises étaient un peu mieux construites, mais sans approcher du luxe ordinaire des églises arméniennes. C'était dans leurs tombeaux que les habitantes de *Djoulfa* mettaient leur gloire et leurs richesses.

Passez la seconde muraille qui ferme l'autre extrémité de la ville; est-il rien de plus beau que ces milliers de pierres sépulcrales dressées et pressées à côté les unes des autres comme une riche moisson d'épis et qui recouvrent une grande étendue de terrain le long de l'*Araxe*.

……

Ces monuments sont de deux espèces: les uns consistent en une pierre longue de 7 à 9 pieds, plantée debout à la tête d'un bloc long et carré, qui représente le tombeau. Sur cette longue pierre, vous voyez dans un cadre orné d'arabesques superbement travaillées, des croix de différentes grandeurs chargées d'ornements: le haut de la pierre est orné quelquefois de bas-reliefs représentant des scènes de la Bible, ou un saint Georges, ou un sphinx persan, à double corps, avec une tête d'homme au milieu. Chaque pierre a son inscription en arménien, avec la date. Beaucoup sont admirables par le fini du travail qui les rendrait dignes de figurer dans un musée. Quelques-unes des plus belles étaient recouvertes par de petites chapelles murées.

L'autre espèce de monument, moins nombreuse, consistait en béliers couvert d'inscriptions et de quelque reliefs. J'ai dessiné le plus beau et le plus curieux de tous. Ker-Porter en parle dans son Voyage. On y voit un cavalier armé d'une lance qui porte en croupe un enfant prisonnier; il lui a passé autour du cou une corde à laquelle sont attachés trois autres prisonniers qu'il traîne après lui. Plus loin, le même personnage, sans doute, est représenté assis devant une table; d'un côté, un esclave à genoux lui sert à boire, tandis que, de l'autre, un second esclave pince de la guitare; le sphinx n'y manque pas non plus. L'inscription nous apprend que: «Ici repose la cendre de *Manouk Nazar*, décédé l'an 1037 de l'ère arménienne (1578 de J.C.)».

Lycklama a Nijeholt T.M, ***Voyage en Russie, au Caucase et en Perse, dans la Mésopotemie, le Kurdistan, la Syrie, la Palestine et la Turquie exécuté pendant les années 1865, 1866, 1867 et 1868*****, Paris—Amsterdam, 1873,** Tome II, pp. 26–27, 33, 36.

[Le passage de l'Araxe] s'effectue au moyen d'un bac retenu par un câble transversal, les ponts qui existaient autrefois ayant été rompus, et la Perse comme la Russie ne

montrant aucun désir de les rétablir. Je parle au pluriel, car s'il me faut croire les anciens voyageurs et les historiens persans, outre le pont de *Djoulfa*, on en avait construit un ou deux autres plus ou moins éloignés, soit au-dessus soit au-dessous de cette ville. Chardin et Tavernier avaient pareillement été obligés de passer l'*Araxe* en bateau, et on a vu que ce dernier attribue à *Abbas-le-Grand* la destruction du pont qui reliait Djoulfa à la rive posée, et dont on aperçoit encore les ruines au milieu du fleuve.

Chéryf-êd-Dyn … historien persan en fait une description vraiment imposante. Il était construit de quartiers de rocher parfaitement joints ensemble. Parmi ses arches, il y en avait des immenses, l'une de soixante *guez* ou coudées persanes, l'autre de cinquante. On avait pratiqué un caravanseraï sous une des arches qui restait à sec en tout temps, et les deux extrémités de ce pont étaient fortifiées par des portes construites, comme le pont même, avec des quartiers de rocher … Quoique l'*Araxe* soit en cet endroit d'une assez grande largeur et surtout fort rapide, avec les moyens que la science possède il serait aisé aujourd'hui de construire un nouveau pont qui servirait de trait d'union entre la Russie et la Perse. Mais peut-être, que ni l'une ni l'autre de ces puissances ne se soucie d'établir là une communication trop facile.

……

[Avant Marande] … on voit encore les restes d'un vieux caravanseraï du temps d'*Abbas-le-Grand*, remarquable par son architecture imposante et ses proportions monumentales. Déjà, du temps de Tavernier, cet établissement si utile aux voyageurs, qui trouvaient, dans la chaleur du jour et pour la nuit, un abri assuré et gratuit, avait cessé d'être entretenu avec soin: «on ne trouve là – dit-il -, qu'un grand caravanseraï abandonné, quoiqu'on y ait fait de la dépense, et qu'il soit bâti de belles pierres de taille qu'il a fallu apporter de fort loin». Aujourd'hui cette construction du plus illustre souverain de la Perse moderne, dont on rencontre les traces à chaque pas, est devenue une ruine … La partie la mieux conservée est la porte d'entrée, ornée d'un revêtement de marbre et de plusieurs inscriptions.

……

[A Sofian on voit] les restes d'un nouveau caravanseraï à peu près dans le même état que celui que j'avais rencontré la veille. A en juger par ce qu'on voit encore, il devait être d'une grande beauté. Sa destructions ne peut remonter fort loin, car Tavernier en parle comme d'un édifice existant dans son entier, vers le milieu du dix-septième siècle.

Madame B.Chantre, *A travers l'Armenie Russe*, Paris, Hachette et C., 1883, pp. 174–175, 175–176, 177, 178–182 (Fig. 6).

La route qui nous mène à *Djoulfa* franchit un col, et débouche sur un superbe plateau bien cultivé, situé à 1250 mètres d'altitude, d'où la vue plonge sur la vallée de l'*Araxe* au sud-ouest. A partir de ce point, on descend: le sol se couvre de pierres et se transforme en un vrai désert, envahi par des sauterelles … A neuf heures on rejoint la route de poste d'*Ordoubat* à *Djoulfa.* A cet endroit (900 mètres d'altitude) se présente une station de poste. Dès lors nous quittons la route pour nous diriger à travers le désert, vers une ville appelée *Karabaghilan.* Elles s'élèvent en amphithéâtre

MADAME B. CHANTRE

A TRAVERS

L'ARMÉNIE RUSSE

OUVRAGE CONTENANT

CENT CINQUANTE ET UNE ILLUSTRATIONS

GRAVÉES D'APRÈS LES PHOTOGRAPHIES PRISES PAR M. CHANTRE

ET DEUX CARTES

PARIS

LIBRAIRIE HACHETTE ET CIE

79, BOULEVARD SAINT-GERMAIN, 79

1893

Fig. 6 Title page of the volume of Madame B. Chantre, *A travers l'Arménie Russe* (Chantre 1893)

sur deux collines recouvertes de dépôts morainiques, et entre lesquelles s'étend un vaste ravin.

Rien ne peut donner une idée de l'aridité et de la solitude de cette cité morte: ni source, ni ruisseau, ni verdure, mais partout des pierres sèches et de l'argile calcinée par un soleil brûlant. C'était sans doute un bourg, analogue comme importance à celui de *Magri*, et dont la ruine et l'abandon doivent être l'ouvre du farouche *Chah*

Abbas, qui ravagea toute la contrée au XVII siècle … Sur le versant droit du ravin se dresse une sorte de forteresse avec tours en partie rasées. L'une d'elles pourtant encore 7 à 8 mètres de hauteur, et il semble qu'elle était en forme de cône tronqué. Au fond du ravin un énorme barrage de 40 à 50 mètres de longueur sur 10 de largueur devait retenir les eaux de pluie, si rares pendant une partie de l'année dans ces pays, et former un bassin d'alimentation. En avançant toujours, nous découvrons une construction octogonale, surmontée d'un toit piramidal: ce monument, d'assez petites dimensions, recouvrait un caveau funéraire.

……

Toute cette partie de la vallée de l'*Araxe* est empreinte d'une poésie sauvage. La symphonie en rouge formée par le sol, les eaux et les monts que viennent seules troubler les neiges étincelantes et grandiose, d'autant plus que les escarpements qui bordent la vallée affectent souvent les formes les plus bizarres, les plus tourmentées. Ce sont des schistes et des grès, passant du rouge au jaune par toutes les gammes possibles, soulevés parfois verticalement, et quelquefois même renversés. Ça et là de belles taches d'or formées par les champs de blé mûr ajoutent encore à l'intensité de couleur du paysage.

……

Voici enfin la station de douane de *Djoulfa*, dont les bâtiments s'élèvent dans la plaine poussiéreuse juste en face de la douane persane et au bord de l'*Araxe*, qui atteint ici sa courbe la plus méridionale.

……

19 juin. – Après avoir été sur le territoire persan … et traversé deux fois l'*Araxe*, en bac et à dos de mulet, j'assiste avec intérêt au passage incessant qui se fait, entre les deux rives, de gens, de colis et de troupeaux, car c'est ici que passe la route la plus importante reliant la Perse à la Transcaucasie. D'immenses caravanes chargées de produits de la Perse et de l'*Aderbeïdjan* y passent d'un bout de l'année à l'autre … Tout ce rivage de l'*Araxe* n'est qu'un désert de sable, visité par le vent pendant une partie de l'été. Point de bonne eau, pas un arbre, pas de légumes, pas de fruits: les employés là mènent une existence bien triste. Le village de *Djoulfa* proprement dit se trouve à trois verstes plus loin.

……

Laissant le village actuel à droite, nous atteignons l'*Araxe*, qui forme un coude avant de s'engager dans une sorte de défilé sauvage et solitaire, bordé d'immenses escarpements en forme de pics et d'aiguilles couleur de sang. En longeant la rive droite pendant quelques instants, on arrive sur un plateau légèrement élevé au-dessous de la rivière, et couvert de ruines … Aujourd'hui, sur l'emplacement de cette grande et riche cité règne un désert de ruines … Le sol est couvert de décombres sur une longueur de plusieurs verstes. Les matériaux étaient en pierres de grès bigarré, liées avec de l'argile rouge, et rien d'architectural ne s'offre à nos yeux. Seules quelques églises se distinguent ça et là, encore ne présentent-elles aucune trace de sculpture ni d'ornement. Les monuments les plus importants ayant été brulés, puis rasés, il est difficile aujourd'hui de reconstituer par la pensée l'aspect de la ville … On arriva alors devant la nécropole, si vaste que ses stèles hautes et rapprochées la font rassembler à une forêt minuscule pétrifiée … Chaque tombe se compose de deux pierres: l'une posée sur le sol, l'autre dressée à la tête et portant presque toujours des sculptures et des inscriptions. On remarque au milieu de ces pierres dressées, ornées de croix à profusion, un certain nombre de tombes dont les stèles affectent la forme d'animaux, notamment de béliers. Sur cette dernière catégorie de monuments se voient des sculptures représentant des scènes allégoriques et même des inscriptions, mais aucune croix ni aucun autre symbole chrétien.

Le plus curieux de tous ces béliers, signalé déjà par Ker-Porter et Dubois de Momtpéreux, présente un cavalier en croupe. A sa suite marchent trois autres captifs, attachés par une corde. Le même cavalier est représenté ailleurs assis devant une table, tandis que deux esclaves se tiennent près de lui, l'un à genoux qui sert à boire, l'autre, un musicien, pinçant de la guitare. L'inscription gravée sur ce curieux monument apprend qu'il fut érigé en 1578 de l'ère chrétienne sur la tombe de *Manok Nazar* … Les stèles dressées ont en moyenne 2 mètres de hauteur sur 50 à 70 centimètres de largeur, et ont été taillées dans un même grès rouge. Leur état de conservation est si parfait qu'on les croirait érigées de la veille. Au milieu des croix se voient des personnages représentant les saints, les apôtres; des scènes de la Bible s'y développent en bas-reliefs, ainsi que des animaux fantastiques tels que des chimères ailées, des oiseaux et fréquemment le sphinx persan, à double corps, avec une tête d'homme au milieu. A cela il faut joindre une infinité de dessins géométriques, d'arabesques, d'entrelaces qui courent un peu partout, enlaçant gracieusement et complétant cette décoration d'un goût bien original. Presque toutes sont dignes de figurer dans un musée, tant le fini du travail est admirable.

Sur la roche nue de la montagne au pied de laquelle s'allonge le cimetière, tranche une seule tache de verdure. Cette végétation est entretenue par une source, et cache en partie une petite chapelle solitaire … Sur un certain point on voit encore les débris du fameux pont de *Djoulfa* dont la construction au-dessous de l'*Araxe* fut longtemps regardée comme une merveille.

2 Descriptions of the City of Khoya in the Relations of Western Travellers in Persia

1271–95

Marco Polo, in: Marsden, William (Ed): Re-Edited By Thomas Wright, The Travels of Marco Polo: The Venetian, Published by Doubleday, New York, 1948, BOOK FIRST—chap. IX. Of the noble City of Tauris, in Irak, and of its Commercial and others Inhabitants, p. 5.

Tauris is a large and very noble city belonging to the province of Irak, which contains many other cities and fortified places, but this is the most eminent and most populous. The inhabitants support themselves principally by commerce and manufactures, which latter consist of various kinds of silk, some of them interwoven with gold, and of high price. It is so advantageously situated for trade, that merchants from India, from *Baldach*, *Mosul*, *Cremessor*, as well as from different parts of Europe, resort thither to purchase and to sell a number of articles. Precious stones and pearls in abundance may be procured at this place. The merchants concerned in foreign commerce acquire considerable wealth, but the inhabitants in general are poor. They consist of a mixture of various nations and sects, Nestorians, Armenians, Jacobites, Georgians, Persians, and the followers of Mahomet, who form the bulk of the population, and are those properly called *Taurisians*. Each description of people have their peculiar language. The city is surrounded with delightful gardens, producing the finest fruits. The Mahometan inhabitants are treacherous and unprincipled. According to their doctrine, whatever is stolen or plundered from others of a different faith, is properly taken, and the theft is no crime.

10–13th century

Guy Le Strange, *The lands of the eastern caliphate. Mesopotamia, Persia and Central Asia from the Moslem conquest to the time of Timur*, University Press, Cambridge, 1905, p. 166.

On the northern shore of the lake was the town called *Tarûi* or *Tasûj*, which is apparently identical with the modern *Tursah. Mustawfi*, as already said, often speaks of the Salt Lake of *Tasuj* or *Taruj*, and the town therefore shared with *Urmiyah* the honour of giving its name to this sheet of water. In the 8th *[14th A. D.]* century *Tasuj* must have been an important place, it was warmer than *Tabrîz* and damper, being so near the lake, and it was surrounded by gardens and orchards. To the north-east of *Salmâs* lies *Khawî* pronounced Khoi, on a stream that flows north to the river Aras (Araxes). *Khawî* was a strongly fortified and flourishing town according to *Yâkût* and *Kazwînî*, surrounded by fertile lands and famous for its excellent brocades. There was also a spring here which had the reputation of being hot in winter and cold in summer. *Mustawfi* says that the *enceinte* of its town walls measured 6500 paces, and that its people were a white skinned race like the *Khatâi* (Chinese); eighty villages were of its dependencies.

1403-6

Ruy Gonzales de Clavijo, *Embassy to Tamerlane—1403–1406*, Translated from Spanish by Guy Lestrange, George Routledge and Sons, London, 1928, p. 148.

On Thursday the 5th of June at midday we reached that same city *[Khoy]*, which lies in a plain, encircled by many orchards and corn lands, these again being surrounded by extensive grazing pastures irrigated by many streams and conduits of water. The city of Khoy itself is defended by a wall of baked bricks, with many towers and barbicans: and at this city the province of Upper Armenia has its limit, the frontier into Persia being passed here, but in Khoy itself the population for the most part is of the Armenian race.

1471-72

Caterino Zeno, *Commentarii del Viaggio in Persia*, in Charles Grey, *A narrative of Italian travels in Persia in the Fifteenth and Sixteenth Centuries*, Printed for the Hakluyt Society, London, 1873, pp. 59–60.

[Published for the first time in 1588 but written after 1473].

... Amarbei and Semper marched ahead, and hearing that *Selim* had crossed the Euphrates and was advancing by forced marches, retreated to *Coi*, where *Ismail*, who had come from *Tauris*, was in person.

Being informed of the large forces *Selim* was bringing with him on this enterprise, he caused his army to be strongly entrenched, and returned to Tauris to collect more troops, and then to show front to the enemy.

Coi (*Khoi*, the capital of a rich district, with a considerable trade between Turkey and Persia; it has a population of twenty-five Thousand, and is a well-built, handsome town, on the *Ak Schai*, a tributary of the Aras) is a city which they say was built on the ruins of the ancient *Artasata* (the site of *Artakarta* is fixed on the Aras, a little to the south of *Erivan*).

1473-79

Josaphat Barbaro, *Viaggio dello istesso Messer Josaphat Barbaro in Persia*, in *Viaggi Fatti da Vinetia, alla Tana, in Persia, in India, et in Costantinopoli ...*, Venezia, 1543, p. 55.

[si va a Coi] cinque giornate lontano della quale si truova una campagna, nella quale è una grande città altre volte destrutta per il *Zamberlan*. Truovasi etiandio molte ville, et dietro ad esse un altro lago lungo miglia ducento, et largo miglia trenta, nel qual ci sono alcune isole habitate. Finalmente si truova poi due città, *Tessu* e *Zerister*; le quali tra tutte e due fanno da fuochi tremila: altre cose memorabili, non habbiamo vedute in questi luoghi, salvo che in tutti si fanno lavori di cottone, di tele, di canapi, di grisi, et di schiavine assai; et qualche poco di lavoro di seta. Hanno carne assai massimamente di castroni, vini, et altri frutti assai; i quali essi conducono in mar maggiore [Mar Nero] nelle terre che sono dintorno.

[A/T:...we reached Coi ... five days far from which, in a plain, a large town is located, that was once destroyed by Tamerlane. Moreover, we found many villages, and behind them another lake two hundred miles long and thirty miles large, where are some inhabited islands. At last we reached two towns, Tessu and Zerister, that together amount three thousand hearth. We didn't see other memorable facts in those places, except for an intense manufacturing in cotton, canvas, ropes, grisailles, and cloaks with sleeves and hood; and few silkwork. They have a lot of mutton meat, wine, and fruits at large, which they send to the countries surrounding the Black Sea.]

1490

Giovan Maria Angiolello, ***A Short Narrative of the Life and Acts of the King Ussun Cassano*****, in: Charles Grey,** ***A Narrative of Italian Travels in Persia in the Fifteenth and Sixteenth Centuries*****, Printed for the Hakluyt Society, London, 1873, pp. 54–55.**

[Published for the first time in 1583 but probably written in 1490]

Coi, which in ancient times was a large city, as can be seen by the great space occupied by the ruins. In this place, it was anciently the custom (which is still observed) for the troops to assemble when the King of Persia was about to take the field with his army. This city, a short time ago, was in ruins; but when *Sciech Ismael* succeeded to the throne he began to rebuild it, and has now restored the greater part. And, among other things, a large palace has been built, which in the Persian tongue is called *Douler Ghana (Duulct Khaueh)*, signifying "pleasant abode". This palace is all walled round with bricks, and is of great extent, with an *Arim (Harem)* all together; within there are many halls and chambers, and it is built in one vault—that is to say, with one flooring; and it has a large and magnificent garden. It has two gates, with two fine courts, beautifully decorated, and these entrances are like two cloisters of a convent of friars. Before the gate which looks west are three round turrets, each of them eight yards in circumference, and about fifteen or sixteen high. These turrets are built of the horns of Namphroni stags, and it is considered that there are none like them in the world. The Persians also consider these things very magnificent. Therefore for show they have built these three turrets of the horns of these animals, as the mountains are rocky and full of game. And *Sultan Sciech* Ismael boasts that he and his lords have killed all these animals. And truly *Sciech Ismael* takes the greatest pleasure in the chase; and to show that he is a skilful hunter he has had these three turrets built, and takes more delight in living in this place than in *Tauris*, as this country is well adapted for hunting.

In this city, they also make much crimson dye, by using some red roots, which they dig out of the ground with spades and hoes, and then take to *Ormus*, and they are employed as red dyes in many parts of India.

Sec. XVI

Anonymous (probably Francesco Romano), ***Viaggio di un Mercante che fu nella Persia*****,** in **Ramusio, Giovanni Battista,** ***Navigazioni e Viaggi*****, Giulio Einaudi editore 1978–88, collana I Millenni, Ed. Marica Milanesi, 6 voll.,** cap.7, pp. 1142–1143.

Da questo luogo *[Elatamedia, vicino a Van]* camminando tre altre giornate si truova *Merent*, ch'anticamente fu gran città, come si vede per gli edificii antichi, ed è posta in una bellissima pianura, con molti fiumicelli e giardini assai, e dentro v'è solamente un borgo con un bazzarro. E scorrendo piú oltre tre giornate si vede una bella e gran pianura, circondata da gran montagne, nel mezo della quale è una gran terra nominata *Coi*, che ne' tempi antichi fu una gran città, come pel circuito di molti edificii si vede.

In questo luogo anticamente (e oggidí ancora s'osserva) era costume di ragunar le genti, quando li re persiani volevan uscir con esercito in campagna.

Questa città prima era rovinata, ma poi che *sciech Ismael* è successo nel regno, egli ha cominciato a rifabricarla e n'ha rifatta una gran parte: e fra l'altre cose è stato fatto un gran palagio, il quale con vocabolo persiano è detto *Doulet chana*, che vuol significare la casa graziosa. Questo palagio è tutto murato di mattoni, grandissimo, con un *arin* tutt'insieme; dentro vi sono molte sale e camere, ed è fatto in un volto, come sarebbe dire in un solaro, e ha un bellissimo e gran giardino. Ha poi due porte con due magnifiche corti degnamente fabricate, e quest'entrate sono simili a due chiostri di convento di frati; avanti la porta che sta verso ponente vi sono tre torrioni fabricati in tondo, e ciascuno d'essi volge otto passa, e d'altezza sono da 15 o 16 passa. Questi torrioni sono fatti di corna di nanfroni cervi, e si giudica che nel mondo non ne siano altretanti; e appresso i Persiani queste cose sono riputate molto magnifiche, onde per magnificenza hanno delle corna di quelli animali murato tutti questi tre torrioni, però che tutte quelle montagne sono alpestre e piene di salvaticine; e sultan *sciech Ismael* porta il vanto co' suoi baroni d'aver ammazzati tutti li detti animali. E veramente sciech Ismael piglia grandissimo piacere delle caccie, e per mostrar ch'egli è valente cacciatore, ha fatto fabricare le dette tre torri, e sta molto piú volentieri in quel luogo e con molto maggior dilettazione che in *Tauris*, per esservi luoghi molto accommodati alle caccie. In questa città si fanno anche assaissimi cremesini, per esservi alcune radici rosse, che si cavano dalla terra con vanghe e con zappe, e poi sono portate in Ormus, e le adoperano in far tinta rossa in molti luoghi dell'India.

[A/T: From that place [Elatamedia, near Van] in three more walking days, Merent is located, which in ancient times was a great town as shown by its old buildings. It's placed in a gorgeous plain, with many little channels and gardens, and within the town is just one borough with a bazaar. In three more days journey you meet a beautiful and large plain, surrounded by high mountains, in the middle of which is a great place called Coi, which in ancient times was a big town, as shown by many rows of buildings. In old times (and same you can see nowadays) the people used to be gathered when the Kings of Persia were about to take the field with their army. This town was previously in ruins but from the time scheich Ismael is ruling he started rebuilding and a large parte of it was done. Among others, a big palace was built, which in the tongue of Persia is called Doulet Chana, that means pretty house.

The palace has all brick walls, is very big and includes an arin. Inside of it are many rooms and bedrooms, and has a single vaulted ceiling, and a beautifully large garden. Moreover, it has two gates leading to two magnificent and well constructed courtyards. Those entrances are similar to friars' convent cloisters; in front of the Westward door are three round towers eight paces in circumference and fifteen or sixteen paces high. Those towers are made with the horns of Nanphroni stags and none alike are deemed to exist elsewhere in the world. Persians consider such things very magnificent, thence they built in the walls the horns of those animals, because their mountains are all alpine and full of game. Sultan scheich Ismael boasts, together with his barons, to have killed alle the said animals. And indeed scheich Ismael is much pleased in hunting, and in order to show how good he is at hunting he had those towers built, and stays much more willingly and pleasantly there than in Tauris.

There he can enjoy places very fit for hunting. In this town they also make a lot of crimson dye, because of the red roots they dig with spades and hoes from the ground and then bring to Ormus to make crimson dye in many parts of India.]

1603

E. Denison Ross and Eileen Power, *Don Juan of Persia*[Uruch Beg]. *A Shi'ah Catholic (1560–1604),*Harper and Brothers New York and London, 1925 (first edition 1603), pp. 43, 117.

The people of *Azerbayján* are very warlike and ever given to fighting, the nobles more especially. The province has an extent of 150 leagues across.

Further to the south-westward come three provinces of *Kurdistán*, lying in the direction of Arabian *Iráq* or *Babylonia.* The nearer of the three provinces of *Kurdistán*, the one of which we now speak, marches with the province of *Azerbayján*, and its capital towns are the cities of Khoy and *Salmás*, whose combined population may amount to 30.000 householders (or 135.000 souls). In the outlying parts of this province there are 15 other towns. The people here are not very civilized, and being warlike are given to constant fighting. They live half the year in their tents camped in the open country, after the fashion of the Moors (of Morocco), though otherwise they are quite unlike these latter folk …

1633 around

Joannes Laet (de), *Persia seu Regni Persici Status. Variaquae Itinera in atque per Persiam; cum aliquot iconibus incolarum.* Secunda editio priori auctior, Lugdunum Bataviorum, 1633, p.68.

[The cities in Azerbaijan he cites are: Salmas, Choy, Ourmya, Merent, Nahziovan and Maraga].

Chouvvy sive Choy oppidum, quod Vlughbeeckius et alii Arabes collocant ad longitudinem LXXIX grad. Et 40 serupul. Latitudinem autem XXXVII grad. Et 40 serupol. Meminerunt illius itineraria in parte secunda; paginis 198, 207 et 219. Putant veteribus dictam Artaxata; et distare à Taurisio octaginta milliaribus Asiaticis vel Romanis.

[A/T: Chouvvy or Choy town, which Vlughbeeckius, and others Arabs place to the length of79.40°, breadth of37.40°. They mentioned in the second part of this itinerary (pages 198, 207, and 219). They believe that was named by Ancients, Artaxata; far fromTaurisio eighty Asian or Roman Miles.]

1655

Evliya Celebi, *Narrative of Travels in Europe, Asia, and Africa in the seventeenth century by Evliya Efendi*, Translated from the Turkish by the Ritter Joseph von Hammer, London, Oriental Translation Fund of Great Britain and Ireland, 1834, pp. 130, 148–149.

From *Tabriz* the Sultan returned, and laid waste the countries to the right and left of *Azerbaijan*, such as *Khoi, Manand, TesU, Barud, Dumboli, Rumieh*, and after a few days arrived safe and sound at the castle of *Kotur*. This castle, one of the strongest belonging to the Persians, though fiercely attacked, did not surrender, and as winter was approaching they abandoned it.

……

Our way now led northward through marshy grounds, and we pitched our tents on the borders of *Wishlechai*. This river issues from the mountains of *Nahkshivan* and joins the *Aras*. We suffered much from dust here, and after two hours traveling reached the castle of *Khui*, a fine town of *Azerbeijan*, the seat of a *Sultan*, who commands a thousand men; the public authorities are a judge, *Muftì, Darogha, Kelenter, Nakib, Munshi, Kuruji* and *Dizchoken Agha*. The castle is built in a square form on the plain, its circumference is seven hundred paces; the ditch is not very deep. Of its two gates one leads to the south, the other to the west; it contains a hundred houses and a mosque. Its builder was *Ferhad Pasha*, but the old town was built by *Shah Haider*, and it has been sacked more than once since. This suburb or outer town consists of seven thousand houses with terraces, seventy mosques, eleven of which are *Jamis*, two baths, seven khans, and a thousand shops with gardens. I and my boys measured the circumference, which is ten thousand paces. The air is rather warm and favourable therefore to the culture of rice. The river issues from Mount *Selmas*, and flows into the *Araxes*. The fruits are famous, above all the Prophet-pears, which have no equal in sweetness and delicious taste. Owing to the mildness of the climate, the inhabitants are all white, and the women are innumerable. Some historians call this town *Asharistan, Iranistan* or *Turkistan*. Its districts reckon one hundred and eighty village, the inhabitants of which are for the most part *Sunnites, Shafiites*, who paid a capitation tax to *Shah Ismail* for the privilege of letting their beards grow *[Sakal Tuli]*, but have been exempted since the time of *Shah Sefi*.

1805–1806

Pierre-Amédée Jaubert, *Voyage en Arménie et Perse fait dans les années 1805 et 1806*, Paris, Pélicier et Nepveu, 1821, pp. 148–149 (Fig. 7).

La ville de *Khoi*, qui n'est pas fort ancienne, a des fortifications régulières. On n'y voit pas un grand nombre de mosquées ni de maisons considérables; mais les rues sont ombragées d'arbres, et l'on y trouve un assez beau caravanserai spécialement réservé pour les marchands. On peut évaluer à vingt-cinq mille âmes la population de cette ville, dont les habitants se disent aussi d'origine tartare, ce qui a valu à la contrée le surnom du *Turkestan* de la Perse. Les trubles qui ensanglantèrent le règne d'*Aga-Mehemed-Khan* furent nuisibles à *Khoi*, enforçant un grand nombre de familles à s'expatrier.

Le lendemain de mon arrivée, j'allai visiter *Hussein-Khan*, gouverneur de la province. Cet officier me reçut dans un pavillon élevé au milieu d'un jardin, où je vis, non sans surprise, des arbres: symétriquement alignés, des bassins de marbre et des jets d'eau. Ce pavillon était un peu exhaussé, ouvert par-devant, et éclairé au moyen de vitraux de couleur.

Fig. 7 Pierre Amédée Emilien Probe Jaubert (1779–1847) (E. Ducrocq editeur—Wikimedia Commons)

1807

Paul Gardane, *Journal d'un voyage dans la Turquie d'Asie et la Perse, fait en 1807 et 1808*, Paris, Le Normant, 1809, pp. 52–53.

Neuf heures et demie de *Zorava* à *Qu-oye*: 5000 âmes. La route est sur des montagnes de marbre et de porphyre. … Nous dinons chez le *Khan* dans une salle fort ornée et dont les vitrés sont de diverses couleurs. … Qu-oye est entouré de murailles et de tours, et rassemble exactement aux gravures de Jericho que l'on voit dans les Bibles. Cette ville est sujette aux tremblements de terre. La plupart des maisons sont crevassées. Fabrication de sabres, de serges et d'indiennes. Les portes ne ferment point ici avec des serrures, mais avec des cadenas. Si vous avez besoin d'une table ou d'un banc, les menuisiers apportent chez vous un billot et leurs outils, et travaillent sous vos yeux. Douze heures un quart de *Qu-oye* à *Tersoucht*. Cette ville avait deux mille maisons renversées, il y a deux mois, par un tremblement de terre; il n'en reste que cent. Les plaines sont cultivées. On y voit l'arbuste du coton dont le produit enrichit l'industrie et le commerce. Toutes les terres que nous voyons sont imprégnées de sel, et il est en si grand quantité, que le matin nous aurions pu le prendre pour de la giboulée. La terre résonne sous les pieds des chevaux.

1807

Joseph Michel Tancoigne, ***A narrative of a Journey into Persia …*** **London 1820,** vol. I, p. 67 and vol. II, pp. 169–170.

Vol. I

Khoi is situated in the middle of a well cultivated plain, covered with trees and gardens; it is surrounded with high brick walls, and all the streets have little channels of fresh water, which produce an agreeable coolness; they are also bordered with plane trees or poplars: the houses are built of clay, and have generally but one story. This town suffered much some years ago, from an earthquake, which completely ruined several of its streets: however, it still contains twenty-five thousand inhabitants. Were it not for this disaster, the beauty of its situation, and fertility of its soil (one of the most productive in all Persia), would have, no, doubt, attracted a much greater number.

Vol. II.

[En 1807] *Khoi* venoit d'éprouver un nouveau tremblement de terre qui lui avoit causé les plus grands dommages, et les habitants étoient à peine revenus de la stupeur où les avoit jetés ce désastre si commun en Perse.

1807–1809

Adrien Dupré, ***Voyages en Perse fait dans les années 1807, 1808 e 1809…*****, tome II, Paris, 1819,** pp. 241–250.

Nous laissons au bas des collines qui l'appuient, le village de *Charab-Khânéh* avec les arbres et les vignes qui l'environnent, et à droite, près de ce lac, celui d'*Ef-Tchesmè*. Non loin d'un lieu où l'on payé un droit de péage, un chemin conduit à *Selmas*, chef-lieu d'un district de soixante à quatre-vingt villages, à douze lieues au S. S. O. de *Khoi*. Ce bourg, situé dans une plaine entourée de montagnes calcaires, qui contiennent des carrières de marbre, renferme cent familles chaldéennes.

catholiques, et dépend du gouvernement de *Khoi*. Ces chrétiens sont sous la direction d'un archevêque nommé *Jean Guriel*, qui a étudié à Rome au collège de la Propagande. Le plus grand et le plus beau de ces villages catholiques est à trois-quarts de lieue au S. E. de *Selmas*: il s'appelle *Kosrew*.

Le chemin est constamment beau, quoique le terrain ait quelques inégalités. Au-delà de ruines considérables et de champs cultivés arrosés par divers ruisseaux qui affluent au lac nous atteignons *Tessuch.* Cet endroit, autrefois important et peuplé, comme l'attestent des décombres immenses, n'est plus qu'un village renfermant à peine une centaine de feux. Les causes de sa ruine ont été les incursions des Ottomans et leurs guerres avec les Persans. Quelques familles Turcomanes sont venues depuis peu de temps s'établir dans ce village abondant en fruits et en raisins.

Le 21 nous traversons, en sortant de *Tessuch,* une eau qui coule de droite à gauche près d'une mosquée. Au pied de la chaine du *Yam,* sont divers villages plantés d'arbres fruitiers et de peupliers. Nous franchissons plusieurs petits ruisseaux qui vont se perdre dans le lac, après avoir arrosé des champs cultivés. Au-delà de quelques jardins situés sur la route, nous pénétrons dans une gorge, laissant au fond d'un ravin un village environné d'arbres. Un ruisseau coule devant nous, et quand nous l'avons passé, le terrain devient pierreux. Sur la gauche, presque au bas des collines, est le village de *Kizyldji.* Après avoir traversé un faible cours d'eau, nous commençons à gravir le mont *Yam,* et mettons une heure à le franchir par un chemin étroit et pénible. Nous marchons sur la rive gauche d'un ruisseau qui coule devant nous un second, que nous passons, avait son cours de l'E. à l'O., et allait se joindre au précédent. Nous entrons bientôt après dans la plaine de *Khoi.* A gauche, est une colline détachée sur laquelle s'appuie un pont jeté sur la rivière de *Cotour.* On le traverse en prenant une autre route de *Tessuch* à *Khoi.* Nous passâmes à gué la rivière, que la fonte des neiges avait grossie et rendue très-rapide. Cette rivière prend sa source près de *Mahmoudi,* à trois journées au S.O. de *Khoi,* passe au village de *Cotour,* qui lui donne son nom, mêle ses eaux à celles de *l'Ak-chai,* non loin de *Meraguier,* à une journée à l'E. de cette ville, et se jette dans l'*Araxe.*

Nous laissons sur la droite *Zoraw,* village arménien, et après avoir traversé divers filets d'eau détournés du *Cotour,* nous entrons à *Khoi.*

Cette ville, bâtie sous *Kérim-Khân,* semble avoir tiré son nom de ses salines (*khoi* en Kurd signifie sel). Elle est dans une plaine longue de six lieues, large de trois, bien arrosée, abondante en fruits de toute espèce, en plantes céréales et en coton. Environnée de murailles et d'un fossé profond, elle a quatre portes en pierres, savoir: *Tchors kapoussy* au N., *Gullù-Boulak* id. à l'E., *Mahalè* id. à l'O., *Selmas* id. au S.

Comme dans toutes les villes de la Perse, on y voit un nombre infini d'édifices ruinés. Ravagée dans les dernières guerres qui eurent lieu entre l'Empire ottoman et l'Empire persan, avant et depuis *Amurath IV*, elle a beaucoup plus souffert en 1800. Son gouverneur, *Djaffer-Kouli-khân* leva l'étendard de la révolte contre le roi actuel *Fcth-Ali-Châh.* Après s'être défendu pendant quelque temps, il se réfugia chez les Russes, qui l'élevèrent au grade d'officier supérieur, et lui confièrent le commandement de *Chuzzè.*

La population de *Khoi* consiste en cinq mille familles, dont deux cent cinquante sont arméniennes schismatiques. Tous ceux des habitants qui sont persans appartiennent à la tribu *Dombeli. Khoi* et la province dont elle est la capitale, sont gouvernées par un *khân-khâdjàr*, nommé *Nedjef-Kouli.* Ce khân relève du prince *Abbas-Mirza,* à qui il paye une redevance annuelle d'environ soixante mille tumans. La ville ne renferme aucun édifice remarquable. Les rues sont traversées par des canaux bordés de peupliers et de saules. L'industrie est presque nulle, si ce n'est quelques fabriques de toile. Mais *Khoi* est le passage de toutes les caravanes qui vont de *Tawriz* à *Arz-roûm* et d'*Arz-roûm* à *Tawriz*. Le bazar est petit et couvert. Les marchandises ne sont sujettes qu'à un droit de péage, qui est de trois à quatre riales par charge, suivant la valeur des objets.

1808–09

John Macdonald Kinneir, ***Geographical Memoir of the Persian Empire*** **...,London 1813**, pp. 154, 327.

There is no town in Persia better built or more beautiful than *Khoee*: the walls are in good repair; the streets are regular, shaded with avenues of trees; and the ceilings of many of the houses are painted with infinite taste. These paintings, however, are not modern; and as the immediate predecessors of *Shah Ismael* frequently held their court in this city, they were probably executed about that period.

...

The city is surrounded by a wall and ramparts: it has two churches and three mosques; and the monastery of *Karu Killeesa* is famous for the beauty of its architecture, its antiquity, and grandeur.

1810

James Morier, ***A Journey through Persia, Armenia*** **...London, 1812,** pp. 297–299.

June 3. The distance from *Tasouj* to *Khoi* is called eight *fursungs [= parasang, a Persian unit of measurement for how far a man can walk in a day]*; we were however nine hours on the road, and calculated the journey at thirty-six miles. The general direction was N. 30 W. Our course for the first ten miles, to the foot of the range, (which encloses the plain and lake of *Shahee*) bore nearly West; then we suddenly turned to the North through the mountains and, for ten miles more, wound among them through some very narrow defiles, and by some sharp ascents and descents, till we reached on the opposite side the plain of *Kiwi [Khoy]*. Towards the lake the mountains are mostly of an argillaceous soil, but change into fine earth as they approach the plain of *Kiwi*.

In this direction they are green to their very summits, and their intervening valleys are covered with the finest pastures.

We had left *Tasouy* by moon-light: we could not therefore discover with any accuracy the nature of the country, which we traversed in the first part of our route; though we discerned indistinctly groves of trees, and heard the falling cascade in the recesses of the valleys. The first view of the plain of *Kiwi*, from the summit of the pass in the mountains, is sublime. The city and its more immediate territory are seen on the N. but separated from the rest of the plain by a border of green hills, which seem to divide the expanse into two parts. At the distance of two *fursungs* from *Kiwi*, we passed on the right the village of *Disajiz*, surrounded by fields of wheat and barley. On the left of the plain are some more villages; and one curious mound of red soil, crowned by a hillock of salt, besides several other white mounds, which are described as entirely of the same substance. We passed the small range of hills, and came all at once upon the more circumscribed plain of *Kiwi*, which is opened by a seven-arched bridge, bordered on each side by rocks, and forming with the fine stream below a complete picture. The river is called the *Otour*, and flows from W. to E. falling into the *Arras* or *Araxes*, about twelve *fursungs* further to the Eastward.

The plain of *Kiwi* (in breadth from N. to S. five miles, and in length ten) was the richest tract that we had seen. It was covered with corn, broken only here and there by the foliage of enclosed gardens. Of these gardens we ventured to enter one, which was renowned all over the country for its beauty and fruitfulness. It stands on the left of the road about two miles from the walls of *Kiwi*, and was made by *Hossein Khan*, Governor of the city in the time of *Aga Mohamed Khan*; but it has now become the property of the government. It consists of a fine alley of *chenar* trees, which leads up to a pleasure-house, now falling into decay, built on the elevation of six terraces, from each of which falls a beautiful cascade, conducted by *kanauts* from the neighbouring mountains. On the right and left is a wood of fruit trees of every sort and description, with a fine crop of grass at their roots. From the pleasure-house is seen, through the alleys of *chenars*, the whole territory of *Khoi*, one of the most lively landscapes that we found in Persia. The *chenar* is really a delightful tree; its bole is of a fine white and smooth bark, and its foliage, which grows in a tuft at the summit, is of a bright green. Those in the garden had not attained their full growth. Their trunks are everywhere carved with the invocation of "*Ya Ali*" proceeding probably from the ecstasies of those, who visit this little Persian paradise.

Khoi is surrounded with a wall, and with towers of a different construction to any which we had remarked in other fortified towns of Persia. They are triangular in front, with a species of connecting work behind them. There are four gates, which are of stone, and very superior to most of those that I had noticed elsewhere. Within the walls are twenty mosques and six baths. There are said to be ten thousand houses, and a population of fifty thousand persons, of which the larger proportion are Armenians. The Mussulmans live in a parish or *Mahalé* of their own. The territory is so extremely fertile, that *Kiwi*, with the surrounding villages, pays annually to the public treasure the sum of one hundred thousand *tomauns*. *Khoi* is much warmer, from its local situation, than Tabriz. Roses here were in full flower, whereas a little opening bud was reckoned a rarity at Tabriz; and probably in twenty days from the date of our visit, the plain lost it verdure, and assumed the beautiful gilding of a ripe corn-field.

Six *fursungs* South from *Kiwi* is an equally large and populous town called *Salmas*; where, as I afterwards learnt at *Arzroum*, are "sculptured rocks and many ruins."My informer added, that one of the subjects represented two men, of whom one, looking over his left shoulder, pointed with his hand to a spot which the people of the neighbourhood affirm to contain a hidden treasure, though they admit that the deposit has escaped all research.

1816

James Morier, *A second journey through Persia, Armenia, and Asia Minor, to Constantinople, between the years 1810 and 1816*, London, Longman, Hurst, Rees, Orme and Brown, 1818, pp. 304–305.

On the 1st of June we travelled over a continuation of the mountains until we descended gradually into the vale of *Khoi*, which for richness of cultivation, water, pasture and population is perhaps not to be surpassed by any spot of similar extent, either in Persia or any other country. It is nearly oval. About fifteen miles in length and ten in breadth, and is surrounded by mountains, though as usual, unadorned with wood, yet beautiful in their shapes and distances. It produces great quantities of corn, besides cotton and rice. The soil in so stiff that requires sometimes even ten pairs of buffaloes to drag the plough-share through it. When the plough is at work, two or three men according to the length of the team are seated upon the yokes, exciting their cattle by a loud song, which at a distance in the stillness of the morning was very agreeable. Their plough is an instrument of more mechanism here than in the southern parts of Persia, and furrows the earth much more effectually. The corn we observed grew thicker and better than in other parts, which doubtless proceeds in some degree, from the great quantity of water with which this plain is blessed.

Most of the villages are peopled by Armenians. In passing through one of them we observed a board hung between sticks, placed conspicuously on the top of a house. It answers the purposes of a bell, for when beaten it emits a sound that is heard at a distance, and calls the poor folks together to prayers.

Our tents were placed near the walls of the town, and before we reached it we were again greeted by infantry dressed in red.

……

Attempts have been made to fortify *Khoi* according to rule, upon a plan drawn out by one of General Gardanne's officers. It has cost the Prince large sums of money, more from the instability of the workmanship and materials employed upon it, than from any other cause.

He now has determined to complete the fortifications, and we found some hundreds of workmen employed under the direction of a young Persian, by name *Mirza Baiker*, who acquired a superficial knowledge of engineering, from both the French an English officers.

1812–1813

Gaspard Drouville, *Voyage en Perse fait en 1812 et 1813*, Paris, 1825, Tome I pp. 53–54 and Tome II, p. 221 (Figs. 8 and 9).

Tome I

Les villes sont fortifiées de la même manière et plus ou moins régulières, mais en général elles sont défendues par une grande quantité de tours, distantes d'une demi-porté de fusils les unes des autres, et assez souvent entourées de fossés larges et profonds.

Il n'existe dans toute la Perse que deux villes qui soient fortifiées dans un système européen, *Khoï* et *Abas-Abad*; il est impossible de trouver rien de plus mauvais. Tout ce qu'on y a fait, consiste en de mauvaises courtines, très-imparfaitement flanquées par de petits bastions si étranglés, et d'un saillant si aigu, qu'on aurait de la peine à y manœuvrer deux pièces de canons. Ajoutez à cela que le tout est sans revêtements, à une toise près, sans chemins couverts, ni glacis, ni palissades. Ce que l'on a bien

Fig. 8 Portrait of Fatey Ali Shah from G. Drouville, *Voyage en Perse, fait en 1812 et 1813* (Drouville 1819)

voulu décorer du nom de contrescarpe, n'est d'aucune défense, et l'on n'aurait pas beaucoup plus de peine à la descendre que les degrés d'un escalier un peu élevé. Ces mauvais ouvrages furent construits par un officier anglais, soit-disant ingénieur et qui était au service de la compagnie des Indes. Peut-étre fut-il, faute d'autres, considéré comme le Vauban de l'Asie.

Tome II.

Khoéy grande et magnifique ville de l'*Azerbidian*, située au milieu d'une vaste plaine, sur l'*Otour* qui la traverse et dont on à tiré des petites ruisseaux qui coulent au milieu des rues à travers deux rangées d'arbres.

Elle est en grande partie habitée par des Arméniens; elle était autrefois fortifiée des hautes murailles flanquées des grosses tours. Mais le prince royal, vu l'importance de sa position l'a récemment fait fortifier à l'européenne.

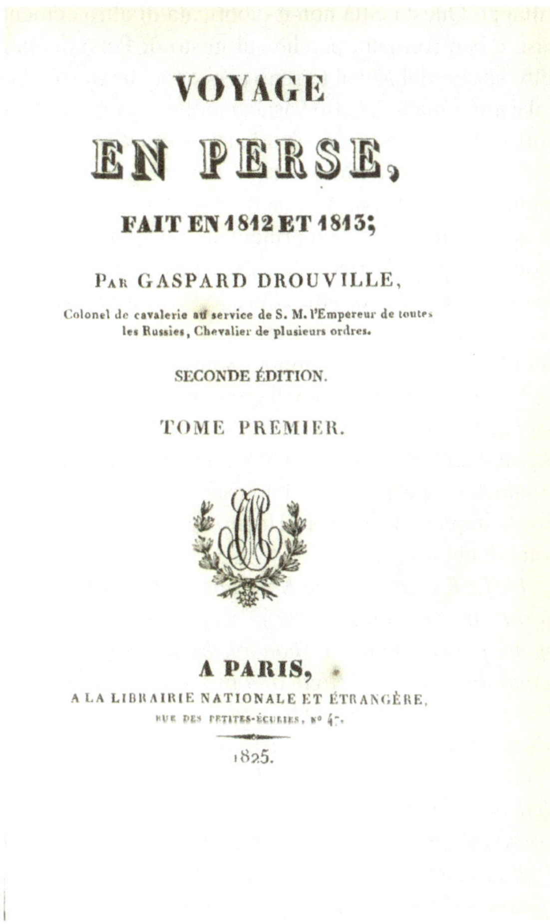
VOYAGE

EN PERSE,

FAIT EN 1812 ET 1813;

PAR GASPARD DROUVILLE,

Colonel de cavalerie au service de S. M. l'Empereur de toutes les Russies, Chevalier de plusieurs ordres.

SECONDE ÉDITION.

TOME PREMIER.

A PARIS,

A LA LIBRAIRIE NATIONALE ET ÉTRANGÈRE,

RUE DES PETITES-ÉCURIES, N° 47.

1825.

Fig. 9 Title page of the volume of G. Drouville, *Voyage en Perse, fait en 1812 et 1813* (Drouville 1819)

Lorsque j'ai quitté la Perse elle avait déjà un front totalement terminé. La place doit former un pentagone irrégulier. Elle est gouverné par un *khan* de première classe qui a le titre de *beglerbey*, mais qui est vassal du prince royal; l'autorité des gouverneurs de *Khoé* s'étend sur plusieurs districts du *Kurdistan*, tributaires de la vice-royauté de l'*Azerbidjan*.

1818 around

Giuseppe Campanile, *Storia della Regione del Kurdistan e delle sette di religione ivi esistenti*, Napoli, Fratelli Fernandes, 1818, pp. 41–42.

Il *Koi*, detto ancora *Koi-Sangiàk* è la capitale di una provincia del *Kurdistan* detta *Soràn*. Ivi fu la residenza del bascia, il quale è riconosciuto, e protetto, per mezzo della pelliccia *[an Ottoman honor]*, dal Visir di Bagdad. Il dominio del *Koi* si stende circa a tre giorni di lunghezza, e spesso a due di larghezza. Ha sotto di se dugento cinquanta villaggi. Questa città non è fabbricata di altro cemento, che di loto. L'interno delle case è ben formato, perché sul gusto di Persia colla quale confina, non essendovi altro spazio dal *Koi* al terreno persiano, che di soli due giorni. Il linguaggio comune è il *Kurdo*; ma il dialetto varia in molte voci da quello del *Badinan*, ed è di questo più goffo. Quasi tutti sanno, e parlano ancora il persiano, ed il turco. La città è circondata d'alti monti per lo più o sterili o incolti; ed è irrigata da varj ruscelletti di acque non molto buone a bersi. Alcune di esse sono acide, altre amarette. Il traffico principale di questa città, e del suo principato è il tabacco da fumo. Portano ancora dalle indie molte, ma piccole pietre di agata. Quivi sono molti negozianti esteri. I nazionali non lasciano di esser torpidi, come lo sono tutti i *Kurdi*; mostransi però alquando più svelti e civili con i forestieri. Vestono ancora più galanti degli altri *Kurdi*. Questa città fa circa dieci mila anime. Sono tutti turchi, e pochi ebrei.

Un'ora distante ha un villaggio nominato *Armutta* ove sono tutti cristiani cattolici caldei, che fa circa quattro cento anime. Vivono col lavorare i terreni, e con i mulini ad acqua. I terreni abbondano di grano, orzo, legumi, robbia, poche frutta, e molte erbe botaniche. I pascoli sono buonissimi, perché sempre bagnati d'acque che scorrono divise in piccioli rivoletti. Hanno de' buoni formaggi, e del miglior butiro. Le carni sono di buon gusto.

[A/T.: Koi, also called Koi-Sangiàk is capital town to a Kurdistan province called Soràn. It was home to the pasha, who is recognizable, and protected, by means of the fur [an Ottoman honor], by the Visir of Baghdad. Koi's domain stretches about three days in length and two in breadth. It consists two hundred fifty villages. This town is made of no else concrete, than mud. The interiors of the houses are very pleasant, according to the taste of Persia, to which boundaries it's very near. Actually Persian territory is just two days' walk far. Kurdish tongue is the most common; but dialects vary in many words from the Badinan's, and is clumsier than it. Almost everybody knows and still speaks Persian and Turkish. The town is surrounded with high mountains, mostly sterile and non grown; and is is watered by a lot of small channels the water of which is not very good to drink. Some of them are sour, other ones are bitter. The main deal of this town and its princedomis tobacco. From Indies they still bring many but small agatha stones. Here thrive many foreign traders. The national people instead are lazy, like all Kurds are; although they are quite faster and civil with foreigners. They are more elegantly dressed than other Kurds. This town counts about ten thousand souls, all Turkish with few Hebrews.

At one hour distance is a village named Armutta where all the people are Christian Catholic Chaldeians; it counts about four hundred souls. They live of farming, and with water mills. The ground is rich in wheat, barley, legumes, madder, few fruit, and many botanical herbs.

Pasturage is very good because constantly crossed by small water channels. They have good cheeses and even better butter. The meats have good taste].

1826–30

Robert Macdonald, ***Personal narrative of Military Travel and Adventure in Turkey and Persia*****, Edinburgh, Adam and Charkes Black, 1869**, pp. 124–130.

The first town of any consequence that we came to was *Khay*. The fields in the vicinity of this town were covered luxuriantly with fine water-melons and sweet melons. To look these fruits was quite enough to put one's teeth on edge. I never saw anything so tempting; but the look was nothing to the taste. They were so exquisitely delicious and succulent, that they might have even been presented to the gods on Olympus in preference to their nectar. The town had a brown and monotonous appearance, and was enclosed and defended by a mud wall, flanked with towers, and an entrenchment. The streets were narrow and dirty, and the houses flat-roofed, and built of clay.

......

The governor's mansion, although it was only built of mud, had not by any means an inelegant appearance. The way it was lacquered and painted over with various colors gave it a showy and stylish look. It had gardens, with clear and beautiful fountains, in front and rear; but the gardens were kept in very bad order and taste, though there was no deficiency of trees and rose-bushes. The mansion, the gardens, and the whole premises were enclosed round with a high mud wall. The various apartments were void of any kind of furniture in the shape of tables and chairs; in fact, these articles are unknown in the country. The floors, however, were covered with rich and beautiful carpets; on these, both stranger and native have to squat. Bedsteads are things unknown, cushions laid on the carpet to recline on forming the bed of a Persian.

......

We left *Khay* pretty early in the morning, and were escorted by the kind and hospitable governor to a beautiful garden about a mile and half outside of the town, were we had an excellent breakfast prepared for us. On entering the gate of this garden, in front of us lay a fine broad avenue, at the top of which stood a showy and fairy-looking palace. On each side of the avenue were rows of high and splendid rose-bushes, which were covered over with roses of the most beautiful hue that I ever beheld. Well might the trees be proud of such lovely-looking brides by their sides; but they had every cause to be jealous of them, for the *bulbuls*, or nightingales, appeared to share a great deal of their favors. The nightingales are very numerous here, and the rose-bushes are their favorite haunts. In front of the palace was a delightful clear fountain; in front of this again were a few enchanting and gently-falling cascades. These had a most refreshing and pleasant effect upon the eye, and added greatly to the beauty of the scene. This garden, I was told, was once considered the finest in Persia; but it is now very much neglected, being covered over with weeds, and the palace also appeared falling into ruin.

1829

Thomas Alcock, *Travels_in_Russia_Persia_Turkey_and_Greece in 1828–29*, London, Clarke and Son, 1831, pp. 102–104.

In the course of our march, not many hours before we got to *Hoey*, we fell in with a body of between two and three hundred Armenians ... The party consisted of about forty Christian families, who had obtained permission to emigrate, and they were then on their road to Russia ... The poor people appeared overcome with delight that they were leaving the iron rule of Mahometans, and were going to colonize in a Christian land; but I have heard it disputed whether the delegated authority in Russia, especially in the remote parts of the empire, is not equally severe as the government of Turkey, or of any other state.

Hoey was in the hands of the Russians during the latter part of the war with Persia in 1827, and many of the houses bear evident marks of the dilapidations committed by the troops whilst they were there, for the principal buildings were made use of as barracks during their occupation of the town. At this place we felt, for the first time in our lives, the shock of an earthquake … The house rocked from side to side; but before we had time to collect our senses, or consider the cause of our alarm, all was composed. On enquiring with earnestness what damage had been done, we were calmly answered that these occurrences are very frequent, but that they seldom do much injury … Upon this occasion no damage appeared to have been done, although the shock was so violent that the side of our room seemed to move five or six inches out of its perpendicular.

1829

George Fowler, *Three Years in Persia with travelling adventures in Koordistan*, London, Henry Colburn Publisher, 1841, vol. I, pp. 242–243.

Being detained at *Khoie* a whole day to obtain horses, I rambled through the bazaars, and saw a good deal of this large city, the gates of which are respectable; but the bridges over the dry ditch are of so tumble-down a description as to be highly dangerous; they are built on slight poles, so as to yield in the middle some six or eight inches of level, and miserably propped with poles from below. There is design in this image. Fancy a contumacious governor within; he cuts down the bridges in five minutes, and defies the outside authority.

…

The plain of *Khoie* is most extensive, and richly dotted with villages; but we must not tarry amongst them, being now within two days of *Tabreez*.

…

I have seen districts depopulated in this way, particularly in the neighborhood of *Khoie*; and thousands of them, I understand, found an early grave in Russia. Nothing grieved *Abbas Meerza* more than to see so many valuable *"rayats"*, or tributary subjects, leaving their native soil, which had been manured with their industry, and rendered flourishing by their occupancy. The Armenian villages were always the most flourishing, and yielded much greater revenues than the Persian villages. One of the objects of the war, on the part of Russia, seems to have been that of acquiring population …

T. B. Armstrong, *Journal of Travels in the seat of War, during the last Campaigns of Russia and Turkey*, London, A.Seguin, 1831, pp. 164–167.

… In four hours … we arrived at the neat town of *Tasse,* within a mile of the lake; its environs form a complete garden-verdure is everywhere to be seen in this country; and a vast population supported by its productions. After refreshing we proceeded;

a heavy rain and fog obscured the lake, with its islands, from our view; but during three hours' ride to *Almanserai* it cleared up. When we found ourselves clos on the smooth waters of *Ouroomia.* The circumference of this lake is considered to be about two hundred and fifty miles; its length, from north to south, ninety; and breadth, thirty-two; the scenary was most beautiful. In the centre of this immense lake several singular-formed islands are dispersed, and the opposite mountains of *Koordistan* bound the prospect: the *tout ensemble* is highly romantic, and reminded us of the scenary of Switzerland: the waters of the lake, we were told, is so salt, that no fish can live in it. *[They lodge in Alamanserai]* ... Got away early, and commenced crossing a rocky steep pass in the mountains; the road proved rather indifferent and rugged; in two hours we arrived at the summit: we descended through a narrow glen, and in three quarters of an hour passed a village, and commenced crossing a wide open plain, which continues for two hours. When within three quarters of an hour of *Khoye*, we fell in with a caravan of Armenians, with their goods and cattle, emigrating to *Erivan*; we approached their camp, when they came out to meet us, perceiving we were Christians, by our dress. They complained the bad treatment they met with from the *Mussulmen* where-ever they passed, and pointed out a village where the inhabitants, not content with exacting twenty *tomains* from the poor wretches before allowing them to pass a bridge, had stolen from them two children, and the whole camp were waiting to-day, in hopes of recovering them ... Soon after leaving them we passed a small river, and entered an avenue of poplar trees, which continue to the gates of *Khoye*, a distance of about a mile. The situation of this city is interesting; it is in a plain, surrounded on three sides by mountains; the environs, for miles, are laid out in gardens, which at this time of the year give a fresh and beautiful appearance. We entered across draw-bridges, through a double set of fortified walls. This place was taken by the Russians in the late war and has only been evacuated by them about a month. We were escorted by a soldier from the gate to the governor's house, where we were invited to remain.

We stayed here to-day, for the purpose of disposing of greater part of our mules and horses, previous to entering Turkey. The khan treated us with much civility: he amuses himself in ornamenting the *Koran*, and his son is very clever in drawing flowers. About noon we experienced a sharp shock of earthquake, one of the domestic, standing by a stream opposite the house, was shaken from his position into the water ... We were told such occurrences were frequent, and the inhabitants thought very little of them.

1836–37

James Ballie Fraser, *A Winter's Journey (Tatar) from Constantinople to Tehran with travels through various parts of Persia,* London, Richard Bentley, 1838, vol. I., pp. 347–348, 352.

Our approach to the city was through a long succession of gardens, and there were great tracts of ice, occasioned by the bursting and overflow of the water-courses. We had been told on the way that the city gates would be shut at sun-down, but when we reached the place there was no one to whom to proclaim our quality. The gate was not only shut but abandoned; at least, no one would answer to our appeals, though made with abundant vociferation and noise: so, after roaring ourselves hoarse for a quarter of an hour, we were forced to retreat and look for a caravanserai without the walls.

……

Khoee, one of the few fortification in Persia that approaches regularity, is a town of considerable size, and rather better laid out than usual; but the quantity of ice and snow choked its streets, prevented me from forming any correct idea either of its populousness or condition. I observed many good houses in it, however, and though there was no want of ruins, the aspect of the place upon the whole was less mean than that of most towns in Persia. The bazaar, through which I took a turn, though dull and rather ill-filled, was extensive, and kept in tolerable repair.

1836

Lieutenant-Col. Justin Shiel, *Notes on a journey from Tabriz, through Kurdistan, via Van, Bitlis, Se'ert and Erbil, to Suleimaniyeh, in August, 1836,* in The London and Paris Observer, Paris, 11 March 1838, p.158.

After a tedious march of nine hours, in a N.N.W. direction, and various ascents and descents among hills, we arrived at the district and village of *Kotur*, which are attached to the governorship of *Khoi* in Persia. Overlooking this village, of about 60 houses, is a tolerable mud-fort on a mound, yet it was plundered in October, 1835, and held possession of during a month, till all the grain was carried off by a branch of the tribe of *Hekkari*, nominally under the *Pashalik* of Van. We were well received at the village, which is chiefly inhabited by Armenians; the chief was extremely civil, and supplied us with the best house, which belonged to an Armenian … Everybody was filthy and in rags; but their poverty was more apparent than real, for the house we lodged in was crowded, like the generality of the other houses, with cows, buffaloes, calves, etc. The houses were not much higher than four feet outside, but as there was a considerable descent at the entrance, the height was much greater within … We left *Kotur*, and crossed to the left bank of the *Kotur Chai*, which flows with a very rapid stream. We entered a wide defile, named *Balanjik*, twelve or fourteen miles in length, with the above stream flowing through it. Our general course was west: at about the fifteenth mile we entered an immense meadow, and at the twenty-fifth mile halted, at the ruined village of *Mulla Hasan*, on the bank of a small stream flowing from the south … This is the most unsafe part of the road between *Khoi* and Van, from the vicinity of the tribe of *Khan Mahmud*, who has taken possession of this meadow.

1836–38

Horatio Southgate, *Narrative of a Tour through Armenia, Kurdistan, Persia and Mesopotamia…*, New York, D. Appleton & Co, 1840, vol. I., pp. 325–328, 332.

We passed, between dawn and sunrise, several hills from which salt is taken in nearly a pure state, and struck the road between Khoy and Tebriz, where a good bridge crossed a stream that was now reduced to a mere thread. The rest of our way lay over the plain of Khoy, along a road which was the best that I saw in the East, excepting only the post-road of the Sultan, between Constantinople and Nicomedia, and a portion of another in the direction of Adrianople.

Excepting these, and the poor causeways in Asiatic Turkey, constructed during the reign of the late Sultan, for the transportation of cannon, the reader may generally understand, when I speak of roads, simple footpaths, such as may be seen across pasture grounds in New-England. The country over which they pass is, for the most part, in the state in which nature left it. Sometimes among the mountains, the difficulties of the way have been relieved by the labor of man, where they would otherwise have proved impassable, but, seldom is this affected to such a degree as to make a commodious path. It is singular to meet, on such roads as these, excellent bridges of stone or brick. Often, however, they are wanting, and the traveller is compelled to make his way by fording, which, at certain seasons of the year, is not always an easy or a safe task.

The road across the plain of *Khoy* was bordered by willows, which must once have been beautiful, but are now in a sadly neglected state. At short distance the road was cut through by canals, over which there was no passage but little pathways, three or four feet wide. The plain itself, which is of nearly a circular form, and, as I judged by the eye, about twelve miles in diameter, is pronounced by Morier, who traversed the greater part of Persia, the finest in the country. Morier, however, did not visit the plain of *Ourmiah*, which far surpasses it, both in natural beauty and in the variety of its products.

……

Our muleteer conducted us to an excellent caravanserai in *Khoy*, where I was pleased to find myself not so great an object of curiosity as on the plains of *Salmas* and *Ourmiah*. I had, on entering Persia, divested myself, as far as possible, of my Turkish garb, but, having no other covering fo my head, I was obliged to retain the fez until I reached *Tebriz*, where I exchanged it for the black lamb-skin cap of the Persians. Khoy lying so near the border of Turkey, and on the great route from Erzurum, the red cap and turban of the Turk was not so strange in its streets as in the villages of *Salmas* and *Ourmiah*.

……

Our caravanserai was of one story and of brick, the principal particulars in which a Persian *caravanserai* differs from a Turkish *khan*, the words being only two names of the same thing, in different languages. In Persia, also, the building is often of two stories, especially when it is designed to afford rooms for merchants or tradesmen. When there is but one story, the building is generally elevated immediately over the arched entrance, thus affording space for an airy and comfortable room above. We had wished to occupy it in the present instance, but John assured us that it was intended as a place for the lodgers to say their prayers …We concluded, however, that, being ourselves lodgers, our privilege was the same as others; accordingly, we occupied it for an hour, on Sunday morning, in reading the daily service of the American church.

The next day, while walking in the bazars, I was accosted by a well-dressed Persian and invited to sit down with him: a piece of politeness much more common in Persia than in Turkey. He offered me the *kalioun* (water-pipe), and made many inquiries concerning my object in travelling. He could not understand me when I told him it was for purpose of observation, and could not conceive of any man, in his senses, travelling for any other reason than necessity.

……

The most remarkable objects I observed, were the bazars, which were built of brick and vaulted, with domes of the same material. They extended in a single line from the gateway to the heart of the town, with branches, at short intervals, running down the caravanserais. I walked from my room to the gate of the city, a distance of three eighths of a mile, under this brick canopy. The brick edifices are the most remarkable in Persia, and every traveler observes with astonishment the great beauty and durability of the arches and domes constructed with no other materials than burnt clay and mortar. Khoy, I was told, contains 8000 families. But, although I received the same information from several individuals, it appeared to me, from other indications, exaggerated.

1837

E.D'Arcy Todd, *Itinerary from Tabriz to Tehran,* via*Ahar, Miskhia, Ardabil, Talish, Gilan and Kazwin, in 1837,*Journal of the Geographical Society, 1838, vol. 8, pp. 29–30.

On quitting *Tabriz*, the road for about a mile runs along the right bank of the *Bosmiyeh* river … The village of *Khójeh* contains about sixty or seventy houses … After leaving *Khójeh*, the country is covered with small isolated green and red hills … The village of *Kizil Kapán* … has suffered dreadfully from a late earthquake, which has shaken every house in the village … The road gradually descends from *Kizil Kapán* into a valley. At 3 miles a ruined *káravánseráï* marks the commencement of a gentle ascent, about a mile in length, by the side of a small stream; the descent for three miles is more abrupt. A ruined *káravánseráï* stands at the foot of the pass. One mile beyond, the road turns to the eastward, and follows the cause of the *Ahar* river on its right bank. A range of hills to the north sprinkled with snow. The road, after clearing the pass, is excellent, being hard, level and free from stones. Much cultivation: wheat and barley on both sides of the river.

1837

Lieut.-Colonel Charles Stuart, *Journal of a residence in Northern Persia and the adjacent provinces of Turkey,*London, 1854, pp. 126–131.

We descended through a narrow valley into the rich plain of Khoi, which is well wooded in some places, and bounded by a finely broken range of hills.

…

We arrived at two o'clock at the pretty village of Pereh, at the western extremity of the plain of Khoi. Its houses, built of mud, or of unburnt brick, are divided by orchards and embosomed in trees, chiefly walnuts. Ellis's quarter has a broad verandah in front, and three windows with wooden lattices -much too summery an abode for the time of year. I am nearly starved in a cottage within the same enclosure; it has doors or windows which cannot be closed on every side, besides several holes in the roof, which admitted this afternoon much more than was agreeable of a thunder storm.

Our grand entrée into Khoi, about six miles from Pereh. The fertility of the plain, chiefly produced by artificial irrigation, is most refreshing to the eye after passing through the "deserts idle" of Kurdistan. It is covered with corn-fields, orchards, melon-grounds, and gardens interspersed with villages and groves of lofty chenars. I observed a mulberry-tree by the roadside as large as a good-sized English oak.

A large escort left Pereh with us, and parties which joined us on the road augmented the procession to at least 200. The Governor, Hajee Hateem Khan, met us half way, preceded by ferashes, (carpet spreaders, tent pitchers, petty executioners, &c;) carrying long peeled sticks, which are anything but idle badges of office in their hands, and accompanied by another nobleman. There was a grand scene of galloping about the plain, and a waste of powder, such as I have before described.

.....

A deputation of citizens on foot met us outside the gates of the town. Khòi is surrounded by a double enceinte of mud fortifications, both in a sad state of decay. The outer enceinte consists of a line of curtains, redans, and bastions, with a glacis, ditch, and covered way, which were traced, I believe, not many years since, by Colonel Monteith, of the Madras army: the inner is a high wall with flanking towers: the space between is occupied by gardens and hovels; streams of water run through the principal streets of the town, and rows of willow-trees are planted beside them. Most of the inhabited houses are built of mud, but I remarked some unfinished brick buildings in ruin. Our appearance astonished the Khoites: the gentlemen with white staves were quite necessary to keep off the crowd. We are logged in a suite of apartments with large windows of painted glass, opening upon an enclosed garden (without flowers). This is the "anderoon" (harem) of a palace which once belonged to Jehangueer Meerza, one of Abbas Meerza's younger sons.

......

We found the floor of this apartment covered with piles of beautiful fruits, melons, grapes, pears, &c. and every variety of execrable sweetmeats -a present from the governor.

......

The luxury of having a clean room to oneself is not to be described.

We halted all yesterday. Khoi is said to contain between 4000 and 7000 families, of whom 100, inhabiting a distinct suburb, are Armenians. Dr. R——tells me that the bazaars are the finest in Persia, except those in Sheeraz. They are substantially built of brick, and are lofty and capacious. The ceiling is a succession of arched domes, lighted by square holes at the top. The principal alley is occupied by cap-makers; another, by sellers of cloths and linen; worsted sock makers and copper-smiths have likewise extensive quarters. Khoi is celebrated for the two latter articles, the worsted socks are striped and variegated with many colours. The principal merchants have small rooms and shops in the caravanserais: one of these, to the right of the cap-makers' street, is an octagon, very neatly built and cleanly kept, with a fountain in the centre. It was pleasant to see some old fellows seated on their heels at the doors of their shops, "with spectacles on nose," poring over large quartos [large books] which they held on their knees. M. Garibaldi, a Geonese merchant, now a tenant of one of the pigeon holes in the caravanserai, dined with us yesterday.

1839

Charles Texier, *Description de l'Armenie, la Perse et la Mesopotamie…*, Vol. II, Paris, 1852, chapts. XV–XVI, pp. 27–28, 28–29, 29–31 (Fig. 10).

Après avoir franchi [le] passage, qui forme le point réel de séparation entre la Turquie et la Perse proprement dite, nous entrons dans la plaine de *Khoie*. L'aspect du pays change totalement. Nous nous trouvons comme par enchantement transportés au milieu des jardins tracés avec art, des champs bien arrosés, d'une population active, industrieuse et qui parait au-dessus du besoin. Ce premier sentiment fut tout à fait en faveur des Persans. Les pays que nous venions de quitter, ravagés par des guerres continuelles et presque entièrement dépouillés par les Russes dans leur dernière campagne, habités par des tribus qui ne connaissent d'autre richesse que

Fig. 10 Title page of the volume of CharlesTexier, *L'Armenie, la Perse et la Mesopotamie* (Texier 1842)

l'indépendance, nous avaient présenté le tableau d'un peuple qui n'a pas encore su faire un choix entre l'anarchie la plus désordonnée et le plus dur despotisme.

Nous arrivions dans l'*Aderbeidjam*, la plus riche et la plus fertile province de la Perse, la seule dont les habitants, ayant une communauté d'origine avec les Turcs, dont ils parlent la langue, n'ont pas pour les étrangers cette antipathie qui caractérise les provinces méridionales.

Depuis *Elam*, le costume avait complétement changée: nous n'apercevions plus un seul turban; mais tout le monde, les paysans comme les grands personnages, était coiffé du bonnet de peau de mouton noire, qui est devenu la coiffure de toute la nation, depuis que les princes de la dynastie des *Cadjars* ont monté sur le trône.

Ayant fait halte près d'une fontaine, on mit le couvert pour notre déjeuner. Pendant que nous étions assis, arriva un soldat qui se reposa près de la fontaine. Nous l'invitâmes à manger avec nos gens, mais il refusa avec une sorte de dédain.

Jamais un Turc n'eut agi aussi; mais quoique les habitants de cette province soient infiniment moins fanatiques que ceux du sud, et fréquentent plus volontiers les étrangers qui ne sont pas de leur secte, jamais ils n'oseraient manger en public avec eaux.

Les travaux de la récolte n'étaient pas encore achevées; nous remarquions de grands chariot trainés par de bœufs, dont l'encolure relevée leur donnait un aspect particulier. On rentrait le blé, mais le ricin et le coton étaient encore sur pied.

……

XVI pp. 28–29.

L'aspect de la ville de *Khoie* ne démentit pas l'idée que nous nous étions faite à la vue de ses environs. Elle est entourée d'un retranchement en terre, construit d'après les lois de la fortification moderne. Les plans furent, dit-on, donnés par un officier attaché à l'ambassade du général Gardanne. Après avoir franchi le pont-levis, nous entrâmes dans une rue bien alignée, et plantée des deux côtés d'une allée de platanes qui sont constamment rafraichis par des ruisseaux d'eau vive. Mais ce qui rendra toujours les rue des villes de Perse extrêmement tristes pour un Européen, c'est que nulle part les maisons n'ont de fenêtres sur la rue, et qu'on ne voit de part et d'autre que de longs murs de briques dans lesquels sont percées de petites portes.

Si l'on juge d'après ses monuments, la ville de *Khoie* n'est pas très-ancienne. Il n'y a pas un édifice auquel on puisse attribuer une antiquité de trois siècles bien constatés. Lors la Perse et la Turquie étaient deux Etats florissants, cette ville était un des entrepôts les plus considérables du commerce entre les deux peuples. Aujourd'hui elle parait peu disposée à déchoir de son rang. Il y a peu de villes en Orient qui soient aussi commerçantes.

En arrivant à *Khoie*, nous fûmes reçus avec une grande politesse par le maître de la maison, et tout le reste du jour se passa en visites de la part des principaux habitants de la ville. Bientôt arrivèrent les présents. On nous apporta plusieurs plateaux chargés de fruits admirables, de pastèques d'une grosseur inouïe, des melons parfumés, et surtout de ces raisins qui n'ont pas leurs semblables en Europe, le grain en est petit, transparent comme l'ambre, et sans aucun pépin. Nous fûmes véritablement charmés de l'accueil qu'on nous fit et de la manière cordiale avec laquelle les habitants venaient causer avec nous. Le lendemain, on nous conduisit dans l'intérieur de la ville. Je fus surpris de l'activité qui régnait dans le bazar et ce qui me frappa surtout, c'est que pas une des cinq cents boutiques qui le composent n'était à louer, tandis qu'en Turquie, dans les bazars des villes les plus commerçantes, il y a toujours six boutiques sur dix, et quelquefois plus.

……

pp. 29–31.

L'industrie dans laquelle excellent particulièrement les habitants de *Khoie* est la fabrication des ustensiles de cuivre. On ne saurait voir rien de plus varié ni de meilleur goût que les vases de mille formes diverses qu'ils fabriquent avec une matière de premier choix.

Les tissus de laine, et surtout les bas de diverses couleurs, se fabriquent aussi avec une grande perfection. Ces tissus se consomment en Perse. Une partie du maroquin qu'on emploie pour la sellerie se tire de Russie; mais celui de Perse sert généralement pour les ouvrages plus délicats.

La population de *Khoie* ne dépasse pas, *intramuros*, une vingtaine de mille âmes. Quelques voyageurs ont, il est vrai, porté ce chiffre jusqu' à trente mille; mais si l'on songe qu'une grande partie de la ville est occupée par des places spacieuses, par des bazars dont les marchands habitent différents quartiers, par des caravansérais destinés aux étrangers, et que les maisons n'ont qu'un étage, on aura bien de la peine à placer trente mille âmes dans l'enceinte des murailles de *Khoïe*. C'est, du reste, le défaut commun à presque tous les voyageurs, d'exagérer les chiffres, par la raison que les Orientaux, n'étant pas fixés eux-mêmes sur les détails statistiques, ont les mêmes dispositions. La population de *Khoïe* se compose d'Arméniens, qui sont tous commerçants et qui paraissent vivre dans l'aisance, et de musulmans, presque tous de la secte d'*Ali*, et qui, outre leur commerce, peuvent être appelés aux fonctions publiques.

La ville est entourée de faubourgs, qui sont distants d'environ un demi mille, et qui sont tous entourés de jardins et d'arbres fruitiers. Après avoir parcouru la ville, nous témoignâmes à notre hôte o le désir de rendre visite au prince *Méhémet-Rhaïm-Mirza*, frère du schah régnant, et gouverneur de la province de *Khoïe*. Il n'en est pas en Perse comme en Turquie pour la facilité d'aborder les grands personnages. Il faut, dans une visite officielle, que tout soit réglé d'avance, qu'on détermine la distance à laquelle on commencera à saluer le prince, et celle à laquelle il rendra son salut. Nous passâmes par tout le cérémonial, car nous tenions peu à ce que notre salut nous fût rendu quelques pas plus loin ou plus près.

Le prince nous attendait à onze heures nous nous rendîmes au palais, et nous fûmes reçus par un des principaux officiers, qui nous offrit le *kalioun*. Après une pose d'un quart d'heure, le *férach-bachi* vint nous prendre pour nous introduire près du prince. Nous entrâmes par un kiosque orné de peintures dans une grande cour oblongue, plantée de rosiers et d'arbustes, et arrosée par des canaux d'eau courante. A l'autre extrémité de la cour s'élevait un kiosque dont la façade, supportée par des colonnes, était ouverte sur le jardin. Nous étions encore complètement cachés par les arbres et nous n'avions aperçu personne, lorsque notre guide, se tournant vers nous, nous fit signe de saluer, et en même temps il fît une inclination si profonde, avec un air si sérieux, que nous commençâmes à prêter une grande attention à la cérémonie. Il fît ensuite vingt pas, nous le suivîmes en ordre et toujours au milieu des rosiers. On fît une nouvelle halte, et notre guide nous dit encore une fois: *Sélam*! Encore un salut général. Après avoir fait le tour du bassin, nous commençâmes à apercevoir le prince. Il était assis sur l'angle de la fenêtre, et vêtu d'une robe de soie jaune, il portait la coiffure habituelle des Persans, et, quoique jeune, il avait déjà cette longue barbe noire qui est la principale beauté des Persans, et qui réellement distingue leur race de tous les autres peuples du monde.

XVII.

La province d'*Aderbeidjan* est une des plus septentrionales de la Perse: elle confine à l'Arménie au nord-ouest, à l'ouest au Kurdistan, au nord au *Ghilam*, et à la province d'Irak au midi.

Son climat est assez variable, car les étés y sont très chauds, et l'hiver le thermomètre y descend quelquefois jusqu'à 20° centigrades. Le changement de la température est dû à l'élévation du plateau au-dessus du niveau de la mer, élévation qui, en moyenne, n'est pas moindre de quinze à seize cents mètres. Mais ce qui fait la beauté du climat c'est la régularité des printemps, qui ne sont jamais attristés par les brouillards humides si fréquents dans les contrées du nord.

Dès que les premiers beaux jours sont venus et que les pluies de l'équinoxe ont cessé, la végétation se développe avec rapidité sans que des gelées tardives viennent anéantir l'espoir de la récolte. Parmi les fruits que le territoire de *Khoïe* produit avec tant d'abondance, je ne dois pas oublier les abricots, le seul fruit qui, dans son pays natal, soit préférable à celui que nous cultivons dans nos climats, car la pêche en Perse, et la cerise à *Cérazonte*, sont bien inférieures à celles que la culture nous donne.

En sortant de *Khoïe*, nous allâmes visiter les ruines d'une mosquée qui a souvent attiré l'attention des étrangers. Il ne reste plus de cet édifice que deux minarets qui étaient à l'extérieur entièrement revêtus de cornes d'antilope. Ce monument, remarquable par son extrême bizarrerie, n'offre plus qu'une masse presque informe. Nous nous mîmes cependant à le dessiner. Comme les minarets sont dans un village, nous nous attendions d'être obsédés par les curieux, mais ils nous laissèrent parfaitement tranquilles, et nous pûmes nous livrer à nos opérations. Nous sortîmes de *Khoïe* par la Porte du sud, devant laquelle s'étend une avenue plantée d'arbres qui a plus de 3 kilom. de long: à droite et à gauche sont de vastes jardins et des maisons de plaisance qui pour la plupart tombent en ruine. Nous longeons la chaîne de montagnes que nous avions franchi les jours précédents.

Peu de temps après avoir quitté la ville, nous traversons, sur un pont de six arches, une rivière peu profonde que l'on appelle *Atour*. Des deux côtés de la rive s'étendent des collines peu élevées qui séparent le bassin de l'*Araxe* de celui de la mer de *Chahi* ou lac d'*Ouronmiah*. Nous nous dirigions sur *Dilmen*, avec l'intention de visiter toute la côte occidentale du lac et ses innombrables villages, habités en grande partie par des catholiques. De nombreuses caravanes partaient en même temps que nous, les unes pour *Senna* dans le sud, les outres pour Tabriz. J'avais été surpris à *Khoïe* de rencontrer si peu de femmes dans les rues, mais je vis bien en route que les Persans sont infiniment plus attentifs que les Turcs pour cacher jusqu'à la moindre apparence d'une forme, en enveloppant les femmes dans de véritables sacs qui les couvrent de la tête aux pieds.

1839–40

Jules Pichon, *Journal d'une mission militaire en Perse (1839–1840) par le lieutenant Jules Pinchon*, Paris, L.Gougy, 1900, pp. 45–47.

Étant sur les portes de la ville de *Khoi*, nous n'apercevions encore ce qu'on apercevait de bien loin. Avant d'entrer en ville, il fallut tourner et retourner mille chemins couverts que nous aurions donné à tous les diables; à cause du grand froid que nous éprouvions. Enfin nous passâmes sur un petit pont et nous ne tardâmes pas à entrer dans la ville. Il y avait beaucoup de boue et les rues étroites de cette ville, rendues, pour les voyageurs monté et en nombre plus étroites encore à cause des saules qui sont plantés au milieu des rues. Nécessitèrent une longue suite, puisque nous qui ouvrions la marche étions obligés de marcher sur un rang. Enfin nous arrivâmes au lieu de notre logement. Nous fûmes logés près de *Hussein-Khan*, dans de passables appartements. ... Dans l'après- midi nous fûmes visiter les remparts, qui ne sont autre chose que deux murs de circonvallation, flanqués de tours de distance en distance. Ces murs sont très élevés sans avoir de bonnes bases; aussi n'est-il pas rare à *Khoi*, comme dans les autres villes de la Perse, de voir tomber lambeaux par lambeaux les choses dans lesquelles surtout ils devraient mettre tout leur salut. Les fortifications de *Khoi*, qui sont bien renommées en Perse, me firent la plus grande pitié. Nous fûmes visiter ensuite la caserne, vieux amas de ruines, où se sont retirées quelques femmes depuis que les soldats logent chez eux. Sans les déranger nous désirâmes visiter l'extérieur. Nous aperçûmes à la basse-cour une pièce de canon de vieux modèle, dont la pareille se trouvait à côté sous une casemate, et quelque soldats, excessivement mal accoutrés formaient à l'entrée la garde d'un poste.

......

Je fus réellement étonné des bazars que nous visitâmes. Il sont grands et sont sous de vastes voûtes assez régulièrement construites. La hardiesse de ces voûtes m'étonna beaucoup et était de beaucoup supérieure à celles d'*Erzeroum*. Voici donc comment se composent les bazars d'Orient et ici ceux de *Khoi*: ce sont des vastes galeries, sous des voûtes très élevées et ayant à la partie supérieure et sur les côtés de dessins persans, de distance en distance, dessins du reste grossièrement travaillés. Les boutiques sont sur la droite et sur la gauche. Elles ne se ferment point. Les marchandises sont dans des rayons et le maître est assis à côté, sur ses talons. Les marchandises de même espèce se suivent assez ordinairement et sont sur chaque partie d'un bazar. De distance en distance sont les caravansérails, où les marchands Gèorgiens s'établissent d'ordinaire. Cette manière de placer par quartiers les marchandises d'une même nature n'est pas chose générale et de rigueur, cependant cela existe à *Khoi*.

......

Nous remarquâmes ... un très mauvais entretien dans cette ville plus encore que dans les autres villes de la Perse, dont nous fûmes à même de juger plus tard; il y avait même, dans certains endroits, du danger à ne pas regarder ses pieds, car nous aperçûmes plusieurs trous sur le milieu des rues qui avoisinaient les bazars. Nous sortîmes de *Khoi* le lendemain. Nous remarquâmes une avenue charmante à l'autre côté de la ville, quelques maisons de campagnes, ainsi quelques arbres sur notre droite et notre gauche; cela nous fit un vif plaisir.

1840–41

Eugène Flandin, Pascal Coste, *Voyage en Perse de MM. Eugène Flandin, Peintre, et Pascal Coste, Architecte, attachés à l'Ambassade de France pendant les années 1840 et 1841*, Paris, Gide et J.Baudry, 1854., pp. 135–137 (Figs. 11 and 12).

Tout en regardant à droite et à gauche ces fantasias dont j'avais eu un avant-goût en Algérie, je cherchais des yeux *Khoi* qu'on nous avait annoncée comme une grande ville. Nous étions dans une plaine magnifique, bornée au loin par d'immenses et

Fig. 11 Jean-Baptiste Eugène Napoléon Flandin (1809–1889) (BNF. Gallica—Wikimedia Commons)

Fig. 12 Xavier Pascal Coste (1787–1879) (Wikimedia Commons)

belles chaînes de montagnes. Devant nous s'étendait une longue muraille crénelée dont le ton brun tranchait fortement sur la blancheur de la neige; beaucoup d'arbres placés derrière surmontaient ses créneaux, mais, du reste, aucune maison, ni même aucun minaret n'accusaient là l'existence d'une ville. Nous traversâmes pourtant de larges fossés sur des ponts que dominaient d'un air assez martial des ouvrages de défense qui portaient l'empreinte de l'art européen.

En effet, lors de l'ambassade du général Gardanne, des officiers qui étaient à sa suite, et parmi lesquels figuraient MM.Trezel, Fahvier, Lami, avaient proposé des plans de fortifications au gouvernement persan. Ils ne furent pas adoptés de suite, mais plus tard, avec son envie de bien faire, et, en même temps, son ignorance de la science militaire, le Prince *Abbas-Mirza*, gouverneur de l'*Azerbaîdjân*, voulant défendre cette province contre les Russes, les fit exécuter en les modifiant. Il en était résulté que ces travaux, tout en valant mieux que ceux qui sont dus aux simples maçons du pays, étaient insuffisants pour protéger la ville contre un siège régulier fait par une armée européenne.

Nous passâmes une première porte, puis une seconde en retour, et nous nous trouvâmes dans les rues de la ville, où notre étonnement de ne point l'avoir aperçue de loin cessa quand nous pûmes voir de près les constructions. Toutes les maisons sont basses, elles n'ont qu'un rez-de-chaussée, et aucune ne s'élève même à la hauteur des murs d'enceinte. Contrairement à ce qui se voit dans les villes turques, les mosquées de celle-ci n'ont pas de minarets, et leurs coupoles s'élèvent à peine au-dessus des plus modestes maisons.

Peu faits encore à l'usage persan de bâtir en briques, nous étions frappés de la monotonie et de la tristesse de l'aspect des rues dans lesquelles ne se voyaient d'ailleurs aucune fenêtre. Ces rues sont fort étroites, à l'exception de quelques-unes au milieu desquelles coulent des ruisseaux d'eau vive, ombragés par des saules dans toute leur longueur.

Autant qu'il est possible d'évaluer le chiffre d'une population en Orient, il faut porter celle de *Khoi* à environ vingt mille âmes. Elle a été, nous a-t-on dit, plus considérable jadis; mais les guerres civiles l'ont considérablement diminuée notamment lors de l'usurpation des *Kadjârs*. Les habitants passent pour être d'origine tartare, comme que tous ceux du Nord de la Perse; mais parmi eux il y a un grand nombre d'Arméniens.

Le territoire de *Khoi*, que nous pouvions juger à cette époque de l'année, est un des plus fertiles du royaume, aussi est-ce un de ceux qui fournit au trésor du *Châh* un des plus fort tributs.

1841–42

De Vecchi Felice, *Giornale di carovana. Brano di un viaggio nell'Armenia, Persia, Arabia e Indostan, fatto negli anni 1841–42*, Luni Editrice, Milano, 2016 (two volumes), pp. 27–28, 33, English translation from: Francesco Surdich, Azerbaijan in the account of the trip to the East by Felice De Vecchi (1841–1842), in: Geopolitical, Social Security and Freedom Journal, Vol. 1, issue 2, 2018, pp. 40–41.

... the city of *Khoi*, situated on the edge of a plain cheered up by pleasant countryside, by an affair of pleasant and green pastures where between the blonde of the ears, the plants of cotton, flax, castor-oil, tobacco, the intricate indulgence of the large watermelon plantations sprang up like many pawns of a variegated chessboard, the thick houses, which gave the appearance of simple and industrious life ... *[describes a plough]* of a form that elsewhere I have never seen ... An enchanted oasis, where birds hovered with feathers vaguely mottled by the no longer intense warble and insects with very bright colours and peregrines ... for which the entomologist following them "went to the broth".

...

[In Khoi they have the opportunity to visit the caravanserai] There is nothing that reminds us of our clean shops (...): but weeds close to the walls or through the streets; so you feel like walking in a labyrinth; they were creaking under a mountain

of manufactures stacked as God wants; all trivial stuff and strange in form and use; strange at least for us, people of Europe.

1841–42

Gaetano Osculati, ***Note d'un viaggio nella Persia e nelle Indie Orientali negli anni 1841, 1842*****, Monza, Tipografia Di Luca Corbetta, 1844**, pp. 11–12 (Fig. 13).

Koi è città di forma quadrata, ha due miglia di circonferenza, ed è difesa da profonda fossa che gira attorno a doppi baluardi, costrutti in mattoni cotti al sole, ed intonacati esternamente di creta e paglia sminuzzata, come lo sono in Persia tutte le case. Un lungo bazar coperto occupa presso che tutto l'interno della città, ed è di forma

Fig. 13 Gaetano Osculati (1808–1894) (from G.Osculati, *Esplorazione delle regioni equatoriali lungo il Napo ed il fiume delle Amazzoni,* 1854—Wikimedia Commons)

elegante. Vi sono inoltre molte fabbriche di tessuti in lana. Nella popolazione contansi molti armeni d'origine tartara. Sonovi venti moschee e molti bagni. Per l'abito europeo che indossavamo una turba di gente ci seguiva con molesta curiosità, sì che anche col soccorso della frusta ci era penoso l'aprirci la strada. Non appena l'indimani uscimmo, e trasportammo i nostri effetti dalla cella dove avevamo passata la notte essa venne subito inondata d'acqua per purificarla, come essi dicevano, dall'immondo contatto dei cani infedeli. La strada era fiancheggiata da bei pioppi, e molti eleganti *Kioski* ne abbellivano i dintorni; attraversato il fiume *Otour*, giunsimo la sera a *Sayd-Agy* accompagnati nella giornata da gagliardo vento e pioggia.

[A/T.: Khoy is a square shaped town, has a two miles perimeter and is protected with a deep ditch surrounding double bulwarks built with sunburned bricks, plastered with clay and chopped straw, as all buildings are in Persia. A long roofed bazaar occupies nearly all the inside town, and has an elegant shape. Moreover, there are many wool textile manufacturers. Among the inhabitants are many Armenians of Tatar origin. There are twenty mosques and many baths. Because of our European dresses, a crowd of curious people kept following us annoyingly, to an extent that we hardly could make our way even whipping. The next day, as soon as we left the room where we stayed overnight and moved our belongings, it was soon flooded and purified, as they said, from the filthy contact with the unfaithful dogs. The road was flanked with beautiful poplars, and many elegant Kiosks made the environs even better. After crossing river Otour we reached Sayd-Agy at evening, along with strong wind and rain all day.]

1841 e 1853.

Moritz Wagner, *Travels in Persia, Georgia and Koordistan; with Sketches of the Cossacks and the Caucasus*, London, Hurst and Balchett, 1856, in Three volumes, vol. III, pp. 75–78 and 83–84.

The high table land of *Choi* deserves its name. It consists of a spacious plain, sixty miles in length, and almost equalling that of *Erzeroum* in dimensions. Its system of natural irrigation, and its natural fertility, are inferior to those of the Armenian highlands, but the ingenuity and industry of the inhabitants, who are almost a match for Englishmen in agriculture and horticulture, amply compensate for other deficencies. Dams and other artificial channels divert the waters of the streams in all directions, especially near populous villages, and the natural aridity of the plain, is relieved by a series of verdant oases. The gardens and fields yield abundant crops, wherever human industry succeeds in irrigating the land. But the plain would become a desert without this distribution of water.

The table-land of *Choi* presents a very inviting appearance to the eye of the European traveller, just arriving from the barren highlands of Armenia. The eye is relieved by the numerous gardens and shrubberies surrounding the different villages like a green curtain, and decorating the little houses with natural verandahs. The sight of this charming foliage gently agitated by a cool breeze after a recent storm, combined with the exhilarating influence of the air, filled our whole party with a happy genial mood, in which the very horses seemed to share, and we all moved on to the town with light and elastic steps. The verdant meadows and artificial rivulets were fringed with silver poplars of large dimensions, which formed also a hedge round the numerous gardens, and in some places thick clumps, which reminded me of the groves in Schleswig–Holstein. The eye was gratified also, by the beauty and diversity of the fruit trees, including apple, pear, apricot, cherry, walnut, and expecially mulberry trees. The latter were loaded with large fruit exceeding in flavour, any others that I have ever tasted. Mulberries are the most abundant fruit produced and consumed in *Aserbeidschan.* The black species are smaller, and not so sweet, but larger and more juicy than the white kind, and they are regarded in the larger cities, as the most savoury fruit next to grapes and apricots. No idea of the beauty of the Persian mulberry can be formed in Europe, where the same tree only yields small and indifferent fruit. The contrast is quite as great as that presented by the sour, shrivelled oranges of Hyères, and the golden globes of the groves at Blidah in Algeria. Most European vegetables are raised in the gardens surrounding *Choi,* which is likewise encompassed by a complete zone of flowerbeds, producing a fine crop of roses, for which all Persians have as strong a partiality, as that which they feel for iced condiments. Wheat and barley are the predominant species of cereals in this district.

The town of *Choi* is almost buried under the growth of the surrounding foliage, and you are scarcely aware of its existence, till you are within the place, which is encompassed by an indifferent mud wall, an effectual protection against *Koordish* forays, but absolutely useless against European field-pieces. The town is entered by two gates and the architecture of the houses presents much analogy to that prevalent in Russian Armenia. The dwellings consist of low houses with mud walls, like those of *Erivan,* and the streets are unpaved. The two most conspicuous objects of the town are the spacious bazaar, and the Great Caravanserai, consisting of two large square courts, one of which is surrounded with fruit trees and adorned with a fountain. The shops and guest's chambers surrounding this court are spacious and cleanly kept. They answer as quarters to the poor tradesmen and lower classes, whilst' the caravan cattle fill the court. The bazaar, which is of very unusual size in proportion to the town, is situated between the two caravanserai courts, and is the whole day full of life and bustle …*Choi* cannot be numbered among the largest or most splendid cities of the East, but its propinquity to the Russian and Turkish border gives it considerable commercial importance, without mentioning the flourishing smuggling trade carried on with Russian Trans-Caucasia, since the introduction of the new Russian customs. I admit that the display of merchandize in the bazaar of *Choi* is not nearly so splendid, or varied, as that at *Tabris*, but very considerable, in proportion to the size of the town, and the population of the vicinity. There was a respectable selection of Persian silks, shawls carpets, peltries, and leather, besides the most usual European articles. The tradesmen in the bazaar, who seldom move from their shops in Turkey, were most pressing here, begging me to inspect their goods. Many of the merchants stopping at the caravanserai, brought me sundry articles, which they thought calculated to interest me, including Roman and Persian coins, modern coins, with effigies of saints, crosses of artificial gold, cornelian, and agate, on which talismans were engraven.

1847–48

Xavier Hommaire de Hell, *Voyage en Turquie et en Perse*, Paris, P.Bertrand, 1855, Tome premier, deuxième partie, pp. 522–529.

En sortant du village *[Bélatchik]*, nous nous engageons dans une gorge, par laquelle s'écoule la rivière qui passe à *Khoï*, et dont le nom est *Khotour*. Les montagnes arides qui bordent cette gorge ne présentent le plus souvent que de hautes escarpes; mais plus loin la gorge se transforme en une large plaine où se trouve *Khotour*, village kurde. On ne peut m'indiquer d'une manière positive les limites exactes des deux pays. *Khotour*, où apparaissent les bonnets pointus des Persans, est grand, relativement aux villages précédents. En le quittant, la vallée de la rivière se rétrécit, et présente de temps à autre des petites plaines où sont installés des campements kurdes, dont les tentes, plus soignées qu'en Turquie, annoncent plus d'aisance et de bien-être ... La vallée s'élargit insensiblement, et le chemin devient très facile. La culture commence à reparaître. Après un nouveau rétrécissement de la vallée, on sort tout à fait des montagnes, ayant devant soi une immense plaine traversée par la *Khotour*. Çà et là des masses d'arbres annoncent des villages assez rapprochés les uns des autres. Le fond de la plaine est bordé par une chaîne élevée, mais où l'on n'aperçoit aucune trace de neige. ... Toute cette plaine de la *Khotour* est très bien cultivée. A leur sortie de la montagne, les eaux de la rivière, recueillies dans de nombreaux canaux étagés les uns au dessus des autres, vont fertiliser tout le sol de la plaine. Rien de plus intéressant que de voir ce système d'irrigation ... Avant d'entrer à *Khoï*, nous traversons plusieurs villages que l'on peut presque considérer comme les faubourgs de la ville. Tous les chemins, à cette heure du soir, sont couverts de champagnards venant de *Khoï* et se rendant chez eux, ce qui répand une agréable animation sur la contrée. Sauf l'étrangeté des costumes, on pourrait se croire aux abords d'une ville européenne, tant les conversations sont animées, les enfants gais et turbulents. Ce qui me frappe surtout, est la propreté des maisons, de même que leur nouveau mode de construction. Quoique bâties de terre, la plupart possèdent sur leur façade de vastes vitraux de couleur, dont les châssis, d'une grande légèreté, sont très élégants de forme et de dessin. Nous nous installons dans un beau *konak*, au fond d'un jardin, dont la pièce principale, ornée d'une cheminée, s'ouvre par un grand vitrage de couleur sur le jardin. La façade opposée présente un enfoncement rectangulaire dont le fond n'est également qu'un vitrage donnant sur un jardin intérieur.

...

Khoï présente une enceinte de fortification à l'européenne, construites en terre; mais, vu son délabrement actuel, elle ne saurait offrir aucune résistance sérieuse. L'intérieur de la ville est un assemblage irrégulier de maisons et de rues plus ou moins tortueuses, s'éloignant complètement de la physionomie de nos cités. Les maisons ont rarement une façade sur la rue, étant précédées d'une cour, qui les isole de la voie publique. Beaucoup sont en ruines, et cela tient à un grand tremblement de terre qui eut lieu il y a cinq ans, et qui ne laissa pas une seule maison intacte. Un millier de personnes périrent sous les décombres de leurs demeures. Aujourd'hui on est en train de reconstruire la ville, mais avec peu d'activité. On ne voit aucun important édifice à *Khoï*: dans ce moment on bâtit une grande mosquée de briques, de la même forme, à peu de chose près, que celle de la vieille mosquée de *Diarbékir*. Les travaux sont déjà élevés jusqu'à la naissance de la couple; la construction, parfaitement conçue et soignée, fait honneur aux architectes persans.

Khoï compte 2.800 maisons contenues dans l'intérieur de l'enceinte, et 3.700 dans les faubourgs situés à l'entrée de la ville, du côté de l'ouest. Toute la population est persane, sauf une cinquantaine de familles arméniennes établies en dehors de la ville. Près de 4.000 hommes, tant infanterie que cavalerie, et 400 canonniers, forment la garnison ordinaire de *Khoï*; mais aujourd'hui les casernes sont presque vides, par suite de l'expédition du *Khoraçan*, et des troubles qui ont éclaté à *Kirmanchah*. Les portes de la ville ne sont même pas gardées par une sentinelle; je n'ai aperçu qu'un seul soldat au bazar, dans une tenue peu respectable.

Ces bazars, la plupart bâtis de briques, et voûtés, sont très vastes, mais on voit qu'ils ont également souffert du tremblement de terre. L'animation qui y règne prouve qu'on est au milieu d'un peuple essentiellement industrieux, ne reculant devant aucun métier. Je remarque, entre autres, les chaudronniers, dignes d'admiration par leur habilité à couvrir de ciselures les vases qu'ils fabriquent. L'industrie spéciale de *Khoï* consiste dans la fabrication des chaussettes de laine … Les environs de la ville produisent du sel gemme, qui est exporté en grande quantité dans toutes les contrées voisines.

Au milieu du jour, je vais faire ma visite au gouverneur, qui me reçoit dans une pièce dont la disposition est tout semblable à celle de notre *konak*; seulement le vitrage est ici remplacé par un cheminée à côté de laquelle se tient Son Excellence, couchée sur un petit matelas. Les murs sont d'un gris rouge, ornés de baguettes blanches de plâtre. Tous les coins des vitraux formant les croisées offrent de charmants dessins de fleurs, en verres de couleur, faits avec une rare perfection.

……

Au coucher du soleil, le tambour annonce la fermeture du bazar, et deux heures plus tard, le même roulement se fait entendre trois fois, à de courts intervalles, pour annoncer que tout le monde doit rentrer.

……

Malgré la pluie fine qui ne cesse de tomber depuis l'aube du jour, nous montons à cheval et gagnons une belle route bordée de fossés et coupant en ligne droite une plaine admirablement cultivée, qui aboutit à la *Khotour*, sur laquelle est un pont de sept arches, dont les piles, jusqu'à une certaine hauteur, sont de pierres de taille et le reste de briques; entre les arches sont des ouvertures placées directement au-dessus des piles; quant à la forme des arches, elle présente une ogive de figure triangulaire, généralement adoptée dans toutes les constructions persanes.

1849–52

Mary Leonora Woulfe Sheil, *Glimpses of Life and Manners in Persia, by Lady Sheil. with notes on Russia, Koords, Toorkomans, Nestorians, Khiva, and Persia*, London, John Murray, 1856, pp. 287–288.

Khoee was the first town of importance after *Tabreez*. On our way to it we passed close to the Lake of *Ooroomeya*, otherwise called *Shahee*. On ascending the high pass leading to *Khoee* the lake lay at our feet. We had a fine view over the expanse of that silent water, the Dead Sea of Persia, which contains no living thing. The islands of the lake, the mountains of *Maragha*, those of the *Mikree Koords* at the south of the lake near the *Jaghata'i*, part of the district of *Ooroomeya* with the *Koordistan* mountains behind it, revealed themselves with the rising sun. It was a fine panorama. We only stayed to rest and breakfast at *Khoee*, that important and defenceless city, open and ready to be seized upon by any invader.

1855–58

Arthur De Gobineau, *Trois Ans en Asie (de 1855 à 1858)*, Paris, Librairie de L.Hachette et C., 1859, p. 510 (Fig. 14).

Khoy est une ville charmante; sa belle porte, construite en marbre noir et blanc, dans le goût des édifices du Caire, et comme je n'en ai vu nulle part ailleurs en Perse, donne immédiatement entrée dans le bazar. *Khoy* est riche, marchande et célèbre par

Fig. 14 Joseph Arthur de Gobineau (1816–1882) (Wikimedia Commons)

ses fabriques de chaussons de laine tricotés ou tissés avec beaucoup de goût. On voit autour de la ville de très-belles plantations d'arbres et, ce qui est trop rare pour ne pas être dit, le gouverneur actuel, prince de la famille royale, est le créateur de ces plantations. C'est un administrateur intelligent et un soldat d'une grande bravoure, comme il a souvent lieu de le prouver dans ses rapports avec les *Kurdes* de la frontière.

En quittant Khoy nous allâmes à *Zourabad*. Je m'aperçus que nous commencions à abandonner la Perse, car, la nuit, on voulut forcer la porte de ma chambre pour voir ce que j'emportais avec moi. Une escorte fut jugée nécessaire et nous prîmes trois Kurdes, déterminés coquins, afin de paralyser leur vaillance ... Nous ne trouvions plus désormais cet air jovial et poli auquel nous étions habitués; mais un aspect sombre, ou des figures patibulaires, ignobles dans leur déférence; par-dessus le marché, des maisons ou plutôt des trous creusés en terre, dont toute propreté était absente.

1860

John Ussher, *A journey from London to Persepolis; including Wanderings in Daghestan, Georgia, Armenia, Kurdistan, Mesopotamia and Persia,* London, Hurst and Blackett, 1865, p. 661.

"*Khoi*, which is now a small town, bears, like all other Persian cities, numerous traces of its former greatness. It is still surrounded by walls and towers built of mud, and contains a bazaar which seemed populous and busy. A large manufacture of socks, called *"tifteck"*, which are much esteemed in Persia, is carried on, and as we rode through the streets, the shops seemed filled with both native and foreign goods. On most of the latter that were exposed for sale Russian letters were visible, showing them to have been sent from that country. A considerable number of small villages were dotted over the plain in the vicinity. The town was surrounded by gardens and orchards, and, on the whole, had a more prosperous and thriving appearance than any place we had seen for a long time".

1870

Edwin Lord Weeks, *Persia and India*, Harper and Brothers Publishers, 1896, pp. 37–39.

Khoi.—A large walled city, with moat and drawbridge, sloping walls and battlements of rose-tinted mud. We ride down a shady but dusty avenue crowded with citizens who are looking on at an Armenian religious procession, with priests and banners. We halt for lunch at a caravansary, just outside the city gate, and, climbing up to the broad wooden balcony just over the entrance, we find several of our Persian fellow-travellers, who are already installed. After a long resistance we have at length capitulated to the Persian watermelon, and begin to believe that there is no harm in him. In a country where the brackish water is impossible to drink, the filtered and sweetened juice in the heart of a melon seems to be nature's own substitute; but our dragoman, who had become sadly intemperate in the matter of melons and sliced cucumbres, now began to show the disastrous results of his indulgence. His face, which was round and ruddy at the outset, had become elongated and haggard, and his flabby cheeks hung in wrinkled folds. In vain we physicked him and dosed him with cholera mixture and quinine; we invariably caught him the next day after an indisposition surreptitiously devouring forbidden fruit. While we are eating on the balcony, during the space of an hour or two, thirteen bodies are deposited in the cemetery across the way. Plainly there must be something wrong about the sanitary condition of this place. For a short distance beyond *Khoi* we follow a well-made carriage road shaded by great trees, which ends suddenly at the bank of a river, and we then strike across the hills again.

1879–82

Sir C. W. Wilson (ed.), *Handbook for travellers in Asia Minor, Transcaucasia, Persia,* etc., London, J. Murray, 1895, p. 226.

From *Bayezid* a road, skirting the marshes, runs to the Persian frontier post *Surp-khân*, whence there is a track, via *Moku*, to the *Tabriz* road. It is better, however, to rejoin the caravan road at *Kizil-diza* (3 h.), the Turkish frontier post, which is 6 1/2 h. from *Diadin* (p. 228). 2 h. after leaving *Kizil-diza*, we pass *Teperis K*. where the road to *Van* ... turns off to the right, and 1 mile further we cross the *Kazli Gedik Bel*, a low col. where the Turco-Persian boundary is marked by a ruined tower. Descending to *Ovajik*, the Persian frontier post, we afterwards pass through *Kilisse Kend* (Armenian), *Arab-diza* (Armenian), in the centre of a plain, and *Ak-diza*, to the large village of *Kara-aïneh* (7 and half hrs.). Thence the road runs through *Tegnesh*, *Kirk- kendi*, and *Zeiva* to *Zorava* (7 h.). We now cross some low hills to *Pire* (7 h.), in the *Kara-dasht Dere*, whence there is an easy rood to *Kotur* (p. 235).

Khoi (4 h.), an important commercial town, situated in a rich well-cultivated valley about 15 miles by 10. The town is surrounded by a double *enceinte* of mud fortifications, and is entered by two gates. Canals from the *Khoi Chai*, bordered by willows, run through the principal streets. The *bazârs*, built of brick, are amongst the finest in Persia. The plain, which is shut in by mountains, produces large quantities of corn, cotton, and rice, and it is studded with Armenian villages. The population of the town, 20,000, is, with the exception of about 100 Armenian families, Persian. From *Khoi* there are roads to *Julfa, Van, Urmia*. There are two roads, both running through rich, well-cultivated districts, from *Khoi* to *Tabriz*. One via *Marand*; the other by *Tasuj* (9 *farsang*), and *Alishah* (8 *fars.*), to *Tabriz* (6 h.).

1880–81

Colonel Charles E. Stewart (edited from his diaries by Basil Stewart), Through Persia in disguise, With reminiscences of the Indian mutiny, London, George Routledge and Sons, 1911, pp. 188–191.

At five miles short of *Khoi* we halted at a caravanserai, situated in a very flourishing village, to give the horses a feed, and we did not reach *Khoi*, which proved to be a large town, until two in the afternoon. *Abdullah*, the young Persian who had invited me to his house and who had accompanied me from *Tassili Khoi*, took me in. I found his house a small but very pretty one, standing in a nice garden, with cherry and pear-trees in full blossom. Flowers were coming up, and a grapevine was trailing over the trellis-work. The front of the house was towards the garden, the back showing a blank wall without any windows. Downstairs there were, I believe, four rooms, two for the men, and two others, separated by a passage, for the women. Upstairs there was only one room, the whole front of which was open, and looked on the garden. The room was very prettily furnished, and the front was closed with a lattice-work … They were a middle-class Persians family, fairly well off, but not rich, as they kept no servant, but did everything for themselves … About five o'clock Abdullah and I went to the public bath to have a Turkish bath. It was very enlightened of the people here to let a Christian enter their bath, a favour not usually granted in Persia. *Khoi*, however, being near Turkey, they follow the Turkish customs, which allow Christians to go to the public baths.

……

On the 15th May, about 8 a.m., I went to call upon the Prince-Governor of *Khoi*, and found it was a long distance to the palace, *Khoi* being a large town. A small stream of running water flows down the centre of the streets, and is bordered by poplars and willows. After riding about a mile we came to a large fort, built of unburnt brick, and partially constructed according to modern ideas, which would render it very difficult for anyone to take without modern artillery. The fort is also surrounded by a wide and deep ditch. Part of the town is contained within the fort, and the Prince's palace is situated in a central position in it. On turning out of the street, one first enter a large courtyard where a few old-fashioned brass guns, about twelve pounds, were parked. They were field guns, and would no doubt be used as field artillery, but I do not think they had any horses or other means of moving them.

……

Round the square were quarters for soldiers; a long passage led out of the square, and we went down this to a door, where a guard of infantry was stationed, armed with breach-loaders of Schneider pattern, but their rifles were fearfully dirty and neglected … I left my horse near the guard, and walked through the guard-room into a courtyard where was an open room with an awning projecting from the front. I was admitted into this room, and was met at the door by a very nice-looking Persian; the *Kajar* Princes are, as a rule, very good-looking men.

1887–1889

Edward Granville Browne, *A Year amongst the Persians. Impressions as to Life, Character, and Thought of the People of Persia, received during twelve months residence in that Country in the years 1887–1889*, London, Adam and Charles Black, 1893, pp. 50–56.

Next day (26th October) we found to our delight that our escort was reduced to two, who still continued their attempts to scare us with alarms of robbers. Whether the road was indeed dangerous I do not know, but it was certainly amazingly bad. About mid-day, on emerging from a very fine gorge, we saw at our feet a wide and cultivated plain, surrounded almost entirely by mountains, except to the right, in the direction of *Urumiyyé*. In this plain lay the beautiful little city of *Khuy*, and, somewhat nearer to us, the suburb of *Piré-both* surrounded by a mass of gardens. The latter we reached in about an hour, and here we rested for a while. Thence onwards to the very walls of *Khuy* (appropriately styled *"Daru'ṣ-ṣafa"* ["the Abode of Delight"]) our way lay through pleasant gardens of poplars, willows, and fruit-trees, and fields planted with cotton. At 3.30 p.m. we entered the town, and put up at a clean and well-constructed caravansary.

While the baggage was being unloaded, I perceived that we were undergoing an attentive scrutiny on the part of a magnificent-looking dervish, who wore on his head a green turban, of which one end depended over his shoulder, and carried in his hand a shining battle-axe. Presently he began to address enquiries to *'Ali,* and, on learning from him that I spoke Persian, approached me and entered into conversation.

He proved to be a native of *Kirmán, Mir Jalálu'd - Din* by name; and his extraordinary fertility of imagination, which often carried him far beyond the bounds, not only of the probable, but of the possible, rendered him a very amusing companion, if not a very reliable informant. He at once constituted himself our guide, philosopher, and friend, and hardly quitted us during the three days which we spent at *Khuy*, declaring that he perceived us to be excellent fellows, worthy of his society and conversation. He assured us that he had travelled much, and had thrice visited London, once in company with the *Shah*; that he had instructed members of the Russian royal family in Persian; and that besides this, his native tongue, he was conversant with no less than ten languages, including Kurdish, Russian, and the dialect of *Sistan* on the eastern frontier of Persia. Having given us these details about himself, he began to question us as to our destination, and, on learning that we were bound for *Tabriz*, told us that we must on no account omit to visit the towns of *Salmás*, *Khusraváhád*, and *Dilmaghán*, more especially the latter, in which, as he declared, there were no less than a thousand English residents, who, through converse with dervishes and Sufis, had become enlightened and philosophical.

While we were engaged in conversation, a man entered the room to enquire our names and whence we came, the object for which this information was sought being, as *Mir Jalálu 'd-Din* informed us with perfect gravity, that it might be inserted in the newspapers of *Tabriz*! His imagination being now temporarily exhausted, our worthy friend bade us good-night; and, promising to be with us betimes in the morning, and to show us something of the town, left us to repose.

Our first business on awaking in the morning was to make enquiries as to the possibility of obtaining a bath in the adjacent *hammám*, and this indulgence was without difficulty accorded to us. On our return we found our friend the dervish awaiting our arrival.

...

About 11 a.m. we accompanied him for a stroll through the town. He first took us to a neighbouring caravansary and introduced us to a Syrian Christian of *Urumiyye*, named Simon Abraham, who practised the trade of a photographer, and spoke English (which he had learned from the missionaries settled at that place) very well. He, in his turn, introduced us to another Syrian Christian, called Dr. Samuel, who kept a dispensary at the opposite side of the caravansary, and who likewise possessed a good knowledge of English. Both received us very cordially, and did much to render pleasant our sojourn at *Khuy*.

In the afternoon we were taken by the indefatigable *Mir Jalálu 'd-Din* to visit a *tekhé,* or retreat for dervishes, situated near the walls of the town. The dervishes, who were a most heterogeneous crew, including, besides Persians, Kurds and negroes, received us very hospitably, and gave us tea.

……

We next visited the dispensary of Dr. Samuel, whither *H–* had already preceded us. Here for the first time I was able to appreciate the difficulties incidental to the practice of medicine amongst a people whose curiosity prompts them to hover round the physician long after their own cases have been dealt with, and who are only too eager to throw out hints on diagnosis and treatment whenever they get the opportunity.

……

It was with much regret that on the following morning (29th October) we prepared to quit *Khuy.*

……

Quitting the town by a gate opposite to that by which we had entered it, we passed through a long avenue of poplars, and shortly afterwards reached a point where the road bifurcated, one branch running southwards in the direction of *Urumiyye*, and the other, which we pursued, eastwards towards the hills which we must cross to reach *Tabriz*. Near the summit of one of these hills was a small *imámzádé*, or shrine, which, as Farach informed us, was reputed most efficacious in curing persons afflicted with hydrophobia, or bitten by a serpent. After a short stage of four hours we reached a little village called *Seyyid Táju 'd-Din*, where we halted for the night.

Next day we continued to ascend for about two hours, until we reached the top of the pass. From this we had a magnificent view of the great salt lake of *Urumiyye*, glittering in the sun, and studded with numerous rocky islands, which, as an effect of the mirage, appeared deeply indented at the base. Descending by the dry bed of a river which did duty for a road, we soon entered the plain which skirts the lake on this its northern side.

……

Before 2 *p.m.*. we reached our halting-place, *Tasuch*, a large but uninteresting village distant about a mile from the shore of the lake.

1888–90

Paul Müller-Simonis, *Du Caucase au Golfe Persique*, Université Catholique d'Amérique, 1892, pp. 114–115, 117.

L'absence presque totale de minarets et le peu d'élévation des maisons de *Khoï*, fait qu'on n'aperçoit presque point la ville avant d'y arriver. On ne voit qu'une longue muraille, et par derrière un rideau d'arbres. Cette muraille, élevée par Mirza Abbas il y a une cinquantaine d'années, produit de loin un effet imposant; une contre-escarpe, des créneaux vigoureusement dessinés, lui donnent un air respectable. De loin c'est quelque chose, mais de près ce n'est rien; le tout est en pisé et, faute d'entretien, tombe en ruines.

Le *caravanseraï* est assez primitivement installé: des portes ridiculement basses où l'on se cogne régulièrement la tête; des fenêtres dont les châssis seuls subsistent rendent l'endroit des moins confortables. La cour est remplie de chevaux; tout est bondé de voyageurs, car demain part la grande caravane des pèlerins *chiites* qui vont à *Kerbéla* vénérer le tombeau de *Hussein.*

……

Khoï, située à environ 1136 mètres d'altitude, passe pour une des plus belles villes de Perse; ses rues, assez larges et régulières, sont arrosées par des canaux et plantées d'arbres. Les mosquées y sont rares, comme d'ailleurs toutes les constructions monumentales, car les tremblements de terre y causent parfois de grands dégâts.

Lorsque la Perse et la Turquie étaient deux États puissants, cette ville était l'un des entrepôts les plus considérables du commerce entre les deux peuples. Aujourd'hui, comme place frontière, elle garde encore une certaine importance; elle est tête de ligne du chemin d'*Erzéroum*, par *Bayazid*;[8] de celui de *Van* par *Kotour*; et de celui d'*Erivan* par *Nakhitchévan.* Le bazar est très actif; l'industrie dans laquelle excellent particulièrement les habitants de *Khoï*, est la fabrication des ustensiles de cuivre qu'ils savent faire fort variés et du meilleur goût; la matière en est, paraît-il, excellente.

La population de Khoï est de 20 à 30,000 âmes; la majorité des habitants se dit d'origine tartare, et l'élément musulman y passe pour fanatique.

……

Le chemin au sortir de *Khoï* est, pendant trois quarts d'heure environ, une large avenue plantée d'arbres; selon la règle, rien n'est entretenu et les ponceaux, qui avaient été fort soigneusement construits, commencent à crouler. L'avenue aboutit au *Kotour-Tchaï*, que franchit un grand pont. Naturellement il menace ruine; personne ne s'en sert et l'on passe la rivière à gué!

1890

Albert Develay, *D'Arménie en Perse à travers le Kurdistan*, in Journal des Voyages et des Aventures de Terre et de Mer, n. 989, 21 juin 1898, p.54.

… nous arrivons à *Khotour*, dont la vallée, encaissée et souvent resserrée entre des parois de rochers, est une des voies de communication entre la Perse et la Turquie. Dans cette région, les Kurdes bataillent entre eux de chaque côté de la frontière. Il convient de dire que ces Kurdes joignent, à la bravoure, un côté chevaleresque séduisant et qu'ils pratiquent largement l'hospitalité, sans cupidité éhontée; il n'en est pas de même des Kurdes que nous rencontrerons, plus au Sud, entre *Souleïmanieh* et *Sinha.*

Le *hakim* de *Khotour* nous fait le meilleur accueil, et c'est sous l'escorte de deux gendarmes persans que nous entrons à *Khoï*. La neige a disparu bien avant que nous n'apercevions la ville, située à 1.2000 mètres d'altitude seulement dans une plaine basse et marécageuse. *Khoï*, dont la population doit dépasser 20.000 habitants, a l'aspect de toutes les villes persanes: une méchante ébauche en plâtre qui ne connaîtra jamais ni le bronze ni le marbre.

1889–90

Samuel G. Wilson, *Persia Western Mission*, Philadelphia, Presbyterian Board of Publication, 1896, pp. 126–127, 159–160, 300–301, 309, 313.

… The other festival celebrates the assumption of Mary, bodily to heaven. It is preceded by a week of fasting. The celebration was described in an essay by a scholar in our Tabriz school in the following manner:

"There is a church of the Mother of God in a village called *Sayidabad*, near *Khoi*. On September 25th at an hour after noon, the pilgrims follow the road toward the shrine that they may present their thanks for their preservation in the past year or deliverance for any special trouble that may have befallen them.

If the one who has vowed is rich, he takes with him a ship to sacrifice, but if poor he must at least take twenty or thirty candles to burn before the picture of Mary the Mother of God. The Turks seeing the Armenians gathering from the villages, load their donkeys with watermelons and take them to sell to the pilgrims who will remain overnight in the churchyard. For those who are paying a vow must remain all night and engage in prayer. They buy and eat of the watermelons and so pass the night.

When morning arrives the *Dĕr-Dĕr* or priest of the village comes to the church to sacrifice the offering. The animal must be killed with his head toward the East, to which they look in prayer. The skin and one quarter must be given to the priest, but the head, feet, liver and tallow to the sexton. Then they rub the sacrifice with salt, cut in pieces and put it in a kettle to boil while the people enter the church for prayers. There is a confuse multitude. The brides and girls are praying, the men are burning candles, some are praying earnestly, but the women are talking so incessantly that they neither hear anything nor allow anybody else to hear.

……

After eating they enter the church, kneel several times before the pictures of the Mother of God, burn a candle and in the cool of the evening return home.

……

Sunday following the feast of the Assumption is the day for the blessing of the fruits. Many Armenians will not eat grapes until that day. Then a quantity is brought to the church and prayer is made for a blessing not only on the grapes, but on all the fruits of the year. These are then distributed, and they go and enjoy their vineyards. This feast accomplishes one good thing in that it keeps some Armenians from eating unripe grapes as the Mussulmans commonly do. But in some seasons and places, the grapes have ripened several weeks before the festival.

……

Some interesting experiences happened to us at *Khoi*. In the suburbs we have a school and an evangelist. We appointed preaching services every night in the week, and a good attendance aroused the priest to oppose. He preached against us in the church, saying, "By going to the Protestants you will become contaminated in two months and become lepers." (They nickname us "*Prots*" and the word for leper is "*Barod*"). "They are wolves in sheep's clothing. They will steal your hearts. I will have to shut up the church." … The priest in the other ward cursed us strongly and threatened the people that he would not marry, nor bury them, baptize their children, nor give them the sacrament if they came to us, and his daughter came and turned back the women from our meeting. In a village on the *Khoi* plain, opposition was severe because an Armenian inclined to Protestantism, had "eaten" his fast … In the village, one place of meeting was the stables. The *sakki* or platform of the stable is a great winter place of congregating. We took an earthen castor-oil lamp by which to read. The heat from the oxen, donkeys and buffaloes supplied the place of a stove ... The air soon became steamy and stifling, so that one was compelled to break through the conventionalities, and take off his coat. The people, from forty to one hundred, crowded around so that they had not sitting room. Generally the service ended in a free discussion. Thus the opportunity for seed-sowing is abundant, but success arouses opposition. A man may be a drunkard, impure or profane, and no one has a word to say: but if he becomes evangelical, he has denied the faith and is subject to ostracism. A merchant told me, "I believe Protestantism is the best form of Christianity, but if my brother becomes a Protestant, I will beat him". Under such difficulties, work grows slowly, but the people are surely becoming enlightened.

......

Cholera began in *Tabriz* before July 29th. The firs known cases were among the Mohammedans ... Meanwhile the disease was spreading rapidly. The polluted water supply, public baths, ceremonial ablutions and other causes promoted the disease. Ignorance of sanity laws, and disbelief in infection or the much spoken of microbes, assisted its spread. It was not many days until people were dying by hundreds in every quarter.

At the first alarm, multitudes escaped the infected city ... The Armenians fled to *Muzhumbar* and *Karadagh. Mussulmans* scattered themselves everywhere, many returning to their native villages.

......

Of the Armenians who remained in the city, mostly the poorer class, one in eight died ... Many sacrifices of sheep were made and distributed. In a village thirty sheep were sacrificed at the festival of Mary "the Mother of God" ... Processions with certain rituals and prayers were made around the churches.

......

In Salmas the cholera was light ... In *Urmia*, Dr. Cochran, in anticipation of its coming, had prepared a pamphlet, setting forth the nature and cause of cholera and giving advice as to its prevention and treatment. It was of great value and its suggestions were acted on by the nobles and people. The cholera came in its lightest form.

1892

Samuel G.Wilson, *Persian life and customs with scenes and incidents of residence and travel in the land of the lion and the sun*, Fleming H. Revell Company, New York, Chicago, Toronto, 1895, pp. 82–85.

The journey around Lake *Urumia* is one which few travellers make, yet it leads through some of the pleasantest regions of Persia. In going northward one sets out over the great caravan route from Tabriz to *Trebizond*. The telegraph and post routes go to *Marend*, thence over the mountains to *Khoi*. The caravan route lies over the *Myan* plain – the difficult part of the journey. In spring it is muddy and slippery; in winter it is flooded over and frozen, with a deep covering of snow. The ice breaks under the horses' hoofs. The sky is sometimes cloudy, and the cold penetrates even through an overcoat and a Persian *aba*. Often, too, the river is impassable. In summer the heat and glare of the sun are reflected from the glistening sand, and make it extremely uncomfortable for the traveller. For hours of the journey, moreover, only brackish water can be obtained. No place shows better than this the contrast between summer and winter in *Azerbijan*.

The first station is *Ali Shah* … On the third day we crossed the Pass of *Sayid-Taj-i-Din*. It is steep and rocky, and the road bears the marks of so energetic a hand that one is led to suspect the work of the Roman. For a long distance a path has been cut in the solid rock, and there the horse must walk with wary foot. From the other side the *Khoi* plain opened up before us. It is well watered by the *Kutur* and *Aland* rivers, both of which rise on the Turkish border and, uniting, flow into the *Aras*. The approach to the city is through a wide avenue, two miles in length, bordered by watercourses and double rows of willows, fenceless gardens extending on both sides. At its farther end the walls and battlements of the city frown upon us. A double wall, with towers (having picket holes for cannon and musketry), eleven feet thick, built of sun dried brick, and encircled by a moat twenty feet deep, surrounds the city. Five large and ponderous gates of wood, sheeted with iron, lead out to the chief highways. Across the moat are slight bridges, easily removed on the approach of an enemy. It is a frontier fortress but the time is long past when such breastworks could withstand the cannon of either Russia or Turkey. Against Kurds, however, the fortress would still afford protection. Many of the streets of the city are of unusual width for the Orient, with rows of trees and running water in the centre. Low garden walls give glimpses of the interior. The bazaars are poor, though caravansaries are numerous. There are several noted mosques and an armory. Considerable copperware is manufactured in the town. The population of twenty thousand within the walls is entirely Mussulman.

……

Mussulman superstition in *Khoi* is in no way more strikingly manifested than in the reverence shown to the mosque over the tomb of the *Hadji Mir Yacob*. This man, a descendant of the Prophet, was a meek villager, to whom doors were supposed to open of their own accord, and by whose blessing bandages received power to dispel fever. When he died, not in the olden times of legend, but just forty years ago, they erected a mosque on his grave, which has become a shrine and a refuge. I was permitted to look into the interior. The mausoleum was decorated with shawls, embroideries, mirrors, chandeliers, and other offerings of the faithful. The sick are brought and laid within its precincts for the operation of the faith-cure. A hundred swords and dirks were hung around, consecrated by criminals fleeing to the shrine with vows of repentance. The governor, however, noticed that it was only discovered criminals who professed repentance, so he abolished its privilege as a sanctuary. Here, they say, the Imam *Husain* manifested himself when the Turks forbade access to *Kerbela*. A deformed and lame boy was healed, though unbelievers say there was a collusion between a surgeon and the *mollas*. Once a camel, mistaking it for a caravansary, entered the place. The faithful immediately exclaimed, "Even the dumb beast recognizes the presence of the Prophet!" Decked in gorgeous array, they paraded it through the streets, each one taking some of the hair of the animal as a blessed talisman. The camel became an attaché of the mosque.

At *Khoi* there is a curious structure called the Tower of *Shan Ismiel*, he who fought with Sultan Selim I on this historic plain in 1514. The tower is circular, and is about sixty feet high, with innumerable horns of wild goats worked into the masonry, the points projecting outward. Some say that the horns were the spoils of one day's hunting by the royal party. This story not being marvellous enough, another has been invented to the effect that a holy man from *Tabriz*, a miracle-worker, called all the gazelles together from the country round about. These were slain and consumed in a single feast, and the horns were then built into a tower as a memorial.

The Armenians of *Khoi* live in three wards outside of the walls and in the surrounding villages. They are shopkeepers, farmers, and carpenters. They are financially prosperous, but intellectually and spiritually destitute. The church is said to be six hundred years old. The windows are very high up and small, like port-holes. In one corner is a small stone basin for the immersion of infants. One of the pictures is of Constantine and Helena, with Greek inscriptions. Another Armenian church stands in a Mohammedan village. Many sick ones are brought to its shrine, and for two days in the year pilgrimages are made to it from the Armenian villages. Superstition preserves the church from destruction. One Mohammedan took a stone from it for his stable. The same day he ran the tongs into his eye. He brought back the stone and made an offering to the offended saint.

From *Khoi* northward to *Maku* is about seventy miles. The district is mountainous. Much of it is dame, that is, not artificially irrigated. Its inhabitants are of the tribe of *Kara Goyunlis*, or the "Black Sheep," who once ruled *Azerbijan*.

1906–7

Eugène Aubin, *La Perse d'aujourd'hui. Iran-Mésopotamie*, Librairie Armand Colin, Paris, 1908, pp. 51 65.

A *Navahi*, la rivière ronge le pied de la montagne et la magnifique plaine de *Khoï* apparaît.

Les conduites d'irrigations se répandent en tous sens, des lignes de saules bordent les champs; des masses de verdure recouvrent les villages; melons, courges et pastèques, mûrs en cette saison, s'alignent dans les melonnières. Au fond, la ligne des grandes montagnes du Kurdistan, d'où descend la *Kotour Tchaï*.

Khoï est une ville fort. Elle se trouvait naguère sur le chemin des envasions turques, qui, d'Erzeroum, contournaient le massif de l'Ararat; elle reste toujours sous la menace des Kurdes. Aussi gardent-elle des murs crénelés, des bastions en terre, des fossés, des remblais, des ponts-levis, de doubles portes, tout un système de défenses moins efficace que pittoresque. La ville a 14.000 habitants; 6.000 seulement tiennent dans ses murs; faute d'espace, les maisons, privées de jardins, se serrent les unes contre les autres; à peine quelques saules réussissent-ils à pousser auprès des ruisseaux. Le reste de la population occupe trois quartiers extérieurs, par de là les cimetières: *Chahanagh, Imamzadé, Robat*.

……

Le commerce de *Khoï* n'a guère de relations qu'avec Costantinople par la voie de *Trébizonde*, accidentellment avec *Tauris*; cotons, fruits et peaux gagnent la Russie par l'*Araxe*.

Les gens de *Khoï* gardent jalousement les terres voisines; à peine quelque Armeniens est il rendu proprietaire.

……

En dehors du Belge, directeur des Douanes, il n'y a d'étrangers que le gouverneur, à la fois civil et militaire, qui est envoyé de *Tauris*. Residant dans une place forte, il fait également métier de *Reïs-el-Fowdj* et commande la garnison: 2 régiments, 300 cavaliers, 200 artilleurs. Il joint à son gouvernement de *Khoï* district de *Salmas*, où l'ont presentement appellé le incursions *kurdes*.

……

[Le] plus grand propriétaire du lieu, *Hâdjit Haïdar Khân* … possède des villages, des bains, des boutiques, quatre caravansérails. Son jardin de *Daghbaghi*, sur la route de *Salmas*, est le plus beau de la plaine; les eaux y tombent en cascades, entre deux superbes rangées de platanes; sa maison est la meilleure de la ville.

……

La ville de *Khoï* se confit en dévotion. Elle entretient quarante mosquées, un peuple de *mollahs* et de *seyyeds* … Une ville aussi pieuse ne pouvait se passer d'un tombeau miraculeux. Hadji Mir Yakoub a bien voulu le lui fournir … Le tombeau est couvert de draperies, enclos de grilles; des lustres en cristal, des sabres, des poignards pendent au plafond … Si le tombeau de Hadji Mir Yakoub est le plus lucratif qui soit a *Khoï*, il n'en est pourtant pas le plus illustre. Au quartier de *Robat*, un vieux minaret émerge des arbres; le balcon et la lanterne en sont fort délabrés; une multitude de cornes de mouflons pointent entre les briques. La légende veut qu'un roi, venu à *Khoï*, ait tué en un seul jour le nombre de mouflons nécessaires à la décoration du monument. Le respect populaire y vénère ce qui reste du tombeau de *Chems-é-Tabriz* [philosophe, fondeur de l'ordre des *Mevlevis*].

……

L'enceinte fortifiée de *Khoï* appartient aux seuls musulmans; 110 familles arméniennes vivent dans les quartiers extérieurs. Il existe encore une très vieille église grégorienne, construite en pierres jusqu'à mi-hauteur; l'intérieur est sombre et voûté. On dit que toute la communauté arménienne y trouva refuge, lors du passage de Tamerlan. Parmi eux s'est établie la *Deutsche Orients Mission*. Luthéranienne, elle fut fondée, il y a huit ans, à la suite des massacres d'Arménie, par le docteur Lepsius, fils du Cèlèbre égyptologue. Son quartier général est à *Ourfa*, en Mésopotamie; elle possède trois stations en Perse, *Khoï*, *Ourmiah* et *Saudj-Boulak*. Celle de *Khoï* est de beaucoup las plus important.

3 Descriptions of the Region and the City of Salmas in the Relations of Travellers in Persia Tenth Century

Ebn Haukal, *The Oriental Geography of Ebn Haukal, an Arabian Traveller of the Tenth Century, translated by Sir William Ouseley*, London, Oriental Press, 1800, pp. 157, 162, 164–165.

Armenia is an extensive and fertile region, bounded by the sea, and full of delightful situations: the towns are *Misan*, *Khounah*, *Bervanan*, *Khoui*, *Selmas*, *Neshoui*, *Marend*, *Tabriz*, *Bezerend*, *Derban*, *Moukan*, and *Khaberan*; and several smaller towns.

......

There is a lake in *Azerbaijan* called the Lake of *Armia*: the water is salt or bitter, and contains not any living creature. All round this lake are villages and buildings: from the lake *Meraghah* is a distance of three *farsang [= parasang, a Persian unit of measurement for how far a man can walk in a day]*; to Armi two *farsang*. The length of this lake is five days journey, by land; and by water, with a fair wind, a person may traverse it in the space of one night.

......

From *Ardebil* to *Meraghah*, the road is this: from *Ardebil* to *Mianeh*, twenty *farsang*; from *Mianeh* to *KhouneH*, eight *farsang*; from *Meraghah* to *Deir Kherkan*, two *merhileh*; from that to *Tabriz*, two *merhileh*; from *Merend* to *Selmas*, two *merhileh*; from *Selmas* to *Hhoui*, eight *farsang*; from *Khoui* to *Berkeri*, thirty *farsang* ... from *Armi* to *Selmas*, fourteen *farsang*; from *Selmas* to *Khoui*, eight *farsang*; from *Khoui* to *Nakhjevan*, three *merhileh*; from *Nakhjevan* to *Deinel*, four *merhileh* ...

[A/N: Units of measurement of the lengths mentioned:

Farsang: Different authors report different measurements. As places and times change. In general it can be estimated that a farsang corresponds to about 6 km.

Merhileh: another vague measure. It could refer to a day trip on foot (10–13th century].

GuyLe Strange, *The lands of the eastern caliphate. Mesopotamia, Persia and Central Asia from the Moslem conquest to the time of Timur*, University Press, Cambridge, 1905, p. 166.

On the high road north of *Urmiyah*, and at some distance back from the north-western corner of the lake, is Salmas. *Mukaddasi* describes this as a fine town with good markets and a Friday Mosque built of stone; the population of the place in the 4th [10th] century was of *Kurd* origin. *Yâkût* says that in the 7th [13th] century Salmas lay for the most part in ruin; but the *Wazir 'Ali Shah*, *Mustawfi* writes, rebuilt its walls 8000 paces in circuit during the following century, in the reign of *Ghazan Khan*, the Mongol, and the town had then regained its former importance. Its climate was cold, and a river which rose in the mountains to the west passed through it to the lake.

1633

Joannes Laet (de), ***Persia seu Regni Persici Status. Variaquae Itinera in atque per Persiam; cum aliquot iconibus incolarum.*** **Seunda editio priori auctior, Lugdunum Bataviorum, 1633**, p. 68.

[The cities in Azerbaijan he cites are: Salmas, Choy, Ourmya, Merent, Nahziovan and Maraga].

Salmas; de terra inquit Armeniae: Salmas urbs est à stagno Kannudhan remota; stagnum vero istud salsum est caretque omni animantuim et piscium genere. Distat idem stagnum à Dachercan XII mill. Pass. Habetque in longitudine stationes quatuor, inter Septemtrionem et meridiem; eius autem latitudo protenditur inter Maragham et Armeniam spacio fere LX mill. Pass. Hyberno tempore maxime excitantur in eo procellae, adeo ut cymbae naustagium patiantur.

[A/T: Salmas; the land of Armenia begins: the city of Salmas is located at a distance from Lake Kannudhan [Urmia Lake]; this lake is salty and devoid of any animals and fish in general. Distance from Dachercan [a town located in ancient maps near Maraghe, SE of Urmia Lake] 12,000 paces. It has four stages in longitude, between north and south; its width also extends between Maragham and the Armenian land about 60,000 paces. In winter, huge storms develop there, to the point that the small boats are wrecked].

1812

Antoine-Jean Saint-Martin, ***Mémoires Historiques et géographiques sur l'Arménie*****, Paris, De l'Imprimerie Royale, 1818**, pp. 178–179 (Fig. 15).

La Persarménie étoit située à l'orient de la province de *Gordjaik'h* et au midi du *Vasbouragan.* Il est très-difficile d'en déterminer d'une manière positive l'étendue et la situation; les Arméniens ne nous fournissent que fort peu de lumières sur ces objets. Il paroît qu'elle occupoit les rives occidentales et méridionales du lac d'*Ourmiah*, et qu'elle étoit formée aux dépens de l'*Aderbaidjan*, avec lequel elle fut confondue après la destruction du royame d'Arménie. Elle étoit dans l'antiquité, partagée en neuf petits cantons, sur lesquels nous n'avons pas plus de renseignemens que sur ceux de la *Gordjaik 'h.* Les plus connus de ces cantons sont ceux de *Her* et de *Zarevant*, ou *Zarovant* qui étoient limitrophes et qui étoient très-probablement situés sur le penchant des montagnes des *Kurdes*, vers l'orient, du côté de la ville d'*Ourmiah.* On trouvoit dans les environs de ces deux cantons, un pays assez étendu, qui paroît devoir être placé du côté du nord, sur les bords du lac d'*Ourmiah*, qui se nommoit *Rhodog* et qui étoit quelquefois placé dans la province de *Gordjai'k*, mais qu'il faut cependant attribuer à la Persarménie, puisqu'il comprenoit la ville de *Salmas* et meme les cantons de *Her* et de *Zarevent*. La Persarménie est actuellement comprise dans l'*Aderbaidjan* et soumise aux Persans. La principale ville qui se trouvoit dans l'Arménie Persane, étoit celle qui est nommée actuellement par les Persans *Selmas.*

Fig. 15 Antoine-Jean Saint-Martin (1791–1832) (Wikimedia Commons)

Elle est située à l'occident du lac d'*Ourmiah*, sur les bords d'une petite rivière qui vient des montagnes des *Kurdes*, qui sont à une petite distance, aussi vers l'occident; elle fait partie de l'*Aderbaidjan* et elle est soumise à la domination Persane. Cette ville est fort ancienne; elle existoit déjà au milieu du quatrième siècle. Les Arméniens la placent généralment dans le petit canton de *Her*; selon d'autres, elle étoit dans le canton de *Rhodog*; comme elle étoit dans le voisinage de la province de *Gordjaik'h*, elle y a aussi été placée quelquefois. Les Arméniens appeloient autrefois cette ville *Saghmas* ou *Saghamasd*; ils l'appellent actuellement *Salmasd* ou *Salamasd*.

1817–20

Sir Robert Ker Porter, *Travels in Georgia, Persia, Armenia, Ancient Babylonia during the years 1817, 1818, 1819 and 1820*, London, Printed for Longman, Hurst, Rees, Orme and Brown, 1822, pp. 597–599.

High mountains bounded it to the north; but all between, seemed rich, verdant country: and, on enquiry, I found I was now entering the wealthy district of *Salmos.* Having traversed its cultivated and level ground for nearly four miles … we drew towards a narrow, but level dell, formed and divided by detached masses of the rocky mountain stretching out into the plain. Our course lay within it for about twenty minutes, when we arrived at the entrance of a fine rural-looking valley to the south-west, and riding up it, soon reached the village of *Tamar* … and not far distant, the people told me I might see a curious group of men cut in stone, on the face of one of the mountains … About two miles from the village, and along the great plain of *Salmos*, rise several detached rocky mountains. On the southern face of one of these, and a good way up the precipice, the rock has been smoothed for the length of sixteen feet, and there I found a very rude piece of workmanship, representing four figures, two of which are equestrian. Rough as it is, the style is similar to that of the *Sassanian* dynasty, mentioned in a former part of my journal.

This bas-relief consists of what may be called two groups, though evidently connected in one point of history. The group to our right, (when standing before it), consists of a personage on horseback, whose horse trappings and own apparel shew the *Sassanian* fashion before described. His hair flows in the full bushy curls, a globular shape surmounts the helmet, which here has no other ornament, excepting the royal streamers. His left hand holds the reins, appearing at the same time to rest on the hilt of a long straight sword. Besides the waving ends from his head and neck, others stream from behind his back, as if attached to the robe which clasps over his chest. Similar appendages float from his heels. With his extended right arm he grasps a ring or chaplet, either giving or receiving il from a person on foot, who stands at the head of the horse. This latter figure is bare-headed, his hair parted over the forehead, and thence devolving on each side in full and abundant curls. He is dressed in a short tunic reaching to his knees, and fastened round the waist with a band and a sword-belt. Loose trousers, apparently fastened around his ankles, complete all that is now discernible in his *appareil.* The second group, but in order of march it precedes this, represents a similar action; and, indeed, the chief difference in its details lie in the dress of the equestrian personage and the decorations of his horse. The caparison of this animal appears more splendid than the other's, and the head of the rider without the usual *Sassanian* complement of hair, a singular circumstance in bas-reliefs of that era, and seems to mark him to be a prince rather in alliance with those sovereigns, than one of their race. The drapery which incloses his leg and thigh is precisely like that on the figure in Plate XXIV, at *Nackshi Roustam*; whereas the other cavalier has his limb covered in the style of *Shapour*, in Plate XXII.

This performance, on the whole, is very poorly executed, and ought to be called a very low relief, since it scarcely rises four inches above the face of the rock, the detailed forms being made out upon it by the simple line of the chisel, in the manner an engraver draws on copper. Therefore it is surprising to me how any part of so slighty elevated a surface should have stood the wear of so many centuries, considering the changes of seasons, and other inclemencies, which must have passed over it. Great mutilation has taken place, yet enough remains, to shew the original rude execution of this indeed curious specimen of the state of the arts of sculpture in Armenia at that day, for I should be inclined to suppose its date early in the fourth century of our era; and a miserable idea it gives of the then Armenian taste in such works; a contempt, indeed, of these imitative branches of the fine arts, manifested also in the coins of the country, all that I have seen being particularly barbarous. I say contempt, for their artists must have known better models, from the constant communication between that country and those of Greece and Rome. This is the only relic of antiquity I could hear of on the plain …

1824–29

Joseph Wolff, *Missionary Journal and Memoir of the Rev. Joseph Wolff*, 3 vols, 1824, 1828, 1829, vol. III, pp. 131–145 (Fig. 16).

On my arrival at *Khosrowa,* I met Don Isaiah, a Chaldean Catholic, who had left the Propaganda just one year before I entered the Propaganda; and it was determined … that I should join him in his mission, after my being ordained Priest of the Church of Rome.

……

he is a Catholic Missionary, and I an Evangelical Missionary [Wolff, born of a Jewish family, had converted to Catholicism at a young age, had entered the College of Propaganda Fidae with the intention of becoming a missionary; however in 1821 he abandoned Catholicism for the Anglican Church].

I asked him whether I could lodge in his house, for which favour I pay; he received me readily. (p. 131).

[A/N: A religious discussion follows, addressing to the ease with which Don Isaiah worked the conversion of the Nestorians to Catholicism, essentially based on contributions in money and exemption from taxes. This is not the place to address this aspect of the competition between Catholics and Protestants in Persia and reference is made to the original text for further information. Here and there in the story we can grasp various pieces of information that deserve to be reported below].

…

it must first of all be observed, that there are 160 Chaldean Catholics residing at *Khosrowa.* Who have been converted, eighty years ago … (p. 131).

…

A young man … Ithniel, by name, had copied the Syrian translation of the Pentateuch, with Nestorian, or rather Chaldean characters, for the use of the Chaldean Catholics, and Nestorians scattered throughout *Curdistan*; and Ithniel intended to sell it to the Propaganda of Rome, in order that it might be printed there, and then sold to the Nestorian and Chaldean nations throughout *Curdistan* (pp. 134–135).

Fig. 16 Joseph Wolff (1795–1862) (https://archive.org/stream/researchesmissio00wolf#page/n5/mode/2up—Wikimedia Commons)

… From Teheran down to *Salmast* and *Ooromea* the Jews live more at ease and comfort than their brethren at *Shiraz* and *Isphahan.* The Jews in the districts of *Salmast* and *Ooromea* speak a corrupt Chaldean, as the Chaldeans Christians do, and which they call *Gebilah*: they speak Hebrew in their colleges (p. 138).

…

Jews in the district of *Salmast* are in the following places, and following number:

	Families
1. Shahar	50
2. Bashkala	50
3. Bajerka	40
Whole amount	140
Souls	700
(p. 139)	

…

I was, therefore, very glad to meet at *Khosrowa* with Alexander Mirza, the son of the late king of Georgia, whose sceptre was taken, and the lawgiver from between his feet, by Catherine II, Empress of Russia. Alexander Mirza swore eternal enmity to the Russians, and found an asylum in Persia, where Abbas Mirza made him governor of the village called *Khosrova* (p.140).

…

(Tab. pp. 145–146):

Number of Jews and Christians in the district of *Salmast*				
Shahar	Jews	50 Families, 3 rabbis, and 2 synagogues		
		Families	Churches	Priests
Patawar	Armenians	200	2	4
	Chaldeans Cath.	50	1	0
Sawra	Armenians	400	1	1
Khosrova	Armenians	50	1	1
	Chaldeans Cath.	160	2	2
Pakajoj	Armenians	250	1	2
Kulusur	Armenians	40	1	0
Sarna	Armenians	120	1	2
Melam	Armenians	120	1	2
	Chaldeans Cath.		1	
Gulisan	Armenians	12	1	
	Chaldeans Cath.	5		
	Nestorians	6		

… It is general opinion among Christians in Europe, that almost all Christians in the East are Nestorians; but this is erroneous idea, although encouraged by some travellers. There are only 250.000 Nestorians existing now on earth, and these live in the borders, and in the mountains of *Curdistan* … the Nestorians have among them more of primitive Christianity than he, with all the Cardinals, and all the other Christian denominations of the East have.

1831

Eli Smith and Harrison Gray Otis Dwight, *Missionary researches in Armenia: including a journey through Asia Minor, and into Georgia and Persia, with a visit to the Nestorian and Chaldean Christians of Oormiah and Salmas*, London, George Wightman, 1834, pp. 314–315 and 356–358 (Fig. 17).

Our morning's ride to Khoy was three *fúrsakhs* … Almost the whole distance was a plain, and at one place a copious spring of warm salt water boiled up in a small pond by the side of the path. As we approached the city, the fertility of the soil increased,

Fig. 17 Eli Smith (1801–1857) (Missouri History Museum—Wikimedia Commons)

cultivation became general, and villages were frequent. We viewed the plain of Khoy under more advantageous circumstances, on our return from *Tebriz*. The whole basin, enclosed by the surrounding mountains, is of great extent. Much of it is broken up by hills of a dry soil, and not remarkable for fertility; though they, as well as the mountains, are destitute of the sun-burnt and arid aspect of the scenery of *Tebriz*, and of the valley of the *Aras*, and afford, upon their sloping sides and swelling summits, verdant pastures for flocks and herds. In the vicinity of the city, the plain is extremely level, and watered almost to excess; and its fame for fertility has not exceeded the reality. Besides numerous gardens which beautify it, grain and cotton are extensively cultivated.

……

We spent the day at Khoy, and were almost confined, by the cold and dampness of the weather, and my feeble health, to the caravanserai. In our subsequent visit we had a better opportunity to see the city. In former times, it belonged to the Armenian province of *Vasbooragan*; now it is one of the cities of *Aderbaijan*. Though its walls, as well as it houses, are of mud or unburnt bricks, broad and regular streets, intersected by canals with shade trees upon their banks, gave it an airness and regularity unknown in almost every other oriental city I have seen. It has a lofty and well-built *bazar*, too, that would hardly disgrace Constantinople. The extreme moisture of the plain, upon the level surface of which it is situated, combined with the warmth of the atmosphere in the summer months, must subject its inhabitants to frequent fevers. A son of the prince royal, named *Jihangir Mirza*, now resides here, and governs for his father the district of which it is the capital. Its moslem population we heard estimated at from 4000 to 7000 families. A distinct suburb contains about 100 families of Armenians, who have two churches and two priests. In the villages around are a few more, but nearly all have emigrated north of the *Aras*.

A tolerably well-made and perfectly straight road conducted us from Khoy, for about two miles in the direction of *Tebriz*, to the principal river that waters the plain. A double ditch of running water lines each side, and waters rows of the *sinjid*, or silver-leafed willow, which form an avenue the whole distance. The *sinjid* is a common tree in all the plains of *Aderbaijan* which we visited, and being invariably planted along the margin of the canals, never failed to remind us forcibly of the imagery of the first Psalm. It produces an insipid fruit, resembling the date in form, which abounds in every bazar. About half way from the city to the river, a side avenue conducts from the road to the country seat of a *khan*, in an extensive and luxuriant garden of fruit trees upon the side of a hill, which carried our minds away to other countries, where protecting governments allow private wealth thus publicly to exhibit itself more frequently than in Persia. The river is crossed by an arched bridge of stone.

…

Taking a Chaldean guide from *Khosrova*, we made an excursion this morning to Old Salmas, now generally called simply *Shehir*, or the city. The plain continued of the same character as was noticed at the first village on entering the district. Level almost as the floor of a house, with a soil of rich light loam, and irrigated throughout with canals, it presented a landscape truly charming. Cotton and rice do not flourish here, and only grain is cultivated; but with that almost every spot was green. Our guide assured us that it produces from eight to twelve, and in some places 15-fold. Numerous villages appeared, and all were surrounded, like *Khosrova*, with forests of fruit trees, which seem to flourish here with almost unequalled exuberance. The apple, pear, quince, peach, apricot, walnut, and *sinjid*, were the principal.

Trough gardens thickly set with these, we worked our way into *Saoora*, a village on the road. Our guide assured us that five or six hundred families of Armenians had emigrated from it to Georgia, and that only a few of that nation, with a few moslems, were left. We found the Chaldean church in the keeping of a solitary family of that sect. Its door was a mere eliptical hole cut through a single stone, and so small as to admit one's body with difficulty. Within were three small chapels under the same roof, dark and dirty, and without furniture, except a few old shawls and Romish pictures suspended upon the walls. No books even were to be found, and the keeper said the clergy brought them from *Khósrova* when they came to say mass on the great festivals.

Salmás is mentioned by ancient writers as a town in the Armenian province of Persarmenia. The modern village is scattered over a considerable space, on the site of the old one. Its houses are poor, and it has an aspect of decay. The only remains of antiquity we saw, were two or three cylindrical monuments or towers of an order similar to that at *Shamkór*, but much inferior in height. They were constructed of brick, and marked with inscriptions in the Arabic character, betraying a moslem origin. The plain extends hardly more than a mile south-westward from the town, and then commence the semi-independent mountains of the *Kurds*. Though so near that lawless people, however, the district is never disturbed by them, except when the Persian government is in a weak and disorganized state. The entrance to the Chaldean church was like that just mentioned at *Saoora*, and its interior was in a similar state, except that it contained three or four old books, one of which was written in the Estrangelo character upon parchment. There being no Chaldean or Nestorian in town, the key was kept by an old Armenian. Still, the clergy come from *Khosrova* to say mass in it upon the great festivals. Among other ancient inscriptions upon the stones in the external face of its walls, we noticed one in a character entirely unknown to us. We found the Jewish synagogue neater and better carpeted than the churches; and the Jews better dressed than the Christians. Their copies of the law were beautiful, and we tried to purchase one, but in vain. They reckon 30 or 35 families here; in *Oormiah* they amount to 300 families; and in *Khoy*, I believe, there are a few; but elsewhere in *Aderbaijan* we heard of none. In *Teheran*, *Kashan*, and *lsfahán*, they are more numerous. The priest at *Khósrova* said that these of *Salmás* are doubtless much oppressed, but less so than his own townsmen; for not being cultivators of the soil, their property is not so tangible. Their appearance seemed to justified his opinion. Thee Jews of Persia generally are the most demoralized, and oppressed part of the community. They are said to have neither tradition nor history to inform them when their ancestors came into the country. We naturally look among them for the remains of the ten tribes; but if such where their origin, all traces of it have been effaced. They now resemble their brethren elsewhere, except that their reverence for the Talmud is perhaps somewhat less, and there is some doubt whether they have all the books of the Old Testament. This resemblance may have been produced by long intercourse with Jerusalem Rabbiess, who often pass by them on their way to the north. We found one at *Salmás*, at the time of our visit. They speak the languages of the country; but respecting the common use of a vulgar Hebrew among them, we received contradictory statements. The Armenian church bore marks of great poverty, and was probably stripped of its furniture and ornaments at the time of the emigration. Upon that occasion, we were assured, 200 families left the town. Only 29 remained, and 40 have since joined them, so that their present number is 60. In the whole district there are now but about 400 families of Armenians. They have no school, nor had they any before the war.

1833–35

Rev. Justin Perkins, A *Residence of eight years in Persia among the Nestorian Christians, with Notices of the Muhammedans*, New York, Allen, Morril and Wardwell, 1843, p. 171.

In the morning, we visited several of the gardens, near the village. *Khosrova* is the residence of a priest and a monk, educated in the Propaganda at Rome. The Catholic bishop, whom Messrs. Smith and Dwight saw here, is dead. The priest had now gone to *Tabréez* to perform mass for the Russian ambassador, who is a Catholic, there being no Papal ecclesiastic in that city. The monk stood by the road-side, as we were passing to the gardens; but knowing us to be Lütrans [Protestants], and seeing a number of the villagers accompanying us in a friendly manner, he turned his back upon us and retained that attitude until we had passed by. We little regretted the loss of his acquaintance. Just at evening, we visited the old Armenian town of *Salmas*, situated two miles west of *Khosrova*, quite under the mountain. It has a very venerable appearance, but now contains only about three thousand inhabitants, the majority of whom are Muhammedans and the rest Jews and Armenians. There are two lofty brick towers near the town, that bear the marks of considerable antiquity, on which are inscriptions in the Arabic character.

1836

Justin Shiel, *Notes on a journey from Tabriz, through Kurdistan,* via *Van, Bitlis, Se'ert and Erbil, to Suleimaniyeh, in August, 1836*, in "The London and Paris Observer", Paris, 11 March 1838, pp. 54–56.

At *Tabriz* two routes were proposed to me for reaching the Turkish camp, which was supposed to be situated to the northeast of *Mosul*. One was by proceeding to *Júlámerik*, an independent Chiefship in the mountains of *Kurdistan*, bordering on Persia, and from thence through *Tiyárí*, the territory of the *Kaldani [Chaldrean]* or Nestorian Christians, whose almost impracticable country joined to their own warlike character, enables them to avoid rendering obedience or tribute to Turk, Kurd, or Persian, and still corresponds with Xenophon's character of the *Chaldreans*: "they are said to be a free people and warlike." The objections to this route were the necessity of assuming the character of a *Dervish*, that is of travelling under the appearance of great poverty, as my informant said that otherwise there could be no security, and of performing a part of the journey on foot, a portion of the road being totally impracticable for cattle. I therefore selected the route by *Van*, although more distant.

July 15th.—We left Tabriz and proceeding in a western direction arrived on the evening of the 17th at the town of *Dilmán*, situated in the large and fertile valley of *Selmas*, which is bounded by the Lake of *Urumiyah*, on the east, and by the mountains of *Kurdistan* on the west. In this district Armenian Christians are very numerous, and exclusively occupy whole villages; there are also some villages inhabited by Roman Catholics, under the spiritual guidance of a *Khalifah* or Superintendent, appointed by the Vicar General in Baghdad. Like the Nestorians, these last call themselves *Kaldánis*, which is probably a national, not a religious designation, while they also style themselves *Katóliks*, a name by which they are known all over *Kurdistan*. A strong mutual antipathy exists between them and the Nestorians. I remember a bishop of the latter church once telling me, that the Armenians were not very good, the Musselmáns were much worse, but that the vilest of all mankind were the *Kátólíks*. The contentions of the fifth century are not yet forgotten, and they still retain the doctrine that separated them from the Roman Catholic Church.

Yet the Nestorians are generally tolerant; they are anxious for instruction, and they have given a welcome reception to some American missionaries who have established themselves in *Urumiyah*. The Roman Catholic bishops of *Selmas* are generally educated at Rome.

……

Selmas is also a home for many of the Russian deserters in the service of Persia. Here they marry and settle when they are worn out and unfit for service and form a sort of colony. There are also many *Leks* established in *Selmas*. This is a tribe of the ancient Persian race which dwells chiefly in the south of Persia.

They were brought here by *Nadir Shah*, but they have forgotten their language, and now speak only Turkish.

Dilmán, commonly called "the City," is a new town built by *Amir Khan Khájar*, a relation of *Fath'Ali Sháh*. There is an old city of the same name placed one *farsakh* to the west, now almost in ruins; the only reason I could discover for removal to the present site, is its greater distance from the *Kurds*, and therefore greater security. The new town is of considerable extent, and is said to contain 15,000 inhabitants. Like almost all the villages in the district of *Selmas*, *Dilmán* is surrounded by gardens; the streets are clean, but the *bázárs* are poor and ill supplied. Caravans are sent from *Dilmán* to *Van*, *Júlámerik*, *Tiflis*, and *Erzrúm*.

…

We passed through the old town of *Dilmán*, a large portion of which is in ruins. From the number of mounds in the neighbourhood it has the appearance of having been once of considerable extent, and it is described by the Orientalist, St. Martin, as being a very ancient Armenian city. Near the town we saw two very high minarets standing alone in the plain, the mosques to which they were probably attached having fallen to decay; they were said to have been built by the '*Osmánlis*, who possessed this part of Persia upwards of a century ago.

My inquiries respecting some ancient sculpture on the face of a mountain near the old town of *Dilmán* were answered with difficulty: after a long search I discovered that it was half a *farsakh* to the S.E.; our road was west, and my party having preceded me, together with its being near sunset, obliged me to abandon my intention of visiting it. I learnt that, in the range of hills separating *Selmás* from the district of *Urumiyah*, there was some sculpture of a similar description, but whether it has been examined by any European I know not".

1836

James Ballie Fraser, ***A Winter's Journey (Tatar) from Costantinople to Tehran with travels through various parts of Persia,*** **London, Richard Bentley, 1838,** pp. 47–49).

[Selmâs plain]

Nevertheless, the district of *Selmâs* is still rich and productive, and in passing through it you see a number of fine villages, particularly towards the upper end near the mountains. But I was surprised to find that the river *Selmâs*, instead of being, as I had heard, a stream navigable for boat several miles from the lake, is at this season a miserable rivulet which scarcely wetted our horses' feet while crossing it.

The plain of *Selmâs* itself, which I should think extended thirty miles inland from the lake, and which may in some parts be ten miles across, has at one time obviously been covered by the lake. Indeed, the gradual recession of the waters of this inland sea is a phenomenon which affords curious matter of speculation. Like all other salt lakes in this part of Asia, and some fresh, it has no outlet, although it is itself a receptacle for the waters of many and considerable streams which flow from the mountains round it. That these were once more abundant sources of supply than at present, appears probable, since, if report on the subject can at all be trusted, the lake, within the memory of man, was much larger and fuller than it now is. Its only drain at present can be from evaporation, unless an increased demand of water for irrigation were supposed; but as there is every reason to believe that cultivation has rather diminished than increased, that cannot be the case. However the diminution of waters may be caused, there is no doubt that a gradual increase of land takes place on the margin of the lake, and land of great richness, as is proved both by the crops it yields and by the natural grass it produces, the flocks and herds of the inhabitants being unable to consume the store of herbage in the meadows.

1837

Captain Richard Wilbraham, ***Travels in the Transcaucasian provinces of Russia and along the Southern shore of the Lakes of Van and Urumiah, in the Autumn and Winter of 1837*****, London, John Murray, 1839**, pp. 367–370 (Fig. 18).

Towards evening we arrived at the fertile and well-cultivated district of *Salmas*, which lies between the lake of *Urumiah* and the mountains of *Koordistan*. It is thickly studded with villages, inhabited chiefly by Armenians. The Principal town of the district is called *Dilman*, and is the residence of the governor.

Fig. 18 General Sir Richard Wilbraham (1811–1900) (Royal Archives, Windsor Collection Wikimedia Commons)

I passed it at some distance to my left, and took up my quarters in a miserable village built among the ruins of an ancient Armenian city. On the face of the neighbouring mountains are some rude sculptures hewn out of the solid rock, apparently of great antiquity.

…

On the following day, after crossing the plain of *Salmas*, and passing through numerous straggling villages with large but neglected gardens, we traversed successive ranges of low hills, the snow at every step being less deep, until in about five hours we came within view of the lake of *Urumiah*. The first glimpse of the broad sheet of water was very beautiful. Towards the north-east, in the direction of *Tabreez*, rose a very bold chain of mountains, now clothed with snow; but to our right the shores were low, and every sign of winter had disappeared. Large tracts of swampy land, overgrown with tall reeds intervened between the base of the hills and the shores of the lake. At our feet lay the small village of *Kolingi*, where I intended halting for the night, being half way between *Salmas* and the city of *Urumiah*.

The waters of the lake are so salt that no fish can exist in them: several large streams of bitter, brackish water flow into the lake from the northward, and it doubtless contains numerous subaqueous saline springs.

Soon after leaving *Kolingi*, we quitted the shores of the lake, and struck across the wide plain of *Urumiah*. The rich loamy soil saturated with melted snow, rendered the road almost impassable. In summer this must be a beautiful country; and I have been told by many who have visited it in that season, that it reminded them of English scenery. The numerous villages are surrounded by orchards; and the rows of tall trees which line the banks of the brooks, by which the plain is intersected in every direction, give it the appearance of an enclosed country.

1836–38

Horatio Southgate, *Narrative of a Tour through Armenia, Kurdistan, Persia and Mesopotamia, New York, 1840, vol. I,* pp. 288–289, 291–292, 299–302, 313–314, 321–328.

We passed several villages. Near one, called *Derik*, inhabited by Persians, were several warm mineral springs of Sulphur and iron. Two of them were enclosed, and used by the villagers for bathing. Another was an Armenian village entirely deserted. Its lonely church still crowned the top of the hill on which the village stood. The sight of it brought vividly to mind the sanctuaries of God in a far distant land, where I knew thousands would gather upon this holy day for a pure and spiritual worship.

……

Our course turned more Southerly as we approached *Salmas*, until, at length, the town itself broke upon us with its verdant gardens, its wide-spread plain, and, in the distance, the North west corner of the lake of *Ourmiah*. The scene was entirely Persian, and we felt at a once that we had entered a new country. The contrast was equally striking in the reception that we met. In Turkey, on arriving at a town, the traveler has only to present his firman to the governor, to secure for himself respectable lodgings; in Persia, firmans are unknown, and the traveler is often glad to find a shelter for himself in the meanest caravanserai.

In Turkish towns there was always someone to help us to dismount, and to draw off our unwieldy travelling-boots; in Salmas, a crowd gathered about us to gaze and inquire, but no one would render us any service. In Turkey, the arrival of a traveler excites no stir, no man rises from his place to ask after him, no impertinent throng collects about him to annoy him with comments and questions; we had hardly dismounted in Salmas, before half the town were gathered in the court of the caravanserai, where they stared at us and remarked upon us as freely as if they were republicans. When we endeavored to rid ourselves of them, they laughed at us and made sport of us.

The caravanserai at which we had alighted, was a mud-building, some twelve feet high, erected, as usual, about a court, but containing no other accommodations than rows of cells about six feet square, opening upon the court, without doors, and entirely unfurnished within. I inquired for the *Khan*, or governor of the place but he was residing in *Dilman*, a few miles distant on the plain.

......

All my efforts to obtain more commodious lodgings being unsuccessful, I betook myself from the burning sun, to one of the cells of the caravanserai, and set down upon the ground in a corner, to spend Sunday.

......

Salmas was formerly the chief town of the district in which il lies, but the residence of the *Khan* is now removed to *Dilman*, and *Salmas*, deserted by the greater part of its population, who were Armenians, and emigrated to Russia after the last war with that country is no more than a village. Some remains in the vicinity of the town attest its antiquity and its former importance.

......

Dilman appears to have risen into importance since the decline of Salmas. It is entirely a Mussulman town, and, with a population of about 2000, has several mosques and four schools. The plain contains numerous villages, with a population chiefly Christian. Their number was stated to me to be sixty-two, but, as my authority was not of the best kind, I cannot vouch for the accuracy of the information. The Christian population consists of Armenians, Nestorians, and Chaldeans, the last being proselytes from the Nestorians to the Church of Rome. These are, so far as my information extends, the only adherents of that Church among the Nestorians of Persia. They have a Bishop at *Khosrova*, who is a zealous and efficient promotor of the Papal interests. The bazars of *Dilman* are small, and the place, upon the whole, is one of humble pretensions.

...

We left *Dilman* before day. The full moon, declining in the West, shed her light over the plain, and on the distant lake. Our course lay Southerly, through ripe fields of grain, which diffused around the fragrance of harvest-time. At early dawn we observed husbandmen at work. It was now the 18th of July, and they were gathering in their crops. The sun came up gloriously in the East, just as we entered the mountains.

1839–40

Eugène Borè, *Correspondance et Mémoires d'un voyageur en Orient*, Tome II, Paris, Olivier-Fulgence, 1840,

§. III. - *Comment la foi catholique fut apportée au pays de Salmas, l'ancienne Médie*, pp. 256–257.

Il y a un siècle, vivait un jeune *Chaldéen* du *Diarbekre*, converti par le zèle des Dominicains, qui évangélisaient cette partie de l'Asie occidentale. Après avoir franchi les hautes montagnes du *Curdistan*, il vint au village de *Kosrova* exercer sa profession de teinturier. Il était très ignorant, selon le monde, mais la grâce lui avait départi une science préférable à celle qui, malgré ses ténèbres et son insuffisance, nous pousse à l'orgueil. Il savait aimer Dieu et son prochain, pratiquant ainsi toute la loi, qui se résume dans ce double précepte. La nature de la vérité, rayon de la lumière increée, est de se répandre et de briller au dehors, en communiquant à tout ce qui l'entoure ses ardeurs secrètes. Aussi le jeune artisan devint-il bientôt l'apôtre des jeunes apprentis qui l'avaient choisi pour maître. Ses instructions, et plus encore ses bons exemples, opérèrent leur conversion. A ces prosélytes se joignit un homme veuf, doué de quelque instruction, et qui fut jugé capable d'être le père spirituel de la société naissante. Il l'envoya près du patriarche de *Mossoul* pour être ordonné. Lorsqu'il revint, sa maison servit de chapelle aux catholiques. L'intolérance des nestoriens, au milieu desquels ils vivaient, les obligeait au secret et ils le gardèrent si religieusement que, durant vingt années consécutives, leur Église put se consolider et s'étendre, à l'insu de tous les profanes. Enfin l'évêque nestorien, Mar-Isaïe, découvre le mystère et, l'heureux changement opéré dans son village lui ouvrant les yeux, il va dans la Géorgie, à *Akaltsiké*, faire son abjuration entre les mains des missionnaires catholiques, puis retourne à *Khosrova* convertir le reste de ses ouailles. Le patriarche de *Mossoul*, apprenant cette joyeuse nouvelle, lui envoya quelques Dominicains, dont les instructions éclairèrent et affermirent ces nouveaux frères. Le successeur de *Mar-Isaie*, Jean Guriel, élevé au collège de la Propagande, rapporta de ce centre glorieux du catholicisme la science de la foi qu'il avait recouvrée, et vint à son tour la propager dans les villages environnants. *Pataura*, peu distant de *Khosrova*, fut reconquis par son zèle et cette église s'agrandit encore journellement, grâce à l'activité pastorale de monseigneur *Mar-Michael*, disciple, comme lui, de la Propagande, qui vient d'être promu tout récemment à l'importante dignité de patriarche des *Chaldéens*.

§ IV.—*La foi de la plaine de Salmas passe au pays d'Ourmi, malgré l'opposition des musulmans et des nestoriens,* p. 258.

De *Khosrova* le catholicisme s'étendit dans le pays adjacent d'*Ourmi*. Cette vallée, bordée à l'est par le grand lac bleu et salé du même nom, et resserrée au sud et au couchant par les montagnes des *Curdes*, est la terre promise de la Perse. Le pur froment, le riz, la vigne, le tabac, le coton et toutes les variétés d'arbres fruitiers y croissent avec une heureuse abondance et leurs fruits tirent du terroir une qualité qui les rend supérieurs à ceux des pays circonvoisins. Les souvenirs les plus vénérables consacrent ces lieux, où l'on a placé le berceau de *Zoroastre*, réformateur du culte des constellations célestes et de la lumière. En ce, sur la rive opposée du lac, est *Maraga*.

1839

Charles Texier, *Description de l'Armenie, la Perse et la Mesopotamie*, Paris, 1852, pp. 31–34.

……

Toute cette plaine de terrains est peu accidentée, aussi n'avons-nous à traverser que de faibles ruisseaux qui portent leurs eaux au lac. Pendant que je contemplais la chaine de montagnes du *Kurdistan*, qui s'élève comme un mur infranchissable, les nuages s'étaient accumulées, le vent, qui avait soufflé avec assez d'intensité, s'étant calmé tout à coup, une grêle épouvantable tomba sur la plaine. Le bruit que faisaient les grêlons dans les nuages ressemblait à des sacs de noix qu'on agitait dans l'air ils étaient du calibre d'une balle de munition. Lorsque le temps s'éclaircit, nous aperçûmes les montagnes parfaitement blanches. Ces orages de grêle ne paraissent pas plus fréquents en Perse que dans nos climat, les habitants étaient aussi étonnés que nous de la grosseur des grains de grêle. Nous logeâmes à *Dilmen* chez un habitant. J'ai toujours, autant que possible, évité de demeurer dans les *caravansérais*; car j'ai remarqué qu'on y est en butte à une indiscrète curiosité. Les informations que nous prîmes à *Dilmen* ne nous éclairèrent pas beaucoup sur le point principal de notre excursion.

Les bas-réliefs que nous voulions voir sont célèbres jusque dans le *Kurdistan*, paraissaient parfaitement inconnus aux habitants du village. Le lendemain, 17 septembre, nous nous décidâmes à nous rendre à *Selmas*, espérant être plus heureux. La plaine, de deux lieues d'étendue, qui sépare ces deux petites villes, est assez bien cultivée; mais, au moment où nous passions, toutes les récoltes étaient rentrées. Nous faisons lever à chaque pas des troupes nombreuses de francolins et d'une certaine espèce de perdix très-commune dans ces contrées.

XVIII

Les villes de Perse, qui de loin présentent un aspect assez agréable, offrent toujours de près le plus triste tableau. *Selmas* se présentait à nous ombragée par un rideau de hauts peupliers, ayant à droite un vaste cimetière, indice d'une population nombreuse, quelques minarets, et en avant de la ville une grosse tour de briques ornée d'émaux de différentes couleurs, qui me paraissait un monument remarquable. Je me détournai de ma route pour aller examiner ce monument, qui me semblait avoir quelque analogie avec la tour d'*Érivan*; mais, en approchant, je n'eus plus devant les yeux que une ruine plus dégradée encore par la main des hommes que par le temps, et dont l'intérieur n'est plus qu'un vide immense sans toiture. La corniche extérieure de la tour était formée d'un ajustement en polygone, avec des émaux bleus et verts. L'intérieur de la maçonnerie est en briques cuites d'une assez bonne qualité, ce qui indique l'époque d'une construction assez reculée, et antérieure même à celle de *Chah-Abas*, car du temps de ce prince on commençait déjà à négliger la bonne exécution de ces matériaux de première nécessité, tout en s'efforçant de conserver l'usage des ornements d'émail, seule décoration possible dans un pays où le marbre est chose presque inconnue, et où la pierre à bâtir est si rare. L'intérieur de la tour était divisé en deux étages, et on pénètre maintenant par une porte basse dans un ancien caveau qui a tous les caractères d'un lieu de sépulture. Cette tour est parfaitement isolée de toute parts, et n'a dans son voisinage la trace d'aucun édifice auquel elle aurait pu se rattacher.

La ville de *Dilmen* est assez moderne; elle fut bâtie, au dire des habitants, par *Émir Khan-Kadjar*; les ruines que l'on observe aux environs appartenaient à l'ancienne *Dilmen*, qui fut détruite par les guerres des Mongols et de *Timour*.

Sous le règne des princes arméniens, les contrées qui font aujourd'hui partie de l'*Aderbeidjam* étaient des annexes de leur royaume et le canton de *Selmas* appatenait à la principauté de Persarménie, qui s'étendait depuis les montagnes qui nous venons de traverser jusqu' à l'autre rive du lac, de l'est à l'ouest. Du côté du nord, ses limites ne dépassent pas sans doute la chaine de montagnes qui soutient le plateau de l'Arménie supérieure. Au-delà de ces monts, on trouve une portion considérable de pays inculte et inhabité. Cette province, du côté du sud, s'étendait jusqu'à *Meraghalt*, qui était déjà une ville importante au commencement du IV siècle. On ignore à quelle époque les princes arméniens ont créé la circonscription de la Persarménie; mais cette province fut de nouveau confondue avec l'*Aderbeidjam* lorsque les Mongols s'en furent emparés.

Selmas passait pour la capitale de cette province. Son antiquité ne remonte pas au-delà du III siècle du Christianisme; elle fut une des premières qui se convertirent à la foi chrétienne, et encore aujourd'hui elle est de toutes les villes de Perse celle qui renferme le plus grand nombre de catholiques, qui sont sous le gouvernement spirituel d'un préfet apostolique, suffragant du vicaire général de *Bagdad*. La langue chaldéenne est encore en usage parmi les membres du haut clergé.

Il existe aussi un certain nombre de nestoriens et d'Arméniens hérétiques qui sont du ressort du patriarcat d'*Etchmiadzin*. M. Eugène Boré a passé quelque temps dans cette ville, pour chercher à mettre un peu d'ordre parmi les membres épars de l'Église latine, et en même temps pour établir des écoles où les enfants des catholiques puissent trouver une instruction qui leur manque. Ces pacifiques et utiles occupations ont attiré sur lui la haine et les injures des membres des autres sectes, mais on doit dire qu'ils ont été excités par quelques missionnaires étranger, plus jaloux que les Orientaux de l'accueil qui était fait à notre compatriote.

En entrant dans la ville de *Selmas*, nous trouvâmes une grande troupe de cavaliers assemblés pour la chasse au faucon. C'était jour de marché. Nous nous arrêtâmes un moment pour demander quelques indications: nous devînmes bientôt l'objet d'une curiosité telle, que la foule qui nous entourait empêchait nos chevaux de faire un pas. Un paysan se chargea enfin de nous conduire au lieu que nous voulions visiter; il nous promit en outre de nous faire voir dans les rochers une tête sculptée, objet, disait-il, extrêmement curieux.

Nous nous dirigeâmes vers l'est, du côté du lac, marchant non sans peine au milieu des canaux nombreux qui sillonnent la plaine, les pluies de la veille ayant détrempé le terrain, de sorte que nos chevaux entraient jusqu'au jarret dans l'argile dont le sol est formé. L'horizon se termine, du côté de l'ouest, par une chaine de montagnes qui nous cachait la vue du lac. C'est dans un groupe de rochers détachés de la chaine que se trouve cette sculpture; elle est si peu apparente, qu'arrivé au pied de la roche, notre guide ne parvenait pas à la retrouver. Nous y arrivâmes cependant, et je reconnus, dans le tableau que j'avais devant les yeux, l'exactitude de la description que m'avait faite un *Kurde* qui connaissait ce monument. «Il y avait, disait-il, deux *padichah* (deux rois) qui étaient venus à *Selmas* accompagnés d'un écuyer, pour chercher le trésor qui est caché dans les rochers. Ils arrivèrent presque en même temps sur la place. Une querelle s'éleva entre les écuyers: chacun d'eux voulait indiquer le premier à son maitre la place du rocher, et arrêter le coursier de l'autre roi, lorsque apparut un magicien qui les changea en pierres».

XIX

Ces rochers sont formés d'un calcaire bleu, un peu schisteux, et par conséquent difficile à travailler, s'exfoliant à l'air, en un mot, très-peu propre à recevoir une sculpture. Aussi le bas-relief de *Selmas* a-t-il sensiblement souffert. On peut néanmoins reconnaitre que ce monument date de l'époque des princes sassanides. Son exécution est assez grossière: les figures manquent principalement de modelé, elles sont seulement découpées sur la pierre en silhouette. Le sujet représenté dans ce bas-relief est le même que ceux qui sont retracés dans la plupart des monuments de cette époque: c'est toujours le triomphe d'un des princes sassanides sur des ennemis vaincus. Cependant ici il y a une variante qui aurait besoin d'un éclaircissement. On remarque deux cavaliers, vêtus l'un et l'autre de la même manière, montant des chevaux harnachés de même, et portant tous deux sur la tête ce globe qui est un des indices de la puissance royal: il semblerait que c'est le même personnage que l'on a voulu représenter deux fois: une figure à pied, qui se tient devant le cheval, semble offrir au cavalier des captifs, on doit regarder ce monument comme relatif à quelque conquête des princes perses sur les rois d'Arménie, car ce ne sont pas des Romains qui sont représentés.

Les cavaliers sont armés d'une longue épée, qui s'attache au ceinturon d'une manière fixe; elle est remarquable par sa longue poignée. Le cavalier de droite est coiffé d'un casque arrondi, mais ne porte pas la couronne. Derrière lui flottent quatre bandeaux qui se rattachent à sa coiffure. Deux masses énormes de cheveux frisés encadrent sa figure. Il porte un justaucorps serré ce une tunique qui couvre une partie de la selle. On n'a remarqué aucune inscription autour de ce monument qui puisse aider à l'expliquer d'une manière satisfaisante.

En quittant ce lieu, nous invitâmes notre guide à tenir sa promesse: il nous conduisit à travers la plaine vers un monticule volcanique qui porte le nom de *Kara-Tépé*, sur le sommet duquel est bâtie une petite église qui appartient aux Arméniens. La prétendue tête qu'il devait nous faire voir n'était autre chose qu'une croix grossièrement sculptée sur les rochers, et semblable à celles que nous avions vues aux environs de *Van*.

Les villes de *Dilmen* et de *Selmas* contiennent plusieurs familles catholiques qui devaient être visitées par le préfet apostolique que nous avions rencontré à *Erzéroum*. Les chrétiens, dans celle contrée, semblent vivre en assez bonne intelligence avec les musulmans. Je suis porté à croire que l'identité du costume entre pour beaucoup dans la sécurité qu'ils trouvent au dehors, les Arméniens n'étant pas obligés, comme les rajas de Turquie, de porter un vêtement d'une forme et d'une couleur désignées qui appelle sur eux le mépris public. Les Arméniens qui se trouvent en Perse n'y sont pas comme peuple conquis, mais en vertu de traités que leurs ancêtres ont faits avec les anciens *schahs*, et dont les clauses sont encore exécutées. N'ayant trouvé à *Selmas* ni l'évêque ni les autres Arméniens que nous espérions rencontrer, nous ne séjournâmes point dans cette ville et retourmâmes le même soir à *Dilmen*, en recommandant aux muletiers de se tenir prèts au point du jour, afin de nous rendre à *Tabriz*.

1840

William Ainsworth, *An account of a visit to the Chaldeans, Inhabiting Central Kurdistán: And of an Ascent of thr Peak of Rowándiz in 1840,* Journal of the Royal Geographical Society of London, vol. 11, 1841, pp. 58–60.

We crossed over a ridge of trachytes and descended by *Khâni Berin*, re-ascending amid hills of conglomerate and igneous rocks, from whence we obtained our first view of the fertile plain of *Salamast* bearing due magnetic E., with the lake of *Urumiyah* beyond. There are moments which never slip from a traveller's memory, when, after a long journey on a heated or monotonous plain, a range of mountains with their anticipations of cool waters and refreshing breezes come into sight, or when, fatigued with mountain-toil and travel, a plain, smiling with gardens and villages, and full of promises of delicious repose, presents itself to his delighted vision.

Our descent to the plain from hence occupied us, however, 3 long hours, when we reached some basaltic cliffs, which led directly to the cultivated plain. On one of them were the foundations of a castle constructed of stones of large dimensions: to the S. was also a bold rock of limestone, which protruded out of the plain, bearing the ruins of *Karnawi* or *Marandos* castle; and before us rose a small hill, the last of the basaltic knolls, with a small Christian church. Pits were dug in the bed of a river close to us to obtain gravel, which is sifted, and then sprinkled over the land, to adapt it for growing water-melons. Two more hours amid villages and gardens brought us to the Chaldean village of *Khosrowa*, the place of *Khosroes* (the *Khusrue* of Monteith's map), where we had an introduction to a relative of *Mar Shim'on*, and were well received and hospitably entertained.

The district of *Salamast* is covered with villages, as may be seen by a glance at Colonel Monteith's map, which appears, in what regards this district and that of *Urumiyah*, to be founded upon actual survey. These numerous villages have, as in many parts of the E., a common market, where is also the residence of the governor, and the whole is inclosed like a fort. This place is designated sometimes *Salamast*, sometimes *Dilman*, but is generally known in ordinary parlance as *Shahr*, "the town", simply. It is the same with the district of *Urumiyah*. In all this part of Persia a bad Turkish is the language generally spoken: the better classes alone are acquainted with Persian. The Christians all look to the Russians as their protectors; but the Persians have imbibed a notion that *Mohammed 'Ali*, or rather *Ibrahim Pasha*, is destined to be the great bulwark of Islamism, and the conqueror of the Christian foe; and Major Rawlinson also remarks the same thing of *Soldúz*.

Friday, June 26th.—We rode by *U'la*, where the American missionaries have a school, and *Turmel*, to the hills which advance in bold rocks, bearing two castles over the lake of Urumiyah, and which are designated *Kará Básh*, or Black Head. But they have a culminating point westward, which, had still a few patches of snow on its hoary head, and which, rising about a thousand feet above the level of the lake, is called *Zendasht Tágh*, or *Túr Zendasht* by the Chaldeans. These hills gave me much hard work, for their structure was very varied; the results, however, may be given in a few words: the fundamental rock was large-grained granular hornblende and feldspar, and the same mineral small grained passing into basalt. This rock became large-grained lamellar as in gneiss, or small-grained schistose as in certain chlorite schists. Superimposed were a breccia of limestone with fragments of hornblende rocks and limestone rocks often saccharoidal. A second series of rocks presented feldspar and black mica, large grained and lamellar, passing into black mica schists, and common mica schists much waved and contorted, and these into clay schists of various colours, red, green, and gray. Associated with these was a third series, consisting of quartz rocks, generally with a waxy lustre, and passing into jade. We crossed this range of hills and stopped in the gardens of the Chaldean village of *Gawalán*, to the N. of which is a larger Christian village called *Jemalawah* by the Chaldean residents, but *Jelálábád* by the Persians.

1843–45

Rev. Joseph Wolff, *Travels and Adventures of the Rev. Joseph Wolff*, London, Saunders, Otley and Co., 1861, p. 224.

On their way to *Ooroomia*, both the Bishop and Wolff came to *Salmast*, where they met with *Alexander Mirza*, a relation of the late King of Georgia – the Georgian kings having their genealogy from King David.

……

The Georgian king had been expelled from his capital, at *Tiflis*, by the Russians, who took possession of his country; and he retired into Persia to *Salmast*, where he died.

1862–65

Giuseppe Anaclerio, *La Persia descritta. Relazione di un viaggio per Giuseppe Anaclerio*, Napoli, Vincenzo Marchese, 1868, pp. 58–59.

[I Lazzaristi francesi] stabilirono la prima casa di missione in Khosrova nell'Aderbeijan, villaggio ora intieramente caldeo cattolico, dove risiede un Arcivescovo di questo rito. Il paese ora è piccolissimo non contenendo più d'un cinquecento in seicento abitanti. Fu in questa città che si conchiuse la celebre pace tra i Persiani e Costantino, e dal nome del loro re la città venne chiamata Khosrova. Fu in questa città medesima che i persiani restituirono il sacro legno della Croce, e che Costantino riportò in trionfo nell'Occidente. Questo fatto è di una tradizione costante fra i Caldei, i quali ogni anno celebrano perciò una solenne festa in memoria di cotesto avvenimento. Coll'andar del tempo ed a seconda che il cattolicismo si stabiliva in questa provincia, si credè necessario di fondare un piccolo seminario per i giovani caldei, onde formarne poi dei preti per la conversione dei loro fratelli nestoriani; questo piccolo seminario è diretto dai RR.PP Lazzaristi. … Nel 1865 questo seminario non conteneva che una ventina di giovani, il numero più grande che possa contenere quel locale. La cattedrale era rovinata allorquando i missionari fondarono la missione: grazie però alle premure ed alle cure dei Cristiani, una nuova cattedrale è stata costruita secondo l'architettura europea. I missionari in numero di quattro, sono occupati alla predicazione ed all'istruzione del popolo.

[A/T: (French Lazarists) established their first mission house in Khosrova – Aderbeijan, a village currently entirely Chaldean Catholic, home to an Archbishop of this rite. Today the village is very small, consisting of no more than five/six hundred inhabitants. Right in this town the famous peace was made between the Persians and Constantine, and from the name of their king the town was called Khosrova. It also was right in this town that the Persians gave back the holy wood of the Cross, which Constantine brought back triumphantly to Occident. Among the Chaldeans

this is by then a constant tradition: every year they celebrate a solemn festivity in memory of that evenience. On time passing and catholicism more and more settling in that province, the foundation of a small seminar for young Chaldeans was deemed necessary, in order to train priests for converting their Nestorian brothers; the small seminar is guided by Lazarist Fathers. ... RR.PP Lazarists. ... In 1865 the seminar could give room to but twenty young men, the largest number that room might contain. The cathedral was in ruins when missionaries founded the mission: but thanks to the care of the Christians a new cathedral was built according to European architecture. The missionaries are in number of four, and are committed to preaching and education of the people].

1885

Henry Binder, *Au Kurdistan, en Mésopotamie et en Perse*, Paris, Maison Quantin, 1887, pp. 82–85 (Fig. 19).

Un chemin poussiéreux, en plein soleil. Nous conduit à travers les champs à *Diliman*. La ville est entournée de murs crénelés en terre qui s'effondent peu à peu sous l'action des pluies et dont les débris comblent le fossé. Il n'y à pas de plus grande ville à soixante kilomètres à la ronde, les rues et les bazars sont fort animés; le nombre des habitants, musulmans, arméniens, chaldéens et juifs, est de plus de vingt mille.

Nous traversons à grand peine la place du marché pleine de monde, et, suivant les bazars, nous atteignons le caravanseraï, où se trouve la poste. Nous demandons des chevaux; il n'y en a pas; c'est aujourd'hui le seul jour de courrier de la semaine et tous les chevaux sont en route. Néanmoins, en montrant la lettre que m'avait remise le *Velyed*, le maître de poste devient plus amiable et nous promet de nous en procurer; il va en demander au bazar. En attendant, nous nous installons du côté de l'ombre dans une des petites chambres qui entournent le caravanseraï. Un caravanseraï est un grand bâtiment carré avec una seule porte et sans fenêtres extérieures; une grande cour carrée à l'intérieur et tout autour, comme creusées dans l'épaisseur d'un mur énorme, des cases à un mètre du sol, larges d'environ deux mètres, profondes de quatre et hautes de trois. Tous les caravanseraïs sont à peu près pareils: quelques-uns ont un étage; les chambres du premier ferment avec des portes; les voyageurs, qui emportent tout avec eux: lits, cuisine, etc., se logent dans une de ces niches où ils se trouvent aussi confortablement que nous dans nos meilleurs hôtels européens; les animaux sont parqués dans la cour attachés à un piquet.

......

On nous regarde comme des phénomènes, il faut fermer la porte pour que le caravanseraï ne soit pas envahi par la foule. En partant nous traversons une partie de la ville qu'arrose un ruisseau; on sort des remparts par une porte, qu'il serait, je crois, impossible de fermer tant le bois en est vermoulu.

......

Après bien de péripéties, des chutes, des passages où nous nous laissons guider par l'instinct de nos montures, nous arrivons à *Guiavilen*, petit village de Chaldéens catholiques. Le prêtre parle un peu français; nous lui demandons l'hospitalité et il

Fig. 19 Henry Binder (Binder 1887)

nous accueille d'une façon très amiable. Nous sommes tout étonnés de l'entendre parler de sa femme et de ses enfants.

1886–91

Arthur John Maclean, William Henry Browne, *The Catholicos of the East and his People, Eastern Syrian Christians of Kurdistan and Northern Persia (known also as Nestorians*), London, Society for Promoting Christian Knowledge, 1892, pp. 77–78.

It may be as well to explain the nature of the country in which the Persians Syrians live. Between the Turkish frontier and the city of *Urmi* are, first, a chain of upland plains which run parallel to the frontier ... ant then a range of grass-covered hills. Eastward there lie other series of plains, which form the western margin of the Sea of *Urmi*, and extend for about a hundred miles in length ... In the centre lies the great plain of *Urmi*, which is about twenty miles broad in its widest part, and is watered by the three rivers known as the *Nazlu* river, the City river, and the *Barandus* river. These flow into the Sea of *Urmi*. To the north lies the little plain of *Gavilan*, which has but little water; its inhabitants have partly the habits of mountaineers, and partly the habits of plain-men. At the north end of the sea is the plain of *Salmas*, watered by a small river of its own and separated from the other plains by a range of mountains. The Christian inhabitants are chiefly Armenians or Syrian Roman Catholicos (the so-called *Chaldeans*). The dialect is very different from that of *Urmi*, and more resembles that of *Qudshanis*, the patriarch's village ...

All these plains are well watered by the short rivers which rise in the frontier mountains and flow down eastwards to the lake. But, by the time they reach the plains, they are most of them dry in summer, as the water is drawn off into artificial channels for irrigation purposes, and even in winter they are fordable in most places, except after a flood. As a consequence of the abundance of water, these plains are very productive, and might be made much more so under a good government. At present a large area near the mouths of the rivers is a great swamp, which causes malaria and retards locomotion.

1888–90

Paul Müller-Simonis, *Du Caucase au Golfe Persique*, Université Catholique d'Amérique, 1892, pp. 119–122, 129.

À la nuit tombante nous débouchons enfin dans la plaine de Salmas. La terre glaiseuse a été détrempée par l'orage; une foule de ruisseaux se croisent en tous sens et nous manquons à chaque pas tomber dans des fondrières.

......

Khosrâva, l'un des villages de population chaldéenne qui sont disséminés au milieu des localités arméniennes ou musulmanes du pays de Salmas, est bâti dans la vallée du *Tcharra-tchaï*: son terroir est composé d'une argile claire, disposée en couches profondes, que l'irrigation rend extrêmement fertile.

À une heure environ à l'Ouest de *Khosrâva* se trouve la petite ville de Salmas que l'on appelle dans le pays la «vieille ville»: ce fut autrefois un centre important, qui donne son nom à tout le canton.

Au temps de la plus grande extension du royaume d'Arménie, les contrées qui font aujourd'hui partie de l'Azerbaïdjan étaient des annexes de ce royaume, et le canton de Salmas appartenait à la principauté de Persarménie; Salmas en était la capitale. Son antiquité ne remonte pas au delà du IIIe siècle de notre ère; elle fut parmi les premières villes à se convertir à la foi chrétienne, et aujourd'hui encore, le pays de Salmas est, de toute la Perse, celui qui, proportionnellement à sa population, renferme le plus grand nombre de Chrétiens.[11] Les Arméniens du canton sont en majorité schismatiques; mais les Chaldéens sont presque tous catholiques.

Ceux-ci ont partagé autrefois les vicissitudes du schisme nestorien; mais ils l'abandonnèrent au XIIIe siècle, grâce aux prédications d'un jeune Chaldéen, originaire du *Diarbekr*. Il avait été, dans sa patrie, converti par les missionnaires dominicains.

……

La mission de *Khosràva* fut fondée par les Lazaristes en 1844; … Les établissements des Lazaristes sont groupés autour de la place du village, tout près de l'église chaldéenne. Comme toute demeure orientale dont la vie est concentrée à l'intérieur, ils ne présentent au dehors qu'un grand mur de pisé d'aspect triste. Lorsque l'on a franchi le porche, la première pièce que l'on rencontre, faisant face à la loge du portier, c'est le *Diwan-Khaneh*, ou salon de réception (littéralement, *place du jugement,* le mot s'appliquait d'abord au *diwan*-tribunal des hauts fonctionnaires; mais il a passé complètement dans la vie ordinaire). Du *diwan* l'on pénètre dans une grande cour où donnent les communs; cette cour sert aux séminaristes de préau de récréation pendant l'hiver, et c'est là que se trouve la chapelle de communauté, isolée des autres bâtiments.

Au fond de cette cour, faisant face à l'entrée, le bâtiment principal de la mission; c'est une grande maison de forme rectangulaire, à un étage sur rez-de-chaussée, solidement bâtie et dont les maçonneries en briques crues ont été consolidées par un revêtement de briques cuites. Tout y est simple; les chambres sont blanchies à la chaux et les meubles sont très modestes; mais comme la mission est ancienne, l'indispensable s'y trouve, et pour des voyageurs sortant de *caravansérais*, tout paraît magnifique.

Derrière la maison s'étend un assez grand jardin potager avec un promenoir abrité par une vigne grimpante et du houblon. Le tout a un aspect reposant et hospitalier.

L'œuvre principale de la mission est le séminaire: 15 à 20 jeunes gens y complètent leurs études commencées à la mission d'*Ourmiah*, et ceux qui ont la vocation ecclésiastique y font leur théologie.

Cette œuvre, qui date de l'établissement de la mission, est fort importante pour l'avenir de celle-ci, mais elle a toujours été difficile et ingrate, car les jeunes gens manquent en général de constance et se découragent souvent avant la fin de leurs études; depuis sa fondation elle a à peine fourni une vingtaine de prêtres; cependant la dépense est lourde pour le budget de la mission qui doit nourrir et habiller les élèves.

À côté de ce séminaire, la mission a encore une école pour les garçons de *Khosràva*; elle est dirigée sous la surveillance des missionnaires, par des professeurs recrutés dans le village et par quelques-uns des séminaristes.

Il semble que *Khosràva* soit assez aux mains des Lazaristes; ils y ont apporté la civilisation et se dépensent avec la plus grande générosité pour les habitants; leur influence n'est donc qu'une juste récompense de leur zèle.

Le nombre des missionnaires varie généralement de 4 à 5; ils sont aidés par trois prêtres chaldéens qui s'occupent plus spécialement du ministère paroissial.

L'établissement des sœurs de charité, dont les bâtiments sont assez embrouillés comme plan, donne, lui aussi, sur la place de l'église. Les sœurs ont un orphelinat, un asile, une école où tout est gratuit, et où elles prodiguent leur dévouement habituel; elles sont au nombre de sept.

1892

Rev. Samuel G.Wilson, *Persian life and customs with scenes and incidents of residence and travel in the land of the lion and the sun*, Fleming H. Revell Company, New York, Chicago, Toronto, 1895, pp. 89–91.

Salmas is a well-watered and fruitful plain, surrounded on three sides by mountains, and on the fourth by Lake *Urumia*. Through its center flows the *Zolah* River, rising in the Zagros. Its surface, of about one hundred square miles, contains four-score villages. *Dilman* and *Kuhna Shahr*, or "Old City", are the chief towns, which are of five thousand inhabitants each. The former is the capital, and contains the court-house, custom-house, post and telegraph offices, and chief bazaars. The markets are open two days in each place. Sunday is a special market-day, and is called here, as in Turkey, "*bazaar*-day".

A *Sassanian* sculpture in bas-relief. It represents two horsemen and a figure on foot, of mammoth size, supposed to be *Ardeshir* and *Shapur I* receiving the submission of the Armenians.

……

The inhabitants of the plain are mostly Turks, with some Kurds, six thousand Armenians, three thousand Chaldeans and Assyrians, and five hundred Jews. The Jews reside in *Kuhna Shahar*, the Christians in a score or more villages. The Armenians are thrifty and prosperous. They were free from oppression until of late years, when quarrels with other Christians—especially Catholics—and political agitations have given occasion to some Persian officials to fine and maltreat some of them. The Armenian villages of *Salamas* are the best built of any Persia, and farmers and tradesmen alike have a comparatively high standard of living. The chief Armenian merchant had an electric bell to ring at his gate; one of his sons was an amateur photographer, another dabbled in telegraphy.

At *Haftdewan* is a curious old church built of stone, with arched dome and large altar. As usual, it was without seats, and the back gallery was assigned to the women. Its walls we found frescoed and hung with old pictures, representing scenes in the life of Christ end the apostles, the canvas almost falling to pieces from age. One fresco represents the judgment. A male angel, with black hair, holds a pair of balances. The one scale is filled with light, the other with darkness, which is overweighted, though the devil and a group of imps are pulling with all their might on the side of darkness. On the side of light are groups sitting peaceably in Paradise; on the other is hell's gate, and behind it souls in misery. In the church is a dark well, in which many sick persons, and especially lunatics, are confined in the hope of being healed. Mussulmans, while rejecting much of the good they might see in the Armenians, are fully ready to adopt this superstition. I met a straggling procession of about thirty women, with their babies sniffing and coughing. They were Mussulmans from

Dilman coming to the Christian shrine for the healing of their children. Above the Old city is a monastery, to which pilgrimages are made, and where sacrifices are offered. It also has a ceazy-pit for the faith-cure *[in ancient times a belief was shared that a sick person could be cured by spending one night in a sacred place, in this case a pit in a monastery].* The monastery walls were lately strengthened and heightened. The government, hearing that a fortification was built, ordered the walls to be razed. Afterward permission was given to build low walls. The shrine and supposed grave of the Apostle Bartholomew is a short distance off in Turkey, at *Albak (Arebanes).* A numerous caravan of merit-seeking pilgrims yearly visits the shrine with great feasting. The estimated expense for one year of the congregated Salmas pilgrims was thirty thousand day's wages.

The American Presbyterian missionaires reside in *Haftdewan*; the French Lazarists have for many years had the head-quarters of their work at *Khosrova.*

1892

Rev. Samuel. G. Wilson, *Persian Western Missions*1896, pp. 162–164, 167.

… About this time the attention of the mission was specially drawn toward the need of evangelizing *Salmas.* Preachers and teachers from *Urmia* had been in several villages for a number of years. The Armenians seemed friendly. When I was itinerating there, many of them requested us to come and open up schools among them. In some of their schools the Bible in the modern version was being used. At one place the school room was thrown open to us for a meeting. The priest sent out and invited his flock and they listened with approval the truth … At another village they confessed with sorrow their utter ignorance of spiritual things and lack of true Christian experience. The need was evident, and the people seemed accessible.

It was decided to establish a new station, including in its field *Salmas, Kotur, Khoi* and *Maku,* with a population of over one hundred thousand, ten thousand of them being Armenians … A house was rented in *Haftdewan,* the largest and most central Armenian town on the plain of *Salmas.* It has a population of two thousand. Its streets, though crooked, are of good width and have water course running through them lined with large willows. The walls of the yards are unusually high and betoken the prosperity of the inhabitants.

The governor of the crown villages, of which *Haftdewan* is one, determined to prevent our occupying the house unless he received a large fee. Some of the villagers were opposed to our coming. A prolonged contest arose … While I was in *Khoi,* I interviewed the *khan,* endeavoring to have him withdraw his opposition. He refused, saying that even if the *shah* ordered it, we must arrange with him, and informed us indirectly that the sum he desired was £100. I told him we must then refer the matter to Teheran … Later a telegram came, "Government orders house to be given, and *khan* to report in person to Teheran … Shortly afterwards Rev. J.N. Wright was transferred from *Tabriz,* and took up his residence in *Haftdewan* (1884). Active opposition ceased, but the Armenian bishop wrote, enjoining the priest to prevent anyone from coming to us. Some bound themselves with an oath not to have anything to do with the missionaries. But the common people heard us gladly, and teachers and preachers were requested by them in some of the villages.

……

The work among the Armenians of *Azerbaijan* has been a difficult and trying one. One great difficulty which indeed manifests itself among all the races of Persia, is the mercenary idea of religion. The idea seems deep-rooted that religion is a matter of worldly policy, and is to be valued in proportion to its temporal benefits. In Islam it is the established custom to render the convert financial aid.

1903

Abraham Valentine Williams Jackson, *Persia past and present. A book of travel and research*, New York, The Macmillan Company, 1906, pp. 78–80.

The stages of my journey as far as *Dilman* in the Plain of *Salmas*, northwest of the lake, were slow, averaging not more than twenty-five or thirty miles a day, with stops for the night at the villages of *Dizah-Khalil* and *Tazvich. Dilman* was reached in the afternoon of Sunday, March, 22. This town is one of the largest places in the plain, but its *manzil* seemed little more than a shelter, nor was I attracted by an invitation to visit the tea-house near-by, which I later learned was an opium den. The carcass of a dead horse lying in the stream that runs through the town seemed characteristic of the place, but the people appeared to be kindly disposed to a stranger and showed something of a holiday spirit, as they had put on their best clothes to celebrate the season of *No-Ruz* and also Sunday, which is always a special "*bazar* day", the regular *Mohammedan Sabbath* being Friday.

I left *Dilman* the next morning at eight, with the weather bleak and dreary. In about two hours we came to an interesting old Armenian cemetery which is situated on a hillock near a small village. To reach this graveyard I had to leave the wagon and made through snow and water for nearly half a mile. An inscription on a large monument near the summit ascribes the founding of the burying-ground to the *Mamikonians*, an ancient heroic family of Armenia, and I observed a number of grave-stones rudely shaped like a ram, a common image in old Armenian burial-places. I noticed also one long Syriac inscription, but the letters were almost illegible.

After a short stay on the hillock I proceeded on my journey to see the Sasanian bas-relief of the horsemen, which I knew was carved on the side of a rocky hill called *Surat Daghi*, 'Picture Mountain', somewhere on the road between *Dilman* and *Guchi*. Shortly before noon we reached a tiny hamlet and found that we had passed the hill of the sculptures, having met no one on the deserted plain who knew precisely where the carvings were located. Accordingly I took a guide and returned on foot through mud and snow for a distance of nearly three miles in order to examine the bas-reliefs.

The sculptures are carved about a hundred feet above the plain on the face of a somewhat precipitous rock, and are undoubtedly Sasanian in origin, as they present all the characteristic features of the bas-reliefs at *Tak-I Bostan*, *Naksh-I Rustam*, and *Naksh-I Rajab*. The group is composed of four figures, two mounted and two standing … *[see Texier's description]* … The generally accepted identification of the group is that the bas-relief represents *Ardashir Papakan*, the first Sasanian king, and his son *Sapor*, receiving the submission of the Armenians, an event that occurred about A.D. 230, to which period the sculptures approximately belong.

1908

Eugène Aubin, *La Perse d'aujourd'hui. Iran-Mésopotamie*, Libr. Armand Colin, Paris, 1908, pp. 66–67.

Après avoir passé le village arménien de *Diza* et le pont du *Kotour-Tchaï*, la route de *Khoï* à *Salmas* remonte une vallée desséchée pour aborder la montée du *Kara Tépé* (colline noire). Au delà commence la plaine de *Salmas*, verdoyante et arrosée comme celle de *Khoï*, entourée d'un cirque de montagnes grises ... *Diliman*, chef-lieu du district, est musulman; un bourg, enclos de murs, de 4.000 habitants. Les chiites y sont en majorité ... Les sunnites [sont] assez nombreux dans la campagne.

......

Vers le fonde de la plaine, à devant du village de *Kouné Chehr*, se trouve un vieux tombeau comme il en existe plusieurs autour du lac d'*Ourmiah*: ce sont tours en briques surmontées de coupoles; deux portes opposées, décorées d'inscriptions arabes et de mosaïques en faïence bleu, donnent accès dans le tombeau; la coupole est tombée; les motifs décoratifs s'effritent; les inscriptions se lisent à peine: «Construit par *Eminé Khatoun*, fille d'*Oghouz Agha*» et le verset, usité dans les tombes: «Tout le monde doit mourir, sauf Dieu, qui reste éternel».

Kouné Chehr est le seul village de *Salmas* où il y ait des juifs. Les juifs de l'*Azeebaïdjan* sont groupés le long de la chaîne du Kurdistan, de *Salmas* à *Miandouab*. Les premiers vinrent de la vallée du *Tigre*, lors de la captivité de Babylone; ils furent rejoints par le restes de la colonie juive de *Tauris*, jadis florissante, grâce à la tolérance des Mongols, persécutée à la suite de leur conversion à l'islamisme ... Ceux de *Salmas* sont une soixantaine de familles qui font l'usure, le colportage, le petit commerce; quelques-uns possèdent des vignes. De leur sort présent, ils ne se plaignent point; leur village appartient à la princesse héritière, et c'est une garantie suffisante contre les exactions. Mais le pays n'offre point de chances de développement à la jeunesse, qui tend à passer en Russie.

......

Les soixante villages de *Salmas* sont disséminés sur les deux rives du *Zolla-Tchaî*; indiqués par des bouquets d'arbres, séparés par des champs, où se dressent les tours de guet contre les Kurdes: 40.000 habitants, Musulmans, Arméniens et Chaldéens.

Parmi les chrétiens, les Arméniens se trouvent être les plus nombreux: ils forment à peu près le quart de la population, répartis dans 23 villages. *Kelassar* est leur centre commercial, comme *Diliman*, celui des musulmans; mais le véritable chef-lieu de la communauté est *Heftevan*, siège du vicariat grégorien: un gros village de 446 maisons, presque exclusivement arméniennes. Les Habitants l'ont racheté, ainsi que les principaux villages arméniens de la plaine (*Payadjuk*, 350 maisons; *Mehlam*, 200); les musulmans ne gardent le sol que là où se trouvent isolées des familles chrétiennes. Dans chaque village, existent l'école et l'église; pour les soins médicaux, quelques protestants, Chaldéens ou Arméniens, ayant étudié chez les missionnaires américains et pris leur grades aux Etats-Unis.

……

Le village voisin, *Khosrova*, sert de capitale aux Chaldéens; l'évêque catholique de *Salmas*, la mission lazariste y résident. Le diocèse s'étend jusqu'à l'entrée de la plaine d'*Ourmiah*, compte de 5 à 6.000 fidèles; il paraît que, de ce côté l'ensemble de la nation est uni; peu ou point de Chaldéens protestants, nestoriens, orthodoxes. Ces gens sont groupés dans sept villages: la plupart habitent *Khosrova* (400 familles) et *Patavour* (100); quelques-uns à *Gulizan* et *Ougla*; naguère, le village de *Saours* était chaldéen; il est maintenant habité par 168 familles arméniennes, en majorité catholiques.

……

Ces villages chrétiens sont organisés sur le même modèle que les villages persans. Ils élisent annuellement leur maire, chargé de percevoir l'impôt, et leurs «barbes blanches». Les sept paroisses possèdent églises et écoles. La maison lazariste entretient à *Khosrova* un séminaire et un collège; neuf clercs, cent cinquante enfants y apprennent le français et le persan. Les Sœurs de charité desservent un petit hôpital et une école fréquentée par cent soixante-quinze petites filles. Un médecin chaldéen … soigne toute la communauté. A côté de la mission, la maison de l'évêque chaldéen, l'église de *Mar-Ghiverghis* (Saint Georges), forment un bloc religieux au centre de *Khosrova*. Le cimetière est à l'entrée du village: un amas de pierres tombales, noircies par les ans, la plupart allongées en forme de sarcophages, quelques-unes affectant la figure d'un bélier. Selon l'usage de l'Orient, la population vit autour de ses prêtres. Beaucoup vont chercher fortune à Tèhéran ou au Caucase, même à Constantinople.

4 The City and the Lake of Urmia in Some Travel Reports in Persia

Ruy Gonzalez de Clavijo, ***Narrative of the Embassy to the Court of Timour at Samarcand, A.D. 1403–6*****, London, Hakluyt Society, 1859,** pp. 87–88.

[The lake]

On Monday, at noon, they came to a place called *Caza*, which was large and populous. It was situated in a plain, with many fruit gardens, and streams flowing from all parts. Before reaching this place, there is a lake of salt water, a hundred miles round, and in it are three islands, one of which is inhabited (Lake *Urumiyeh*). They passed the night at a place called *Cusacana [Kuzah Kunan]*, which was a large town, but half destroyed. They say that *Toktamish*, Emperor of Tartary, destroyed it; who was afterwards conquered by *Timour Beg*, and driven out of his dominions, and is now deprived of them; as will be related to you presently. In this place there were many Armenians.

On Tuesday they slept at a place called *Chauscad*, in a plain covered with many orchards, vineyards, and fruit trees; and many streams descended from a mountain, which rose up above the plain, and irrigated the gardens. They take quantities of fruit from this town to the city of *Tabreez*, and other places.

They passed that night in a plain, and most of the road they traversed that day, passed through gardens and vineyards; and the road was level, and it seemed very pleasant to pass through these gardens.

On Wednesday, the 11th of June, at the hour of vespers, they arrived at the great city of *Tabreez*.

James Morier, *A Secon Journey through Persia, Armenia, and Asia Minor to Costantinople between the years 1810 and 1816,*London, Longman, Hurst, Rees, Orme and Brown, 1818, pp. 286–289.

[The lake]

A good direction to the ponds is a peak in the adjacent mountains, immediately under which they are situated. They bear from the village of *Shirameen* N.80 W and are distant about two miles from the border of the lake.

From our camp at *Shirameen*, we had an extensive view of the lake, which is bounded on the north by mountains. Near it are situated the town of *Salmas*, and the productive district of *Shebester*. The plain of *Tabriz*, the salt desert, with the hills and vallies of *Uzkoh* inclose ìt to the east, whilst a sublime range of snowy mountains of the *Courdistan* girds it to the west, and it is terminated to the south by *Maragha*, its table lands, and its extensive pastures. It is most commonly known by the name of *Deria Shahee*, or the Royal Sea, on account of the peninsula of that name, that extends itself within it, and is said to be twelve *fursungs* in circumference.

It is also called *Deria Maragha*, and *Deria Oroumieh*, from the two large towns of that name, that are situated in its neighborhood, and frequently only *Deriacheh*, or the Little Sea. According to *Mirkhond*, in his *Ajaib al balladan*, it was included within the limits of Armenia, being called by him *Deria Armenistan.* Eight *fursungs*, or about 280 miles, are mentioned as its circumference; and from all we can learn, it is generally very shallow, its greatest depth not being more than three or four *zer* or cubits, and in some places scarcely one. We got these particulars from a man who had been sent by the Governor of *Maragha* to the Ambassador, expressly to serve as a cicerone and who assured us that he had crossed over to *Oroumieh* in a boat. He, moreover, asserted that there existed a causeway, which was now covered by water, from the village of *Chawan* to *Oroumieh*, but so visible, particularly in the shallow parts as not to be doubted; adding that it might be possible still to cross over the lake without the aid of a boat. He could not speak of the antiquity of this work, as it had never been used by the inhabitants of the present day; nor do I recollect it to be mentioned by any of the old travelers: it may therefore date from remote times, none of the later kings of Persia, except the great *Shah Abbas*, being remarkable works of public utility.

The same fact which appears in the Caspian Sea, the Dead Sea, and many other lakes in the globe, is also to be remarked here: I mean the daily reception of a great quantity of water without any visible increase in the lake itself. No less than fourteen rivers of different sizes discharge themselves into the lake of *Shahee*; and although from the general character of Persian rivers, I should not suppose any of them to be so large as the Jordan, yet still collectively they cannot fail to make up a very large mass of water. Instead of increase, there are many visible signs of diminution of the water, from which we may conclude that the evaporation is greater than the supplies from the rivers. *Shahee*, that was formerly surrounded by water, has for the two or three last year been attached to the main land by a swamp, which is an evident encroachment upon the lake. The plain of *Chawan*, and the district of *Deenab,* upon which are large and conspicuous tracts of salt, have also evidently been abandoned by the water, which, from local tradition, as well as the conformation of the surrounding mountains, probably washed the bases of the nearest highlands. That there has been great variation in the depth of this lake is certain, first from the causeway, which was built when the whole was doubtless one great swamp; and, secondly, from the visible diminutions above mentioned, and this may be accounted for, by the uncertain supplies of the rivers themselves, which depend almost wholly upon the mountain torrents. Persian rivers, in general, are on the one day overflowing their banks, and on the next, can scarcely be called rivulets, on which account, it would be impossible to form an estimate of the proportion of the supply to the evaporation. I should conceive, that the lake must have its largest quantity of water in spring, when the snows melt, and the torrents flow with the greatest violence, an opinion which derives confirmation from the nature of the small salt water lake, seen on the right of the road between *Ojan* and *Tabriz*, which in spring receives its water from the mountain, and in summer is completely dry from evaporation alone. This lake resembles in many things, to what Sandys calls "that cursed lake Asphaltides or the Dead Sea. Like it, its water seems dull and heavy, and the late Mr. Brown found that it contains more salt than that of the sea. We were informed, that as soon as the rivers disgorge any of their fish into it, they immediately die. We saw swans in the lake, near the coast contiguous to *Shirameen*. Like the Dead Sea, it also supplies the adjacent country with a salt of beautiful transparency, although the inhabitants generally prefer the rock salt, which is cut from quarries in the neighborhood of the petrifactions.

It contains four islands of different heights and sizes, exclusive of *Shahee*, which, though in fact a peninsula, is also called an island. They are small and barren, but are very conspicuous from their white cliffs, which, on a clear day, are visible from the top of the mountain above *Tabriz*. We were told, that not long ago a race of wild asses existed upon them, which had originally been turned loose there by the *Seffies*, but which are now destroyed. At present they are tenanted by a family of venomous snakes and other reptiles, and are no otherwise useful than as affording a low brushwood and other material for fuel, to those who will give themselves the trouble to carry it away. *Shahee*, we hear, is inhabited, and contains twelve villages, the inhabitants of which are said to be strangers to the small pox, and live to a good old age without the dread of it; but notwithstanding this blessing, it is looked upon in Persia as the Gyarus of the Romans, for hiter *Abbas Mirza* sometimes sends offenders into exile, a punishment which the Persians hold in great abhorrence.

Fig. 20 *Map of the Ottoman Empire frontiers in the vicinity of Urmia with seven gates*, 1881–88 (Archive of the Documentation and Diplomacy History Center of the Ministry of Foreign Affairs of the Islamic Republic of Iran Document n.R-757)

James Morier, *A journey through Persia, Armenia, and Asia Minor, to Constantinople, in the years 1808 and 1809,*London, Longman, Hurst, Rees, Orme and Brown, 1812, pp. 295–297.

[The lake and its Territory] (Fig. 20).

Shebester is a large town, surrounded by several villages, and by more wood and cultivation, than any spot I had yet seen in Persia. Hitherto indeed the want of trees, either as a shade to the road, or as a relief to the inequalities of the heights, had been constant and uniform. We admired therefore doubly the beauties of our present course. Streams of running water were meandering in every direction amid the numerous willows, poplars, almonds, and other trees, which bordered our road: and at intervals the artificial dikes were opened to admit water into the beds of rice. The greater part of the country was covered with verdure, for the new corn was already well advanced both in maturity and plenty. Peasantry enlivened the fields by the labours of the spade or the plough.

After quitting *Shebester* we came in full view of the delightful lake of *Shahee.* It derives its name from the surrounding *Mahalé*, which may contain twenty villages. I was told that its waters are as salt as the sea, and that the sand over which they flow, produces the salt used at Tabriz. It extended itself N. W. and S. E. before us, and its western extremities were terminated by a stupendous chain of mountains, whose snowy summits, softened by the haze, contrasted admirably with the light azure of the lake. As we proceeded, the long mountain (which I mentioned in the route of yesterday, extending itself and forming a peninsula in the lake) appeared to have no connection whatever with the surrounding lands; and, by a stranger to the real topography, would have been pronounced an island. Its termination (to the south as seen from our road) was in the form of a sugar-loaf.

Near *Shebester* we passed the village of *Misholéh* and, lower down in the plain, those of *Arsaléh* and *Halee*, on the left of the road. Others indeed are seen at every turn, situated at small intervals on either side alternately, all in the *Mahalé* of *Ghuméh*. Among them are *Besh-kefelout* on the left; *Khomyhé*, prettily surrounded with verdure, on the right; *Shinwar*, on the left again; *Khuzec-dunar*, on the right, three *fursungs* before we reached our stage at *Tasouj*; and on the left, about two miles from the borders of the lake, *Alibangolu*, the first place in the *Bolouk* of *Aeenzaub*. In this line we stopt and fed our cattle and ourselves; while a refreshing breeze from the Westward just curled up the waters of the lake, and waved the corn fields which extended themselves on all sides of us.

Our bread and moss was shared by a stranger who was going to *Oroumi*, a large town, distant thirty *fursungs* from *Tabriz;* and situated, by the pointing of his hand, S.50 W. from us, on the left or West side of the lake, which the road continues to skirt through its whole course. On the East of the lake is *Saouk Bolag*, the site of the ancient city of *Sheherivan*. The country, through which we passed in the day, was interesting and picturesque; in every turn of the view enriched by the lake and its surrounding capes and mountains.

From all that I could learn in this region, (and I inquired of many who had travelled repeatedly over this part of *Aderbigian*), there appeared to exist no other lake than this of *Shahee*. And I have as regularly made direct inquiries about the situation of the city of Van and its lake, without obtaining anything like a satisfactory answer. On the contrary, the very existence of such a place, and such a lake, was always denied; I mention this, when the position of Van has been clearly ascertained, to show how general was the ignorance of the people on every subject which was not immediately within their own circumscribed district. Nor was I more successful in my inquiries on the real extent of the lake before them: every one said that it was very large, and that it reached further, than from its appearance we might suppose.

At about five miles from *Tasouj*, there is a village on the left called *Rahdar Khoné*; and then a station of *Rahdars*, or custom-house officers. As we passed it, one of them, a man of a much more respectable appearance than any of the class whom we had seen on other occasions, told us that a driver with seven loaded mules had gone forwards, and refused to pay the duties, alleging that his beasts were carrying part of our baggage; and were therefore in the King's service, and as such exempt from the impost. In fact, however, my *Charwardar* (or conductor of the mules or caravan) had added to my charge this number, above those that were necessary for my purposes; and, having already received a part of their hire from me, was now employing them still more to his own profit, by conveying upon them, duty-free, in my name, the goods of some *Tabriz* merchants. On discovering the fraud, I resigned him into the hands of the officer, with full liberty to exact his dues; a licence, under which he began immediately to cudgel the shoulders of the defaulter. The duties here are high, being five reals on each load. Some miles before we reached *Tasouj*, the lake begins to make all elliptical termination, and the road to turn off on a more Northern angle. We were eight hours in travelling the whole distance from *Ali Shah*, which we reckoned at thirty-two miles, on a bearing of N. 60 W. *Tasouj*, from the great extent of the ruined walls about it appears once to have been a large place, but it is now reduced, by earthquakes, to the denomination of a village. There are remains of domed bazars and mosques, spread in every part of the place.

M. Vivien de Saint-Martin, ***Mémoires Historiques et géographiques sur l'Arménie*****, Paris, De l'Imprimerie Royale, 1818**, pp. 56–61.

[The lake]

A l'orient du lac de Van, on en trouve un autre à-peu-près de la même étendue, qui a cent trente milles de long, sur la moitié environ de large, il est nommé actuellement par les Turcs et les Persans, indifféremment *lac de Tebriz* ou *lac d'Ourmiah*. *Aboul'féda* l'appelloit lac de *Tela* ou d'*Ourmiah*. «le lac de *Tela*, dit-il, est le même que celui d' *Ourmiah*; il est situé entre *Meraghah* et *Selmas*, à l'occident de la première et à l'orient de la seconde de ces villes. *Meragha* est à dix-sept parasanges de *Tebriz*, et depuis la rive nord-est du lac jusqu'à *Meraghah* il y a une journée de chemin. Ce lac avoit environ cent trente milles de longueur de l'occident à l'orient, en inclinant un peu vers le midi, et a-peu-près la moitié de largeur». Le premier de ces noms lui venoit de l'île de *Tela*, qui étoit située au milieu de ses eaux, et qui renfermoit un château très fort, élévée par l'ordre de l'empereur *Moghol Houlakou*, pour qu'il pût y déposer en sûreté ses trésors (*Aboul'féda*). Cette île et le fort qu'elle contient sont appelés actuellement par les Arméniens *Kogartchin ghalasi*. Le lac d'*Ourmiah* est salé comme celui de Van, et il paroit que pour cette raison il porta, comme ce dernier, dans le pays, la dénomination *d'Aghi Dzov*, c'est-à-dire lac salé. Les Arméniens l'appellent encore vulgairement lac d'*Ormi* ou d'*Ourmiah*, ou lac de *Tavrej* ou *Tabriz*.

Il paroît que ce lac est le même que celui que Strabon appelle *Mantiane*, qui étoit situé vers les frontières de l'*Atropatène*, et dont le nom, selon lui signifioit *bleu*. «L'Arménie a aussi de grands lacs, dit ce géographe: l'un d'eux est celui de *Mantiane*, ce qui signifie *bleu*; il est, dit-on, le plus grand qui existe après le *Meotis*. Ses eaux sont salées; il s'étend jusqu'à l'*Atropatie* et il contient des sources salées». Il est probable que ce lac est le même que celui de *Spauta*, qui étoit aussi salé, et dont le même auteur avoit parlé un peu auparavant dans sa description de l'*Atropatene*. Il est impossible de le confondre avec le lac de Van, que Strabon désigne clairement sous le nom *d'Arsene* et de *Thonitis* ou plutôt *Thospitis*, et il doit être celui d'*Ourmiah*, que l'on povuoit attribuer tantôt à l'*Atropatene*, puisqu'il séparoit ces deux pays. Ptolémée le nomme encore lac de *Médie*. En decrivant l'*Atropatene*, Strabon aura donné à ce lac le véritable nom qu'il portoit chez les indigènes et qui signifioit *bleu*, et, dans la description de l'Arménie, il lui aura donné celui de la province qu'il arrosoit de ces eaux. La *Mantiane* ou *Matiane* étoit un pays assez considérable qui s'étendoit à l'orient de l'Arménie du côté de la Mèdie: et il ne seroit pas étonnant qu'on ait confondu son nom avec celui de lac *Spauta*; les rapprochements étymologiques que nous allons faire, ne contribueront pas à éclaircir ce point, et à mettre hors de doute l'opinion que nous venons d'émettre.

Je ne crois pas qu'il existe dans les langues Arménienne et Persane, un seul mot pour désigner la couleur bleue, dont le son puisse avoir une analogie même éloignée avec celui de *Mantiane,* qui, selon Strabon, avoit cette signification. Le nom de *Spauta*, que l'on trouve dans la description de l'*Atropatene* du même auteur, a beaucoup plus de rapport avec celui que les Arméniennes donnoient dans l'antiquité au lac d'*Ourmiah*; et d'ailleurs le sens même du nom *Arménien* confirme pleinement, par son origine étymologique, l'opinion du géographe Grec. Strabon ne s'est donc point trompé sur la signification du véritable nom du lac d'*Ourmiah*, mais il a eu tort de lui donner celui du pays qui se trouvoit sur ses rivages, et de lui attribuer ensuite un sens qui ne pouvoit s'appliquer qu'au nom même du lac, que, par erreur, il a négligé de rapporter dans sa description de l'Arménie.

Les Arméniens appellent le lac d'*Ourmiah*, depuis une époque fort reculée, *K'habodan*, nom qui est évidemment dérivé du mot *gaboid*, qui signifie *bleu* et qui n'en diffère que par une légère nuance de prononciation et par une terminaison qui étoit peut-être particulière au langage usité dans l'*Atropatene.* L'analogie de son qui existe entre le *Spauta* de Strabon et le *K'habodan* ou *Gaboid* des Arméniens, n'est pas très frappante, malgré l'identité de signification; mais nous avons déjà eu occasion de rectifier l'orthographe, d'un grand nombre de noms mentionnés dans ce géographe dont le texte a éprouvé une très-grande quantité d'altérations, et il est très-probable que le nom de *Spauta* est dans le même cas que ceux que nous avons déjà corrigé. Nous avons déjà vu qu'il existoit, près des sources de l'Euphrate, une montagne connue sous le nom de *Capotes*, mot qui signifioit la même chose que le nom du lac en question. Il seroit permis de penser qu'originariement on lisoit dans le texte de Strabon, *Kapauta* au lieu de *Spauta*, qui sera introduit dans les manuscrits par l'ignorance des copistes.

Il paroît que le nom *K'habodan,* en usage chez les Arméniens pour désigner le lac d'*Ourmiah*, a été connu d'une manière assez exacte des étrangeres; car, dans l'abrégé que nous possédons de la Géographie du *scherif Edrisi*, on trouve entre la ville de *Merâgah* et l'Arménie, un lac appelé dans la traduction Latine, *Kanoudan*. Il ne peut être que celui qui nous occupe. Ce nom se trouve déjà mentionné antérieurement dans un ouvrage de *Masoudy*, qui vivoit environ deux siècles avant *Edrisi*. La légère différence qui existe entre l'arabe *Kanoudan* et l'Arménien *K'abodan*, est fort peu de chose, et elle ne surprendra pas les personnes qui connoissent l'alphabet Arabe, et qui savent combien il est facile de se tromper en plaçant les points diacritiques qui servent à déterminer la valeur de lettres qui ont toutes la même forme.

……

Le géographe *Ibn-Haukal*, qui vivoit à-peu-près à la même époque que *Masoudy*, parle aussi de ce lac, qu'il appelle *Kaïoudhan*; ce qui est toujours le même nom différemment altéré par le cause que nous avons déjà indiquée. « Le lac, dit-il, qui est dans l'*Aderbaïdjan* et qui est appellé *Kaïoudhan,* est salé, et ne contient ni poissons, ni reptiles. Une grande quantité de vaisseaux vont et viennent entre *Ourumiah* et *Merâghah*».

Le lac d'*Ourumiah* est environné de tous cotés par de montagnes: au nord-est et à l'est, il a les monts *Sehend* et *Silan*; au sud et à l'ouest, les montagnes de l'Irak Persan, celles du Kurdistan et de *Karah-dagh* ou en persan *Siahkou*, c'est-à-dire, *montagne noire*. Le lac reçoit un grand nombre de rivières qui coulent de ces diverses montagnes et l'alimentent perpétuellement. Les principales de ces rivières sont le fleuve de *Seraw*, ou le *Sourkhab*, qui a quarante lieues de cours et se jette dans le lac, après avoir reçu les rivières de *Serd*, d'*Edounr* et d'*Oudjan*. La rivière de *Tchfteh* a vingt lieues de cours et se jette dans le lac de *Tebriz*, après s'être grossie des eaux du *Safy-roud*, qui passe à *Merâghah*, du Teftou, du *Meraw* etc. Le lac d'*Ourmiah* reçoit en outre une grande quantité d'autres rivières ou torrents qui lui viennent des montagnes du Kurdistan et de l'*Irak*.

Eli Smith and H. G. O. Dwight, *Missionary researches in Armenia: including a journey through Asia Minor, and into Georgia and Persia, with a visit to the Nestorian and Chaldean Christians of Oormiah and Salmas*, London, George Wightman, 1834, pp. 386–387.

[The Countryside]

After leaving the few villages in the vicinity of *Jamálava*, the narrow plain between the mountains and the lake ceased to be cultivated. At the distance of four or five miles, the plain of *Oormiah* opened, projecting for some distance into the lake on the left, and extending up between the mountains on the right. It seemed of great extent, and almost perfectly level. A road led directly to the town of *Ourumiah*, on its south-west side near the foot of the mountains, at a distance ten miles from the lake. We turned to left, and were soon surrounded by marks of a dense population and of extreme fertility. Villages were separated but a little from each other, and the fruit trees and poplars around them resembled a continuous forest. Almost every spot was cultivated with grain or cotton, and the canals which irrigated them (some of them so large as to be used for mill-stream) were so frequent and full, as seriously to impede our progress. Nothing could exceed the apparent fertility of the dark loam which formed the soil. In a word, the native province of Zoroaster (for tradition pretends that Zoroaster was born in *Oormiah*), seemed altogether the finest, the most densely populated and highly cultivated tract we had seen in Asia.

James Ballie Fraser, *Travels in Koordistan, Mesopotamia* etc., vol. I, London, Richerd Bentley, 1840, pp. 50–51; 56–58.

[*The plain*]

Next morning, crossing the plain, we ascended a valley through certain low hills which separate the plain of *Selmâs* from another of about twenty miles in length and ten in breadth, also skirting the lake, and backed by the great mountains of *Koordistan*, many of whose summits were still spotted with snow. This plain was also fertile and well cultivated, and five or six large villages near the hills gave proof that it wanted not its portion of inhabitants. We breakfasted this day in a little valley by a clear stream, disturbing, I am sorry to say, two *Koordish* ladies who had come there for the purpose of washing clothes, a process which they appeared to be performing in the old Scotch fashion, with legs bare to a pretty considerable height.

……

[The agriculture]

Before sunrise we were on our way to the city of *Ooroomia*; and nowhere, certainly not in the East at least, do I remember passing through a richer country. All around us, in whatever direction we cast our eyes, they fell on sheets of varied cultivation, cotton, rice, millet, castor-oil plant, vineyards, orchards, and great fields, from whence already the heavy crops of wheat and barley had been reaped, prettily divided, and fringed here and there with coppice or rows of trees, willow, poplar, *sinjeed* (jujube), and others. Numbers of large villages, each encircled with its own gardens, lay thickly scattered about, and the state of those we passed through was conclusive as to the wealth and comfort of their inhabitants. Thus, for at least twenty miles, did we ride through a perfectly rich and cultivated country to the very walls of the city, which were concealed by trees and gardens.

This day I observed two agricultural improvements which I have seen nowhere else. One was a plough, used to break up the deep soil, of far greater efficacy than the common miserable instrument. It was dragged by eight buffaloes and bullocks, the draught being from an axle with wheels of different sizes at either end; one about twenty inches or two feet in diameter, intended to go in the last-made furrow; the other, not above half as much, to run upon the yet unbroken soil, so that the axle, by this contrivance, was kept upon the level. The other improvement was the adoption of a two-wheeled cart, also drawn by buffaloes—a heavy clumsy machine, it is true, with solid wheels like the old Highland cartwheels, which went creaking and bumping along on their most irregular circumferences, but still effective enough engines for carrying straw, and such like light bulky articles. It is not a little remarkable that these carts, or *arabas*, as they are called, though used in Turkey, Armenia, and Georgia, and even, I believe, at *Khoee*, should not have been adopted even at *Tabreez*, at *Tehran*, or the more eastern and southern provinces of Persia; the pace of improvement is mighty slow indeed here.

Entering the city, I went straight to the house of the *Beglerbeggee*, or governor, *Nejeff Koolee Khan*, an *Affshar* nobleman of very high rank, the King having married a sister of his father, so that some of the Princes are his cousins.

……

[*The Governor's apartments*]

The apartment, in which I am now sitting, is thirty feet long by fifteen in breadth, about twenty feet high, and handsomely vaulted. The walls are fitted up with very tasteful recesses, both arched and square, and the whole, walls and roof, have been painted of a rich white, picked out with blue, red, and gold, except for about three feet from the floor, which is salmon color, ornamented with flowers in deeper tints, all done in oil-colors; and there are some very handsome colored ornaments on the crests of the arches above. One whole end is occupied by a splendid window of painted glass, terminating in a Gothic arch: I wish I could convey to you the pattern. It is indescribably beautiful, and reminds one of the richest figures shown by the kaleidoscope. In the other end, there are two smaller windows of a similar construction, between which is a fire-place: the floor is of tile; but, when suitably dressed up, is first covered with mats, and over them with rich carpets. The palace contains many smaller rooms of equal beauty, and, in fact, I know few things like it in Persia, except at *Ispahan*; for, unlike the buildings of the present day, it is constructed of solid brick and mortar. So for once, at least, I am well lodged.

……

[*The City and its Monuments*]

Of the city and its environs, I must say a few words. It has an aspect of more comfort and solidity than any other I remember in Persia. The gardens, which embrace it to the very walls, unite their foliage with that of many trees within them to produce a rich and verdant aspect; for every house of consideration has its gardens, with rows of chenars and poplars towering above the inclosure. The streets are wider than in most other towns, and have generally a stream of water running in the midst, so that each house may enjoy at will this necessary of life. Few ruins are visible; for, where any exist, they are generally concealed by the wall, which incloses each tenement, like a fort. This mode of rendering each house—that is, those belonging to the great - a sort of stronghold is remarkable, and very characteristic of the state of society which used to subsist in *Ooroomia*. Situated on the borders of *Koordistan*, and in the midst of a "land debateable", frequented by the fierce tribes and clans of that wild country, most commonly at blood-feud with one another, it became a rnatter of necessity for each chief to have at least a place of refuge in case of attack from his enemies.

…… (Fig. 21).

Ooroomia, though an ancient city, and known as such even in the days of the Romans, possesses no show of antiquities. An old tower and mosque of Arab architecture, and built about six or seven hundred years back, can scarcely be regarded as entitled to much consideration. An Armenian church, if we may credit the assertion of its present congregation, has a better claim to attention; for it is said to be one founded by the original Christians of this country. But this claim rests merely on tradition; for, as to the building itself, whatever may have been its original form or size, it now resembles nothing externally so much as a mud hut. The principal part, indeed, is under ground, and we entered it through a low door, scarcely large enough to admit a man into the narrow passage beyond: this led us stooping, into a dark hole, the form of which, even when the eye became reconciled to the obscurity, was hardly to be made out from the quantity of strange hangings that encumbered rather than adorned its walls-offerings of the piety, or superstition, or vanity of its frequenters -suspicious-looking garments, like the shifts of women and the trousers of men, hung from the roof and cornices like the tattered banners of some ancient castle hall or chapel, but quite without their dignity. On one side of what might be the altar, lay a

Fig. 21 Ourmiah view from the house of the Lazarists (Binder 1887)

black bonnet, worn by the priest; on the opposite side, a tattered copy of the Armenian Bible. Beyond was a room, into which none but the priest was permitted to enter, and in this the woman, who showed the church, said there was a stone with some writing in Armenian upon it. The priest was not to be found, so I made no attempt to enter the sanctuary, as, without his aid, I could have made nothing out had I succeed - indeed, even with his aid, I doubt if much could have been done, for these *khaleefahs*, or priests, are very ignorant.

Justin Perkins, *A Residence of eight years in Persia among the Nestorian Christians, with Notices of the Muhammedans*, New York, Allen, Morril and Wardwell, 1843, pp. 177–179 and 186–187.

[The City] (Fig. 22).

The city of *Oròomiah* is quite near the south-western extremity of the plain. It was nearly dark when we reached it. The bishop conducted us directly to the Nestorian corner; and gave us as a room for lodging, one of the porches of the church, taking an adjoining one for himself. The church is a large edifice, built of stone and brick, on, or rather *in*, an elevated spot, all but the roof being imbedded in the ground. The building is divided into several apartments, all of which we entered by extremely small doors. The church itself is considerably larger than the other rooms, and back of it are the baptistery and the *samctum sanctorum* or apartment for consecrating the elements of the communion. We found in the church no images or pictures; but its entire walls were most untastefully -not to say ridiculously- bung with old shawls and pieces of

Fig. 22 View of the city of Urmia (Binder 1887)

calico of every conceivable description and color, for the purpose of ornament. The church is surrounded by an ancient cemetery, some of whose stones -huge oblong blocks of gypsum- are very large and all are inscribed with Syriac characters. In the centre is a beautiful fountain, surrounded by wide spreading shade trees. The city of *Oròomiah* - the ancient *Thebarma,* as already suggested, is renowned as being the birth-place of Zoroaster, *Zerdusht,* as the Persians call him. It is situated on a slight elevation of ground, about ten or twelve miles west of the lake and within two miles of the mountains. It is in nearly 37° 20″ north latitude and 45° 15″ east longitude. Its proximity to the lake and eastern exposure from the mountains, give it a milder climate than that of *Tabréez*; but being so near the lofty *Koordish* ranges, it has considerably more snow in winter than that city. The most interesting remnant of antiquity, in *Oròomiah,* is an immense mosk—the only one in the city that is surmounted with a dome and cupola—which was once a Christian church. lt is an arched structure, finely built of brick and lime. The upper portions have been rebuilt, but the lower walls of the main part appear to have belonged to the original building and bear marks of considerable antiquity. Without the city, about a quarter of a mile from the wall, on the southern side, is an ancient cylindrical tower, perhaps fifty feet high, the origin of which is unknown. On its upper part are figures, which, however, are not sufficiently distinct to enable one to determine whether they are a written character, or merely ornamental appendages. The tumuli of the fire-worshippers, on the plain, will be noticed elsewhere. On every side of the city are gardens of vast extent, surrounded and interspersed with shade-trees of such size and in such numbers as to give to the whole region much the appearance of a great American forest. The city is surrounded by a high mud wall and a ditch, like those of *Tabréez,* but not kept in as good repair. The number of inhabitants is about twenty-five thousand. Vast multitudes, some say, one half of the population -, were swept off by the plague and cholera that raged here in 1829. Of the inhabitants, about two thousand are Jews; six hundred only are Nestorians—the mass of the Nestorians residing in the villages—and the rest are Muhammedans. There were a few Armenians, in the city and the province; but nearly all of them followed the Russians to Georgia. The city within has a very venerable and rather inviting appearance. It has much broader streets, more shade-trees and gardens, and a greater air of general comfort, than most other cities that I have seen in Asia. In the extensive bazars, I observed some of the finest fruit I ever beheld. European goods, as broadcloths, cottons, chintz, etc., are also finding their way to this distant city.

……

We therefore deferred our departure, and in the afternoon visited his gardens. They are about a mile in extent and very tastefully laid out. The rows of thickly studded poplars surround each, with a stream of water running between them. The gardens are also covered with thrifty fruit trees, arranged in rows and squares, and they are ornamented with rosebushes and other flowers, and fountains were here and there, playing in them. We wandered through several other gardens and vineyards beyond, till our ramble brought us to the foot of the *Koordish* mountains. We still proceeded and ascended two or three heights, from which we enjoyed a commanding survey of the most enchanting scenery I ever beheld. We had a perfect view, first of the gardens at the foot of the mountain; next the city; beyond that, the vast plain, gleaming with a grateful harvest and decked with orchards and villages; and finally, the silvery lake, melting away into the blue mountains beyond, and both rising in the distance and apparently meeting the skies. Forgetting, for the moment, the moral night that broods over this beautiful scene, I could hardly resist the impression, as I gazed from the mountain-top, that my eyes were resting upon the garden of paradise.

Laurie Thomas, *Dr. Grant and the Mountain Nestorians*, Boston, Gould and Lincoln, 1853, Chap. V, pp. 59–60.

[The Countryside]

The province of *Oroomiah* lies in the north-western part of Persia. The lofty mountains of Kurdistan look down on it from the west. On the east the beautiful lake of the same name extends, about eighty miles in length and thirty in width. It is four thousand one hundred feet above the level of the sea, and so salt that fish cannot live in its waters, though numerous water-fowl—among which the beautiful flamingo is most conspicuous—enliven its shores. A highly fertile plain gently slopes from the mountains to the lake, comprising an area of about five hundred square miles, and dotted with not less than three hundred villages. These vary in population from one hundred to one thousand inhabitants. The whole region, from Khoy on the north as far south as *Sûldüz*, is a perfect garden. Several rivers from the mountains are almost drained to irrigate the fields on either side. The roads in every direction are shaded by trees, often extending the most delicious fruit within reach of the traveller. The landscape is one of the loveliest in the east, willow, poplar and sycamore trees line the water-courses. Peach, apricot, pear, plum, cherry, quince and apple trees, impart to some portions the appearance of a forest, while the vineyards, and luxuriant fields and gardens of the plain, form a striking contrast to the naked aspect of the mountains. On those not a solitary tree is to be seen. Among them, however, are two hundred villages belonging to the province, half hidden in lovely valleys, or perched high on steep hillsides. One of these last is now the summer residence of the mission, at an elevation of fourteen hundred feet above the lake, while the mountains of Kurdistan rise behind to the height of twelve thousand feet, some of them crowned with snow throughout the year.

Near the centre of the plain, and about four hundred feet above the lake, stands the city of *Oroomiah.* It dates from a distant antiquity, and claims to be the birth-place of Zoroaster. Not far to the east of the city, an artificial mound, about seventy feet high, marks the site of the temple where he fed the sacred fires. It is a walled city, of nearly four miles in circumference, and contains about twenty thousand inhabitants. Of these, six hundred are Nestorians, two thousand Jews, and the rest Mohammedans. The flat-roofed houses are mostly built of mud or sun-burnt bricks, and are enclosed with high walls of the same material. Often the house itself forms one side of the enclosure. The more wealthy have beautiful gardens attached to their dwellings, where the family can enjoy themselves, secure from all intrusion.

Edward L. Cutts, *Under the Crescent in Asia*, London, 1878, pp. 253, 254–257, 260, 262, 265.

[A garden]

Thursday, August 3.—In the morning received visits. In the evening paid a visit to the garden of the *Shahzadah* (the Shah's son, the governor of the city), on the invitation of his chief gardener, who is a Nestorian. The garden is very inferior to our English gardens. It was a large level plot of ground, part vineyard, part orchard, with shrubs and roses planted along the walks. There are two large summer-houses. We were very hospitably entertained with raki, tea, and fruit, and passed a couple of hours pleasantly.

……

[The agriculture]

Just below our tents was the threshing-floor of the village, and we had the opportunity of studying the whole economy. The tenants pay their rent in kind, not a fixed amount, but a proportion of the corn crop, usually, believe, a third. The landlord is secured against dishonesty in the transaction by the law that the corn must be threshed out on his floor. Accordingly we see the people bringing their parcels of wheat on rude, wheel less sledges drawn by oxen and buffaloes, and stacking them in separate heaps around the floor. Every man's corn is threshed in its turn. The process goes on "from morn to noon, from noon to dewy eve." The floor is strewn with a thick layer of sheaves; two buffaloes drag round and round over it a heavy beam, into which a multitude of flints are inserted; both ear and straw are thus cut up into chaff, and the feet of the beasts help to tread the corn out of the husk. Then men with broad wooden shovels throw the chaff up against the wind, the heavy corn falls into one heap, and the lighter straw and husks fall into a second heap close by. A servant of the landlord is on the watch day and night. A little look-out is erected for him, consisting of four poles thrust into the ground, which carry a floor about eight feet high and a roof six feet above that; the sides and roof are loosely fenced with leafy boughs to keep off the sun, but leave interstices through which the watchman can see all round. The landlord's steward comes frequently to take his master's share of the grain.

To sit in front of our tents and watch all this going on in the foreground, with the plain and mountains and distant lake for a back ground, is like a realisation of a Georgic of Virgil.

I may speak here of some other implements which have interested me by their primitive forms. The plough is just like that we see represented in the illuminations of our Saxon MSS. The spade also is of the same primitive fashion; both handle and blade are carved out of one piece of wood; the blade small, rounded at the bottom and shod with iron, with a cross piece of wood a little above the blade for the foot to tread upon. Saxon Bible illuminations of the expulsion of Adam and Eve from Paradise, illustrating the truth that then "Adam delved and Eve span", put into Adam's band a spade of exactly the same shape; and the women here go about twirling a spindle, just like that which the Saxon artist has put into the hands of Eve.

……

[The Church]

The church seems an old one, built of stone, pleasantly situated in an enclosed churchyard planted with trees. The plan of the church is like that at *Kochanes*, with these additions: at the west end is a kind of transept, whose limbs form chapels, in which are placed two handsome tombs of two recent bishops of the line which is called by the dynastic name of *Mar Johannan*. At the east end of the nave is a low screen wall, about four feet in advance of the wall which divides the *sacrarium* from the body of the church. Towards the north end of this screen an embrasure is cut into the wall, where the deacon stands to administer the chalice, while the priest administers the bread from the chancel step. We lingered a little in the churchyard and were introduced to the people; some of the younger women were very nicely dressed, and I was pleased with the appearance and manners of the people (Fig. 23).

There is another ancient church in the city converted into a mosque, but I did not know of its existence till we had left *Oroomiah*, and I did not therefore see it.

After service we went to the house of *Khooda Verdi*, who had invited us to dinner. All the trades here are organised into guilds, and have their officers, as with us in the Middle Ages, and our host is the head of the carpenters' guild. His house I take to be a fair example of the houses of the fairly well-to-do citizens. Out of the hot, dusty, dirty lane of the Christian quarter, we entered through a passage into a courtyard surrounded by buildings - the stable, the workshop, the offices—but planted like a garden. We crossed this into another quadrangle planted still more prettily as a garden, and surrounded by the living rooms. On the further side is the reception-room on the right, and the kitchen on the left, divided by the corridor, which runs through to another garden at the back of the house. I was glad to see so nice a house, everything in excellent order, the garden giving evidence of loving care, and everything wearing an air of substantial comfort.

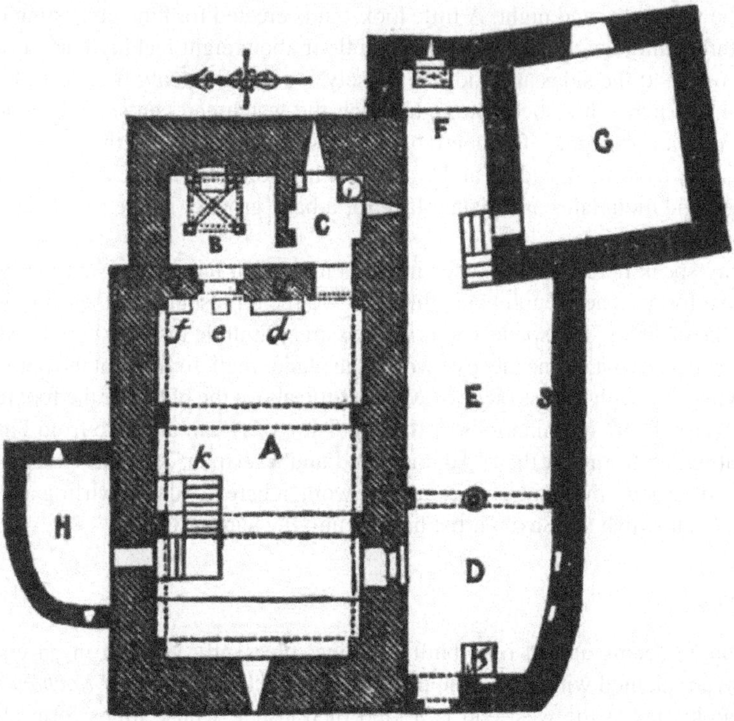

Fig. 23 Ground Plan of the Patriarchal Church, Qudshanis (Maclean and Browne 1892)

……

[The plain]

The plain of *Oroomiah*, stretching about fifty miles north and south, and the salt lake which borders it on the east, are set in a frame of mountains, on some of which patches and lines of snow are still remaining in the middle of August. The plain is very fertile, well-watered and well-wooded; it is very beautiful now, and in spring must deserve its name of "the Paradise of Persia." It is thickly dotted over with more than 1000 villages, of which 104 are inhabited by Nestorians.

Monday, August 7.—The three rivers which rise in the mountains, water the plain, and discharge their waters into the lake, are called the *Nahra-mdeemita* (city river), the *Barandizd*, and the *Nazlu-chai*. These rivers define the jurisdictions of the three bishops. Each takes the Nestorian villages watered by one of the rivers.

……

At several villages the deputations brought us bunches of grapes for our refreshment. In *Oroomiah* they pay great attention to the cultivation of the vine; they say that they have more than a hundred different kinds of grapes on the plain; and that some of the bunches grow so large, that to carry one from the vineyard without brushing off the bloom needs, like the grapes of *Eshcol*, a staff borne by two men. They throw up the soil of the vineyard into ridges, the vine is planted in the side of the ridge and trained over the surface, the grapes lying on the ground, while water is conveyed by a system of irrigation into the intermediate furrows.

……

[*Excursion to the lake*]

After breakfast the bishop had arranged an excursion to the lake for our entertainment. Three carriages and four waited at his door to receive the party. The "carriages" were the great, heavy, rough drays on which the wheat is carried from the field to the threshing-floor; the wheels are formed of several pieces of wood fastened together into a solid wheel, perhaps four feet diameter and two feet thick. They had put poles at the corners, and rigged up a curtain to screen us from the sun and some rugs were spread for us to recline upon. The "four" were draught buffaloes, which are driven without reins. The driver of the shaft-cattle stands in front of the dray, with a huge whip with a leather thong, which he uses very freely; a lad sits on the yoke which is fixed to the horns of the leaders, facing backwards, and keeping up a monotonous chant, without which the cattle would not work.

Mrs. Bishop (Isabella L. Bird), *Journeys in Persia and Kurdistan*, London, J. Murray, 1891, vol. II, pp. 217–220, 235–238, 242, 244.

[The Countryside]

Urmi, October 8. A march over low and much-ploughed hills, an easy descent and a ford brought us down upon the plain of *Urmi*, the "paradise of Persia" and to the pleasant and friendly hamlet of *Turkman*, where I spent the night and made the half-march into *Urmi* yesterday morning. This plain is truly "Paradise" as seen from the hill above it, nor can I say that its charm disappears on more intimate acquaintance. Far from it!

I have travelled now for nine months in Persia and know pretty well what to expect-not to look for surprises of beauty and luxuriance, and to be satisfied with occasional oases of cultivation among brown, rocky, treeless hills, varied by brown Villages with crops and spindly poplars and willows, contrasting with the harsh barrenness of the surrounding gravelly waste.

But beautiful *Urmi*, far as the eye can reach, is one oasis. From *Turkman* onwards the plain becomes more and more attractive, the wood-embosomed villages closer together, the variety of trees greater. Irrigation canals shaded by fruit trees, and irrigation ditches bordered by reeds, carry water in abundance all through the plain. Swampy streams abound. Fair stretches of smooth green sward rejoice the eye. Big buffaloes draw heavy carts laden with the teeming produce of the black, slimy, bountiful soil from the fields into the villages. Wheat, maize, beans, melons, gourds, potatoes, carrots, turnips, beets, capsicum, chilis, bringals, lady's fingers, castor-oil (for burning), cotton, madder, salsify, scorzonera, celery, oil-seeds of various sorts, opium, and tobacco all flourish. The orchards are full of trees which almost merit the epithet noble. Noble indeed are the walnuts, and beautiful are the pomegranates, the apricots, the apples, the peach and plum trees, and glorious are the vineyards with their foliage, which, like that of the cherry and pear, is passing away in scarlet and gold. Nature has perfected her work and rests. It is autumn in its glories, but without its gloom.

Men, women, and children are all busy. Here the wine-press is at work, there girls are laying clusters of grapes on terraces prepared for the purpose, to dry for raisins; women are gathering cotton and castor-oil seeds, little boys are taking buffaloes to bathe, men are driving and loading buffalo-carts, herding mares, ploughing and trenching, and in the innumerable villages the store-houses are being filled; the herbs and chilis are hanging from the roofs to dry, the women are making large cakes of animal fuel (of which they have sufficient for export), and are building it into great conical stacks, the crones are spinning in the sun, and the swaddled infants bound in their cradles are lying in the fields and vineyards, while the mothers are at work. This picture of beauty, fertility, and industry is framed by the Kurdistan mountains on the one side, and on the other by long lines of poplars, through which there are glimpses of the deep blue waters of the *Urmi* Sea. These Kurdistan mountains, a prolongation of the Taurus chain, stern in their character, and dwarfing all the minor ranges, contrast grandly with the luxuriant plains of *Sulduz* and *Urmi*.

As I passed northwards the villages grew thicker, the many tracks converged into a wide road which was thronged with foot passengers, horsemen, camel and horse caravans, and strings of asses loaded with melons and wood. Farther yet the road passes through beautiful orchards with green sward beneath the trees; mud walls are on both sides, and over them droop the graceful boughs and gray-green foliage of an *eleagnus*, with its tresses of auburn fruit.

At the large village of *Geog-tapa* a young horseman overtook me, and said in my native tongue, "Can you speak English?" He proved to be a graduate of the American College at *Urmi*, and a teacher in *Shamasha Khananeshoo's* school (known better to his supporters in England as Deacon Abraham). He told me that I was expected, and shortly afterwards I was greeted by the son of the oldest missionary in *Urmi*, Dr. Labaree.

The remaining four miles were almost entirely under the shade of fine trees, past the city walls and gates, put into tolerable repair after the Kurdish invasion ten years ago, and out into pretty wooded country, with the grand mountains of the frontier seen through the trees, where a fine gateway admitted us into the park in which are the extra-mural buildings of the American Presbyterian Mission, now more than half a century old. These are on high ground, well timbered, and the glimpses through the trees of the mountains and the plain are enchanting. Through the kindness of my friends at *Hamadan*, who had written in advance, I am made welcome in the house of Dr. Shedd, the Principal of the *Urmi* College.

……

[The City]

Very few European travellers visit *Urmi* and its magnificent plain, the "Paradise of Persia," though it is only 112 miles from Tabriz. Gardens come up to the city walls, and the plain, about fifty miles long by eighteen broad, is cultivated throughout, richly wooded, very populous, and bounded on the east not by a desert with its aridity, but by the blue waters of the *Urmi* Sea, and on the west by the magnificent mountains of Kurdistan. The city is some miles to the west of the lake.

Urmi is on the whole very pretty and in good repair. The Christian quarter is almost handsome, well built and substantial, and the houses are generally faced with red bricks. The bazars are large and well supplied, and trade is active. The walls and gateways are in good repair, and so is the deep ditch, which can be filled with water, which surrounds them. Every gate is approached by an avenue of noble *eleagnus* and other fruit trees. The gardens within the walls are very fine, and orchards and vineyards, planes and poplars testify to the abundance of water and the excellent method of its distribution. The altitude is stated at 4400 feet. The estimate of the population varies from 12,000 to 20,000.

Though the Sea of *Urmi* receives fourteen rivers, some of them by no means insignificant, and has no known outlet, it recedes rather steadily, leaving bare a soil of exceeding richness, and acres of dazzling salt. It has very few boats, and none suited for passenger traffic. Its waters are so salt that fish cannot live in them.

The antiquarian interests of *Urmi* consist in the semi-subterranean Syrian church of *Mart-Mariam*, said to have been built by the Magi on their return from Bethlehem! a tower and mosque of Arab architecture seven centuries old, and some great mounds outside the walls, from sixty to one hundred feet in height, composed entirely of ashes, marking the site of the altars at which the rites of one of the purest of the ancient faiths were celebrated. As the birthplace of Zoroaster, and for several subsequent ages the sacred city of the Fire Worshippers and the scene of the restoration of the Mithraic rites, *Urmi* must always remain interesting.

The Christian population of the city is not very large, though it is estimated that there are 20,000 Syrian Christians in the villages of the plain. The city Syrians are mostly well-to-do people, who have come into *Urmi* to practise trades. The best carpenters, as well as the best photographers and tailors, are Syrians, and though in times past the Moslems refused to buy from the Christians on the ground that things made by them are unclean, the prejudice is passing away.

There is a deputy-governor called the *Serperast,* whose duty it is to deal with the Christians. The office seems to have been instituted for their protection at the instigation of the British Government, but the Europeans regard it simply as a means of oppression and extortion, and desire its abolition. Canon Maclean goes so far as to say, "The multiplication of judges in Persia means the multiplication of injustice, and of the number of persons who can extort money from the unfortunate people". The *Serperast* depends chiefly for his living and for keeping up a staff of servants on what he can get out of the Christians in the way of fines and bribes, and consequently he foments quarrels and encourages needless litigation on all hands, the Syrians being by all accounts one of the most litigious of peoples.

I write of the Christians of *Urmi* and its plain as Syrians because that is the name by which they call themselves. We know them at home as *Nestorians,* but this is a nickname given to them by outsiders, and I know of no reason why we should use a nomenclature which attaches to a nation the stigma of an ancient "heresy." They are sometimes called Chaldeans, and the present Archbishop of Canterbury has brought into currency the term "Assyrians," which, however, is never used by themselves, or by any Orientals in speaking of them. The Moslems apply the name *Nasara* (Nazarenes) solely to the Syrian Christians. They claim that Christianity was introduced among them by the Magi on their return from Bethlehem. The highest estimate of their numbers is 120,000, and of these more than 80,000 are in Turkey. The Persian Syrians inhabit the flat country, chiefly the plains of *Urmi* and *Salmas*, where the fertile lands are most carefully cultivated by their industry.

In my last letter I remarked upon the prosperity and garden-like appearance of the *Urmi* Plain. Its 20,000 Syrian inhabitants usually live in separate villages from the Kurds, Persians, and Armenians, and are surrounded on all sides by Moslems of the *Shiah* sect. The landlords or *Aghas* of their villages are generally Moslems, who govern their tenants in something of feudal style. Land is a favorite investment in Persia, and owing to the industrious habits of the Syrians, the "*Agha-ship*" of their villages commands a high price. The *Aghas* often oppress the peasants, but the tenure of houses is fairly secure, and according to Canon Maclean, to whom I am indebited for my information, a system much like the Scotch *feuing* system (though without *feu* charters) is in force. If a man wishes to build a house he takes a present of a few sugar-loaves or a few *krans* with him, and applies to an *Agha* for a site. After it is granted he pays an annual ground rent of 4 s. 9d., but he can build his house as he pleases, and it cannot be taken from him so long as he pays his ground rent. Moreover, he can sell the house and give a title-deed to the purchaser, with the sole restriction that the new possessor must become a vassal of the *Agha*.

In addition to the payment of the ground rent, the tenant is taxed annually by the *Agha* for every female buffalo 2 s., for every cow 1 s., and for every ewe and she-goat 6 cl., after they have begun to bear young.

The *Agha* also receives from each householder annually two fowls, a load of *kiziks*, some eggs, three days' labour or the price of it, and a fee on every occasion of a marriage. Each house pays also a tax of 8d. a year and gives a present of firewood to the *Serperast* of *Urmi*, the Mussulman governor of the Christians. In his turn the *Agha* pays to the *Shah* from a third to a half of the total taxation.

A village house, even when built of sun-dried bricks, rarely costs more than £ 35, and often not the half of that sum. The great feature of a Syrian dwelling is what is called emphatically "the house"; the combined living-room, bedroom, smoking-room, kitchen, bakery, and work-room of one or more families. This room cannot possess a *balakhana*, as its openings for light and air are in the roof. A stable, store-rooms, and granary are attached to it. Vineyards are the chief reliance of the Syrians of the *Urmi* Plain, their produce, whether as grapes, raisins, or wine, being always marketable.

……

On this plain the dress of the men is much assimilated to that of the Persians, but the women wear their national costume. The under-garment is a colored shirt, over which is worn a sleeved waistcoat of a different color, and above this is an open-fronted coat, heavy silver belt are added. The head-dress is very becoming, and consists of a raised cap of cloth or silk, embroidered or jeweled, with a white muslin veil over it and the head, but the face is exposed, except in the case of married women, who draw a part of the veil over the mouth. It is not proper that the hair should be seen.

......

Geog-tapa (*cerulean hill*) possesses one of the largest of the Zoroastrian mounds of ashes. It is a pity that these are not protected, and that the villagers are allowed to carry away the soil for manure, and to break up the walls and cells (?) which are imbedded in them for building materials. This vandalism has brought to notice various curious relics, such as earthenware vessels of small size and unique shape, and a stone tomb containing a human skeleton, with several copper spikes from four to five inches long driven into its skull. In another mound, at some distance from this one, a large earthen sarcophagus was discovered, also containing a skeleton with long nails driven into its skull.

George Curzon, *Persia and the Persian Question*, London 1892, vol. I, pp. 532–536, 548 (Fig. 24).

[The Lake]

Not the least remarkable among the natural features of the mountain system of which Azerbaijan constitutes a part is the cluster of great lakes which are here encountered at a very considerable elevation above the sea. In Russian territory is Lake Gotcha, to the east of *Erivan*; in Turkish territory is Lake Van, to the west of Van. But the largest, to which I now turn, is in Persian territory, and can be seen from the citadel of Tabriz. This lake is commonly called in maps and by Europeans Lake *Urumiah*, from the well-known city, twelve miles distant from its western shore; but this name does not appear to be known to the Persians, who generally call it *Daria-i-Shahi*, or Royal Sea. It is the *Kapauta* of Strabo, his version of the Persian *kabuda*, or blue. The lake is eighty-four miles long, between twenty and thirty miles broad, has a circumference of nearly 300 miles, and an elevation of 4100 feet above the sea. Indented with bays and inlets, studded particularly in the southern part with islands, surrounded by wooded shores and hills, with Mount *Sehend* rising to a height of 11,800 feet from its eastern side, and with the white cone of Ararat piercing the distant clouds on the north, this noble sheet of water presents a fine and delightful prospect. Accounts vary as to its earlier history; for on the one hand it is said to have formerly covered a very much larger area, so much so that the peninsula of *Shahi* or *Shahkuh*, which juts forward into it from the eastern bank, is reported (even as late as by Kinneir) to have been an island twenty-five miles in circumference; on the other hand, local tradition is in favour of expansion, rather than contraction, and there is alleged to have been a causeway for traffic across what is now the bed of the lake to *Urumiah*. Its most peculiar features are its great shallowness, rendering it for the most part little more than a flooded swamp, and its abnormal saline properties, which in salt (of which it contains 22%) and iodine excel even those of the Dead Sea. The bottom of the lake has been proved by soundings to consist of a series of terraces or ledges, the maximum depth being forty-five feet, and the average depth being perhaps fifteen or sixteen feet, though the bather can advance for two miles from the edge without getting out of his depth. The sensations of the latter, if I may

Fig. 24 Portrait of George Nathaniel Curzon (John Singer Sargent 1914) (John Singer Sargent Virtual Gallery—Wikimedia Commons)

judge by the analogous case of the Dead Sea, are not to be envied; for it is impossible to dive or even swim, the limbs being thrown up to the surface, and a thick crust of salt being deposited upon the body, the eyelids, and in the hair. When the wind blows on Lake *Urumiah*, sheets of saline foam are seen scudding along the surface, and the salt is left upon the shore in a solid efflorescence, sometimes several inches thick. No fish or molluscs live in the waters, whose sole living contents are a species of small jelly-fish, which sustain the swans and wild fowl that are occasionally seen. The banks are covered with a thick and treacherous slime, composed partly of salt, partly of decomposed vegetable matter, and emitting a horrible effluvium. Of the sixty islets clustered in a group towards the southern end, three are either cultivated or used as pasture-ground, the largest being five miles in length.

It might be imagined that so extensive a sheet of water, surrounded by such large cities and fertile plains, and said to be singularly free from storms, would have given birth to a busy and profitable navigation, and have been ploughed by the keels of numerous craft. It is not so in Persia. No Persian, not even a Turkish Persian, ever ventured a yard on to the treacherous element if he could possibly help it. The metaphor of 'burning your boats' can have no home in a country where there are no boats to burn. The instincts of lucre alone account for the fact that Lake *Shahi* is navigated at all. In 1838, an uncle of the King, being Governor-General of Azerbaijan, in order to secure a monopoly of the carrying trade, ordered all private boats to be destroyed. The same intelligent policy has been followed by his successors; and at this moment the Governor of *Maragha*, who enjoys the monopoly, allows only three small decked boats of twenty tons burden, which ply between the opposite shores of the lake, and the working of which he sublets to a contractor who pays him 800 l. a year, and makes a substantial profit out of the enterprise. As Colonel Stewart says, what is wanted is a small line of steamers running between the southern and northern extremities, and transporting the grain from the rich cornlands south of the lake where it is plentiful and cheap, to the towns of *Khoi* and Tabriz where it is comparatively clear. But I suppose we must wait for this, as for all good things in Persia.

Near the eastern shore of the lake, and at about six miles from the village of *Dehkharegan*, are the pits or springs from which is extracted the famous semi-transparent marble, sometimes called after the neighboring town of *Maragha*, sometimes after Tabriz. A number of springs, clustered within an area of half a mile in circumference, are constantly bubbling up and precipitating the limestone which they hold in solution. This is deposited in the form of horizontal layers, which are like a thin crust to start with, and can be cracked or broken, but which gradually solidify into hard blocks, with an average thickness of seven or eight inches, the best of which are believed to have been formed when the springs had a much higher temperature than the present (65° Fahr.). When quarried this petrifaction can be sawn either in the thinnest plates, when it is nearly transparent, and is sometimes used for windows, or in more substantial slabs, in which form it is much used for pavements and mural wainscoting. It is a singularly beautiful substance, being of a pink, or greenish, or milk-white colour, streaked with reddish or copper-coloured veins (from the oxide which it contains); and I have seen beautiful samples of it in the palaces and mosques of the East. I have very little doubt that the wainscoting of the *Gur Amir*, or Tomb of Timur, at Samarkand, which I have described in my former work, and which has puzzled all travellers, is composed of this marble, which there is not being more natural than that the great conqueror should have carried home with him at the close of his Persian campaign. The process of petrifaction bears a marked resemblance to that which was in existence till the great eruption of a few years ago at the Pink and White Terraces in New Zealand; and to that which may still be seen at the Mammoth Hot Springs in the Yellowstone Park, in North America, where the induration may be observed through all its stages from a film like frosted sugar to gleaming blocks of snow-white marble.

......

On the other or western side of the lake, and at a distance of twelve miles from its shores, stands the city of *Urumiah* (shortened by the Christians into *Urmia* or *Urmi*), in a plain that is deservedly famous for its abounding fertility. Framed in the Kurdistan mountains, from which descend a multiplicity of perpetual streams, and planted, irrigated, or peopled to the full extent, it has been variously reported to contain 400, 300, and 200 villages (round numbers, which I take to be merely indicative of an unusually extensive population), and has been compared by the fancy or the recollection of different voyagers to the lands at the foot of the Himalayas, to the banks of the Lake of Zurich, and to the wealthy plains of Lombardy. The city, which is situated at an elevation of 4,400 feet above the sea, contains a population of between 30,000 and 40,000, the bulk of whom are *Afshar* Turks, but which comprises a considerable sprinkling of Nestorian, Jewish, and Armenian families. In ancient history *Urumiah* is famous as one of the legendary birthplaces of the scarcely less legendary Zoroaster, and also as one of the burial-places of the Three Magi. The city is enclosed within a wall, with seven gates, and an outer ditch. The only interior structure of any importance is the arsenal, a walled building in the centre of the town, the court of which contains a dozen ancient smooth-bore six-pounders and a single brass howitzer. Until recently, and while the Kurdish terror arising out of the rebellion of *Sheikh Obeidullah* in 1880 prevailed, a garrison of three regiments of regulars, armed with Werndl rifles, was quartered here. To Christian visitors the chief interest of the place will consist in the fact that it is the headquarters of the American, French, and English Missions to the Nestorian populations of the neighbourhood, to which interesting but somewhat intricate subject I now turn.

……

[The Population]

The Armenians being less prolific, less gregarious, and less stay-at-home than the Nestorians, it is recommended to compute their families at an average of six persons, which will give a total of 28,890; or, together with the Nestorians, a grand total of 72,890 for the Christian population of Azerbaijan. The Persian Armenians are a less attractive and an even less reliable people than the mendacious, but peaceable Nestorians. They travel a great deal, and pick up revolutionary ideas, and are disposed to deceit and turbulence. The local head of their church is an arch-bishop at Tabriz, who throws what obstacles he can in the way of the Christian missions; whilst the Catholicos of the entire Armenian church is located not far from the frontier, in Russian territory, at Echmiadzin. The Armenian question is, however, so much a Turkish and so little a Persian one, that I do not feel called upon to say anything more about it here. I shall have occasion to speak of the people again, when dealing with *Julfa*.

Eugène Aubin, *La Perse d'aujourd'hui. Iran-Mésopotamie*, Librairie Armand Colin, Paris, 1908, pp. 69–73.

[The City]

18 kilomètres dans la verdure de la plaine; un pont, orné de colonnettes, traverse le lit desséché du *Nazli-Tchaî*. Les montagnes, où la neige a persisté tout l'été, s'écartent de plus en plus. La route est bordée de saules; de petits murs en terre séparent les

vignobles; les ceps, plantés au creux des fossés, remontent le long des talus; des glacis inclinés servent au séchage des raisins.

Ourmiah est une ville de 30.000 habitants, au centre même de la plaine, fortifiée comme *Khoï* et de même apparence; quelques rues plus larges, ombragée de saules. L'agglomération disparaît sous les peupliers et les platanes. Au dehors, une vieille tour mongole, *Utch Kombaz*—les trois voûtes—avec des restes de faïences et d'inscriptions *koufiques*.

Le commerce est réparti entre 30 caravansérails et 1.500 boutiques. Le raisin sec fait la richesse du pays; il en existe de deux espèces, le *Kichmich*, simplement séché au soleil, le *Sabza*, trempé au préalable dans un bain de potasse. On en exporte une moyenne annuelle de 300.000 *khonkars* [75 kg chacun], pour une valeur d'un million de *tomans*. Ces raisins sont vendus au négociants de *Tauris* ou bien expédiés directement à *Djoulfa* et *Erivan*. Le tabac se consomme dans le Nord de la Perse. Pour l'importation, les marchands d'*Ourmiah* ont surtout affaire avec Constantinople, par la voie de *Trébizonde*, avec *Hamadan* pour les épices et les cotonnades de l'Inde. Par sa position, la rive occidentale du lac échappe davantage à la pénétration russe ... Avant la conquête arabe, le pays était peuplé d'Arméniens, auxquels se mêlaient des Chaldéens et des Juifs. Avec l'invasion mongole arrivèrent les Turcs. Parmi ces sunnites la prédication des Cheikhs ... ne donna point de résultats complets; aucune tribu kurde ne se convertit au chiisme (Fig. 25).

......

Fig. 25 Ourmiah, Persia, Fruit_Market, 1911 (U.S. Library of Congress—Wikimedia Commons)

[The Plain]

La plaine d'*Ourmiah* contient 350 villages; le massif isolé du *Bizau-Dagh* la sépare du lac; elle est entourée par les grandes montagnes où s'ouvre la gorge du *Chehri-Tchaï*. Partout de l'eau et de la verdure. Au travers des champs, les paysans, coiffés d'un bonnet de drap à bordure d'astrakan, conduisent leurs charrettes triangulaires, montées sur deux roues énormes où, à la pointe extrême du triangle, sont attelés les buffles. En bas, les villages sont turcs et chiites; sur les hauteurs, ils deviennent kurds et sunnites. Un appoint considérable d'Arméniens et de Chaldéens s'y joint aux musulmans, si bien qu'*Ourmiah* s'est trouvé le centre naturel des missions diverses, travaillant à l'évangélisation de l'Azerbaïdjan.

Dans la ville les chrétiens habitent les deux quartiers de *Kurdi-Chehr* et de *Mart-Mariam* (Sainte-Marie, du nom d'une vieille église nestorienne). En plaine, les Arméniens sont moins de 6.000, dispersés dans 31 villages; les Chaldéens peuvent atteindre le chiffre de 30.000, et forment le principal noyau de leur nation en Perse. Les sectes concurrentes n'ont plus à se disputer que ce dernier groupe de chrétiens. Rebelle à toute conversion, le musulman persan a ceci de particulier qu'il hésite même à fréquenter les écoles confessionnelles. D'autre part, la religion grégorienne veut être le symbole de la nationalité arménienne ... De ce côté, ni le catholicisme, ni le protestantisme ne peuvent recruter de nouveaux adeptes, et les écoles américaines trouvent maintenant en face d'elles des écoles nationales fortement organisées (il n'existe d'Arméniens catholiques que dans deux villages de la plaine d'*Ourmiah*). La masse chaldéenne reste seule ouverte au travail des missionnaires, qui se la disputent avec acharnement ... Les Lazaristes trouvaient le terrain préparé. Au XVIII siècle, un élève des Dominicains de *Mossoul* ayant converti l'évêque nestorien de *Salmas*, la diocèse entier avait imité son pasteur ... L'ouvre catholique s'est étendue: les Chaldéens comptent aujourd'hui trois diocèses: un archevêché à *Ourmiah*, des évêchés à *Salmas* et à *Senneh (Ardélan)* ... Le séminaire est à *Salmas*: les clercs vont achever leurs études chez les Dominicains de *Mossoul* ou au collège de la Propagande à Rome. L'imprimerie fonctionne à *Ourmiah* pour la publication de livres en langues française, syriaque, chaldéenne et arménienne. 7 Lazaristes, dont plusieurs Chaldéens, desservent la maison d'*Ourmiah*. Le collège a 142 élèves; les Sœurs de charité reçoivent 203 petites filles et tiennent un petit hôpital de 8 lits. Le diocèse a 4.000 catholiques, répartis entre la ville et 45 paroisses; 739 enfants fréquentent les écoles des villages, régulièrement visitées par les religieux français.

……

[The Christians]

Non point que les chrétiens de la Perse soient particulièrement malheureux. Les missions s'accordent à reconnaître la douceur relative du régime persan. Il ne subsiste de ce côté aucune haine entre sunnites et chiites, entre chrétiens et musulmans; tous sont également exposés aux déprédations des Kurdes et à la rapacité des puissants. L'impôt est en-lui-même peu élevé, le loyer des terres raisonnable. Il est généralement admis que les grands seigneurs persans se montrent libéraux à l'égard de leurs paysans; la mal viendrait des intendants et des petits propriétaires.

5 Buildings in Persia

Paul Lucas, *Voyage du sieur Paul Lucas au Levant, premier volume*, Paris 1704, p. 368.

Les maisons y sont faites d'une plaisante manière. L'on fait de grandes fosses en terre que l 'on couvre de pièces de sapin, sur lesquelles on met de la terre, et puis on laisse un trou carré qui sert de porte et de cheminée. Dans ces maisons souterraines le père, la mère, les enfants, toute une famille où il y aura trois ou quatre ménages couchent tous ensemble sur des tapis. Leur bétail, leurs chameaux, leurs chevaux sont dans le mesme endroit.

Gaspard Drouville, *Voyage en Perse, fait en 1812 et 1813*, St. Petersbourg, 1819, Tome I, pp. 52–54, 54–56, 56.

De la construction des bâtiments.

La construction des villes orientales diffère beaucoup de celle des villes d'Europe et c'est surtout en Perse que cette différence devient plus sensible par le manque de pierres.

Les villes persanes sont toutes à peu près bâties de la même manière, c'est-à-dire fort irrégulièrement, et sans ordre, chacun plaçant la façade de sa maison du côté que lui convient, et la faisant de telle dimension qui lui plaît, sans que personne ait le droit d'y trouver à redire.

Les maisons sont bâties en briques séchées au soleil, et liées ensemble avec de la terre délayée, au lieu de mortier. Il y a cependant quelques riches particuliers qui font construire en briques cuites, et même en pierres jointes avec de la chaux, mais ce genre de bâtisse est fort rare et très-cher clans ce pays.

On bâtit en Perse avec une rapidité étonnante, car souvent on voit finie le soir une maison qu'on a vu commencer le matin. J'en excepte cependant les toits qui, étant faits en forme de terrasse, demandant un peu plus de temps pour être consolidés.

Les maçons travaillent en cadence, et c'est en chantant qu'ils demandent les matériaux dont ils ont besoin.

Une grande partie des villages ont la figure de carrés parfaits, entourés de murailles de terre fort hautes, ayant aux quatre angles des tours rondes, percées, ainsi que les murs, de deux ou trois rangs de meurtrières. Un tel village, vu de loin, ressemble d'autant plus à une forteresse, qu'il y-en-a dont les murailles ont plus de cinquante pieds de hauteur.

C'est particulièrement près des frontières de la Turquie que les paysans se ferment de cette manière, pour se garder des Turcs, qui font assez souvent des incursions sur leur territoire, pour enlever les bestiaux et les récoltes.

Les villes sont fortifiées de la même manière et plus ou moins régulières, mais en général elles sont défendues par une grande quantité de tours, distantes d'une demi-portée de fusil les unes des autres, et assez souvent entourées de fossés larges et profonds.

Il n'existe dans toute la Perse que deux villes fortifiées d'après le système européen, ce sont *Khoi* et *Abas-Abad*; il est impossible de trouver rien de plus mauvais. Tout ce qu'on y a fait consiste en de méchantes courtines, flanquées par des bastions si étranglés, et d'un saillant si aigu, qu'on aurait de la peine à manœuvrer deux pièces de canons sur leur terre-plein.

Ajoutez à cela que le tout est sans revêtements, à une toise près, sans chemin couvert, ni glacis, ni palissades. Ce que l'on a bien voulu décorer du nom de contrescarpe n'est d'aucune défense, et l'on n'aurait pas plus de peine à la descendre que les degrés un peu élevés d'un escalier. Ces mauvais ouvrages ont été construits par un officier anglais, soi-disant ingénieur, qui était au service de la compagnie des Indes. Peut-être fut-il, faute d'autres considéré comme le *Vauban* de l'Asie.

Les maisons, dans les villes, sont entourées de murs assez élevés pour cacher totalement les façades, qu'on éloigne toujours beaucoup des rues et qu'on place dans le fond de grandes cours qui les en séparent. On y entre par de petites portes qui ressemblent assez à des guichets de prison. Comme ce sont les seules ouvertures qui se présentent aux yeux, un étranger qui entre pour la première fois dans une ville persane ne sait trop où il se trouve, ne voyant ni édifices, ni façades, mais seulement de hautes et tristes murailles qui forment la trace des rues.

Les maisons sont toutes bâties d'une manière élégante et la distribution des appartements est assez régulière. Elles sont composées de plusieurs chambres et d'une grande salle au centre, qu'on nomme Divan. C'est là que tous les matins les personnes de qualité reçoivent leur clientèle. Cette pièce étant toujours située entre une cour et un jardin, et souvent même entre deux jardins, elle a sur chacun d'eux une large fenêtre, qui va du parquet au plafond, et faite de petites pièces de bois arrangées avec assez d'art en forme de guirlandes et de festons. Les nobles et riches les garnissent de verres de différentes couleurs, mais comme c'est un objet fort rare et très-cher dans ce pays, la classe ordinaire les garnissent avec du papier huilé, qui suffit pour maintenir la chaleur pendant les autres saisons.

Les Persans sont grands amateurs de l'eau, aussi est-il peu de maisons qui n'aient devant leurs fenêtres de larges bassins de marbre blanc ou d'albâtre du milieu desquels jaillissent de jolies fontaines ou jets dont le bruit charme leurs oreilles, quand ils s'abandonnent à la contemplation.

……

Les maisons ont, comme je l'ai dit, d'autres appartements qui consistent en des chambres hautes et basses, construites d'une manière régulière et qu'on nomme *balakoua*. Elles ont de plus des espèces de caves voûtées pour préserver le rez-de-chaussée de l'humidité, et où l'on renferme le bois à brûler et les ustensiles de campagne.

Dans les premières cours, il y a des corps-de-logis latéraux qui renferment d'un côté des chambres pour les étrangers de classe médiocre, et les derviches; de l'autre, les écuries, les magasins à paille et à orge et des chenils pour les chiens de chasse.

……

Les Persans couchent à terre, leurs lits n'étant pas permanents comme les nôtres, sont posés tous les soirs, enlevés tous les matins, et déposés dans des cabinets où ils restent tout le jour. Ils se composent, suivant le rang ou la fortune des individus, d'une courte-pointe d'indienne ouatée en coton, d'un énorme traversin et d'un petit oreiller d'à-peu-près un pied carré pour chaque personne.

Eugène Flandin—Pascal Coste, ***Voyage en Perse de MM. Eugène Flandin, Peintre, et Pascal Coste, Architecte, attachés à l'Ambassade de France pendant les années 1840 et 1841*****, Paris, Gide et J. Baudry, 1854,** pp. 128–129.

Le Tendour

Nous nous couchâmes ce jour-là à *Sophiân*, habité par des Kurdes soumis et tributaires du *Châh*. Ce village avait une meilleure apparence que tous ceux que nous avions encore vus, et nous commencions à espérer que la Perse vaudrait mieux que la Turquie. Une grande quantité de jardins plantés de très-beaux arbres, entouraient les maisons que nous y trouvâmes meilleures et moins sales que celles de l'Arménie, tout y décelait l'aisance et une vie industrieuse dont la culture des terres semblait être la partie importante. Nous y fîmes connaissance avec un nouveau mode de chauffage qui nous parut aussi original qu'économique. Dans la principale pièce de chaque maison est un trou profond de quatre pieds environ, et large de trois. Il est de forme circulaire, et fait à peu près comme une grande jarre. Il n'a qu'un orifice étroit, d'un pied seulement de diamètre. Tout l'intérieur en est mastiqué ou revêtu de pierres noires. Le matin, c'est le four; on le chauffe au moyen de mottes faites avec du fumier, ou avec de petits fagots très-minces, et l'on y cuit le pain de la famille. Après la cuisson, il devient calorifère, on couvre son orifice d'un châssis en bois sur lequel on étende d'amples couvertures, de manière à y concentrer la chaleur. En entrant dans ces maisons, on voit le père, la mère et tous les enfants accroupis, les jambes sous les couvertures ainsi que les mains, et dans l'inaction la plus complète, se laissent vivre dans le torpeur et l'oisiveté. Ces sortes de trous s'appellent *tendours* (Fig. 26).

Filippo De Filippi, ***Note di un viaggio in Persia*****,Milano, 1865,** pp. 139, 140–142.

I mulini

Da quando in quando, lungo i pochi e miseri fiumi della Persia, si incontrano mulini, i quali sono di uno stile tutto particolare. Si presentano come casipole con mura di sassi quasi senza cemento, erboso per l'umidità naturale del luogo; di forma prismatica quadrata, senza finestre, con una sola porticina d'accesso. L'acqua vi è condotta per un piccolo canale di legno che trapassa nel suo mezzo il tetto, formando con questo un angolo di circa 45°. Nulla traspare dal di fuori del meccanismo della macina. Si può facilmente vedere quanto debba costare ai Persiani il trasporto del grano da siti di sua produzione, talvolta per distanze enormi, ai pochi mulini qua e là disseminati; sì pochi da non comprendersi come mai possano bastare ai bisogni della popolazione. Ma ciò che si comprende ancora meno si è come nessuno abbia mai pensato a rivolgere al servizio de' mulini un altro motore naturale, costantemente attivo per tutta l'estensione del paese, voglio dire i venti che soffiano senza posa.

……

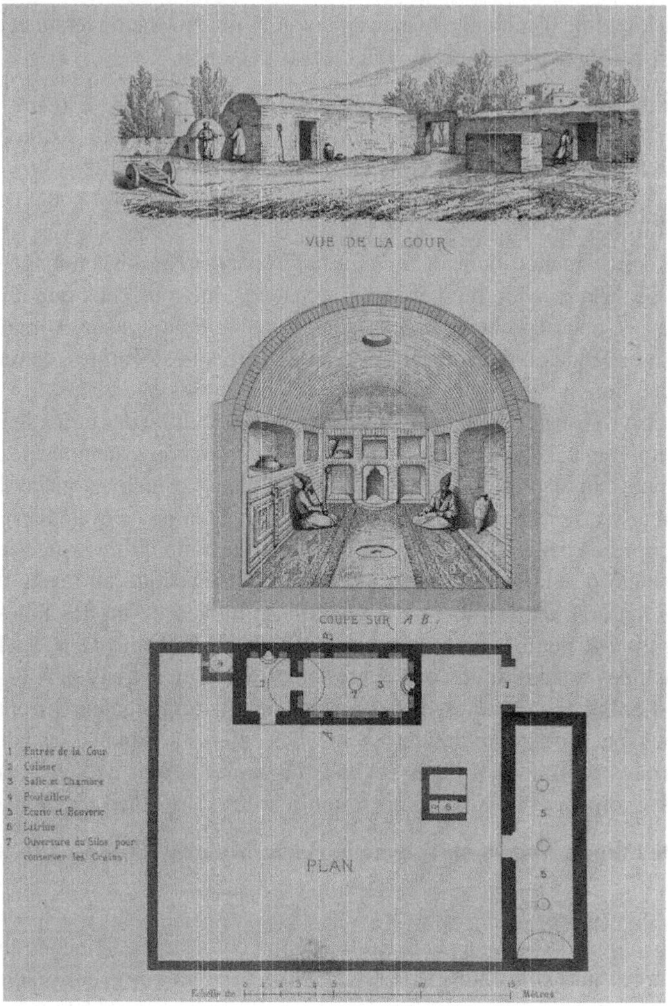

Fig. 26 *Tauris, Maison au village d'Alvar* (Coste 1867)

I villaggi

I villaggi sono labirinti di muriccioli di fango, dai quali appena sporgono gli alberi de' giardini, e le miserabili abitazioni interne, costrutte dal medesimo materiale, fango, con tettoje piane, che nelle notti estive, servono di letto. La struttura di questi sucidi tugurj è studiata in modo da fare il massimo risparmio possibile di legno e di ferro, da escluderli quasi affatto; quindi le porte e le finestre sono misurate per un numero e per dimensioni al più stretto bisogno. È inutile il ripetere che le finestre sono tutte rivolte ne' cortiletti interni o nei giardini. Ad una certa distanza i villaggi passerebbero inosservati, confusi col colore generale del terreno, ove la loro presenza non fosse invece tanto chiaramente indicata dalle piantagioni d'alberi. La grande scarsità di combustibile, fa sì che il cemento calcareo ed i mattoni cotti siano cose di lusso, fin nelle città, riserbate ai bazar ed alle moschee. Una strana singolarità è pur questa, che mentre in Europa i villaggi più floridi sono lungo le principali vie di comunicazione, in Persia è perfettamente il contrario. Lungo quelle linee, che si dicono strade principali, perché più battute dai viandanti, i villaggi hanno in generale l'aspetto della più squallida miseria, mentre negli angoli più segregati da queste linee, si trovano i villaggi più popolosi e fiorenti, per quanto la floridezza è conciliabile colla natura del paese. La ragione di questa dissonanza è semplicissima. Le vie delle carovane sono pur quelle più frequentemente battute dalle truppe, dai numerosi cortei dello *Schah* e de' grandi personaggi, i quali, lungi dal lasciare sul loro passaggio i benefizi del commercio e della vita sociale, impongono ai villaggi contribuzioni gravissime. È anzi costume, quando lo *Schah* onora della sua augusta presenza le provincie del suo vasto impero, che gli anziani dei villaggi, prossimi al supposto suo cammino, gli si rechino incontro con offerte anticipate e spontanee, e con felicitazioni di un prospero viaggio contenenti l'implicito voto che i villaggi, de' quali sono rappresentanti, siano risparmiati. Noi stessi non siamo sicuri di non aver portato con noi qualche piccolo carico di imprecazioni dai villaggi ove il nostro *mehmendar* doveva fare le proviste per la nostra carovana. Rimando ad altra occasione qualche aneddoto curioso su questa materia.

La povertà estrema della vegetazione arborea, relativamente all'estensione del paese, toglie si può dire affatto l'uso della legna come combustibile, ed obbliga i Persiani a rivolgere a tal uso lo sterco de' cammelli e de' bovini. Veramente questa industria è comune alla massima parte dell'Asia occidentale, ed all'Egitto, anche in provincie ove non sarebbe tanto strettamente imposta dalla necessità. A tal fine lo sterco, commisto collo strame fracido delle stalle, è impastato a mano sotto la forma di larghe focacce, e queste sono distese ad essiccare sulle aje dei villaggi, e perfino, a crescerne l'ornamento, sui muri delle abitazioni: al qual lavoro attendono particolarmente le donne. Mi fu assicurato che tal sorta di combustibile dà buon fuoco. In pochi luoghi soltanto, come, per esempio a *Tauris*, si accatasta legna da ardere, e quasi esclusivamente per uso dei pochi Europei che vi hanno stanza. La somministrazione della legna al consumo giornaliero nella cucina, non doveva essere la minore delle cure di chi provvedeva ai bisogni della nostra carovana.

Malgrado le tanto sfavorevoli condizioni fisiche della Persia il tenue prezzo del lavoro e l'abbondanza del bestiame, fanno sì che le derrate di prima necessità per la vita materiale vi abbondino, e siano i prezzi molto più miti che in Europa. Frutta, riso, pane, polli, montoni, si trovano ad ogni villaggio di qualche entità, a poco danaro ed in copia. Il pane è sotto forma di grandi fogli che si fanno cuocere prestamente e male, impastandoli sulle pareti dei forni, e si pongono in vendita ne' bazar su grandi piani inclinati, irti di chiodi. Questi fogli ci hanno le tante volte servito di tovagliolo, di piatto e di cibo nel tempo medesimo.

Le esposte condizioni naturali del paese, la sua latitudine, la sua altezza sul livello del mare, rendono ragione del suo clima. L'inverno è lungo, rigidissimo nella parte più elevata, appena raddolcentesi alla posizione di Teheran; e gli abitanti ne approfittano per ammassare provigioni di ghiaccio; conforto inestimabile a chi, al pari di noi, viaggia la Persia nei mesi più caldi dell'anno. La breve primavera è caratterizzata dalle piogge, il cui benefizio però è rapidamente disperso dai venti, i quali alla loro volta rattemperano gli effetti de' raggi solari in estate, e rendono le notti sarei per dire fredde. Chi percorre queste contrade nella bella stagione non deve tanto premunirsi contro il caldo estivo, inferiore alle comuni prevenzioni, quanto contro i salti giornalieri di temperatura, come quello, per esempio, che accadde a noi di provare, passando da un mattino sufficientemente fresco, ad un calore meridiano di 34° R. all'ombra.

I venti e il debole potere irradiante del terreno fanno sì che, malgrado la serenità delle notti, la rugiada sia insensibile, e non dia compenso alcuno alla vegetazione della mancanza completa della pioggia durante la stagione estiva. Questa lunga, costante arsura, ed il carattere mite delle piogge in primavera ed in autunno, sono appunto tali condizioni naturali da permettere ai Persiani di costruire le loro case in fango che altrimenti, all'impeto delle piogge temporalesche sarebbero in breve tempo stemperate. Uno spettacolo, si può dir quotidiano, che l'aridissimo terreno delle steppe percosso da' cocenti raggi solari produceva al nostro sguardo, era la fata morgana, spesse volte di effetto sorprendente.

[A/T: The mills.*]*

Here and there, along the few and mean rivers of Persia, you'll find mills in a very particular shape. They look like almost mortar-free stone walled huts, grassy because of the natural dampness of the place. They have square prism shapes, no windows, and just one small entrance door. Water is brought in with a small wooden canal passing through the roof in the middle at an angle of about 45°. Nothing can be seen from outside, of the grinder engine. It can easily be seen how expensive be for Persans moving the wheat crops from their growing areas, sometimes at very long distance, to the few mills scattered here and there, so few that it's not understandable how they may suffice to the needs of people. But what is even less understandable is why nobody ever thought to turn in service of the mills another natural engine, constantly active all along the country, I mean the restless blowing winds.

……

The villages

Villages are mazes of muddy low walls, from which the garden trees just peep, and mean dwellings inside, built with the same material, mud, and flat roofs serving as bedrooms in summer nights. The structure of those filthy hovels is conceived as to spare wood and iron to an extent as to make almost without, thus doors and windows are numbered and sized as least as possible. It is useless to repeat that the windows are all facing internal courtyards or gardens. At some distance the villages might not be seen, confused with the general color of the ground, if their presence was not so clearly marked by the plantations of trees. Fuel is so scarce there, as to make lime mortar and baked bricks luxury items even in towns, reserved to bazaars and mosques. A strange particularity is also that, while in Europe the most thriving villages are settled along the main roads, in Persia it's perfectly the reverse. Along those lines that are called main roads, since mostly treaded by the wayfarers, the villages generally look like very mean and poor, whereas the most populated and thriving ones are located apart from those lines, in the in the most segregated corners, as far as prosperity may go with the nature of the country. The reason for that is very simple. The caravan roads are also the most frequently crossed by troops, along with numerous processions of the Schah and great personalities who, far from yielding on their crossing any benefit of trading and social life, charge the villages with very heavy contributions. When the Schah honours the provinces of his vast empire with his august presence, instead, the elders from the villages close to his envisaged passage, use to go and meet him bringing anticipated and voluntary offerings, along with votes for a prosperous journey, implicitly hoping the villages they represent will be saved. We ourselves aren't sure we didn't bring with us some small load of curses from the villages where our mehmendar had to stock up for our caravan. I postpone to another occasion some curious anecdote on this subject.

The extreme poverty of the arboreal vegetation, compared to the extension of the country, takes wood almost completely away from any use as fuel and makes it necessary for Persans to use camel and bovine dung. Actually, that is common use in most of Western Asia and Egypt, even where it wouldn't be strictly required by need. On that purpose the dung is melted with the wet straw from the stables and knead by hand in the shape of large cakes which are placed to dry on the farmyards of the villages and even on the walls of the houses as decorations. This job is made by women in particular. They made me sure that kind of fuel gives good fire. In just few places like, for instance, in Tauris, wood for firing is stacked, and almost exclusively for the few Europeans living there. Providing wood for daily kitchen use had not to be the least task of those who were keeping care of the needs of our caravane.

In spite of the very difficult physical conditions of Persia the low cost of working and the abundance of cattle, help having at disposal a great deal of food for basic needs, at prices much lower than in Europe. Fruits, rice, bread, chickens, mutton are in any village of some entity, cheap and in large quantities. Bread comes in shape of large sheets quickly and badly cooked by sticking them to the walls of the ovens, and later laid on tilted planes full of nails, to be sold in bazaars. Those sheets often have served for us as napkins, dishes as well as food.

The exposed natural conditions of the country, its latitude and height on sea level, account for its climate. Winter is long and very cold in the higher areas, and slightly warmer around Teheran; and the inhabitants take advantage of it to conserve ice supplies: invaluable comfort for those—like us—travelling Persia in the hottest months of the year. Spring is short and rainy but the winds, quickly drying up water and grounds, refresh the hot sun rays in Summer making nights nearly cold. Passengers through these lands in beautiful seasons, shouldn't have to take much precautions for Summer heat, which is not hot as commonly thought of, rather than for daily temperature range, like we happened to pass, for example, from a quite cool morning to 34 °C in the afternoon, in the shadow.

The winds and the weak radiating power of the soil, in spite of cloudless nights, make dew helpless, thus preventing the vegetation from being compensated for the absolute lack of rain in Summer. This long, abiding dry heat, and the meek character of the rains in Spring and Summer, make it naturally possible for Persans to build their houses with mud; which wouldn't be with sweeping and stormy rains, at risk of being melted in short time.

And we may say, a daily show, made for our gaze by the driest soil of steppes struck by the very hot sun rays, was the Morgana Fairy, often of surprising effect.]

Mrs. Bishop (Isabella L. Bird), *Journeys* ***in Persia and Kurdistan*****, London 1891,** vol. 1, letter 4, p.149.

Persian peasant's houses

I have waited to describe a Persian peasant's house till I had seen more of them. The yard is an almost unvarying feature, whether a small enclosure with a low wall and a gateway closed at night by a screen of reeds, or a great farmyard like this, with an arched entrance and dwelling-rooms for two or three generations along one or more of the sides.

The house walls are built of mud, not sun-dried brick, and are only one story high. The soil near villages is mostly mud, and by leading water to a given spot, a pit of mortar for building material is at once made. This being dug up, and worked to a proper consistency by the feet of men, is then made into a wall, piece after piece being 149 laid on by hand, till it reaches a height of four feet and a thickness of three—the imperative tradition of the Persian builder. This is allowed a few days for hardening, when another layer of similar height but somewhat narrower is laid upon it, *takchahs* or recesses a foot deep or more being worked into the thickness of the wall, and the process is repeated till the desired height is attained. When the wall is thoroughly dry it is plastered inside and outside with a mixture of mud and chopped straw, and if this plastering is repeated at intervals, the style of construction is very durable.

The oven or *tāndūr* is placed in the floor of one room, at least, and answers for cooking and heating. A peasant's house has no windows, and the roof does not project beyond the wall.

All roofs are flat. Rude rafters of poplar are laid across the walls about two feet apart. In a *ketchuda's* or a wealthier peasant's house, above these are laid in rows peeled poplar rods, two inches apart, then a rush mat, and then the resinous thorns of the tragacanth bush, which are not liable to decay; but in the poorer houses the owner contents himself with a coarse reed mat or a layer of brushwood above the rafters. On this is spread a well-trodden-down layer of mud, then eight or ten inches of dry earth, and the whole is thickly plastered with mixed straw and mud. A slight slope at the back with a long wooden spout carries off the water. Such a roof is impervious to rain except in very severe storms if kept in order, that is, if it be plastered once a year, and well rolled after rain. Few people are so poor as not to have a neatly-made stone roller on their roofs. If this is lacking, the roof must be well tramped after rain by bare feet, and in all cases the snow must be shoveled off.

These roofs, among the peasantry, have no parapets. They are the paradise of dogs, and in hot weather the 150 people take up their beds and sleep there, partly for coolness and partly because the night breeze gives freedom from mosquitos. In simple country life, though the premises of the peasants for the sake of security are contiguous, there are seldom even balustrades to the roofs, though in summer most domestic operations are carried on there. Fifty years ago Persian law sanctioned the stoning without trial or mercy of anyone caught in the act of gazing into the premises of another, unless the gazer were the king.

Upon the courtyard stables, barns, and store-rooms open, but so far I notice that the granary is in the house, and that the six-feet-high clay receptacles for grain are in the living-room.

Looking from above upon a plain, the poplars which surround villages where there is a sufficiency of water attract the eye. At this season they are nothing but a brown patch on the snow. The villages themselves are of light brown mud, and are surrounded usually by square walls with towers at the corners, and all have a great gate. Within the houses or hovels the families are huddled irregularly, with all their appurtenances, and in winter the flocks and herds are in subterranean pens beneath. In summer the animals go forth at sunrise and return at sunset. The walls, which give most of the villages a fortified aspect, used to afford the villagers a degree of protection against the predatory Turkomans, and now give security to the flocks against Lur and other robbers.

Every village has its *ketchuda* or headman, who is answerable for the taxes, the safety of travellers, and other matters.

Paul Muller-Simonis, *Du Caucase au Golfe Persique à travers l'Arménie, Le Kurdistan et la Mesopotamie*, Washington D.C., Université Catholique d'Amerique, 1892, pp. 123–127, 145–146, 162–163.

Le four et la meule

Nous leur faisons visite au moment où elles cuisent le pain. Le four mérite une description détaillée, car ce four ou *tandour*, pour l'appeler par son nom indigène, est le centre de toute la vie domestique en Orient; il sert de four à pain, d'âtre de cuisine et de poêle.

Lorsqu'une maison a plusieurs chambres, le *tandour* est bâti dans la chambre principale, à peu près dans l'axe de la pièce, mais du côté de l'entrée. C'est en somme une amphore en terre cuite, à parois épaisses de trois doigts environ, enterrée dans le sol de la chambre.

L'installation d'un *tandour* est une chose assez délicate: l'amphore se fabrique en dehors de la maison; il ne faut y employer qu'une terre soigneusement pétrie; la hauteur ordinaire est de 1 mètre à 1.20 m.; le diamètre varie suivant les cas. L'on commence par former le fond de l'amphore et on ne monte les parois que successivement, augmentant leur hauteur d'une main, dès que la couche inférieure a acquis suffisamment de consistance. Lorsque l'amphore est ainsi fabriquée et bien sèche, on creuse dans l'appartement un trou dont le diamètre dépasse d'environ un mètre celui du *tandour* lui-même; au fond de ce trou l'on place une pierre plate; puis besogne délicate, on transporte le *tandour* sur cette pierre. Une ouverture, pratiquée à la partie inférieure de la paroi du *tandour*, communique par un conduit oblique avec le niveau du sol, près de la porte; ce conduit sert d'appel d'air. Tout l'espace libre autour du *tandour* est rempli de cendres que l'on recouvre à hauteur du sol d'un gâchis de mortier.

Le *tandour* étant ainsi en place, on y allume du feu et on le remplit entièrement de combustible; ce feu maintient pendant deux ou trois jours une température suffisante pour cuire la terre du *tandour*.

Il est à noter que l'appartement lui-même prend le nom de *tandour*. Toute la famille s'y rassemble en hiver; lorsque le feu est tombé, on se blottit en rond autour du *tandour*; les plus privilégiés ont le droit de laisser pendre leurs jambes dans le four même; les autres les abritent du moins sous une grande couverture qui conserve la chaleur du foyer.

Ce système de chauffage, laissant la chambre froide et sur-chauffant les extrémités du corps peut, à bon droit, passer pour très malsain; mais il n'y a guère de chance de le voir jamais modifié, car il est universel. Pour éviter les courants d'air, on a généralement soin de ne pas placer la porte dans le même axe que le *tandour*, mais près d'un coin de la pièce.

Reste à parler du combustible.

Dans toute l'Arménie et la Perse, le bois est très rare; on le remplace par les fameuses «galettes combustibles».

Une des grandes occupations des femmes de la campagne consiste à ramasser soigneusement la fiente des animaux. Cette fiente est artistiquement pétrie à la main avec les petits fétus de paille, résidus du battage des grains; la galette que l'on forme ainsi est ensuite vigoureusement plaquée contre un mur où elle sèche, gardant l'empreinte des doigts souvent fort délicats qui y ont travaillé. Les galettes sèches sont empilées en forme de grande meule; lorsque celle-ci est terminée, toute sa surface extérieure est recouverte d'un enduit de même nature, et la provision de combustible se trouve toute faite pour l'hiver.

Plus d'une Européenne, voire même plus d'un Européen se révolterait à la pensée de manger le pain sorti d'un four chauffé par de pareils procédés! Qu'on se rassure! ces galettes sont un combustible parfait; après une rapide flambée, pendant laquelle un odorat délicat peut, si l'on veut, être légèrement blessé, il reste une masse incandescente qui brûle sans flamme, très lentement, avec un grand dégagement de chaleur et sans répandre aucune odeur.

Le pain que l'on cuit dans ces fours s'appelle en chaldéen «*Lavasch*». C'est, au demeurant, le pain national de la Perse. Il est en forme de minces galettes. Pour les préparer, la boulangère - toute mère de famille est boulangère - étend légèrement la pâte au moyen d'un rouleau; puis, d'un mouvement rapide et alterné, elle plaque successivement cette pâte sur chacun de ses bras. La galette s'étire à chaque fois, et quand elle a atteint la forme voulue, on l'étend sur une planchette en bois armée d'une poignée. Cette planchette est arrondie, et sa convexité répond à la courbure du four. Au moyen de cette planchette, une autre femme, préposée à la cuisson, applique vigoureusement la galette contre la paroi brûlante du four. La cuisson est extrêmement rapide; généralement elle est incomplète.

Ce pain frais et croustillant est excellent. Il se conserve plusieurs jours. Au moment du repas, les Persans l'humectent assez fortement pour le rendre flexible, et ces galettes remplissent alors toutes sortes d'offices; roulées en cornets, elles deviennent cuillers pour prendre la sauce; étendues, elles servent d'assiettes; elles deviennent même serviettes; mais le pan ainsi humecté perd de sa saveur et devient lourd.

Généralement les *lavaschs* ne sont pas suffisamment salées; et pourtant le sel n'est pas rare! Les soeurs tirent le leur de la montagne de sel que nous avons vue entre *Khoï* et *Khosrava*; il est de couleur fort grise et très compact; on l'emploie à l'état brut. Les blocs de sel sont tout d'abord pilés dans un creuset en pierre avec de gros marteaux de bois; puis on moud les fragments.

Le moulin que l'on emploie à cet effet est uniformément répandu dans tout l'Orient et jusqu'aux Indes; une pierre *A* est fixée au sol; elle porte un axe vertical *C*. La meule proprement dite *B* est percée à son centre d'un trou circulaire de 8 à 10 centimètres de diamètre. Deux femmes, se faisant vis-à-vis, la font tourner au moyen d'un manche en bois *D*, planté obliquement au deux tiers environ du rayon de la meule, vers la circonférence. L'ouverture circulaire de la meule étant assez grande, le mouvement de rotation autour de l'axe *C* se fait toujours en excentrique, ce qui permet de développer une très grande force de friction et d'obtenir un sel très fin.

Pour Hyvernat, *Khosràva* possédait un objet de grand intérêt, je veux dire son cimetière; il est relativement antique et ses inscriptions tombales en langue chaldéenne offrent de l'intérêt aux linguistes. Les tombes sont généralement très simples; la plupart ne sont que des monolithes à peine dégrossis; quelques-unes reproduisent grossièrement la forme d'un bélier. En dehors des anniversaires des défunts, les tombes sont assez délaissées.

……

Les habitations

Les habitations sont, en général, fort simples. Les murs de clôture sont faits en pisé et les maisons elles- mêmes sont généralement en briques crues. La matière est toujours à portée, car ce n'est que de la terre; le premier endroit venu sert de carrière. Les maçons mêlent à la terre une certaine quantité de paille broyée et coupée menue, pour lui donner plus de consistance et empêcher les briques de casser. Cette terre est assez fortement humectée et les maçons la foulent soigneusement avec les pieds; puis elle est coulée dans des moules de bois mince qui ont environ 21 centimètres et demi de long sur 16 centim. de large et 6 centim. et demi de haut. Le moule rempli, le maçon en égalise la surface avec la main, puis le trempe rapidement dans un baquet d'eau mêlée de paille très menue qui se colle à la surface. On retire alors le moule et on laisse la brique commencer à sécher au soleil pendant quelques heures; après quoi les briques ainsi préparées sont rangées de champ les unes contre les autres et achèvent de sécher.

Quant à la couverture des maisons, dans le centre de la Perse, elle est le plus souvent en voûte; mais, dans l'*Aderbeidjàn*, elle est ordinairement composée d'une terrasse plate, supportée par une charpente en bois de peuplier. Sur cette charpente l'on étend un clayonnage serré qui est recouvert ensuite d'une épaisse couche de pisé. (Fig. 27).

La confection de ces terrasses est évidemment la partie délicate de la construction; car il faut à la fois ne pas les faire trop lourdes et ne pas donner de pente assez forte

Fig. 27 Traditional house in Tadwan (Muller-Simonis 1892)

pour exercer des pressions latérales sur les murs peu solides, suffisante cependant pour l'écoulement des eaux. Les pluies ne manquent pas de délayer à chaque fois une bonne partie de la terrasse; aussi bien est-ce la seule partie de l'habitation qui soit vraiment entretenue. Pour empêcher les vers de démolir les toitures, on répand du sel sur les poutrages et entre les différentes couches de pisé; j'ignore quelle est en réalité le degré d'efficacité de ce préservatif, mais je l'ai vu employer aussi bien dans le Kurdistan qu'en Perse.

Quant aux travaux de serrurerie, on peut simplement dire qu'ils n'existent pas dans une maison persane. Les portes sont toutes en bois et tournent sur un pivot de bois fixé dans deux entailles faites, l'une dans le linteau, l'autre dans le seuil de la porte. Pour fermeture un loquet auquel s'adapte une sorte de serrure en bois, parfois assez ingénieusement et savamment disposée. Tandis que des plafonds élégants masquent les clayonnages dans les maisons riches, ces portes grossières se rencontrent dans les plus belles demeures et l'on peut dire que leur usage est universel.

L'agencement des fenêtres est aussi très rudimentaire; des carreaux de papier huilé remplacent généralement les vitres.

Quant au mobilier, il est des plus simples; tout le luxe est dans les tapis. Dans la salle de réception court le long des murs un banc grossier, élevé de 15 à 25 centimètres au-dessus du sol. Ce banc assez large est recouvert de coussins et de tapis souvent fort riches. Les Persans, qui quittent toujours leur chaussure à la porte, s'accroupissent sur ce divan les jambes croisées et y passent ainsi leur journée. Ce soin d'enlever les chaussures permet de conserver très longtemps les tapis; j'ai vu dans une famille un superbe tapis du *Khorassan* qui avait deux siècles d'âge; il semblait que le temps n'eût fait qu'ajouter à la beauté de ses couleurs. Les murs des chambres sont généralement blanchis à la chaux et n'ont pour ornement que des niches à cadres découpés; les maisons «modernes» y ajoutent les plus horribles gravures d'Épinal. Les guéridons à fumer et les services à café sont généralement élégants.

……

Les églises.

Les églises de ces pays sont toutes très pauvres, et on doit louer le zèle des habitants quand elles sont tenues à peu près proprement (Fig. 28).

Toutes ont une porte ridiculement basse où ne peut passer qu'une personne à la fois, en se courbant en deux. L'on donne cet usage comme un signe de l'humilité qui doit caractériser le chrétien et comme une traduction matérielle de la parole du Sauveur: «Efforcez-vous d'entrer par la porte étroite»: c'est possible; mais j'avoue que je crois beaucoup plus à une préoccupation d'un tout autre genre. Dans les villages, l'église est la seule construction un peu solide. Sa terrasse peut servir de camp retranché; les femmes et les enfants peuvent se réfugier dans l'intérieur, et si la porte est bien basse et bien étroite, elle peut être barricadée solidement et à peu de frais. Dans des pays sans cesse exposés aux incursions des brigands, cette opération

Fig. 28 Gharabagh, entrance to the church of S. Petros (Photo M. Boriani)

doit, et surtout a dû se pratiquer assez souvent; c'est à la préoccupation de la faciliter le plus possible que je suis tenté de rapporter l'usage dont je parle.

6 Agricultural Landscape

Filippo De Filippi, ***Note di un viaggio in Persia,*****Milano, 1865,** (pp. 136–137, 138–140).

[Khanats] (Fig. 29).

Fin da' primi nostri passi nelle steppe della Persia dovetti fare attenzione a certi cumuli di terra di forma conica, aventi alla sommità una sorta di cratere, in molti costrutto, in altri aperto, come la bocca di un profondo pozzo, allineati in lungo ordine perdentesi nell'orizzonte. Ognuna di queste linee di coni, non male rassomiglianti a colossali talpiere, dalle radici dei monti si protende nel Piano, e segna la direzione di un canale sotterraneo, che raccoglie le poche filtrazioni profonde per comporne un canaletto, il quale infine, sgorgando dall' estremo cumulo come da un antro, si versa direttamente sui vicini campi. I coni stessi non sono altro che materiale gettato

Fig. 29 A Qanat under construction (Ichhto—Unesco—*The Persian Quanat* 2015)

fuori di tratto in tratto, onde aprire sotterra un corso all'acqua. In alcuni luoghi ho visto lavoratori intenti a questi scavi, traendo coll'argano dalla bocca dei pozzi il materiale che altri, educato alla scuola delle talpe, andava mano a mano esportando nel progresso sotterraneo. La profondità della scarsa vena dell'acqua, e la rozzezza degli strumenti di lavoro alla quale supplisce la pazienza dei Persiani, sembrano le cause che fanno ai Persiani preferire questo sistema a quello dei canali aperti.

Queste linee di coni che si incontrano di quando in quando negli altipiani della Persia, terminano, come ho detto, ad un antro, dal quale esce un rivoletto d' acqua limpida e fresca.

Qualche volta questo rivoletto si versa in un piccolo stagno artificiale, circondato di salici, e dal quale l'acqua si diparte per vari rami alle campagne circostanti. Da questi canali artificiali dipende la vita dei poveri villaggi sparsi nelle steppe, lontani da corsi di acque naturali. Che se, come pure accade, la vena dell'acqua accumulata a· sì gravi stenti e con sì pertinace lavoro, venga a mancare col tempo, ed il canale sotterraneo si renda asciutto, la misera popolazione del villaggio è obbligata a trasportar altrove i suoi penati. Noi abbiamo infatti incontrato sul nostro cammino più di un villaggio affatto deserto e rovinato per siffatta cagione.

[A/T: Since our first steps through the barren lands of Persia I had to make attention to some heaps of ground in a conic shape, with a sort of crater on the top, built in many of them, but open in others like the mouth of a deep water shaft, lining up until to the far horizon. Each of these lines of cones, quite alike huge shelters of moles, extends itself from the bottom of the mountains to the plain marking the direction of an underground channel, that collects the few deep water filtering into a rivulet which at last flows from the last heap like from a cave spilling over the fields all around. The cones themselves are not but soil dug out from time to time in order to open the underground water flow. Elsewhere I saw workers attending to those excavations, by pulling with a winch from the shafts the soil that other digging workers kept extracting from underground. The depth of the scarce water vein, and the roughness of the work tools for which the Persans' patience makes up, seem the reason why Persans prefer that system to the open channels.

Those lines of cones you can find from time to time on the Persan plains, as I said, end into a cave whence a rivulet of pure and fresh water pours out. Sometimes the rivulet pours into a small artificial pond, surrounded with willows, whence water flows through different branches to the fields around. The life of the poor villages scattered through the steppes, far from natural water courses, depends on those channels. And if, as actually happens, the water vein gathered in such severe hardship and pertinacious work came to be missing through time, the unfortunate inhabitants of the village must move and take their penates elsewhere. In fact along our way we met several villages deserted and totally ruined that reason.]

[Agricultural Landscape] (Figs. 30 and 31).

Quasi ogni villaggio è ombreggiato da alberi fronzuti, addensati, in filari, in boschetti, ora all' aperto ora nei chiusi, è circondato da frutteti arruffati ma rigogliosi, da ajuole verdeggianti, da campi ubertosi, bruscamente contornati nel deserto. Gli alberi che maggiormente abbondano sono i pioppi, i salici ed i platani, e tra i fruttiferi, i gelsi,

Fig. 30 Landscape near Sofian, 2017 (Photo M. Boriani)

i pomi, i ciliegi, i pruni, gli albicocchi, i cui rami cadenti al peso delle frutta mature, trovammo quasi dovunque traboccare e pendere all'infuori de' muriccioli di fango che ricingono i giardini. Una pianta estesamente coltivata è pure *l'Eleagnus hortensis*, il cui frutto farinoso, rassomigliante ad una giuggiola, vedesi in gran copia ne' bazar, col nome di sedschit. Lungo il letto dei fiumi si coltiva il riso, ne' campi raramente il sesamo, più comunemente il ricino, ma per assai maggior estensione il frumento e l'orzo; e quest'ultimo serve di principale foraggio ai cavalli. A tal fine, l'orzo mietuto è sul campo stesso tagliuzzato da grandi coltelli circolari, giranti sul loro asse messo in moto da cavalli, dopo di che, raccolto fra reti di corda, si porta all'abitazione. Altre piante diffusamente coltivate in Persia, sono i poponi e specialmente i cocomeri; ma questa coltivazione richiede un terreno bene innaffiato. I cocomeri sono uno de' principali prodotti di questo paese non solo, ma anche delle provincie caspiche della Russia, ed è veramente incredibile il consumo che ne viene fatto. Si può dire che per tre mesi dell'anno, il nutrimento principale de' Persiani consiste di questo frutto, assai più succulento e dolce che in Europa, ed il quale dappertutto si trova in gran copia ed a prezzo vilissimo.

I Persiani sono ghiotti di frutta zuccherine; i loro mercati ne rigurgitano, e sì grande è la loro abbondanza in alcuni luoghi, che se ne lascia perfino infracidire una gran parte sul terreno. Le piantagioni di gelsi, relativamente rare in questa parte della Persia ch'è al sud dell'Elburz, non danno altro prodotto che nei frutti, essi pure assai più dolci che fra noi. Si coltiva anche la vite, ma questa soltanto presso la città

Fig. 31 Orchard at Savra, 2018 (Photo M. Boriani)

od i più grossi villaggi, ed in modo affatto particolare, imposto dalle condizioni del clima e dalla penuria assoluta di legno da pali. I ceppi della pianta sono allineati in aperta campagna sulle scarpe di grandi solchi irrigati con gran cura, ed i tralci si estendono liberamente sul terreno a guisa delle zucche. L'uva che se ne raccoglie è della più bella e della più gustosa che dar si possa, affatto immune da ogni traccia di crittogama (*Oidium*), mentre ne sono invasi, fino alla completa distruzione dei grappoli anche in questo paese, i rari pergolati dei giardini. Non si fa vino in Persia che in assai scarsa quantità, e dai soli Armeni, e questo vino, forte, disaggradevole, non conviene a palati europei, apprezzanti appena, ed allora come cosa eletta, quello rinomatissimo di Shiraz.

Da quando in quando, lungo i pochi e miseri fiumi della Persia, si incontrano mulini, i quali sono di uno stile tutto particolare. Si presentano come casipole con mura di sassi quasi senza cemento, erbosi per l'umidità naturale del luogo, di forma prismatica quadrata, senza finestre, con una sola porticina d'accesso. L'acqua vi è condotta per un piccolo canale di legno che trapassa nel suo mezzo il tetto, formando con questo un angolo di circa 45°. Nulla traspare dal di fuori del meccanismo della macina. Si può facilmente vedere quanto debba costare ai Persiani il trasporto del grano da' siti di sua produzione, talvolta per distanze enormi, ai pochi mulini qua e là disseminati; sì pochi da non comprendersi come mai possano bastare ai bisogni della popolazione. Ma ciò che si comprende ancora meno si è come nessuno abbia mai pensato a rivolgere al servizio de' mulini un altro motore naturale, costantemente attivo per tutta l'estensione del paese, voglio dire i venti che soffiano senza posa.

[A/T: Almost every village is shadowed by leafy trees, thickened in rows, in groves, either in open or in fenced areas; and is surrounded by matted but luxuriant orchards, by verdant flowerbeds, by fertile fields, abruptly surrounded by the desert. Poplars, willows and plane trees are mostly thriving, and among the fruit trees are mulberry, apple, cherry, prune, apricot trees. Their branches we found nearly everywhere, charged with ripe fruits, overflowing and hanging out of the low mud walls which surround the gardens. A largely grown plant is also Eleagnus hortensis, the floury fruit of which, similar to jujube, is very easy to find in bazaars under the name of sedschit. Rice is grown along the river sides, seldom sesame in the fields, more frequently castor, but most of all wheat and barley; and the latter serves as main forage to horses. To that purpose the harvested barley is shredded straight on field with big circle blades turning on their own axes moved by horses. After that it's picked up and set into rope nets and taken home. Other plants largely grown in Persia are especially watermelons; but that requires a well watered soil. Watermelons are among the main products not only in this country, but also in the Caspian provinces of Russia, and it's really incredible how big that consumption. Watermelons can be said the main food for Persians along three month a year. That fruit here is much sweeter and juicy than in Europe, and can be found everywhere at very low cost. Persians are fond of sweet fruit; their bazaars have plenty of to an extent that in some villages a large amount of them is left rot on the ground. Mulberry plantations, fairly rare in this part of Persia south of Elburz, yield no product but fruit, much sweeter than ours too. Vine also are grown, but just near towns or biggest villages, and in a very particular way due to the climate conditions and absolute lack of wood to make poles. The plant stumps are set in rows in the open fields on foot of large furrows very carefully watered, and the branches grow free on the ground like pumpkins. The harvested grapes are the most beautiful and juicy among all, entirely free of any trace of cryptogam (Oidium), which whereas invades the few garden pergolas up to the complete destruction of the bunches. No wine is produced in Persia but in very tiny amount, and just by Armenians; a strong, unpalatable wine unfit to the taste of Europeans who just appreciate as prelibacy the very famous Shiraz.

From time to time mills can be seen, in their very particular shape, along the few and scarce rivers of Persia. They look like small huts with stone walls nearly with no mortar, grassy because of the natural dampness of the place, in a square prysm shape, with no windows and just one small entrance door. Water is conducted by a narrow wooden canal passing through the roof in its middle at aa angle of about 45°. From outside, the millstone mechanism cannot shine through. You may easily see how expensive be for Persians the transport of wheat from its production places, sometimes from enormous distance, to the few mills scattered here and there in so few number as not understand how may they satisfy the people's needs. But even less understandable is how nobody ever thought of turning to service of the mills another naturl engine, constantly active all over the country, I mean the restless winds blowing].

Paul Muller-Simonis, *Du Caucase au Golfe Persique à travers l'Arménie, Le Kurdistan et la Mesopotamie*, Washington D.C., Université Catholique d'Amerique, 1892, pp. 133–134, 134–137.

[Le territoire d'Ourmiah – distribution des eaux]

Ourmiah est située à une vingtaine de kilomètres à l'Ouest de la «mer d'*Ourmiah*», à une altitude d'environ 1310 mètres. La ville compte à peu près 32.000 habitants, dont la majorité sont musulmans.

Je ne décrirai point *Ourmiah*, car toutes les villes d'Orient se rassemblent; et là où il n'y a pas de beaux monuments publics à signaler, toute ville n'est qu'un assemblage de ruelles mal tenues, bordées de murs pleins, derrière lesquels se cache toute la vie.

Le territoire d'*Ourmiah* a une juste réputation de fertilité; le grand élément de la richesse agricole en Orient, l'eau, ne lui manque point. Les Persans ont eu de tout temps un remarquable talent pour l'irrigation des terres et pour l'utilisation la plus judicieuse de cette eau généralement si rare. Il ne parait pas cependant qu'ils aient employé ici, en dehors de la ville ou de ses environs immédiats, le coûteux système des canaux souterrains qui se rencontrent en d'autres endroits; sans doute, ils n'en avaient pas besoin.

La distribution des eaux est réglée par un fonctionnaire spécial appelé *Myr-Ab* (en persan, grand maitre des eaux).

Ce fonctionnaire est nommé par le seigneur de l'endroit; il a sous sa juridiction, soit un seul village, si celui-ci est suffisamment important, soit les différents hameaux situés sur un même cours d'eau. Dans le pays d'*Ourmiah* l'usage des eaux d'irrigation est libre jusqu'au commencement de Mai; à partir de ce moment, chacun irrigue à tour de rôle. Un particulier veut-il arroser hors tour, le *Myr-Ab* lui vend ce droit d'eau supplémentaire. Cette vente ne doit avoir lieu que sur présentation d'un bon signé par le chef du village; mais, sans aucun doute, il est des accommodements. Parfois le *village* comme corps achète à un autre village situé au-dessus de lui le droit d'irrigation pour une période donnée.

Dans les villages à population mixte, le vendredi, jour de repos des Musulmans, est réservé aux Chrétiens et le dimanche aux Musulmans.

……

[Agriculture] (Fig. 32).

Le pays produit beaucoup de très beaux légumes, et nos choux d'Alsace feraient, dit-on, mauvaise figure auprès de ceux d'*Ourmiah*. Les melons, qui sont excellents, sont représentés par plusieurs variétés. Un des plus goûtés, le *Guermek*, est un melon jaune, très doux, qui mûrit en Juin et se conserve jusqu'en hiver. La pastèque ou *kharpous* atteint parfois le poids de 30 à 40 livres; elle mûrit en été et peut se conserver toute l'année. Son goût, si l'on veut, n'est pas fort délicat, mais c'est un fruit rafraichissant et très sain.

Le *schammàm* est un tout petit melon qui n'est pas comestible, mais que l'on conserve pour son parfum délicat. Son nom figure très fréquemment dans la poésie

Phototypie J.-B. Obernetter, Munich.

ARABAH ARMÉNIENNE

Fig. 32 *Arabah* (chariot) *Arménienne* (Muller-Simonis 1892)

persane. Les oisifs tuent le temps en humant le parfum d'un *schammàm*, qu'ils font gravement passer d'une main à l'autre.

Concombres et tomates atteignent de fort belles dimensions. Le pavot, le safran, le tabac, le coton, forment autant de branches importantes de la culture. Quant à l'huile, elle est généralement tirée du ricin, du lin et du colza.

Le blé de ces contrées est fort beau, et l'on en compte trois espèces. L'avoine n'est pas cultivée; elle est remplacée par l'orge (*arpa*). Le riz et le millet jouent un rôle important dans la consommation.

L'élève des vers à soie a donné un assez grand développement à la culture du mùrier.

Le platane, le saule, le peuplier, le mastiquier (celui-ci plus spécialement dans le territoire de *Soldouz*), l'amandier et le chàtaignier ont aussi fort répandus; mais tous ces arbres ne poussent guère qu'à l'état de « culture». Le chêne à noix de galle pousse à l'état sauvage dans les montagnes.

Comme arbres fruitiers, l'abricotier, le grenadier, le poirier, le pommier, le cognassier, le prunier et le pistachier sont, avec le jujubier, les plus répandus.

Ourmiah est la terre classique des raisins secs. Dans les villages nous voyons partout d'immenses quantités de raisins en train de sécher au soleil. Les raisins ordinaires sont mis à sécher sans autre formalité; mais pour certaines espèces, dans le but de leur donner une belle couleur, on emploie un procédé curieux. L'on fait bouillir longtemps dans de l'eau des cendres de sarments de vigne ou bien aussi de certaines herbes des montagnes: le liquide jaunâtre, provenant de cette cuisson, est ensuite soigneusement décanté et versé dans une nouvelle chaudière, où il est une seconde fois porté à l'ébullition. L'on trempe brusquement, pendant une seconde à peine, les raisins dans cette solution bouillante, et, de suite après, on les étend au soleil pour les sécher. L'on a soin, pour ces qualités de raisins, de préparer des surfaces soigneusement égalisées et recouvertes d'un gâchis de terre et de paille.

Le vin d'*Ourmiah* est excellent; le plus répandu est le vin blanc. Malheureusement, il est rare de voir du vin vieux de plus d'un an; les habitants donnent pour motif l'insuffisance de la production et la difficulté de conserver le vin, durant les chaleurs de l'été, dans des maisons sans caves. Tous ces motifs sont réels; mais, à mon avis, le vrai motif est à chercher dans la soif gloutonne de ces mêmes habitants qui boivent sans mesure tant que dure la provision. Lorsque, l'été venu, le vin est épuisé, «pour ne pas boire une eau que les irrigations rendent malsaine», ces pauvres gens se voient contraints d'étancher leur soif avec leur provision d'eau-de-vie! Boisson excellente par 35 degrés de chaleur à l'ombre!

L'usage universel des galettes combustibles porte évidemment un tort sérieux à l'agriculture; aux environs des villes on cherche à y remédier par l'emploi des détritus organiques. Ceux-ci sont soigneusement recueillis, puis mélangés avec de la terre; au bout de deux ans, ce terreau est répandu dans les jardins. La cendre des foyers est aussi employée comme engrais.

Malgré toutes ces richesses naturelles, le pays est pauvre.

La cause première de cet état de misère réside dans les vices du gouvernement; et ces vices eux-mêmes ont leur origine dans l'influence prolongée des principes de l'Islamisme.

Dans les temps anciens, la Perse était beaucoup plus peuplée et plus fertile. Mais dans ces pays, la *nature vierge* n'existe pas; toute la fertilité est un trésor accumulé par l'industrie humaine. Il suffit d'une série de guerres malheureuses et, plus encore, d'une série de mauvais gouvernants, pour dépeupler un pays; et en Perse, tout pays dépeuplé devient immédiatement aride. L'influence européenne sur le gouvernement persan se manifeste surtout dans les choses superficielles; mais elle n'a pas pénétré les parties vitales de son organisme, et tout marche d'après a vieille méthode.

Edward L. Cutts, *Under the Crescent in Asia*, London, 1878, pp. 277–278.

["Hills of ashes"]

Friday, August 11—There are two (if not more) lofty "hills of ashes" in the plain, concerning which the people have a theory that the ancient fire worshippers had stations here, and that these hills are the ashes of their perpetual fires. One of these hills is just outside the village of *Digala*, and we went out this morning to look at it. The people have found that the ashes and soil are capital manure, and have made extensive excavations, which laid the interior of the hill open in several places almost down to the bottom for our inspection. The ashes lie only in thin irregular layers here and there, at various depths, amidst the heap of soil. Towards the bottom of the hill we observed a portion of a wall in situ. Building stones have been found in other places; fragments of pottery, and entire vessels of pottery (one of which I saw), and implements in metal. The whole remains produced on my mind the conviction that the hill was simply the debris of ancient habitations. If the mounds which lie so thickly between Aleppo and Mosul were dug into I should expect to find similar remains. The soil takes to be the remains of the earthen walls, of which nearly all the villages here are built; the thin layers of ashes may be the charred remnants of their roofs.

7 Travelling in Persia

J. B. Tavernier, ***Les six Voyages de Jean Baptiste Tavernier, a noble Man of France now living, through Turky into Persia and the East-Indies, finished in the Year,*** **London, R.L. and M.P., 1678** (First edition in French, 1676), chapt. X, pp. 105–108, 109, 111.

[Caravanseras]

Les *Caravanseras* sont les hostelleries des Levantins, bien différentes des nostres, et qui ne n'ont ni les commodités ni la propreté. Ils sont bastis en quarré à peu près comme des cloîtres, et n'ont d'ordinaire qu'un étage, et il est fort rare d'y en voir deux. Une grande porte donne entrée dans la cour, et au milieu de chacun des trois autres costez, en face, à droite et à gauche il y a une sale ou grande chambre pour les gens plus qualifiez qui peuvent passer. A côté de cette sale sont plusieurs petites chambres où chacun se retire en particulier. Ces logements son relevez comme en parapet le long de la cour de la hauteur de deux ou trois pieds, et les écuries les touchent derrière, où le plus souvent on est aussi bien que dans les chambres. Il y en a plusieurs qui aiment mieux se retire en hyver, parce qu'il fait chaud, ces écuries estans voûtées de mesme que les sales et les chambres. On pratique dans ces écuries devant la teste de chaque cheval una niche avec une petite fenestre qui répond à une chambre, d'où chacun peut voir comme on traite son cheval. Dans chacune de ce niches deux ou trois personnes se peuvent ranger. Et c'est où les valets vont d'ordinaire faire la cuisine.

Il y a deux sortes de *Caravanseras*. Les uns sont rentez, où on est reçu charitablement comme dans nos hôpitaux; les autres ne le sont pas, et on y paye ce qu'on y prend pour la bouche. Il ne s'en voit guère des premiers que depuis *Bude* jusques à Constantinople, et il n'est permis d'en bastir de cette sorte qu'à la mere et aux sœurs du grand Seigneur, ou aux *Vizirs at Bachas* qui se sont trouvez trois fois en bataille contre les Chrestiens. Dans ces sortes de *Caravanseras* qui d'ordinaire sont bastis de legs pieux, on donne honnestement à manger aux passans, et quand ils partent ils n'ont qu'à remercier le Concierge sans rien débourser. Mais depuis Constantinople jusqu'en Perse les Caravanseras ne sont point rentez, et on ne vous offre que les chambres toutes nuës. C'est à vous à vous pourvoir de matelats et d'ustanciles pour la cuisine, et vous achetez à assez bon compte ou du Goncierge ou des paysans qui viennent des villages circonvoisins, des agneaux, des poules, du beurre e des fruits selon la saison. On y trouve aussi de l'orge et de la paille pour les chevaux, à la réserve de quelques lieux que j'ai marquez dans les routes. On en paye rien à la campagne pour le loüage des chambres des Caravanseras; mais on paye dans les villes, et ce qu'on paye est fort peu de chose. D'ordinaire les Caravanes n'y entrent point, parce qu'ils ne pourroient contenir tant d'hommes et de chevaux, et il n'y peut guère loger commodément que cent cavaliers. Dès qu'on est arrivé chacun a droit de prendre sa chambre, le pauvre comme le riche; car on n'a nul regard en ces lieux-là à la qualité des gens. Quelquefois par honnesteté ou par interest un petit mercier cédera la place à un gros marchand; mais il n'est pas permis de débusquer qui que ce soit de la chambre qu'il a prise. La nuit le Concierge ferme la porte, et doit répondre de tout, et il y a toujours quelqu'un de garde autour du Caravansera.

Pour ce qui est des Caravanseras de Perse, j' ay remarqué ailleurs qu'en général ils sont plus commodes et mieux bastis que ceux de Turquie, et que dans de distances raisonnables on en trouve presque par tout le pays. Il est aisé de voir par cette description des Caravanseras, que s'ils ne sont pas si commodes pour les riches que nos hostelleries d'Europe, ils le sont plus pour les pauvres qu'on ne refuse pas là de recevoir, et qu'on ne contraint pas de boire et manger plus qu'ils ne veulent, estant permis à chacun de régler sa dépense selon sa bourse.

……

[Caravanes]

Les Caravanes sont comme des grands convois composez de quantité de marchands, qui s'assemblent en certains temps et en certains lieux pour estre en estat de se défendre contre les voleurs qui courent souvent par grosses bandes dans des pays qu'il faut traverser, et qui la plupart sont déserts. Ces marchands élisent entr'eux un Chef que l'on appelle *Caravan-bachi*, et c'est luy qui ordonne la marche, prescrit les journées, et qui avec les principaux de la Caravane juge les differens qui peuvent survenir sur le chemin. Il n'ya guère d'honneste homme qui ambitionne cette charge, parce que le *Caravan-bachi* devant acquiter de certains petit droits le long de la route, de quelque manière qu'il se conduise il est toujours soupçonné de peu de fidélité. Quand les marchands Turc font le plus grand nombre dans la Caravane le Chef qu'on élit est Turc; et quand il y a plus d'Armeniens que de Turcs le *Caravan-bachi* est Armenien. Deux sortes de Caravanes. Il y en a de chameaux qui sont les plus ordinaires, parce que c'est la voiture qui coûte le moins, les chameaux, comme je diray plus bas estant de peu dépense, et portent la charge, les uns de trois chevaux, les autres de quatre ou cinq. Mais dans ces Caravanes de chameaux, il y a aussi des chevaux et des mules que les marchands achètent pour leurs personnes, la voiture du chameaux pour l'homme estant incommode quand il ne va que le pas; car s'il alloit toûjours le grand trot elle est assez douce. Il y a aussi des Caravanes qui ne sont que de chevaux, et si les marchands n'en veulent pas acheter, ils trouvent des gens dans la caravane qui leur en loüent.

……

C'est la coutume dans le Levant de faire les journées d'une traite, soit qu'on marche en Caravane, soit que l'on voyage seul. Mais ces journées ne sont pas égales, elles sont tantost de six heures de marche, tantost de dix et tantost de douze, et c'est la commodité de l'eau qu'on ne trouve pas partout, qui les doit régler. En tout temps la Caravane marche plus de nuit que de jour; en Esté pour éviter la chaleur, et dans les autres saisons pour arriver en plein jour au lieu où l'on doit camper. Car si on arrivoit aux approches de la nuit, on ne pourroit dans l'obscurité bien disposer toutes choses, dresser les tentes, penser les chevaux, faire la cuisine, et pourvoir à tout ce qui est nécessaire à un campement. Il est vray qu'au cœur de l'hyver et dans les grandes neiges on ne part guère qu'à deux ou trois heures après minuit, et quelquefois mesme on attend jusqu'à la pointe du jour. Mais en Esté selon la traite que l'on a à faire on part à minuit, ou une heure après le Soleil couche.

……

La caravane campe dans les lieux qu'on sçait estre les plus propres, et sur tout proche des eaux. Quand le Soleil est couché, des *Chaoux*, qui sont des pauvres gens ou Turcs ou Arméniens, ont soin de faire la garde autour du camp, et de veiller sur les marchandises. Ils se promenent par tout, et crient l'un après l'autre en Arabe ou en Armenien, *Dieu est un, il est misericordieux*, et de temps en temps il adjoûtent, *Prenez garde à vous*. Quand ils voyent que l'heure s'approche qu'il faut partir, ils en avertissent le *Caravan-bachi,* qui leur donne ordre de crier que l'on selle les chevaux, et demi-heure après ils crient qu'on charge. C'est une chose à admirer qu'au le second cry des *Chaoux* tout est prest en un moment, et la caravane commence à marcher en grand ordre et en grand silence, chacun a soin dès le soir de se tenir prest, parce qu'il est dangereux de demeurer derrière, sur tout dans les pays que les voleurs fréquentent.

G. Drouville, *Voyage en Perse, fait en 1812 et 1813*, Paris, 1825, Seconde éd.,Tome I, pp. 99–102, 104–109.

[Des Caravanserais] (Fig. 33).

[Les] *caravanserais* sont de vastes bâtiments carrés, renfermant une grande quantité de boutiques, où les commerçants étrangers viennent déposer leurs marchandises, et où, moyennant un certain droit payé aux *darogas* (chefs de police), ils peuvent les débiter. Ces lieux ont de plus des chambres hautes, des magasins a l'usage des voyageurs, des écuries pour les chevaux, et de grands hangars pour abriter les chameaux. Ils sont situés dans l'enceinte des bazars, desquels ils font partie, et sont d'un grand rapport, car il y en a qui rendent annuellement trois milles *tomans* (un *toman* vaut vingt livres tournois). Les nobles et les riches en achètent ou en font construire, quand ils peuvent en obtenir la permission et un local propice, ne pouvant placer leur argent à meilleur intérêt. Ils y mettent quelques-uns de leurs esclaves en qualité d'intendants, et ceux-ci sont comptables envers eux des recettes, qu'ils ne leur rendent cependant pas avec trop de fidélité. Ces lieux sont les points de contact des individus de toutes les nations, mais particulièrement d'une grande quantité de juifs, qui sont là ce qu'ils sont partout ailleurs.

Les *caravanserais* forains sont très-différents de ceux des villes, tant par leur construction que par leur usage. Ce n'est pour l'ordinaire qu'un rang d'écuries qui fait le tour de la partie basse de ces bâtiments, ayant au-dessus quelques chambres et niches sans portes ni fenêtres, qui ne servent plus guère qu'à abriter les chameliers

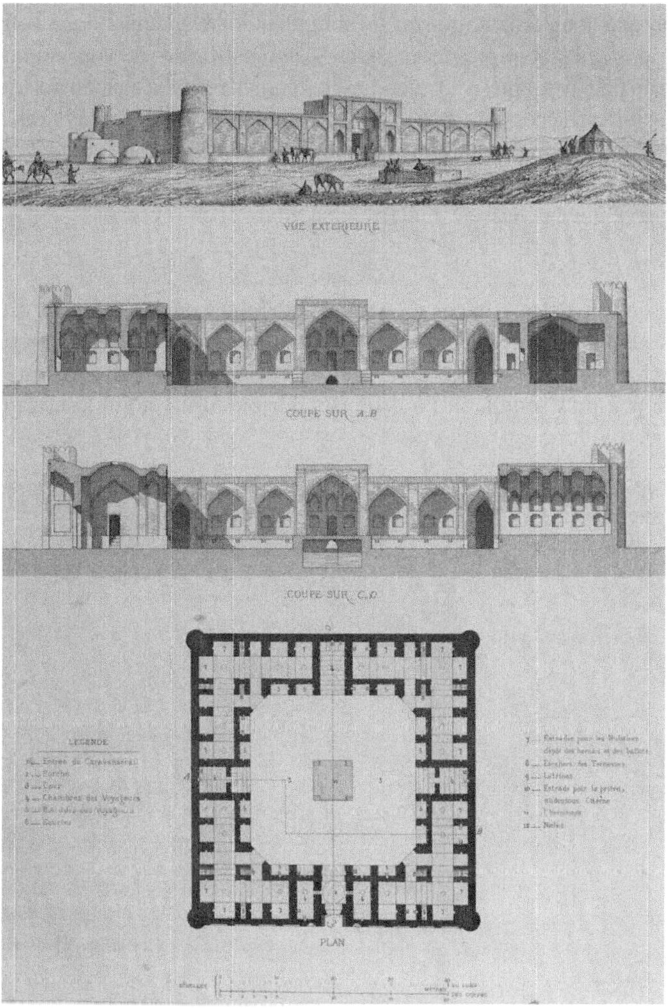

Fig. 33 *Caravanserail à Passengan sur la route de Téhéran* (Coste 1867)

des caravanes, qui s'y réfugient pendant les nuits d'hiver et des saisons pluvieuses. Ces édifices sont pour la plupart très-vieux, délabrés, sans gardiens, et servent assez souvent de retraite aux brigands.

Schah-Abas-le-Grand en avait fait construire de fort beaux sur tous les points de la Perse. Ils furent pendant son règne, et même encore au commencement des guerres civiles, assez bien entretenus, mais depuis ce temps ils sont presque tous tombes en ruine, et comme on néglige beaucoup les réparations en Perse, il est à présumer que dans quelques années il n'en restera plus de vestiges.

Cependant quelques individus riches et dévots en font construire quelques-uns de temps à autre dans les lieux déserts et d'un passage fréquenté. Mais ceux-là même ne subsisteront pas long temps, puisqu'ils sont abandonnés à quiconque veut y entrer, et qu'ils n'ont également pas de gardiens. Aussi est-il rare de voir aujourd'hui des voyageurs un peu distingués s'arrêter dans les caravansérails; chacun porte en voyage tout ce qui lui est nécessaire pour traverser commodément les pays inhabité; cela arrive surtout quand on entreprend quelques pèlerinages de long cours, comme celui de la Mecque, par exemple, pendant lequel on ne profite pas des ressources qu'offre le gibier, parce que la chasse est interdite en allant et en revenant.

[Des Bazards] (Fig. 34).

Les villes, bourgs et les villages un peu considérables ont des *bazards* ou marchés. Ce sont des bâtiments ordinairement situés dans le centre des villes, ou se trouvent réunis tous les marchands et tous les artisans. Ces bazars sont faits en forme de grands corridors à peu près semblables à ceux des dortoirs d'un cloître, mais plus larges,

Fig. 34 *Isfahan, Bazar des tailleurs* (Coste 1867)

et ayant de chaque côté de petites boutiques basses et de peu de capacité, que l'on ouvre à sept heures du matin, et qu'on ferme au coucher du soleil.

Chaque corridor est destiné à une seule espèce de marchandise, ou bien à des artisans d'une même profession, et l'on sait toujours où l'on doit se porter, d'après la nature de ses besoins.

Les *bazards* sont sous la police et juridiction des *darogas*, qui en tyrannisent les marchands et qui commettent toutes sortes d'exactions avec impunité; on peut juger du rapport de cet emploi par le prix qu'on peut y mettre. J'ai vu a *Ourouméa* un particulier de cette ville offrir dix mille *tomans* pour en être revêtu. Chaque corridor à de plus trois ou quatre *alguazils* subalternes qui ne manquent pas de se faire redouter, a l'exemple de leurs patrons, si l'on ne capitule avec eux. Il n'est pas un de ces misérables qui n'ait le talent d'extorquer quelque chose de chacune des boutiques ou ateliers de son arrondissement, contribution qu'on se garde bien de refuser, de peur d'être accusé par eux soit d'avoir vendu avec de faux poids, soit d'avoir ouvert ou fermé sa boutique trop tôt ou trop tard, délits pour lesquels on encourt une punition corporelle et une amende.

Les *bazards* sont, à peu près comme les *caravanserais*, la propriété d'individus riches qui les font construire pour en obtenir des revenus considérables.

……

Personne ne peut coucher dans ces boutiques, car du moment qu'elles sont fermées, elles sont sous la garde immédiate des *darogas*, qui font faire des patrouilles de nuit dans toutes les parties des bazars, par leurs estafiers (ce sont des espèces de lieutenants des *darogas*, que l'on nomme *Mir-aças*, chargés immédiatement de la surveillance de bazars pendant la nuit). Tous les individus qu'on y rencontre passé neuf heures sont arrêtés. S'ils ne sont pas connus, ils reçoivent la bâtonnade; si on leur suppose de mauvaises intentions, on leur coupe le nez ou les oreilles; mais s'il ont volé ils sont aussitôt mis à mort, et leurs têtes roulent devant le palais du gouverneur, pour servir d'exemple à quiconque voudrait les imiter.

……

Les bazars sont des lieux où les marchands étrangers et les oisifs se rassemblent le matin. Comme les rayons du soleil n'y pénètrent pas, la promenade en est très-fraiche en été et assez chaude en hiver. Les dames y viennent aussi quelquefois, soit pour y faire des emplettes, soit pour y rencontrer quelques amies. Mais les femmes du peuple y fourmillent et y restent souvent du matin au soir. Elles parcourent toutes les boutiques pour apprendre des nouvelles dont le débit leur donne un grand relief chez les curieuses du grand monde, dont elles ne sont souvent que les émissaires. J'ai eu fréquemment l'occasion de m'étonner de l'intarissable babil de ces femmes, que je retrouvais au bout de trois heures si échauffées de leur conversation et dans un tel flux de paroles, qu'à peine faisaient-elles attention aux cris répétés de *kabarder!* (garde-à-vous) que font entendre les domestiques qui devancent leurs maîtres en traversant ces sortes d'allées où la foule est sans cesse renouvelée.

De Filippi F., ***Note di un viaggio in Persia,*****Milano, 1865**, pp. 122–124.

[Viaggiare in Persia]

L' ordine che mi sono prefisso in queste note, esige ora che io descriva il modo di viaggiare in Persia. I semplici privati ottengono facilmente dal governatore di una provincia qualunque un firmano e col firmano una piccola scorta della quale è capo un così detto *golam*, sorta di guardia nazionale a cavallo, che provvede per via a tutti i bisogni del viaggiatore, e sovra tutto all' alloggio per la notte in qualche abitazione privata; ma le persone rivestite di un carattere pubblico, gli ambasciatori de' sovrani esteri, hanno a guida un *mehmendar* ed una scorta più numerosa, e devono subire le noje di lunghi cerimoniali e di regali enormemente passivi ad ogni stazione principale. Se l'ambasciata è poco numerosa, come fu dal 1860 al 1861, l'ambasciata prussiana, di sole quattro persone, gli alloggiamenti sono presi ne' villaggi o nelle città che s'incontrano sul cammino; in caso differente, come appunto fu il nostro, il ricovero ad ogni fermata è un piccolo villaggio improvvisato di tende.

Non occorre il dire che solo veicolo possibile in Persia è il cavallo. Ora su tutte le linee più battute si trovano di tratto in tratto stazioni di posta, nel linguaggio del paese chiamate *tschaparkhané*, ove si effettua il cambio dei cavalli, e dove pure si trova disponibile qualche cella oscura e sudicia pel riposo, un samovar ed un *kalian*, per quella forma di ristoro di cui ogni viaggiatore sente maggior bisogno in Persia. Ogni *tschaparkhané* si distingue da un piccolo cortiletto chiuso fra mura di fango, nelle quali mura sono scavate piccole nicchie che servono di mangiatoia pei cavalli, una stanza terrena, ed un'altra superiore alla porta d'entrata costituiscono quasi dovunque i soli quartieri per gli alloggiamenti.

Oltre i *tschaparkhané*, i quali sono d'ordinario ne' villaggi, od a questi molto vicini, s'incontrano in Persia i *caravanserai*, grandi edifizj alla costruzione dei quali hanno proveduto con particolar cura i varj *Schah* che si sono succeduti nel dominio del paese, od alcuni di que' pochi ai quali i *Schah* permettono arricchirsi, e per animo pietoso, o più di soventi, per un contratto interno colla propria coscienza vogliono lasciare ai posteri qualche memoria di sé. Questi edifizj sorgono nella solitudine delle steppe fra le maggiori tratte da un villaggio all' altro e sono una vera providenza pei viandanti. La loro forma di grandi recinti quadrati, con tronchi di torre ai quattro angoli ed un'unica grande porta di accesso, li fa rassomiglianti a piccoli forti.

Ve n'ha che sono costrutti in mattoni cotti, con un certo gusto architettonico, primeggiando quelli edificati da *Abbas* il Grande, ma non ve n'ha alcuno che non porti le tracce del tempo e dell'estrema incuria de' Persiani; e non rari sono quelli rovinati ed abbandonati. Nell'interno di questi *caravanserai*, ai quattro lati del recinto, sono allineati grandi archi a guisa di casematte, che servono di stanza ai viaggiatori (Fig. 35).

……

Sebbene non siano in Persia strade propriamente dette, la natura del paese renderebbe agevole il viaggiare in vettura, ove si accomodassero i burroni che di quando in quando tagliano la solita via delle caravane. Più di una volta ci trovammo nella necessità di far portare le carrozze attraverso passi impraticabili (Fig. 36).

Fig. 35 *Iranian caravanserail between 1851 and 1860* (Luigi Pesce (1827–1864, BNF Gallica—Wikimedia Commons)

[E/T.: [Travelling in Persia] The order I stated for these notes asks for my description of how travelling in Persia is. Private travellers can easily get a firman [a written permission] from the governor of any province, and along with a firman, a small escort under a chief called golam, kind of a national guard on horseback. They look after all the needs of travellers on the road, mainly the overnight accommodation in some private house. However the people with a public character, as the ambassadors of foreign sovereigns, are guided by a mehmendar and travel with a bigger escort, and they have to undergo the trouble of lengthy ceremonials and enormously passive gifts at every main station. If the embassy has few members, like the only four people Prussian embassy of 1860 to 1861, the accommodations are set in the villages or in the towns along the way; otherwise, just as it was for us, the night accommodation uses to be a small improvised village made of tents.

No need saying that in Persia horses are the only possible vehicle. Currently, post stations are placed now and then along the most frequent lines; in the local tongue they are named tschaparkhané, where horses are changed. Here you may also find a dark and dirty rest-room, a samovar and a kalian for that form of refreshment which every traveler feels the greatest need in Persia. Each tschaparkhané is characterized by a small courtyard enclosed with mud walls, where small dug niches serve as horse

Fig. 36 The "road" to Makou (Flandin 1854)

feeder. A ground floor room and an upper one, over the entrance door, are nearly everywhere the only quarters for accommodation.

Besides the tschaparkhané, ordinarily in the villages or very near to them, in Persia the caravanserai can be met, large buildings that the various Schah had built, who dominated for generations the country, or else some of the few that the Schah allow to enrich, and for pityful attitude, or more frequently for a contract with the conscience of their own, wish to hand over memory of theselves. Those buildings arise in the solitude of the steppes among the major stretches from one village to another and are a real providence for travellers. Their large square precincts shape, with tower trunks at the four corners and one big entrance door makes them look like small forts. Some of them are built with baked bricks, and architecturally cured; the ones built by Abbas the Great are the best, but none of them escapes from the marks of time and of the uncarefulness of Persans; lots of them are ruined and abandoned. Inside the caravanserai, at the four sides of the precinct, are rows of high arches alike casematte, that serve as rooms for the travellers.

……

Although in Persia are no proper roads, the nature of the country would make easy travelling by carriage, if only the gaps were filled that from time to time cut the usual roads of the caravanes.

Several times we had to have the carriages brought through impassable passes.]

Edward L. Cutts, *Under the Crescent in Asia*, London, 1878, p. 299.

[Tassach Caravansaray]

Late in the afternoon we halted at the village of *Tassach*; the walls of the village were extensive, and its groves and gardens stretched half across the plain. Our Persians proposed to lodge at the caravanserai, and I was very glad of the opportunity of studying one, and the customs of travelers staying in one. We rode through a great gate-way into a very large oblong courtyard. Over the side through which we entered was a range of rooms; that over the gateway was the best. On the left side of the courtyard was a row of niches in the wall, which served for mangers, and beside each manger was a ring, to which our horses and mules were soon fastened by their halters. Of the right side of the courtyard the further half was occupied with enclosed stabling, the nearer half with small rooms for passengers, with a verandah in front. At the opposite end of the courtyard were the rooms of the keeper of the khan. We took possession of the great room over the entrance, which was approached by an exterior stair in the corner of the courtyard, and had a small anteroom which served as a kitchen. The roof of the verandah formed a "housetop," where carpets were soon spread, and where we ate and slept.

Jane Dieulafoy, *La_Perse la_Chaldée_et_la Susiane Relation de Voyage*, Paris, Hachette et C., 1887, p. 38.

[Caravansaray in ruine] (Figs. 37 and 38).

A moitié de l'étape, la caravane fait une halte de plusieurs heures devant les ruines de l'un des neuf cent quatre-vingt-dix-neuf caravansérails construits sous *shah Abbas*. L'édifice est de forme quadrangulaire, ses murs, bâtis en belle pierre rouge et flanqués de tours défensives, permettaient de l'utiliser comme forteresse en temps d'invasion. La porte, en partie écroulée, est ornée d'une charmante mosaïque de faïence bleue et de briques rosées. Ce caravansérail, comme ses pareils, a servi longtemps de repaire à des bandits, et nos valeureux *tcharvadars* hésitent et tremblent comme la feuille quand Marcel donne l'ordre d'arrêter les chevaux et de décharger l'appareil photographique.

Arthur J. Maclean and William H. Browne, Arthur John Maclean and William Henry Browne, *The Catholicos of the East and his People, Eastern Syrian Christisns of Kurdistan and Northern Persia (known also as Nestorians)*, London, Society for Promoting Christian Knowledge, 1892, pp. 65–67.

[Roads]

The roads are seldom made, and are merely the result of being constantly trodden down by the passing caravans; hence they are generally lower than the neighboring fields, and when the latter are irrigated and the streams overflow, the roads are flooded as a matter of course. The result is an incredible amount of mud, which makes

Fig. 37 *Caravanserail Guilek près de Tauris* (Dubeux 1841—BNF Gallica)

locomotion frequently difficult even in summer, and sometimes impossible in winter. There is an utter absence of public spirit in these matters, and but little compulsion; so that matters go from bad to worse. Indeed, the cultivator often looks upon the road as a convenient place to turn his surplus water into. The stream is a far more important thing than the road. It is necessary that the former should be raised in order to turn the water mills; but it is not the least necessary that the road should be raised to keep a dry means of locomotion! A village is supposed to be responsible for its own roads, but no pressure is put on it, unless the visit of a very great man is expected. In Turkey, indeed, the caravan roads along the rivers, which in winter and early spring are yearly ruined by debris from the mountains and by snow, are repaired in the late spring by forced labour from the neighboring villages, and the bridges are renewed in the same way. But in the plain of *Urmi*, the roads and bridges of the Christian villages are generally worse than those of the Mussulmans. This is because the latter look upon road or bridge making as a work of charity which will secure their entry into Paradise. And, though the short attempts at carriage roads made by some village masters are quite useless, inasmuch as they lead to nowhere, the large bridges over the rivers, of which there are a few in different parts of the country, are a great benefit to the whole community. The bridges over the smaller streams often consist of a single stone or of two poles covered with mud.

Fig. 38 *Caravanserail ruiné sur la route de Marand à Tauris* (Dieulafoy 1887)

Maurizio Boriani Architect, born in Milan, Italy, 1946.

Full professor in Architectural Conservation at the Politecnico di Milano (Department of Architectural Design—School of Civil Architecture)—retired in 2016.

Former Director of Department of Architectural Design at the Politecnico di Milano.

Member of the Teaching College of the Ph.D. in Architectural Conservation at the Politecnico di Milano.

Former President of the Scientific Board of Cultural Heritage Conservation Center of the Politecnico di Milano.

Research Co-ordinator on behalf of the UNESCO, World Bank and European Community, Italian Ministry of Foreign Affairs and other private and public agencies.

UNESCO Consultant for: Albania (2006–2009), Macedonia, Montenegro, Kossovo, Myanmar; expert in international cooperation on behalf of Italian Ministry of Foreign Affairs (China (2009–2012), Vietnam (2017–2019), Armenia (2019–2014), Pakistan (2011–2012, Myanmar (2012–2015).

Research Topics:

History of Architectural Conservation, Conservation and management of historic towns and archeological sites, Cultural landscape preservation and management, Gardens history and conservation, Cultural Heritage and International Cooperation. About those themes he is author or editor of about 240 papers and books.

The manufacturer's authorised representative in the EU is Springer
Nature Customer Service Centre GmbH, Europaplatz 3, 69115 Heidelberg,
Germany. If you have any concerns regarding our products, please
contact ProductSafety@springernature.com

Printed and bound by CPI Group (UK) Ltd, Croydon, CR0 4YY
28/09/2026
02224067-0008